T0349010

Space Microsystems and Micro/Nano Satellites

Space Microsystems and Micro/Nano Satellites

Edited by

Zheng You

Professor, Dean of School of Mechanical Engineering,
Tsinghua University, Beijing, China

Butterworth-Heinemann is an imprint of Elsevier
The Boulevard, Langford Lane, Kidlington, Oxford OX5 1GB, United Kingdom
50 Hampshire Street, 5th Floor, Cambridge, MA 02139, United States

British Library Cataloguing-in-Publication Data
A catalogue record for this book is available from the British Library

Library of Congress Cataloging-in-Publication Data
A catalog record for this book is available from the Library of Congress

ISBN: 978-0-12-812672-1

For Information on all Butterworth-Heinemann publications
visit our website at https://www.elsevier.com/books-and-journals

Publisher: Joe Hayton
Acquisition Editor: Glyn Jones
Editorial Project Manager: Katie Chan
Production Project Manager: Kiruthika Govindaraju
Cover Designer: Greg Harris

Typeset by MPS Limited, Chennai, India

Contents

Preface

Micro/nanosatellites are satellites based on microelectromechanical integrated system (MEMS) technology and a new MEMS-based integrated micro-instrument (ASIM). In multidimensional integration technology large-scale integrated circuit design ideas and manufacturing processes, including the mechanical components such as electronic circuits are integrated, and the sensors, actuators, microprocessors, and other electrical and optical systems are integrated in a very small. Within this space, the formation of mechanical and electrical integration for a specific function of satellite components or systems takes place.

In the late 1980s, microspacecraft, represented by micro/nano-type satellites, became the most active research direction in the field of space, with a new concept and brand-new design. The micro/nano-type satellite has a large number of high-tech advantages, with excellent functional density and technical performance, low investment and operating costs, flexibility, short development cycle, and the advantages of being so small, so that it can not only complete for aerospace tasks, but also in multisatellite network, formation flight, and virtual satellites in completing difficult tasks in space. Micro/nano-type satellites are now widely used in data communications and transmission, ground and space environment monitoring, navigation and positioning, as well as scientific experiments, and many other areas.

In 1994, I wrote a paper entitled "Dual-Purpose Technology for the 21st Century—Micron/Nanotechnology," pointing out that micro/nanotechnology is an important dual-use technology for the 21st century, and its emergence will have a significant impact on economic and national security, in particular, I pointed out that MEMS technology will play a major role in microsatellites.

At Tsinghua University, Professor Youzhe and his team at the time seized development opportunities with the British University of Surrey Space Centre to cooperate in work on advanced technology based on micro/nanotechnology, especially MEMS, with independent innovations developed NS-1/NS-2,

MEMSat, and other micro/nanosatellites, resulting in the reinnovation of micro/nanosatellite technology development. On the one hand, this book explores the use of MEMS technology as the main representative of micron/nanotechnology in the field of space applications, and the use of MEMS technology as in microelectronics, optoelectronics, and micromechanical, ultrafine processing, and other new technologies. This chapter describes several MEMS devices and microcomponents suitable for use in the aerospace field. On the other hand, the authors have developed a new method for the miniaturization and intelligentization of the main functional components of spacecraft, which are suitable for work in the space environment. It is suitable for MEMS devices and microsystems used in the aerospace field and is based on miniaturization, lightweight, low cost, short cycle, and high performance. This book introduces their own research and development of China's first nano-type satellite (NS-1) and MEMS technology based on high-performance miniaturization of satellite functional components and other scientific research. These achievements have won the National Science and Technology Award 2, the National Science and Technology Progress Award 2, and more than 20 national invention patents for the authors' microsatellite technology and MEMS technology, with spatial applications making an important contribution.

It is hoped that the publication of this book will help and inspire scientific and technical personnel who are interested in microsatellite technology, especially MEMS in the field of aerospace applications.

May 2012

CHAPTER 1

Micro/Nano Satellite System Technology

Spacecraft system design technology is closely related to the system design technology according to user requirements in spacecraft and flight processes [1–3]. For micro/nano satellites, the user requirements according to a specific mission are as follow: comprehensive demonstration of its function and the system technical indicators; coordination of the interface and constraints with the rocket, launch site, test and control network, ground application system and other systems; analysis and selection of payload configuration; selection and design of the orbit to achieve the mission; completing the system technical scheme and satellite configuration design; on the basis of the system plan and optimization, determining the technical requirements of each subsystem; structure and mechanism, thermal control, integrated electronics closely related to the general system design and test; system integration scheme determination, final assembly designation, the plan formulation and implementation of system circuit design and performance test after assembling and integration; and determination of components and system-level environmental test conditions, the ground validation test plan, and spacecraft construction rules, etc.

According to the system design of the assignments, tasks, and nature, the system design of a micro/nano satellite is the top design and comprehensive design of the micro/nano satellite. It plays an important role in the realization of the whole satellite mission [4]. The system scheme design determines the spacecraft development direction, the system situation, the scheme, and the design of the subsystem requirements. The quality of the system scheme design will directly affect the overall quality, performance, development cycle, and cost of the satellite.

The key technical problems in system design of the satellite are presented in this chapter. The contents mainly include: micro/nano satellite mission analysis; orbit design and analysis; subsystem scheme selection and demonstration; micro/nano satellite configuration design; analysis and determination of the system performance; characteristic parameter assessment of the satellite

Space Microsystems and Micro/Nano Satellites. DOI: http://dx.doi.org/10.1016/B978-0-12-812672-1.00001-1

system; reliability and safety analysis; and development of the technical processes of the satellite.

1.1 NS-1 NANOSATELLITE TASK ANALYSIS

NS-1 is an exploration test satellite applying new high-end technology, through the research of some key technologies and development, aiming to develop nanosatellite platforms, to carry out critical load-carrying experiments, and to complete the space flight demonstration of new high-end technology. The main task of the test includes the following:

1. CMOS camera tests for Earth imaging [5]: by CMOS camera, shot, storage, and transmission of ground targets, and orbit demonstration experiment of image information processing technology. The CMOS camera field of view has an angle of 12 degrees, and 1024×1024 pixels.
2. Microinertial measurement unit (MIMU) flight experiment: by experiment, to test the performance of the microaccelerometer, microgyros in-orbit, to understand its adaptability to the environment, to test navigation, and attitude determination by the MIMU grouped with other sensors.
3. Small satellite orbit-maintaining and orbit maneuver test: the satellite propulsion system uses liquid ammonia as propellant, the system is relatively simple. To verify satellite orbit maneuver and orbit maintenance workability, provides experience for small satellite networks and formation flights.
4. Satellite program uploading and software test: via satellite system and application programs uploading, the satellite has the ability to update online. At the same time, it reduces the pressure on software development and testing before flight.
5. Some component-carrying tests: arranging for a flight test for some superior performance without flying experience device, to understand its space environment adaptability, provides a reference for subsequent model selection.

1.2 NS-1 NANOSATELLITE SYSTEM SCHEME

The NS-1 nanosatellite is a new high-technology demonstration satellite. In the process of satellite design, lessons were drawn from domestic and foreign satellite technology [6–9]. It combines domestic high-tech achievements, such as microelectronics, with new process technology. Compared with similar foreign satellites, its function is complete and the performance indicators are advanced. The satellite consists of two parts: the payload and service system. The payload consists of a CMOS camera and its control circuit, MIMU, GPS receiver, and new chemical propulsion systems. The service systems consist of the structure, power, thermal control, attitude control, data management, and Telemetry, Tracking and Control (TT&C) communication function module.

NS-1 has adopted an integrated design technology, with the structure, layout, and thermal design according to the payload [10–13]. Adopted electronic integration technology has the onboard computer as the core. To integrate the electronic system, it is necessary for unified management and scheduling of the equipment and resources on the satellite. The computer network has the functions of software uploading and refactoring, which can realize failure recovery. The reliability of the system is high. The system composition block diagram is shown in Fig. 1.1.

The structure subsystem is combined with each subsystem, bearing and passing the dynamic and static load of the carrier rocket, providing a stable work platform for satellite. The main body structure of the system consists of the plate and frame structure. Considering the requirement for weight reduction

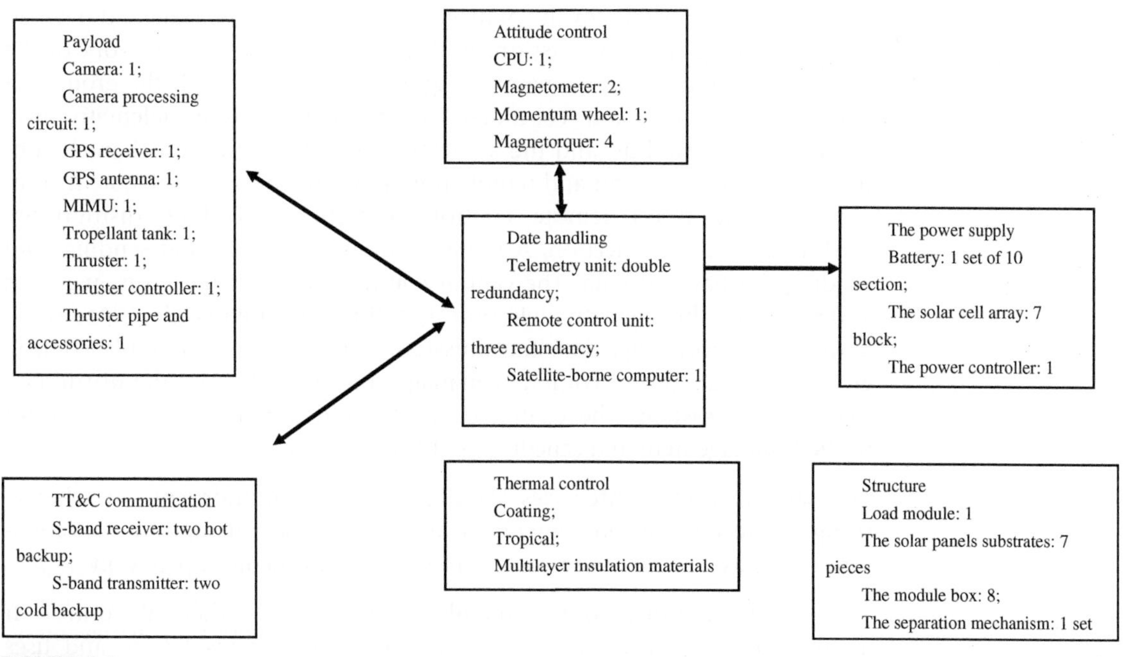

FIGURE 1.1
System composition block diagram.

of the system, an aluminum honeycomb structure is used. The substrate of the solar cell also uses an aluminum honeycomb structure.

The function of the power supply subsystem provides plenty of direct current power for the payload and each of the subsystems during the satellite flight stages, including the free flight phase after satellite–launcher separation and the normal operation stage of the satellite. The satellite power subsystem adopts a high-efficiency gallium arsenide solar cell array and nickel cadmium battery joint as the power supply. In the light area, the solar cell array and battery jointly supply power. At the same time, the solar cell array provides battery charging; when in the shadow area of the orbit, the battery supplies power, ensuring the satellite system works continuously during the eclipse.

Considering the characteristics of small-satellite TT&C communication, includes incorporating a telemetry channel and digital channel, using GPS satellite orbit determination, communicating by an S band transceiver, providing satellite remote sensing, remote control, and an uplink and downlink channel for data injection. The RF channel, in addition to sending and receiving data, also provides the antenna tracking beacon, guiding the antenna for automatic tracking of the satellite and other simple communication functions, forming a multitasking reuse module.

Data/service management is the core of the satellite service management system, and is responsible for the management of the satellite state and payload data processing and transmission. It also maintains the normal work of the satellite and is the core to maintaining effective contact with the ground. It consists of an onboard computer, remote control unit, telemetry unit through CAN data bus, and the asynchronous communication channel connection. The control unit and remote unit as an independent subsystem complete satellite direct remote control instruction decoding, distribution, satellite state data collection, A/D transformation, coding and sending, and sending the payload data. The central control computer monitors, manages and schedules the telemetry, telecontrol, uplink software injection, load operation, data processing, data transmission, data store, and attitude control, etc. The central control computer manages the subsystem by the distributed management system. The central computer and the slave computer adopt unified management and scheduling of the CAN bus.

The satellite uses a three-axes stabilized satellite attitude control scheme, using a three-axes magnetometer for attitude measurement, add a momentum wheel as an actuator to control the attitude of medium precision.

The satellite thermal control mainly applies passive thermal control. It adopts the method of paint and bandage multilayer insulation, and uses active thermal control of the propulsion system. Satellite information flow is shown in Fig. 1.2.

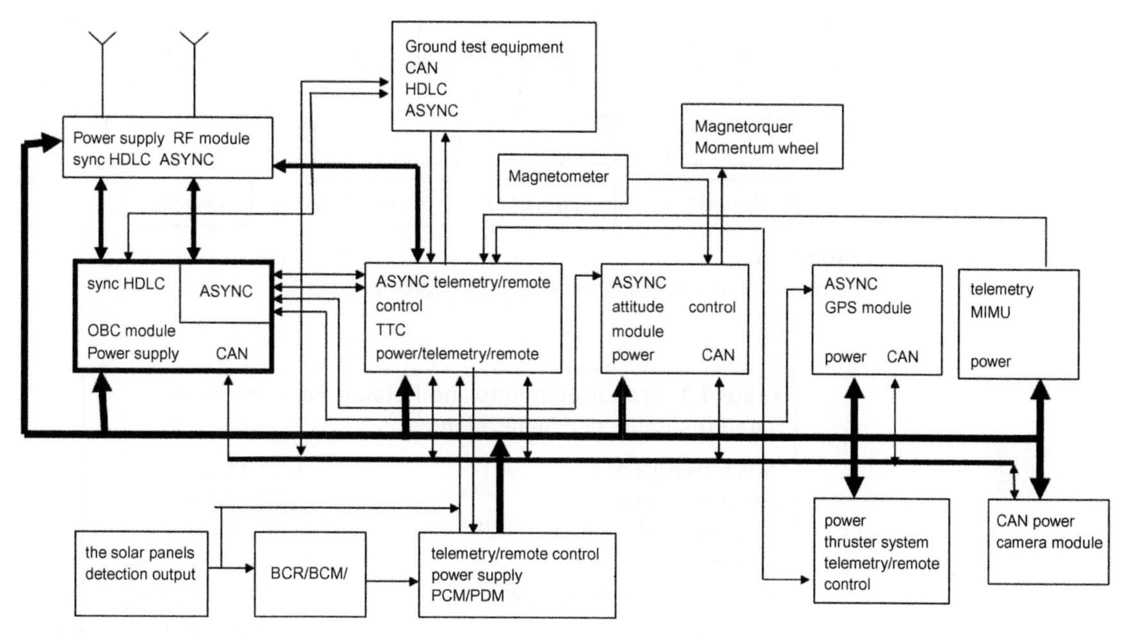

FIGURE 1.2
NS-1 information flow diagram.

After NS-1 satellite–launcher separation, the minimum system automatically starts charging. Once the satellite and launcher have separated and entered orbit, they successively experience free flight, on-orbit test, establishing normal running state and effective load test phase, and satellite specific working mode including the following:

- Orbit mode (minimum system);
- Platform test mode;
- Gesture capture mode;
- Normal mode;
- Load test mode.

Some subsystems of NS-1 have backup capability. Under the condition of local anomalies, they can switch and reconstruct at the component level. Normal and abnormal patterns are shown in Fig. 1.3. The final safe model is the orbit model.

1.3 NS-1 ANALYSIS OF THE SATELLITE INITIAL ORBIT

The NS-1 orbital elements and related parameters in satellite–launcher separation are as indicated in Table 1.1.

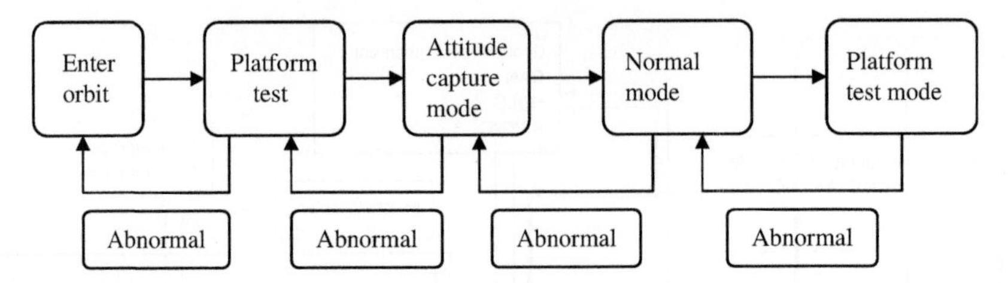

FIGURE 1.3
Relations of various kinds of patterns in normal and abnormal cases.

Table 1.1 Orbital Elements and Related Parameters in Satellite–Launcher Separation

Orbit semimajor axis	6977093.957 m
Orbital eccentricity	0.000166475
Orbital inclination	97.70029 degrees
Perigee argument	263.02387 degrees
Longitude of ascending node	104.04348 degrees
True anomaly	155.46242 degrees
Perigee height	597792.4 m
Apogee altitude	600115.5 m
Absolute speed	7557.28614 m/s
Local absolute velocity azimuth	345.496873 degrees
Absolute speed local angle	0.003962 degree
Geocentric radius vector	6978150.5 m
Longitude	91.716070 degrees
Geocentric latitude	57.653462 degrees
Altitude	615295.5 m
Voyage	3384948.8 m
Perigee geocentric latitude	−79.6237130 degrees
Cumulative departure time after launch	746.9085 s
Flat anomaly	155.45449757 degrees

According to the above parameters, we obtain the initial orbit in Fig. 1.4.

1.3.1 Satellite Orbit Entry Initial Attitude Features

From the rocket flight timing and flight angle, we get the initial attitude at the separation point of the NS-1 satellite and the CZ-2C rocket:

$$\Phi = -62.1736 \text{ degrees}$$
$$\Psi = 0 \text{ degree}$$
$$\Gamma = 0 \text{ degree}$$

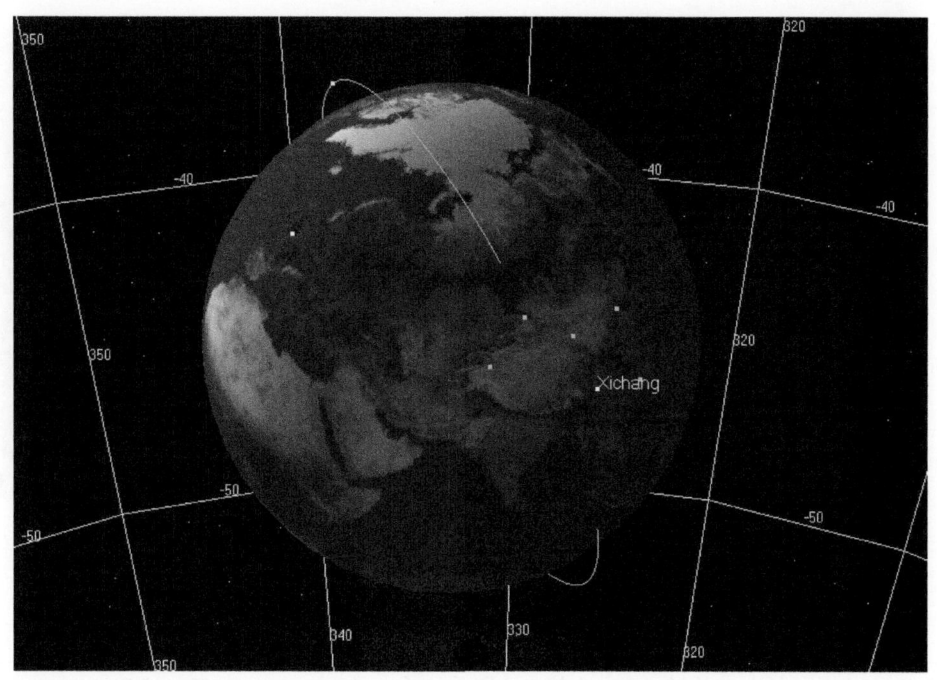

FIGURE 1.4
Initial orbit.

At the satellite–launcher separation moment, the initial attitude angular rate of NS-1 is less than 4 degrees/s around three axes.

1.3.2 Initial Orbit Characteristics Analysis

1.3.2.1 Ground Measurement and Controlability Analysis

1. Main control stations (Beijing) control segment
   According to the initial orbit data of NS-1 satellite and parameters of the main control stations, calculate the control segment of the main ground control stations located in Beijing (Fig. 1.5) within 24 hours. The specific control segment forecast value is shown in Table 1.2. For convenience, the time data in Table 1.2 use UTC, and azimuth is defined as a north direction with clockwise rotation angle.
2. Auxiliary remote sensing segment
   The auxiliary remote sensing segment, according to the initial NS-1 satellite orbit data and the parameters of auxiliary control stations, calculates the remote sensing segment of the NS-1 satellite orbit within

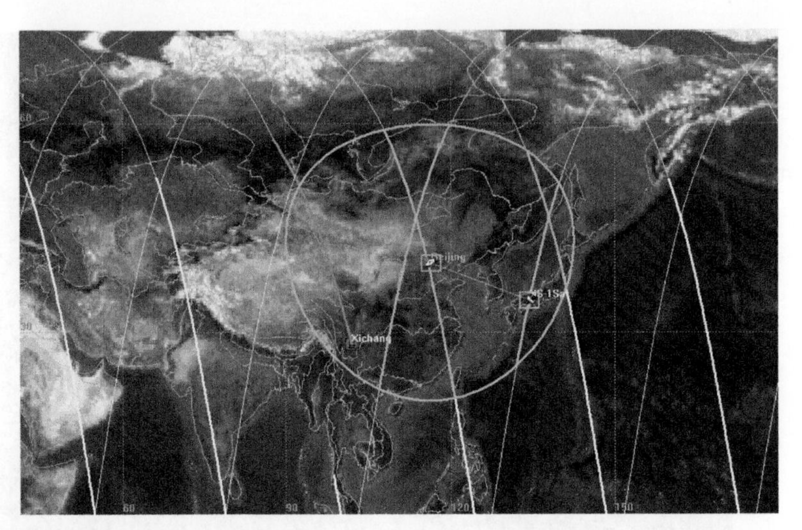

FIGURE 1.5
Control segment of Beijing station within 24 h.

Table 1.2 Calculation Value of Measured Arc in Beijing Station

Order	Time Information		Space Information			Duration Overhead (s)
	Time (UTC)	Event	Azimuth (degree)	Pitch (degree)	Distance (km)	
1	13 Apr 2004 02:03:50.99	Get in	47.411	5.000	2379.134273	427.644
	13 Apr 2004 02:04:50.59	In station	56.804	7.882	2135.761585	
	13 Apr 2004 02:05:50.59	In station	68.459	10.369	1951.462910	
	13 Apr 2004 02:06:50.59	In station	82.104	11.919	1847.357295	
	13 Apr 2004 02:07:50.59	In station	96.686	12.045	1837.799911	
	13 Apr 2004 02:08:50.59	In station	110.618	10.694	1924.444941	
	13 Apr 2004 02:09:50.59	In station	122.686	8.298	2095.250678	
	13 Apr 2004 02:10:50.59	In station	132.517	5.403	2331.337992	
	13 Apr 2004 02:10:58.63	Get out	133.673	5.000	2366.908626	
2	13 Apr 2004 03:38:23.81	Get in	5.098	4.999	2381.154871	620.893
	13 Apr 2004 03:39:23.59	In station	2.850	10.062	1976.386257	
	13 Apr 2004 03:40:23.59	In station	359.195	16.699	1582.391016	
	13 Apr 2004 03:41:23.59	In station	352.506	26.108	1215.022186	
	13 Apr 2004 03:42:23.59	In station	337.591	40.015	909.678572	
	13 Apr 2004 03:43:23.59	In station	297.633	53.934	748.656884	
	13 Apr 2004 03:44:23.59	In station	245.486	46.378	822.409424	
	13 Apr 2004 03:45:23.59	In station	223.631	30.833	1083.802937	

Continued

Table 1.2 Calculation Value of Measured Arc in Beijing Station *Continued*

Order	Time Information		Space Information			Duration Overhead (s)
	Time (UTC)	Event	Azimuth (degree)	Pitch (degree)	Distance (km)	
	13 Apr 2004 03:46:23.59	In station	214.702	19.818	1432.877322	
	13 Apr 2004 03:47:23.59	In station	210.103	12.270	1818.804644	
	13 Apr 2004 03:48:23.59	In station	207.340	6.673	2221.235163	
	13 Apr 2004 03:48:44.70	Get out	206.616	5.000	2364.916387	
3	13 Apr 2004 12:57:24.27	Get in	91.705	5.000	2349.779466	314.554
	13 Apr 2004 12:58:23.59	In station	81.208	7.040	2178.142643	
	13 Apr 2004 12:59:23.59	In station	69.095	8.307	2080.960482	
	13 Apr 2004 13:00:23.59	In station	56.297	8.461	2071.910102	
	13 Apr 2004 13:01:23.59	In station	43.962	7.461	2152.241815	
	13 Apr 2004 13:02:23.59	In station	33.021	5.574	2312.424775	
	13 Apr 2004 13:02:38.82	Get out	30.523	5.000	2363.968953	
4	13 Apr 2004 14:29:58.08	Get in	167.772	4.999	2336.234753	625.578
	13 Apr 2004 14:30:57.59	In station	168.170	10.242	1925.027026	
	13 Apr 2004 14:31:57.59	In station	168.749	17.360	1519.022219	
	13 Apr 2004 14:32:57.59	In station	169.739	28.156	1133.328629	
	13 Apr 2004 14:33:57.59	In station	172.153	47.109	800.611291	
	13 Apr 2004 14:34:57.59	In station	196.256	80.904	616.141945	
	13 Apr 2004 14:35:57.59	In station	340.342	57.961	708.018406	
	13 Apr 2004 14:36:57.59	In station	344.474	34.093	1002.592398	
	13 Apr 2004 14:37:57.59	In station	345.812	21.066	1374.374455	
	13 Apr 2004 14:38:57.59	In station	346.556	12.914	1774.356679	
	13 Apr 2004 14:39:57.59	In station	347.080	7.098	2185.814076	
	13 Apr 2004 14:40:23.65	Get out	347.273	5.000	2366.239406	

24 hours from Guangzhou station, Dongfeng station, Kashi station, Surrey stations, and other auxiliary stations (Figs. 1.6–1.9, respectively), then to obtain the corresponding remote sensing segment forecast values data (data omitted).

According to the ground station control segment forecast value, the NS-1 satellite overhead action sequence is as illustrated in Table 1.3.

It can be seen from the above data that if we can use Kashi and Surrey S-band telemetry signal receiving stations, we will be able to greatly improve the reliability of the telemetry signals from the ground station to the NS-1 satellite early in the orbit.

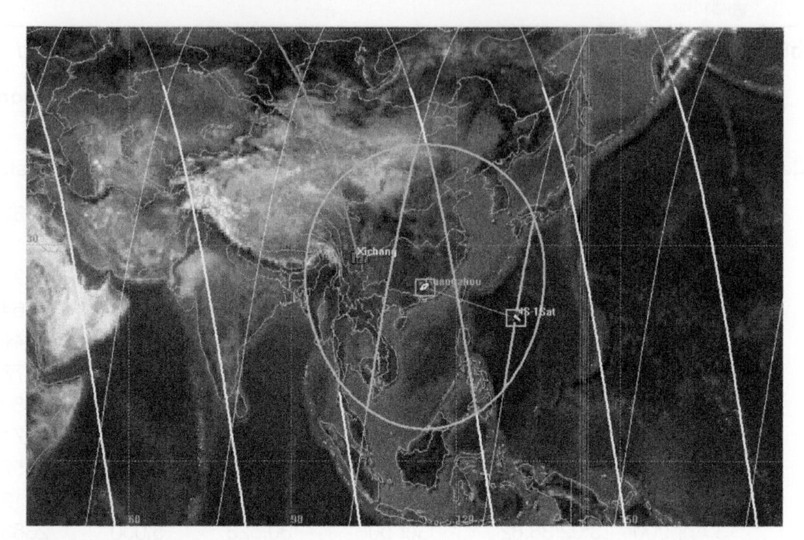

FIGURE 1.6
Control segment of Guangzhou station within 24 h.

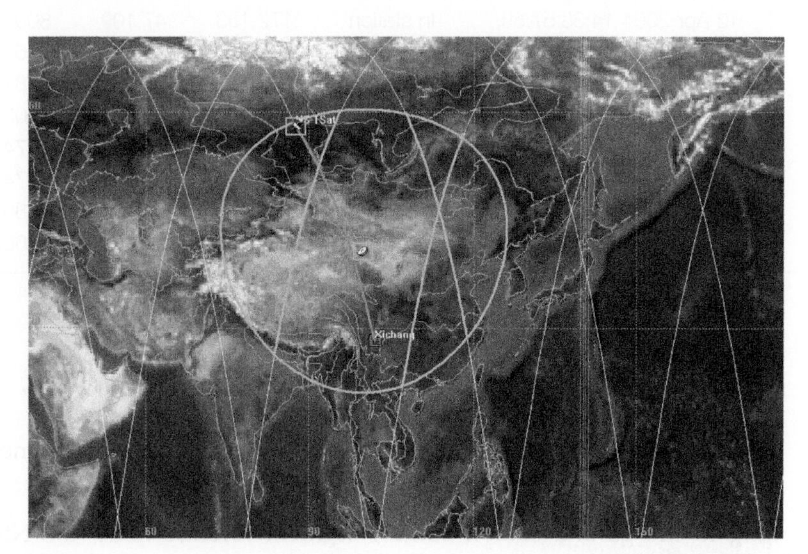

FIGURE 1.7
Control segment of Dongfeng station within 24 h.

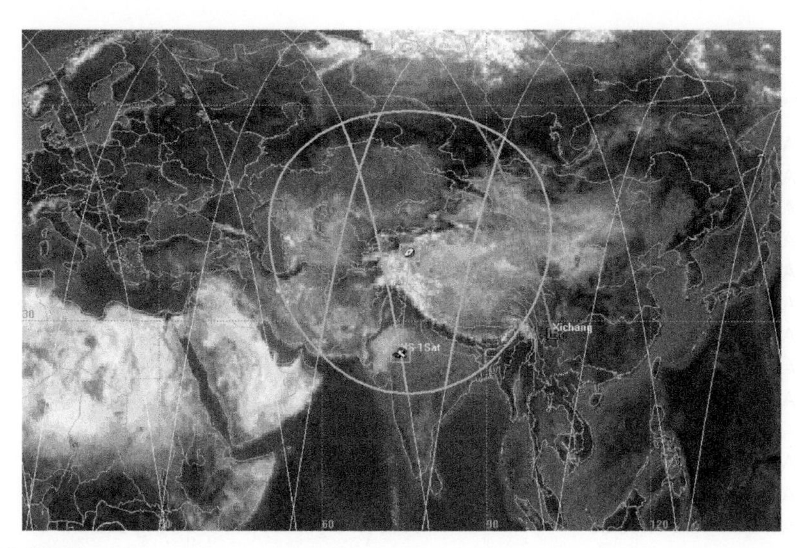

FIGURE 1.8
Control segment of Kashi station within 24 h.

FIGURE 1.9
Control segment of Surrey station within 24 h.

Table 1.3 Satellite Overhead Sequence

Sequence	Time (UTC)	Overhead Action Illustration
1	12 Apr 2004 16:03:49.59	First time getting into Dongfeng station
2	12 Apr 2004 16:04:03.09	First time getting out of Dongfeng station
3	12 Apr 2004 17:30:23.10	First time getting into Kashi station
4	12 Apr 2004 17:40:36.77	First time getting out of Kashi station
5	12 Apr 2004 20:49:29.76	First time getting into Surrey station
6	12 Apr 2004 20:58:10.92	First time getting out of Surrey station
7	12 Apr 2004 22:24:10.81	Second time getting into Surrey station
8	12 Apr 2004 22:34:32.10	Second time getting out of Surrey station
9	13 Apr 2004 00:04:15.31	Third time getting into Surrey station
10	13 Apr 2004 00:08:02.60	Third time getting out of Surrey station
11	13 Apr 2004 02:03:50.99	First time getting into Beijing station
12	13 Apr 2004 02:08:37.00	First time getting into Guangzhou station
13	13 Apr 2004 02:10:58.63	First time getting out of Beijing station
14	13 Apr 2004 02:15:05.08	First time getting out of Guangzhou station
15	13 Apr 2004 03:38:23.81	Second time getting into Beijing station
16	13 Apr 2004 03:39:15.69	Second time getting into Dongfeng station
17	13 Apr 2004 03:42:43.42	Second time getting into Guangzhou station
18	13 Apr 2004 03:48:44.70	Second time getting out of Beijing station
19	13 Apr 2004 03:48:45.58	Second time getting out of Dongfeng station
20	13 Apr 2004 03:52:42.80	Second time getting out of Guangzhou station
21	13 Apr 2004 05:15:05.90	Third time getting into Dongfeng station
22	13 Apr 2004 05:16:23.98	Second time getting into Kashi station
23	13 Apr 2004 05:24:27.36	Third time getting out of Dongfeng station
24	13 Apr 2004 05:25:58.45	Second time getting out of Kashi station
25	13 Apr 2004 06:52:19.42	Third time getting into Kashi station
26	13 Apr 2004 07:01:30.02	Third time getting out of Kashi station
27	13 Apr 2004 10:04:19.50	Fourth time getting into Surrey station
28	13 Apr 2004 10:12:54.98	Fourth time getting out of Surrey station
29	13 Apr 2004 11:39:38.28	Fifth time getting into Surrey station
30	13 Apr 2004 11:50:03.54	Fifth time getting out of Surrey station
31	13 Apr 2004 12:57:24.27	Third time getting into Beijing station
32	13 Apr 2004 13:02:38.82	Third time getting out of Beijing station
33	13 Apr 2004 13:16:51.09	Sixth time getting into Surrey station
34	13 Apr 2004 13:22:48.85	Sixth time getting out of Surrey station
35	13 Apr 2004 14:26:13.00	Third time getting into Guangzhou station
36	13 Apr 2004 14:29:58.08	Fourth time getting into Beijing station
37	13 Apr 2004 14:32:07.10	Fourth time getting into Dongfeng station
38	13 Apr 2004 14:36:14.39	Third time getting out of Guangzhou station
39	13 Apr 2004 14:40:23.65	Fourth time getting out of Beijing station
40	13 Apr 2004 14:40:36.69	Fourth time getting out of Dongfeng station

1.3.3 Orbit Lighting Situation Analysis

The satellite orbit lighting situation determines the satellite power supply capacity, because the NS-1 battery capacity is limited, the orbit lighting conditions have fatal effects on satellite life. Therefore the NS-1 satellite orbit track lighting conditions are analyzed and calculated.

1.3.3.1 Orbit Lighting Time

The NS-1 satellite was in orbit. In the 6 months, the orbital illumination time of each circle varies from 3709.044 to 3732.963 s. In other words, the orbital illumination factor is between 0.6395 and 0.6436. The orbit *lighting time* within 3 days after entering orbit is shown in Table 1.4.

It can be seen from Table 1.4 that NS-1 is in the lighting area before satellite–launcher separation, and stays in the illumination area for more than 1 hour after separation, which is beneficial to the satellite power system.

1.3.3.2 Orbit Solar Angle

Because the NS-1 adopted a nearly noon sun synchronous orbit, the orbit solar angle is kept within a small range. In 6 months, the orbit solar angle fluctuated between 14.494 and 19.497 degrees (as shown in Fig. 1.10). In order to investigate the orbit solar angle and its changes of initial orbit, Fig. 1.11 gives the amplification curve of the orbit solar angle within 3 days after entering orbit.

From the orbit lighting situation analysis results, we can see that within 6 months after launch, the lighting situation of the NS-1 orbit is stable, which is in favor of the satellite power subsystem. But due to the small orbit solar angle, and the certain initial attitude of pitch during the satellite–launcher separation, when making flight procedures, ensuring the side of the solar cell array gets enough sunshine should be taken into consideration.

1.4 NS-1 SUBSYSTEM DESIGN

1.4.1 The Power Subsystem Design

1.4.1.1 Summary of System Function and Principle Block Diagram

The power subsystem of the NS-1 consists of a gallium arsenide solar cell array, a cadmium nickel battery, and a power controller. The power controller consists of a battery charge regulator (BCR), power control module (PCM), power distribution module (PDM), and battery circuit monitoring (BCM). The solar cell array is body-mounted, with solar panels installed on six sides and on top of the array. Each of the solar cell array output diode buses is segregated from the busbar. The solar cell array, as a power device,

Table 1.4 Orbit Lighting Time (Within 3 Days After Entering Orbit)

Lighting Start (UTC)	Lighting Stop (UTC)	Duration (s)
12 Apr 2004 16:03:49.59	12 Apr 2004 17:04:26.08	3636.488
12 Apr 2004 17:39:09.86	12 Apr 2004 18:41:18.25	3728.395
12 Apr 2004 19:16:02.07	12 Apr 2004 20:18:10.39	3728.315
12 Apr 2004 20:52:54.26	12 Apr 2004 21:55:02.43	3728.173
12 Apr 2004 22:29:46.36	12 Apr 2004 23:31:54.45	3728.086
13 Apr 2004 00:06:38.44	13 Apr 2004 01:08:46.46	3728.023
13 Apr 2004 01:43:30.50	13 Apr 2004 02:45:38.51	3728.008
13 Apr 2004 03:20:22.58	13 Apr 2004 04:22:30.67	3728.09
13 Apr 2004 04:57:14.76	13 Apr 2004 05:59:22.89	3728.126
13 Apr 2004 06:34:07.00	13 Apr 2004 07:36:15.02	3728.024
13 Apr 2004 08:10:59.15	13 Apr 2004 09:13:07.06	3727.917
13 Apr 2004 09:47:51.24	13 Apr 2004 10:49:59.09	3727.852
13 Apr 2004 11:24:43.31	13 Apr 2004 12:26:51.08	3727.773
13 Apr 2004 13:01:35.36	13 Apr 2004 14:03:43.09	3727.733
13 Apr 2004 14:38:27.45	13 Apr 2004 15:40:35.20	3727.748
13 Apr 2004 16:15:19.59	13 Apr 2004 17:17:27.38	3727.789
13 Apr 2004 17:52:11.78	13 Apr 2004 18:54:19.53	3727.75
13 Apr 2004 19:29:03.98	13 Apr 2004 20:31:11.64	3727.653
13 Apr 2004 21:05:56.13	13 Apr 2004 22:08:03.65	3727.512
13 Apr 2004 22:42:48.21	13 Apr 2004 23:44:55.65	3727.44
14 Apr 2004 00:19:40.28	14 Apr 2004 01:21:47.65	3727.372
14 Apr 2004 01:56:32.32	14 Apr 2004 02:58:39.68	3727.368
14 Apr 2004 03:33:24.39	14 Apr 2004 04:35:31.85	3727.46
14 Apr 2004 05:10:16.56	14 Apr 2004 06:12:24.04	3727.471
14 Apr 2004 06:47:08.77	14 Apr 2004 07:49:16.14	3727.372
14 Apr 2004 08:24:00.90	14 Apr 2004 09:26:08.17	3727.267
14 Apr 2004 10:00:52.97	14 Apr 2004 11:03:00.17	3727.199
14 Apr 2004 11:37:45.02	14 Apr 2004 12:39:52.14	3727.126
14 Apr 2004 13:14:37.06	14 Apr 2004 14:16:44.15	3727.093
14 Apr 2004 14:51:29.14	14 Apr 2004 15:53:36.25	3727.115
14 Apr 2004 16:28:21.27	14 Apr 2004 17:30:28.42	3727.15
14 Apr 2004 18:05:13.45	14 Apr 2004 19:07:20.55	3727.102
14 Apr 2004 19:42:05.63	14 Apr 2004 20:44:12.62	3726.989
14 Apr 2004 21:18:57.75	14 Apr 2004 22:21:04.60	3726.85
14 Apr 2004 22:55:49.80	14 Apr 2004 23:57:56.59	3726.791
15 Apr 2004 00:32:41.84	15 Apr 2004 01:34:48.57	3726.721
15 Apr 2004 02:09:33.86	15 Apr 2004 03:11:40.59	3726.726
15 Apr 2004 03:46:25.92	15 Apr 2004 04:48:32.75	3726.828
15 Apr 2004 05:23:18.09	15 Apr 2004 06:25:24.91	3726.816
15 Apr 2004 07:00:10.27	15 Apr 2004 08:02:16.99	3726.72
15 Apr 2004 08:37:02.37	15 Apr 2004 09:39:08.99	3726.617
15 Apr 2004 10:13:54.42	15 Apr 2004 11:16:00.96	3726.542
15 Apr 2004 11:50:46.44	15 Apr 2004 12:52:52.92	3726.476
15 Apr 2004 13:27:38.46	15 Apr 2004 14:29:44.92	3726.452
15 Apr 2004 15:04:30.53	15 Apr 2004 16:06:37.01	3726.484

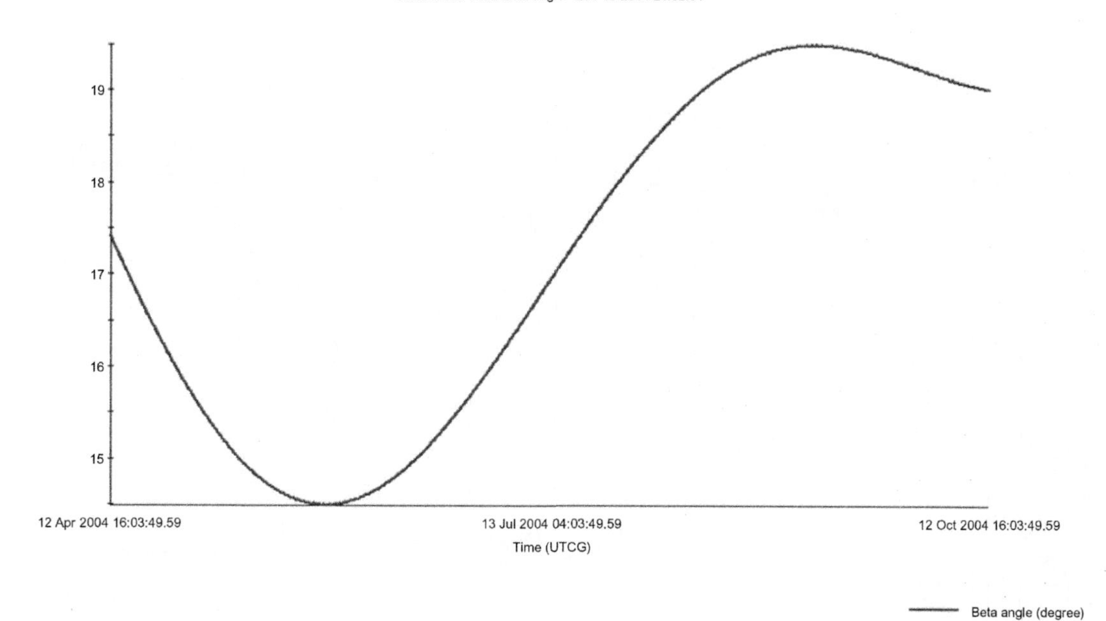

FIGURE 1.10
Orbit solar angle (in 6 months).

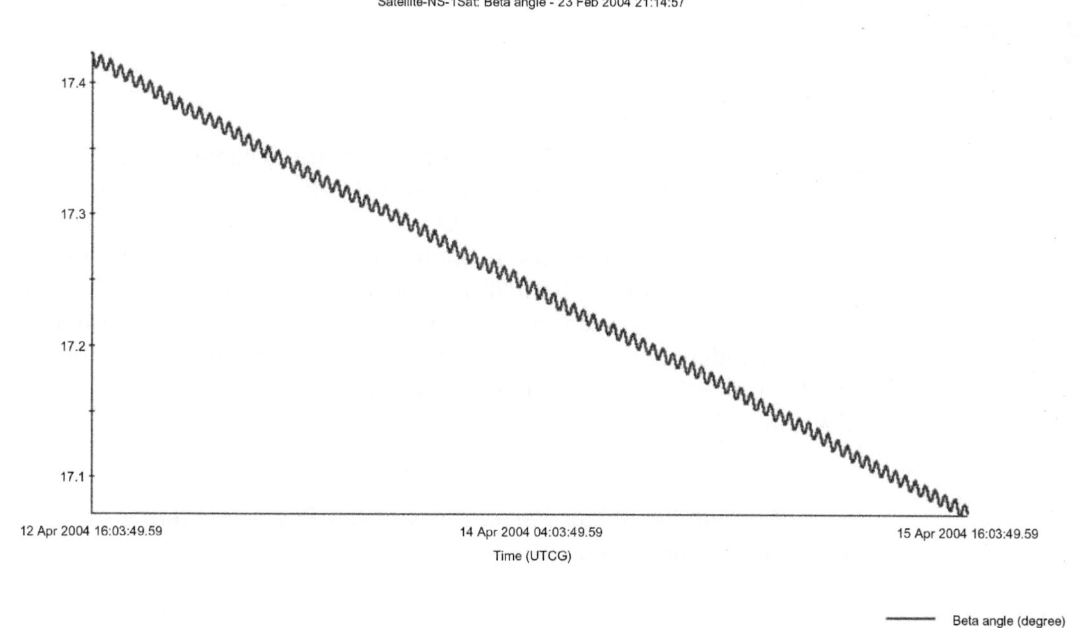

FIGURE 1.11
Orbit solar angle (within 3 days).

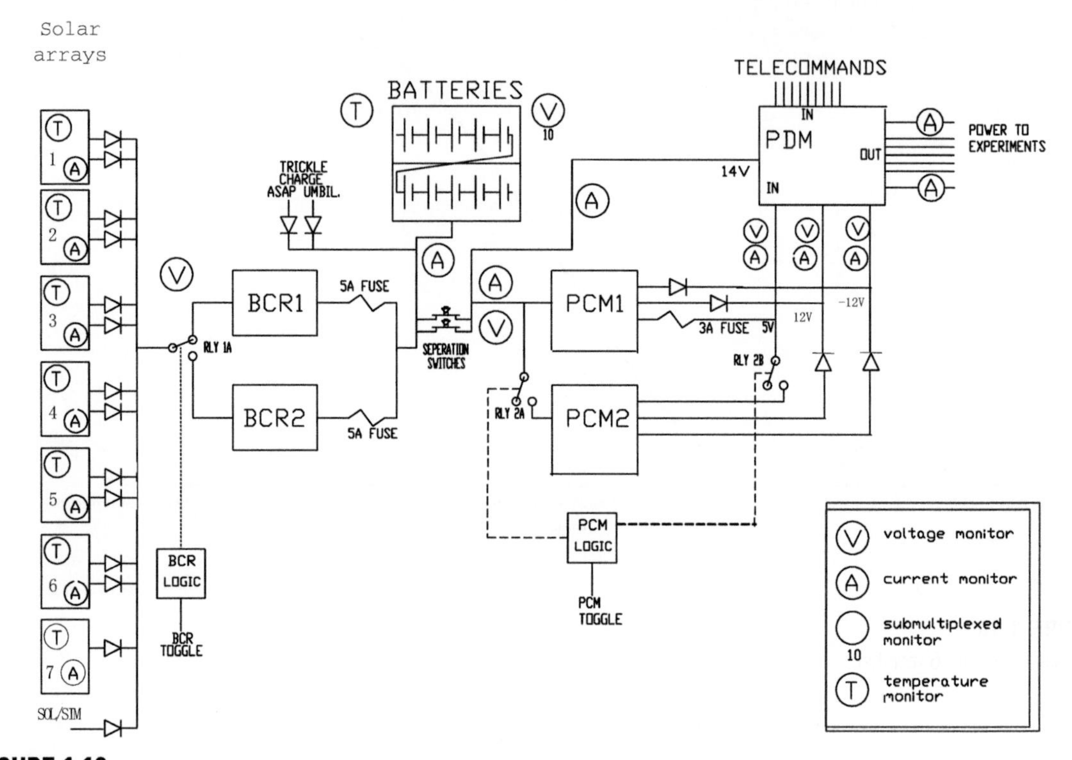

FIGURE 1.12
The power subsystem principle block diagram.

after the BCR, completes power adjustment, battery charging and discharging adjustment and control, and associates with the battery as the satellite primary source. The battery set consists of special test and screening of cadmium nickel batteries, as power supply for the load and platform during the eclipse, and can provide limited peak power to supplement any shortage of the solar cell array output as the primary source supply power. Each subsystem's secondary power supply is produced by the primary source, through a PCM for DC/DC conversion. The power supply subsystem principal block diagram is shown in Fig. 1.12.

1.4.1.2 Main Technical Parameters of the Power System

The main technical parameters of the power system include:

- Average output power: 10 W;
- Primary supply voltage: 14 V;
- Secondary power supply voltage: ±12 V, 5 V.

The power supply performance metrics efficiencies are:

- Efficiency of BCR >85%;
- Efficiency of PCM >70%;
- Efficiency of the battery is about 90%.

The source impedance is low, and a large current pulse load is allowed for a short time. The size is of two standard module boxes (not including the battery).

Temperature ranges are:

- Power controller: −25°C to +55°C;
- Battery: 0–40°C;
- Solar panels: −80°C to +80°C.

The solar cell array is a single-junction gallium arsenide (GaAs) solar cell; with a size of 40 mm × 20 mm × 0.175 mm. The average conversion efficiency is EFF = 18.5% (25°C, AM0), $EFF_{min} \geq 17.0\%$ (25°C, AM0), using solar panels (Table 1.5).

The sides of the solar panels are composed of six pieces of base plate, and the size is 440 mm × 182.5 mm × 10 mm, made from 9.4 mm aluminum honeycombs and each is constructed from 0.3 mm thick aluminum plate, with the up and down surfaces pasted with a 50 mm thick polyimide membrane. Seventy pieces of solar cells are pasted flat on each solar panel. Battery pieces consisted of 35 in series and 2 in parallel. The output voltage of maximum power point of the solar battery array is about 30.1 V (25°C, AMO), within the range of −80 to +80°C, with the maximum power point voltage of about 26.25–37.45 V (AM0). Each panels output power is >13.4 W (AM0, 25°C); each board under the base board is equipped with three temperature sensors, independently measuring the temperature of the solar cells. Among them, two sensors are used for BCR panel temperature compensation, and the other one is for temperature remote sensing.

Table 1.5 Power Supply Metrics

Nominal Voltage (V)	Voltage Range (V)	Ripple Voltage vpp (mV)
14	11–15	150
5	5–5.5	50
12	11.5–12.5	120
−12	−12.5 to −11.5	120

Table 1.6 Cadmium Nickel Battery

Type	Commercial Type of Cadmium Nickel Battery
Sum number	10
Rated capacity	3 Ah
Maximum depth of discharge	40%
Monomer rated voltage	1.2 V

Solar panels have one piece of hexagonal top substrate, whose side length is 182.5 mm (thickness 15 mm). The up and down surfaces are made from 14.0 mm aluminum honeycomb and 0.5 mm thick aluminum plate, with 50 mm thick polyimide membrane pastes on the top surface. Battery pieces consist of 35 in series and 1 in parallel. The panelsoutput power is >6.7 W (AM0, 25°C); panels under the base board are equipped with a temperature sensor, which is used to measure the temperature of the solar cells.

Cadmium nickel battery details are given in Table 1.6.

Monomer batteries experience a series of tests and trials, 10 are screened and matched for encapsulating into the battery set.

1.4.1.3 The Power Supply Reliability Design

The following measures should be taken to protect the power subsystem and power supply equipment.

1. Redundancy design. This is mainly reflected in the backup of PCM and BCR. When a failure occurs, an independent control switch is the priority way, with a remote control switch being complementary. The power conditioning and distribution subsystem is designed as double point and double line system. The design rules out the possibility of a single point failure.
2. Charging system designed with an overcharge protection function. The power supply has an overload protection function, with a power distribution switch limiting flow function. A power circuit series with fuses protects the power supply from being damaged in the case of a load short circuit.
3. Most selected components meet military standards and American 883 standards. Part of the devices meet the industrial- and commercial-grade. The commercial-grade devices are used in Tsinghua I.
4. High equivalent of voltage, inductance, resistance, and impedance are required for the transformers and inductors. Component-level temperature impact test is also needed.

1.4.2 Telemetry and Remote Control System (TTC Subsystem)

1.4.2.1 TTC Subsystem Main Technical Indicators

1. Remote control unit:
   Working voltage: +5 V to +5.5 V. Command channel: 34 way; single chip microcomputer: three pieces of 87C52;
   Communication: code rate of 9.6 kbps;
   RF subsystem interface: 01 decoder and 02 decoder asynchronous serial input. RF subsystem respectively connected to two receivers: RX0 and RX1 asynchronous serial port. 03 decoder asynchronous serial input by multichannel switch by polling the RX0 from radio frequency subsystem, RX1 and OBC, TLM1, ground test equipment of asynchronous serial input signal.
2. Remote sensing unit:
   Working voltage: (1) +5 V to +5.5 V; (2) +11.5 V to 12.5 V; (3) −11.5 V to +12.5 V;
   Telemetry channel: 46 analogue channels, 16 digital channels, 34 remote control instruction channels;
   Normal work of the analogue input voltage range: 0−5 V.
   Analogue maximum allowable range of input voltage: 0.6 V to +11 V.
   Double-machine hot backup: The microcontroller of TLM0, using 87 C52 TLM1 microcontroller with belt CAN SAH-C515C MCU controller.
   Communication interface: the code rate is 9.6 kbps. Asynchronous serial port TLM0, TLM1 telemetry data can be output to the OBC, RF, and ground test equipment. TLM1 CAN can also be received from the CAN bus on remote data frame by asynchronous serial port output to 03 decoder, CAN transmit information through CAN bus communication bus.

Reliability is the main technical problem of the TTC subsystem. In order to achieve high reliability, and provide the necessary design measures in design and to strengthen the weak links, avoid a single point of failure, improve the quality of the components, adopt effective software reliability measures, and ensure the TTC system has high reliability.

The main technology and measures for this include:

- Redundancy design technology;
- Derating design technology;
- PCB electromagnetic compatibility design technology;
- Choose standardized interface circuit design;
- For key parts use military and above specification;
- Hardware watchdog technology.

Software reliability design: as the channel characteristics of tasks, TLM0 will have no interruption telemetry channel 0 programming sequential structure, making TLM0 reliability greatly improved.

1.4.3 Satellite Onboard Computer System (OBC Subsystem)

1.4.3.1 Main Function of OBC Subsystem

The main tasks of the OBC subsystem are to undertake the following:

- Satellite-borne GPS receiving device to receive the label and rail information must be sent to the onboard computer, calculated by the space-borne unit frame, packaging, to send the satellite RF subsystem down to Earth.
- CMOS camera after CAN sent to space-borne computer-generated graphics data, cached by onboard computer, framing and packaging, via satellite RF subsystem down to Earth.
- As the controller of the attitude control system, the control system application software runs on OBC, work attitude control needs a lot of math, this part of the work should be done by onboard computer calculation.
- The attitude of the satellite and other telemetry parameters must be approved by the onboard computer next to the ground, providing measurements of the status of the satellite.
- Indirect remote command received by OBC, analysis, and then via CAN bus, sent to the TTC subsystem; issue instructions by remote control unit.
- Clock management tasks, regardless of the telemetry parameter or remote control command, need the clock to cooperate; an effective task scheduling label also needs to have time.
- Satellite operating system and application should be able to overload, to complete the function of satellite refactoring.

1.4.3.2 OBC System Overall Design

The onboard computer systems use Intel 32-bit RISC processor SA1110 as the CPU, its low power consumption characteristic is suitable for low power applications, and its system operating frequency is 133 MHz/206 MHz. Onboard computer system development is mainly divided into two parts: the hardware system and the software system. The hardware design includes processors, memory, interrupt, RAMDISK, ISCC (integrated serial communication), EDAC error (remedy), bus controller, and remote control remote node. Software development includes the system boot firmware, the underlying drivers, the operating system kernel, application software, etc. Operating system and application software are uploaded on the orbit, for setting up a flexible and fault tolerance satellite system.

1. The satellite's computer hardware system is shown in Fig. 1.13.
   System configuration is as follows:
   - The Intel SA 1110 CPU processor, frequency 206 MHz @ 206 mW;
   - 512K EPROM memory;
   - Program memory with TMR rectifying error detection (1.5 M);
   - 6 M data storage;
   - Timer watchdog technology;
   - CAN bus control node;
   - All landowners and RF S-band in HDLC communication link between the synchronization end up with RF, ADCS, GPS, and TTC asynchronous communication links.
2. The software system of the satellite

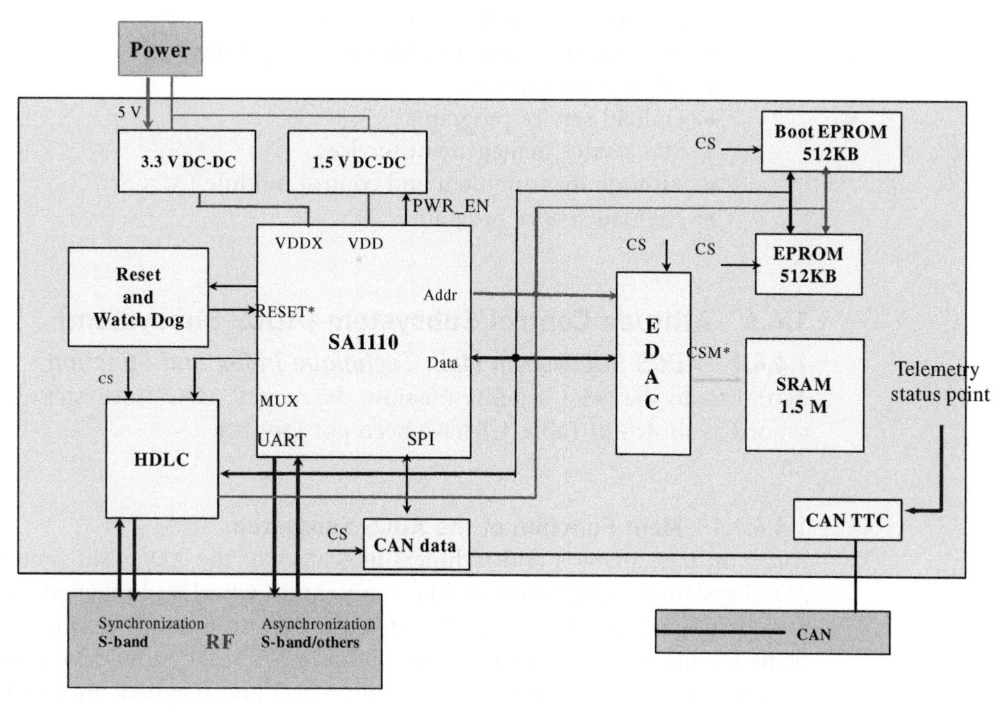

FIGURE 1.13
Schematic view of the computer on NS-1.

The satellite's software system adopts UCOS II operating system, so the development of the entire software system includes operating systems, drivers, and applications in three parts:

- The operating system. Using the UCOS II operating system.
- The driver. The underlying hardware including asynchronous serial port operation, synchronous serial port, and CAN. Driver software system is designed with the upper specific application software, with which the driver software completes specific functions.
- Application software. The application basically covers all the functional requirements of the satellite, but does not support the FAT format file (as a binary data stream). It uses the standard C language application program.

The main software module parts include:

- Bootstrap;
- Operating system of the main program;
- Asynchronous serial port driver;
- Synchronous serial port driver;
- CAN driver;
- QAX. 25 communications services;
- CAN communication protocol services;
- Asynchronous communication protocol services;
- File storage service;
- Upload service program;
- Star service management services;
- Attitude measurement and control module (ADCS);
- Payload service program.

1.4.4 Attitude Control Subsystem (ADCS Subsystem)

1.4.4.1 ADCS Subsystem Main Technique Index and Function

According to the NS-1 satellite mission, the overall ADCS subsystem specifications as shown in Table 1.7 have been put forward.

1.4.4.1.1 Main Function of the ADCS Subsystem

Based on task analysis and technical indicators of the ADCS subsystem, the ADCS system control mode for bias momentum wheel and magnetic control way of triaxial stability control is determined. Attitude-measuring components for the magnetometer and the attitude controller should be borne by the OBC, execution parts for momentum wheel and magnetic moment. This can realize lower cost of directional three-axial stability, and meeting the overall technical indicators.

Table 1.7 Specifications of ADCS on NS-1

Content	Specification
Stability maintenance	3-axis
Gesture measuring accuracy	1 degree
Gesture control accuracy	2 degrees
Gesture stability	0.02 degree/s
Antirotation ability	5 degree/s

Table 1.8 Working Mode of ADCS on NS-1

Rate damping mode	Platform instruments and magnetorquer	Eliminate the large angular rate of separation between satellites and carriers
Spinning mode	Platform instruments and magnetorquer	Establish Y-Thomson spin, inhibit nutation
Antispinning mode	Platform instruments, magnetorquer, and momentum wheel	Starting the womentum wheel, reduce the star racing and the pitch axis angular rate
Three-axes stability	Platform, magnetorquer, and momentum wheel	Establish and maintain a three-axes stable attitude towards ground orientation

The main task of the attitude control system is to guarantee the stability of satellite attitude, meeting the requirements of satellite pointing, and ensuring the collaborative satellite payload flight test task is completed. Due to the satellite using an omnidirectional antenna, the communication system satellite attitude can only carry out basic pointing maintenance. The satellite payload's mission is to take photographs and make a flight test. The flight demonstration experiment task of satellite attitude is controlled without special requirements, therefore the ADCS subsystem will mainly finish the mission to take pictures for design basis, to ensure that the control precision of the ADCS subsystem can meet the normal work of the CMOS camera.

1.4.4.2 ADCS Subsystem Main Working Mode

According to the above analysis, we can determine the NS-1 satellite ADCS subsystem of the main working mode as illustrated in Table 1.8.

Software solutions of ADCS subsystem

The NS-1 satellite ADCS subsystem software consists of two parts:

- ADCS master software (running in OBC);
- ADCS PC software (running on MCU, ATC).

The ADCS master software, running in the OBC, is the core of the ADCS subsystem, and is responsible for the entire ADCS scheduling subsystem of management, large computation algorithm processing, collect the underlying data, receiving ground control instructions, etc. The software of the single-chip processor system running in the microcontroller is responsible for collecting and processing the data directly, control and drive related parts, etc.

1.4.4.3 Subsystem Designation of ADCS

1. Subsystem setup of ADCS

   A schematic view of the ADCS on the NS-1DCS hardware system is shown in Fig. 1.14.

   The hardware system includes the following sections:

   - Two underlying squared MPUs (C515C);
   - Two three-axes magnetometers squared and corresponding processing circuit;
   - Squared four magnetic torque and the corresponding driving circuit;
   - A biased momentum wheel and the corresponding drive and control circuit.

   With the MCU and its peripheral circuit, the magnetometer processing circuit, the magnetic torquer and momentum wheel drive control circuit of the ADCS subsystem of measurement and control

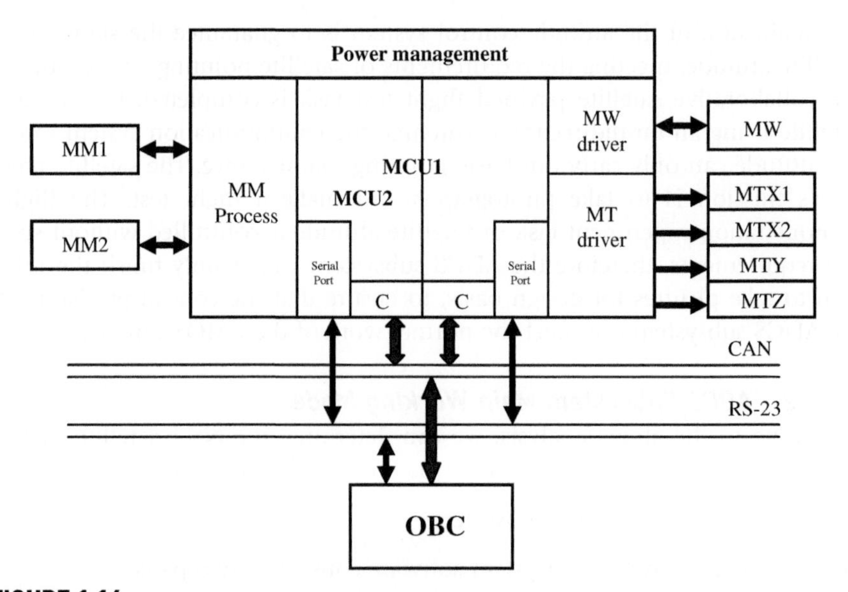

FIGURE 1.14

NS-1 satellite ADCS subsystem hardware structure.

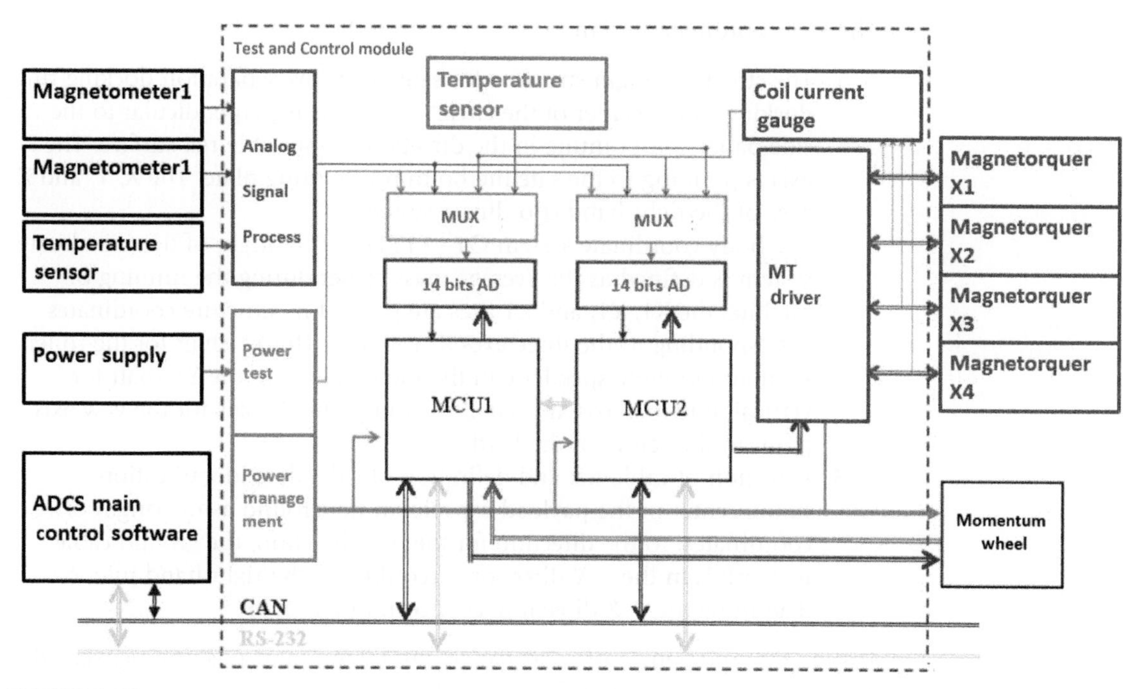

FIGURE 1.15
Signal flow diagram of ADCS.

module. The measurement and control module sends the corresponding data to the OBC ADCS master software, receives master software instructions, and control, and drives the corresponding hardware.

The NS-1 satellite measurement and control module has two microcontrollers, MCU and ATC, respectively. The two microcontrollers CAN through its built-in CAN communicate with another subsystem on the star; the control system works via a serial port (UART) communication between these subsystems.

2. The ADCS subsystem signal flow measurement and control module are as shown in Fig. 1.15.

1.4.5 Structure Subsystem

1.4.5.1 Satellite Coordinate System

During the design process, the satellite needs a variety of coordinate systems, such as the structure of the coordinate system and centroid coordinate system, the coordinate transformation can be done through the origin of coordinate's translation.

The XYZ coordinate system O:

1. The structure origin structure for satellite arrow separation docking ring docking to the center of the circle. The Y axis is perpendicular to the side panel and pointing in the direction of the cable net surface. The Z axis is pointing to the satellite bottom-mounting plate. The X, Y, and Z axes of the right-hand coordinate system.
2. The body coordinate system O1−X1Y1Z1. The origin of the coordinate system is defined as the average mass center during the running of satellite; the X1, Y1, and Z1 axes are parallel in structure coordinates corresponding to the three axes X, Y, and Z. The X1 shaft for the roller shaft and satellite speed are in the same direction; the Y1 shaft for vertical, pitching axis and satellite orbit plane Z1 axis for the yaw axis point to the center of the Earth.
3. Direction of cable net and definition of side plate identification.
   In the center of the payload capsule on the ground as the origin of coordinates, to the direction for the + Z direction, the ground cable network is in the + Y direction, according to the right-hand rule to determine the + X direction; as shown in Fig. 1.16.

Due to the uniqueness of solar panels installation and positioning, six sides and two top surfaces should be defined.

According to the structure coordinate X, Y, and Z definitions, the six-sides are defined as: cable net is in the + Y axis, the remaining five sides from the + Z axis to the Z axis direction run counterclockwise, respectively, as + X + Y + X − Y plane, x−y plane, X−Y plane, X + Y, and on the ground for + Z, as shown in Fig. 1.17.

1.4.5.2 The Satellite Configuration and Layout

The structure subsystem consists of basic module boxes, a load tank, panels, sun sensor, solar panels, and star arrow separation mechanism parts. The satellite function diagram is shown in Fig. 1.18.

The satellite system adopts a modular design—dividing the satellite into payloads and service modules using the advantages of modular design, can result in different modules being run in parallel enabling assembly and testing at the same time, and hence shortening the cycle of assembly and testing. In addition, from the perspective of space, on the same satellite mission, but in respect to different payloads, roughly the same service system module can be used. So that the small satellite platform service module is designed and manufactured. On one hand this helps to facilitate adjustments to the satellite program, and on the other hand, it promotes the adjustments to functions, when developing a new model.

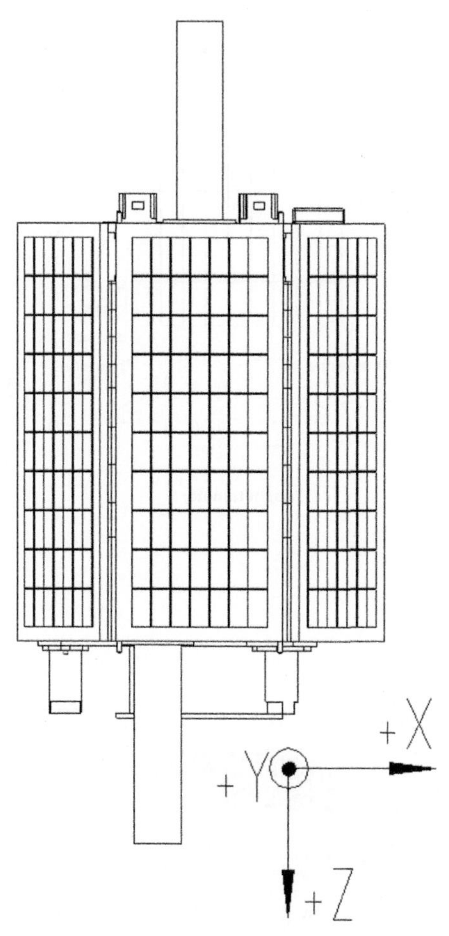

FIGURE 1.16
Definition of coordinates on NS-1.

Satellite platform from top to bottom includes: attitude control, TTC, OBC, GPS, RF1, RF2, the PCM/PDM, 8 BCR modules, with a shell covering the main load-carrying structure of satellite. Each module was hexagon type, with a length of 182.5 mm.

The payload capsule includes a camera module, propulsion module, the momentum wheel, the magnetic torquer 1, storage battery, and MIMU module.

To send and receive signal, two RF antennas are installed on the bottom panel, and other two RF antennas and a GPS antenna are installed on the top panel.

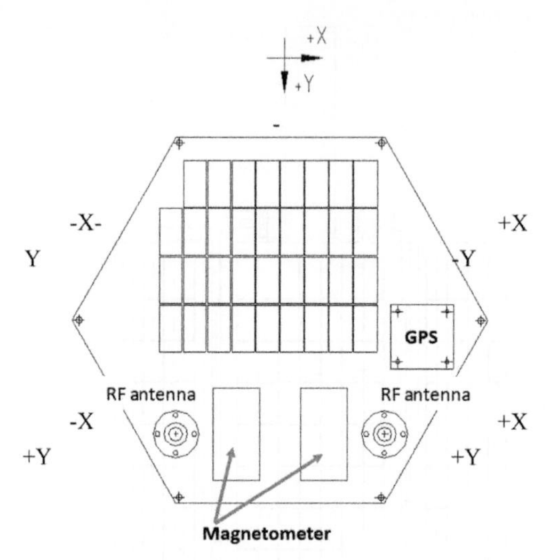

FIGURE 1.17
Definition of faces on NS-1.

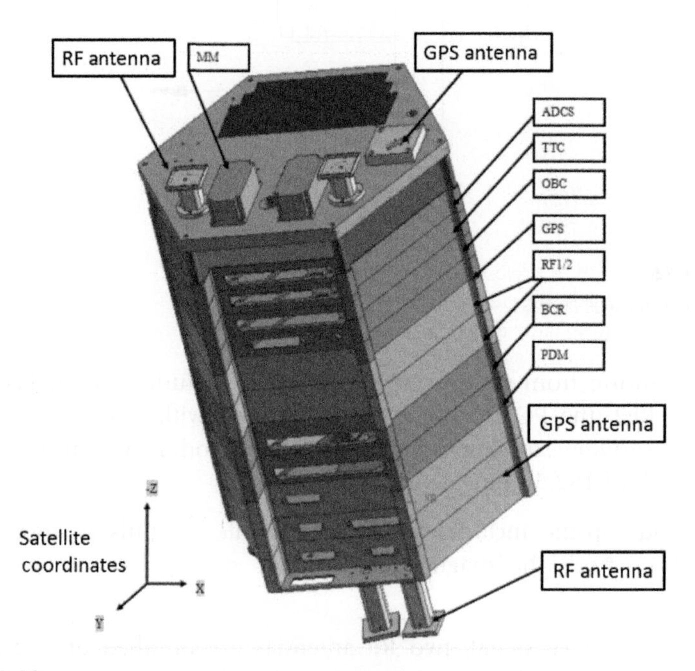

FIGURE 1.18
Exploded view of NS-1.

The solar panels are made up of seven boards, including a satellite roof panel and six side panels.

1.4.6 RF Subsystem

1.4.6.1 RF Subsystem Function and Performance Index

The NS-1 satellite uses a radio frequency subsystem consisting of a receiving channel and a transmission channel, for modulating transmission channel, sending complete remote sensing satellite engineering parameters, forwarding GPS orbit data, storing-and-forwarding communication carrying load data and satellite data, receiving channels to complete the project for the remote control instruction to receiver, demodulating, and managing data from satellite and injected by subsystems. On a satellite RF subsystem, in addition to sending, receiving, and downlinking signals, the transmitter can be used as a beacon, a ground antenna tracking signal, to guide ground capture, and satellite antenna automatic tracking.

Telemetry, remote control and data communication frequency band, modulation system, code rate, and distance requirements are shown in Table 1.9.

The telemetry frame format uses GJB1198.2 standards of PCM telemetry and GJB1198.6 subcontracting telemetry. The remote control frame format uses the GJB1198.1 PCM remote control scheme.

The main technical indicators for a receiving subsystem are listed in Table 1.10.

The main technical indicators of the launch subsystem are shown in Table 1.11.

1.4.6.2 Composition of the RF Subsystem

The nanosatellite's RF subsystem is composed of a receiving antenna (including a duplexer, transmit/receive filters, splitter/combiner), a hot backup frequency receiver with the same frequency, and a cold backup transmitter with the same frequency. The dual-switch transmitter is controlled by the ground control station with upward instruction.

1. Antenna duplexer and network:
   - A set of omnidirectional antenna duplexers mounted on the sky surface and ground surface;

Table 1.9 Basic Characteristics of Telemetry and Remote Control Channel

	Channel	Frequency	Decode	Speed	Distance
Remote test	S	2274.48 MHz	PCM/DPSK	19.2 kbps	2760 km
Remote control	S	2094.417 MHz	PCM/2FSK/FM	9.6 kbps	

Table 1.10 Main Specifications of a Receiving Subsystem

Channel	S Channel
Frequency	2094.417 MHz
Antenna	Gain: ≥ -1.5 dB Directionality: omnidirectional Polarization: circular Impedance: 50 Ω SWR: 1:1.5
Net	Insertion loss: ≤ 5 dB Impedance: 50 Ω SWR: 1:1.5
Decoding method	PCM/CPFSK
Maximum linear frequency offset	5 kHz
G/T	−30 dB/K
Receiver sensitivity (to ensure that the data error rate is better than 10^{-6})	-10^{8} dBm
Dynamic range	50 dB
Frequency tracking range	±110 kHz
Speed	9.6 kbps
Data error rate (under the condition of remote demodulation sensitivity)	10^{-6}
Scrambling polynomial	$1 + X^{12} + X^{17}$
Baseband shaping cosine filter roll-off coefficient α	= 1
Output PCM	NRZ−L
Power consumption (nonrenewable energy)	≤1.4 W
Volume	250 × 240 × 30 mm
Working temperature	−25 to +55°C

- Two diplexers;
- Two front-end RF receivers (receiving band pass filter and low noise amplifier);
- A four-port network;
- Two emission filters.

2. Receiver
   - Two RF receiver modules;
   - Two IF FM receivers;
   - Two FSK baseband processing units.
3. Transmitter
   - Two RF transmitter modules;
   - Two error correction coding and scrambling baseband processing units.

Table 1.11 Main Specifications of the Launch Subsystem

Frequency	2274.48 MHz	
Antenna	Gain: ≥−1.5 dB	
	Directionality: omnidirectional	
	Polarization: circular	
	Impedance: 50 Ω	
	SWR: 1:1.5	
Net	Insertion loss: ≤5 dB	
	Impedance: 50 Ω	
	SWR: 1:1.5	
Decoding method	PCM/DPSK	
Frequency stability	Better than $\pm 1 \times 10^{-5}$	
Frequency accuracy	Better than $\pm 5 \times 10^{-6}$	
Harmonic suppression	≥40 dB	
Noise suppression	≥60 dB	
Impedance	50 Ω	
PCM code	NRZ−L	
Symbol asymmetry	≦10%	
Error correction code	Cascaded convolutional codes (2, 1, 7) and RS codes (255, 223)	
Scrambling polynomial	IESS308	
Error correction encoder input code speed	19.2 kbps	
Baseband shaping cosine filter roll-off coefficient α	= 1	
Phase noise	Δf	Phase noise
	100 Hz	−55 dBc/Hz
	1 kHz	−70 dBc/Hz
	10 kHz	−77 dBc/Hz
	100 kHz	−82 dBc/Hz
	1 MHz	−100 dBc/Hz
Output power of PA	≥ 320 mW	
Maximum power consumption	≤ 3.6 W	
Volume	250 mm × 240 mm × 30 mm	
Working temperature	−25°C to +55°C	

4. Block diagram
   A block diagram of the remote control receiver is shown in Fig. 1.19
5. Telemetry transmitter
   The telemetry transmitter is a dual-system cold backup. A block diagram of a telemetry transmitter receiver is shown in Fig. 1.20.

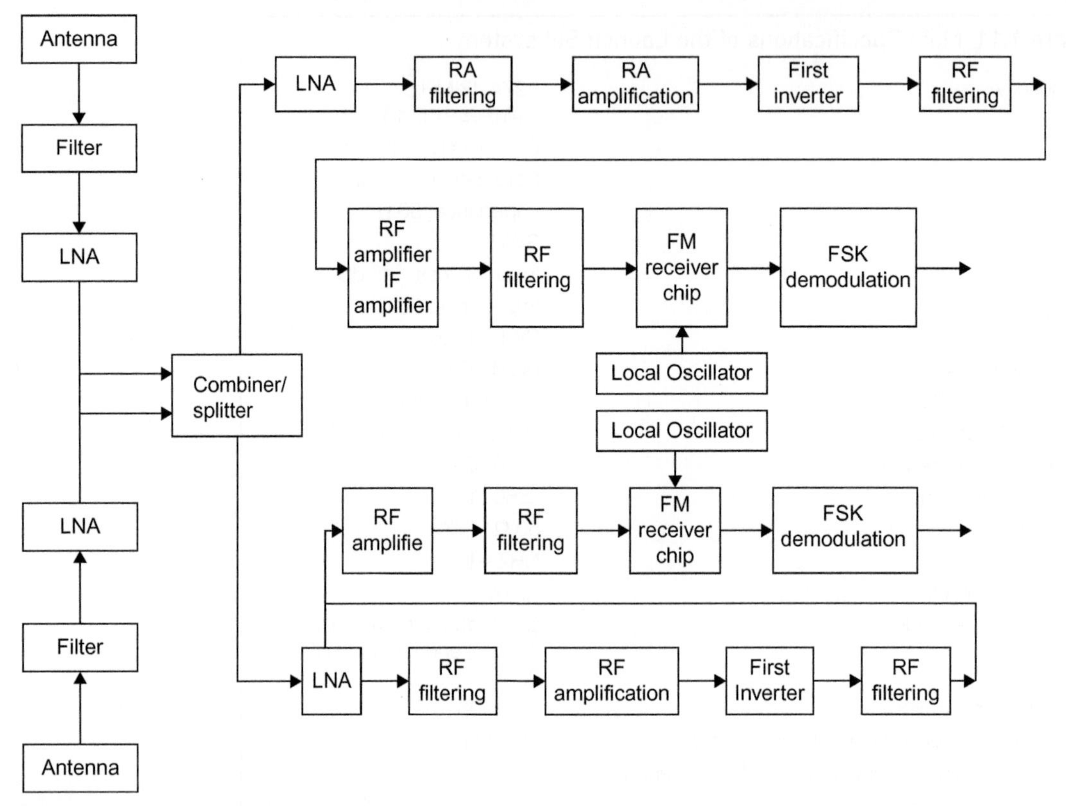

FIGURE 1.19
Block diagram of the receiver.

1.4.7 GPS Subsystem

System functional requirements and environmental uses include the following.

1. Positioning and orbit determination
   The nanosatellite's GPS receiver orbital position measurement and solver give the current satellite position, including three-dimensional position (X, Y, Z), velocity components (V_x, V_y, V_z), and other data.
2. Standard time
   The nanosatellite's GPS receiver simultaneously satisfies the GPS satellite data management subsystem standard time.
3. Software upload
   In order to improve the GPS receiver's orbit measurement reliability, particularly to overcome the consequences caused by space radiation SEU events, the GPS receiver should have the ability to upload software.

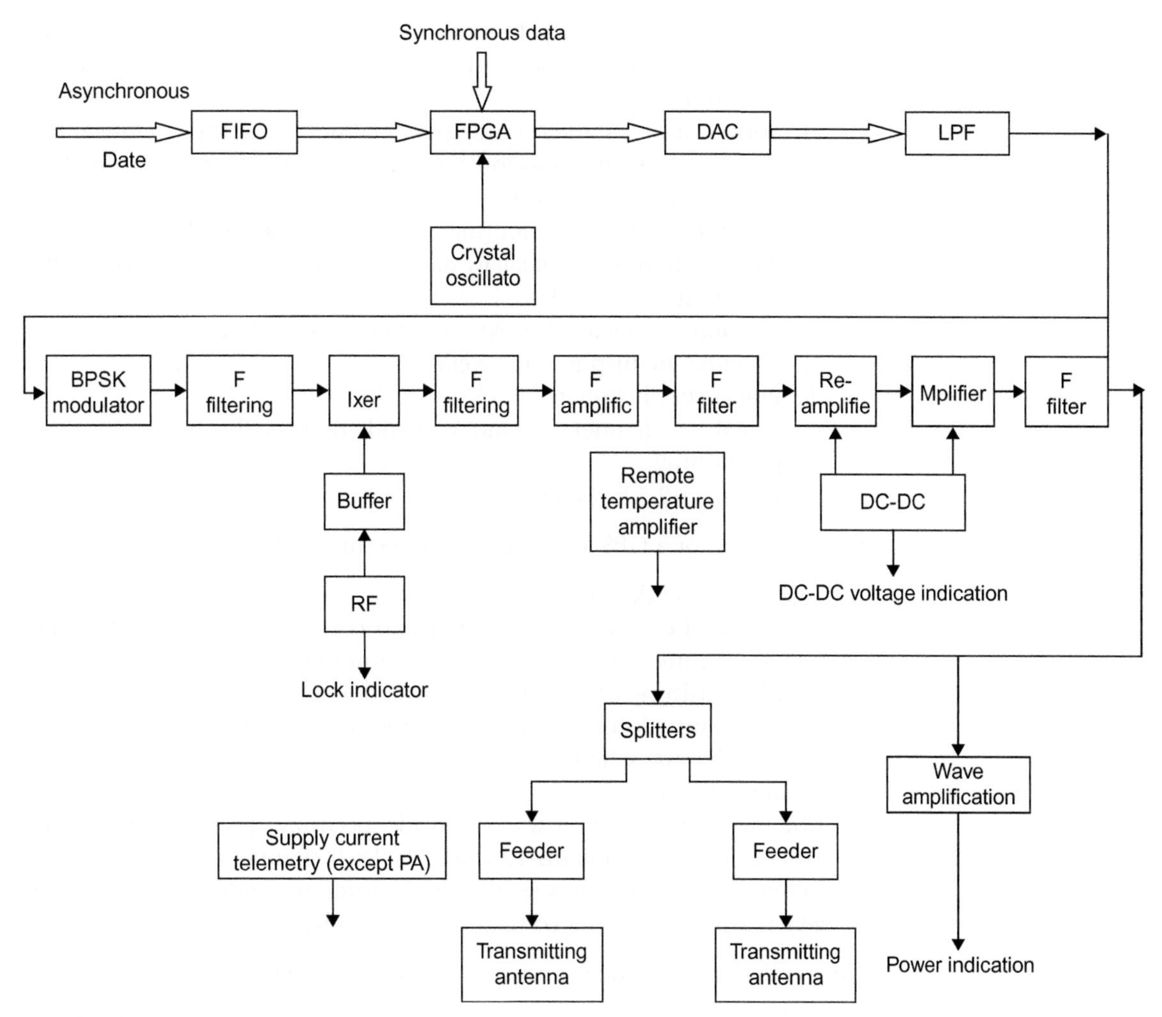

FIGURE 1.20
Block diagram of the transmitter.

4. Environment
   Orbiting speed: 7.8 km/s.
5. The main technical performance and indicators include:
   a. Satellite orbit: 600–1100 km quasi sun-synchronous orbit;
   b. Three-dimensional positioning accuracy (2σ): better than 50 m;
   c. Three-dimensional velocity accuracy (2σ): better than 1.5 m/s;
   d. Time: seconds/pulse, the pulse width is greater than 1 ms, TTL level.

e. Time accuracy (UTC): 1 μs (the GPS receiver calculates the difference between the results and the leading edge of the pulse UTC seconds);
f. Solver output: standard time (seconds/pulse) for the satellite at 84 three-dimensional position coordinates (X, Y, Z) and the velocity components (V_x, V_y, V_z).
g. Acquisition time: ≤3 min (there are circumstances under auxiliary data); ≤40 min (no auxiliary data, and long after the initial boot off);
h. Data update rate: 1 Hz;
i. Dynamic characteristics: speed 7.8 km/s, acceleration 2 g;
j. Antenna: hemispherical coverage;
k. Power supply voltage: +5 V;
l. Signal: 12 channels, L-band, C/A code receiver absolute positioning;
m. Power: ≤2.5 W (single).

A block diagram of a GPS subsystem is shown in Fig. 1.21.

1. Signal channel device
   Signal channel devices are designed to complete amplifying filtering, frequency conversion, and A/D conversion function of the signal, mainly consisting of a reference source, downconvert combination, and a combination of broadband IF and other components. A GPS receiver channel schematic is shown in Fig. 1.22.
2. Channel device
   Channel device comprises a spread spectrum despreading channel processor, which contains 12 channels. Each channel contains the following functional blocks: the NCO carrier, the code NCO, C/A code generator, and a carrier correlator code correlator.

A GPS receiver channel device schematic is shown in Fig. 1.23.

A GPS channel processor internal functional block diagram is shown in Fig. 1.24.

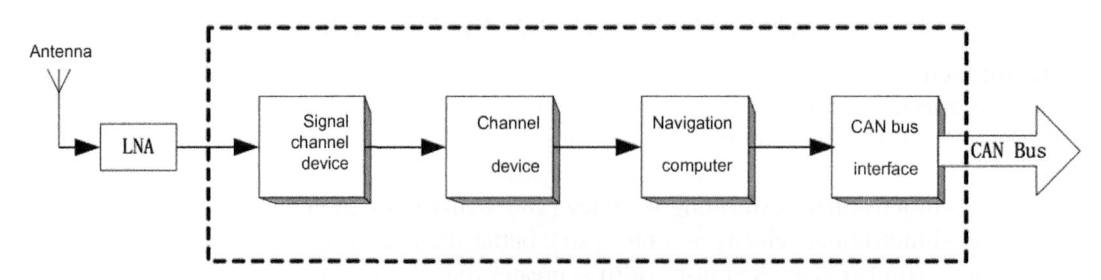

FIGURE 1.21
Block diagram of GPS subsystem.

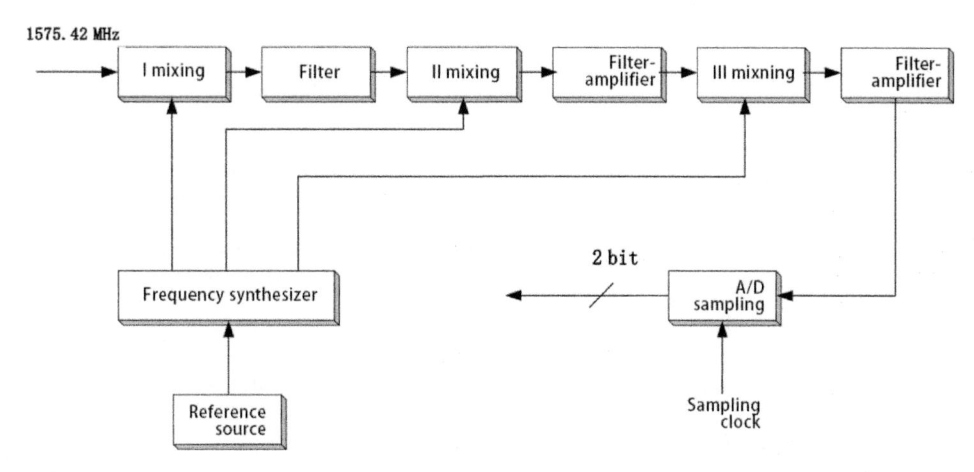

FIGURE 1.22
A GPS receiver signal channel schematic.

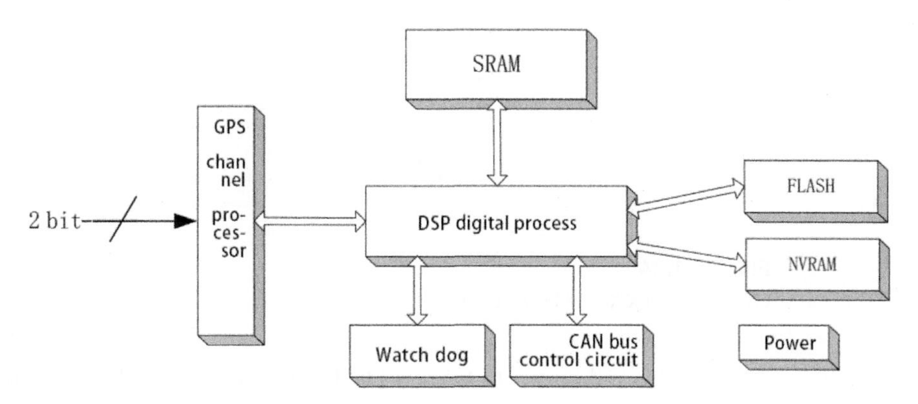

FIGURE 1.23
A GPS receiver channel devices schematic.

1.4.8 Camera Subsystem

The NS-1 satellite has two missions: ground photography and flight demonstration test. The photographic camera payload details for the CMOS NS-1 satellite are listed here.

1. Performance indicators:
   - Focal length: 50 mm;
   - Full field of view angle: 11.7 degrres × 11.7 degrees;
   - Relative aperture: 1:4;
   - Imaging spectrum: 400–800 nm;
   - Ground pixel resolution: 110 m (track height related);

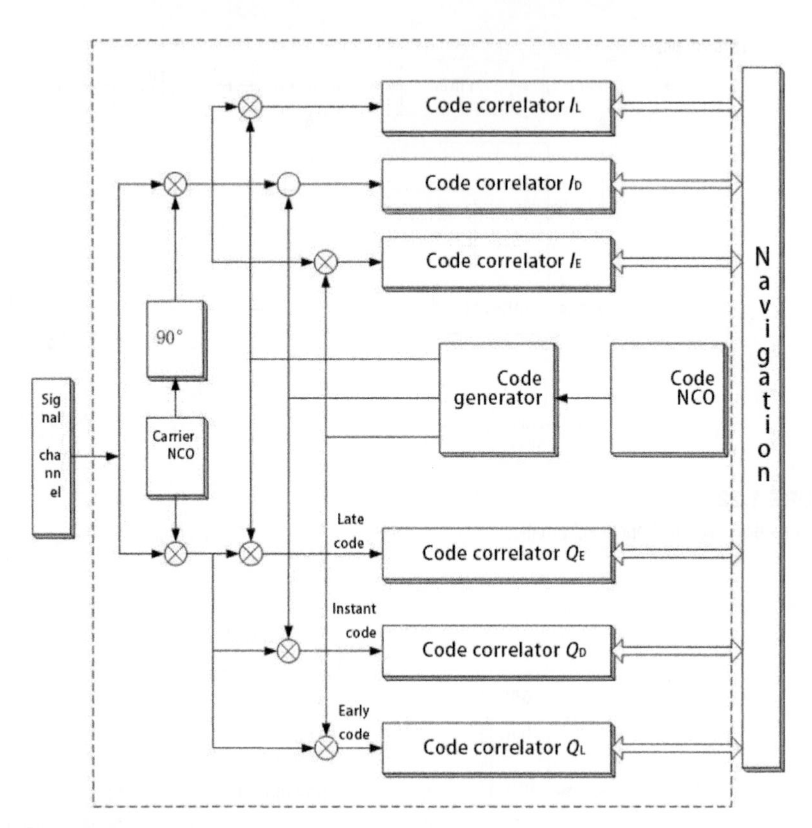

FIGURE 1.24
GPS channel processor internal functional block diagram.

- Ground cover: 113 km × 113 km (about the height of the track);
- Power: <0.7 W;
- Volume: <140 mm × 80 mm × 80 mm.

2. Camera structure:
   Because the nanosatellite has to satisfy the structure's size limit, CEIS volume requirements must be less than 140 mm × 80 mm × 80 mm. The lens parts are about Φ70 mm × 80 mm, where Φ is the diameter of the lens. The size of the electronic system is usually less than 60 mm × 80 mm × 80 mm. Therefore, the circuit board size is not greater than 80 mm × 80 mm. Multiple boards must be stacked together to accomplish this condition. An initial design was for three circuit boards. The overall structure is shown in Fig. 1.25.
3. The optical module:
   The camera lens is the key component of the entire camera. The image quality, structural mechanics, thermodynamic stability, and material composition are all related to the final image quality and the ability to

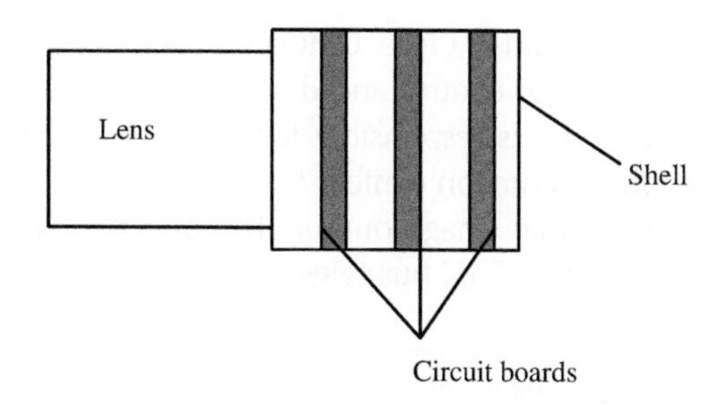

FIGURE 1.25
Overall structure of camera.

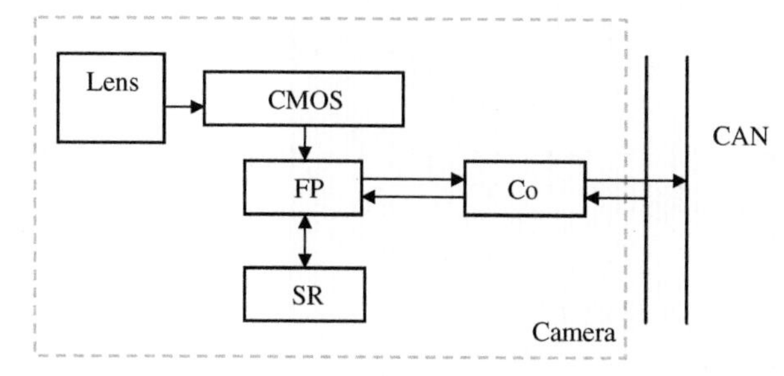

FIGURE 1.26
Electronic system block diagram.

self-focus. Therefore, selection of the most suitable lens from the commercial market is of critical importance. The requirements are:

- Performance parameters and image quality indicators meet the requirements;
- Structure is made of metal;
- Optical parts are of a glass material, that can withstand temperatures in normal operation of −200°C to +500°C;
- Can not have a wear-resistant material between the flexible screw;
- The focusing ring on the lens barrel and screw;
- Weight less than 250 g.

4. Camera electronic system design:
   The nan-satellite's CMOS camera circuit consists of the following modules:
   1. Imaging module: according to the instructions of the microcontroller, under the control of FPGA to complete imaging and image data storage capabilities, and outputs the image data in the TTC controller requirements;
   2. Microcontroller: completing camera system power switch and temperature control, current detection, and communication with the OBC, complete camera control and data transfer functions. The electronic system block diagram is shown in Fig. 1.26.

The entire camera system includes a lens, CMOS sensor, FPGA, SRAM, and a controller. FPGA is responsible for control and data storage output operation of the CMOS sensor. The controller is responsible for receiving external control commands and following the instruction content for the control FPGA operation, and is also responsible for external image output. The interface between Controller and external communication is CAN bus (Figs. 1.27 and 1.28).

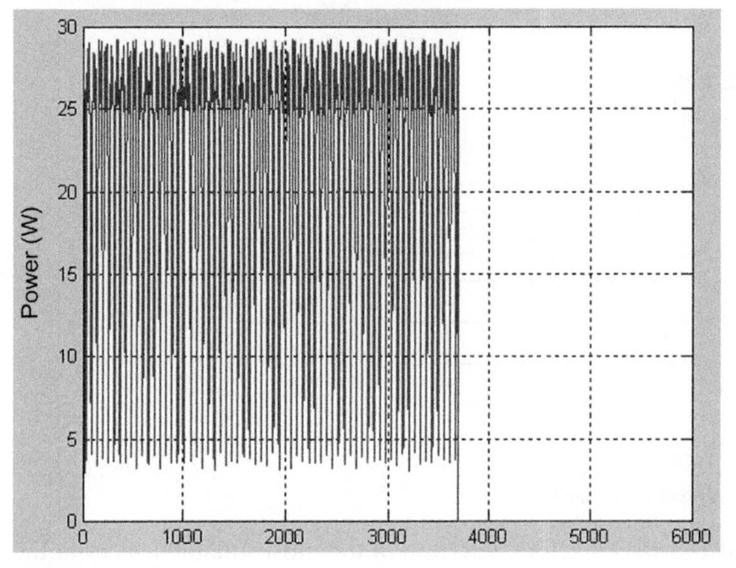

FIGURE 1.27
Free rollover state: the power output of the solar arrays in one orbital period.

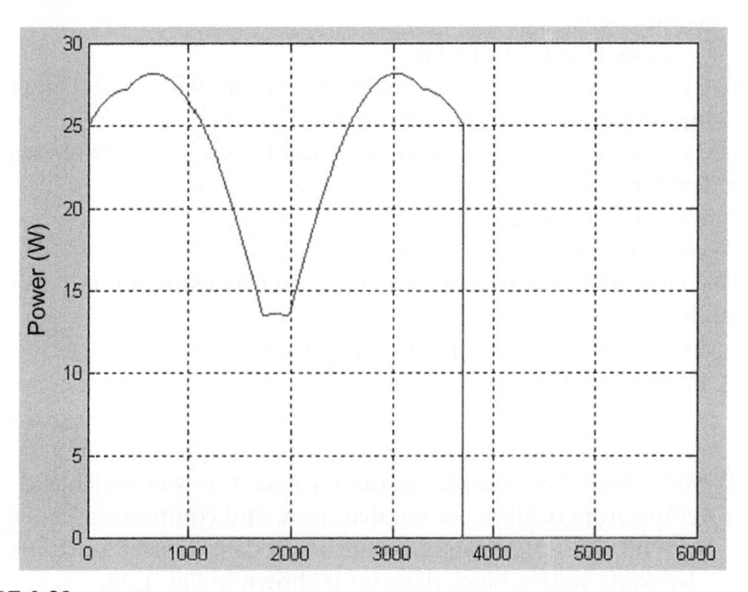

FIGURE 1.28
A three-axis stabilized state: the power output of the solar arrays in one orbital period.

1.4.9 Propulsion Subsystem

The NS-1 propulsion subsystem task is to provide incremental speed in order to maintain or achieve the satellite orbit.

Its performance requirements include:

1. System: liquefaction propulsion system, the system is not less than the total impulse 200 Ns;
2. The quality of the system (including propellant): 1.3 ± 0.1 kg;
3. System configuration: a tank, a combination of a main thruster 1 equipment, a number of other members (a filling valve, a pressure sensor 2, the temperature sensor 2, a filter system isolation valve 2, several pipeline);
4. The thruster: 50–100 mN;
5. System works: pulsed operation, the minimum impulse 0.75 mNs, switching frequency is not less than 1E5; continuous operation (maximum uptime 5 min), the switching frequency is not less than 1E4;
6. Power consumption: average power consumption during operation 2.6 W, peak 6 W;
7. Temperature: $-25°C$ to $+55°C$;
8. Security: fully consider the safety factors, compatibility requirements, having two or more disable functions;
9. Reliability: 0.99.

1.5 EMC DESIGN AND PROPULSION SUBSYSTEM SAFETY DESIGN

1.5.1 EMC Design

Satellite EMC design is aims to ensure that on-board subsystem co-ordination works, meeting one of the important measures of electrical performance. This is done by grounding, shielding, coordination, and control of the interface isolation power output impedance between subsystems. Electromagnetic compatibility testing is an important means of testing the design effect.

1.5.1.1 EMC Design Requirements

1. Ground and isolation
   a. The system mainly uses a secondary power supply. The equipment should distinguish between analogue and digital ground. Both the equipment has to be linked to a single point grounding line in the

device. Then the shortest path to this point is connected to the satellite system ground;

b. Shield cable, usually link to the ground by the tail cover. In special cases could be pinned to the shell's grounding point.

2. Overlapping

   In the metal structure, we often use overlapping to reduce the impedance ground plane.

   The inter star structure impedance should be less than 2.5 mΩ.

   Housing and other electrical equipment should have a potential overlap with the stars and the rest of the structure. Lap resistance should be less than 10 mΩ, socket housing should also ensure that the lap resistance of the structure of the stars is no more than 10 mΩ, and overlap resistance between the solar wings and the structure should be no more than 30 Ω.

   Exposure to the components outside the stars should prevent the space plasma from being charged by the insulating appearance. It can also use lap to ensure mechanical components and thermal control cladding layer and the contact resistance between the star structure meet requirements.

3. Cable laying

   In order to meet the requirements of radiation emission and sensitivity, shielded, twisted-pair cables and correct cables should be used. The cable bundles should be classified as much as possible; these lines should be sorted and bound, separated and segregated if necessary; The internal trace should also separate the signal line from the power cord.

4. Shielded

   Interconnect electronic devices should be shielded to reduce the effects of electromagnetic interference, using a different cable shielding method.

5. Signal interface design requirements

   For receiving and transmitting models of the interface circuit interface circuit model, use different immunity circuitry to ensure minimal interference.

6. Power input filter requirements

   The filter design should consider a variety of factors generated by the ripple voltage and ripple current caused by the interference.

7. Electromagnetic compatibility test

   Electromagnetic compatibility test is related to the test subjects, test items, conducted emissions and sensitivity and radiation. Shooting and sensitivity of the extreme size of the provisions of the relevant requirements.

1.5.2 Promoting Security Subsystem Design

The propulsion subsystem is a high-pressure, toxic system, in order to ensure the safety of person, the satellite, rocket, and satellite launch site security subsystem must be designed with safety.

1. The structure of key components and materials
   tanks, pressure piping, and components must conform to safety requirements. Direct exposure or due to single point of failure exposed to harmful propellant materials, pressure sensors, valves, etc.
   Propellant contact material must be considered for compatibility design requirements.
2. Power off and insurance
   Powered satellites should not produce false starts to the instruction thruster.
   The thruster should not start in the event of electrical interference or malfunction.
   To prevent system failures before the satellite has reached orbit, systems and subsystems must be designed to ensure that there are more than two series of independent insurance (disable functions).
3. Filling
   A propulsion subsystem should extrude propellant gas to avoid it filling between the launch site and to prevent any propellant leak that may arise in the process of filling.

1.6 NS-1's TECHNICAL CHARACTERISTICS AND PARAMETERS OF DISTRIBUTION

1.6.1 The Main Technical Characteristics of Satellites

NS-1's overall technical indicators are listed in Table 1.12.

1.6.2 Satellite Characteristic Parameter Assignment

1.6.2.1 Satellite Mass Distribution and Quality Characteristics

The Carolina star mass distribution is shown in Table 1.13.

1.6.2.2 Power System Dynamic Simulation

The NS-1 configuration is a hexagonal prism, with simulation of the moment of inertia of [0.67 0.68 0.44]. Installation of solar array area:

- Side surface (6): 40 mm × 20 mm × 70 mm;
- Top surface (1): 40 mm × 20 mm × 35 mm.

Table 1.12 Overall Technical Indicators

Track	Types of Height	Sun-synchronous orbit 599 km
	Eccentricity	0.00017
	Cycle	96.66 min
	When descending node local	11:00 a.m.
Emission	Launch vehicle	CZ-2C
	Emission mode	Piggyback launches
Structure	Body size envelope	400 mm × 440 mm (excluding adapter)
Quality and power consumption	Quality	25 ± 1 kg (excluding Xingjian adapter)
	Long-term consumption	About 10 W (normal mode)
Life and reliability	Working life	6 months
	Design and reliability	Greater than 0.65 (end of 6 months, the minimum system)
Camera	Sensor	CMOS
	Spectral range	0.4–0.78 μm
	Number of pixels	1024 × 1024
	Optical lens focal length	50 mm
	Optical F number	2.8–4
	Gradation image	256
	FOV	11.7 degrees
	Star in the transmission mode	CAN bus
GPS receiver	Number of channels	12
	Range accuracy (2σ)	Better than 50 m
	Measurement precision (2σ)	Better than 1.5 m/s
	Time accuracy	1 μs
	Capture time	≤ 3 min (circumstances under auxiliary data)
	Data update rate	1 Hz
Power supply	Power supply system	Solar array—battery-powered joint
	Bus voltage	14 V
	Solar array	Body equipment, a total of seven windsurfing (six sides, one face)
	Monolithic windsurfing size	182.5 mm × 440 mm (side), 182.5 mm (top, hexagonal)
	Solar cell dimensions	40 mm × 20 mm × 0.175 mm
	Type of solar cell sheet	Single-junction GaAs
	The average conversion efficiency of the solar cell sheet	18.5%
	Number of film solar cells	455
	Each output power windsurfing	Side > 13.4 W, top > 6.7 W (AM0, 25°C)
	Battery type	NiCd battery
	The number of single battery	10
	Battery capacity monomer	3 Ah
	Battery discharge depth	40% (max)
	Secondary power supply	+5 V\+12 V\−12 V

Continued

Table 1.12 Overall Technical Indicators *Continued*

Attitude control	Control method	Three-axes stabilized (+stable bias momentum magnetron)
	Attitude measurement accuracy (3σ)	$\pm$1 degree
	Attitude control accuracy (3σ)	$\pm$2 degrees
	Posture stability (3σ)	0.02 degree/s
	Executive body	Momentum wheel (1) and magnetorquer (4)
	Sensor	Magnetometer (2)
Advance	The number of propellant tanks	1
	Thrust size	50–100 mN
	Propellant type	Ammonia
	System quality (including propellant)	1.3 $\pm$ 0.1 kg
	Number of thrusters	1
	Total impulse	Not less than 200 Ns
Data management	TTC subsystem	
	Instruction capacity	
	Analogue telemetry channel count	34 Road
	Number of telemetry channels	46 Road
	Analogue measurement accuracy	16
	OBC	1%
	CPU	SA1110
	Clocked	133 MHz
	EPROM	512K
	RAM	6 M (data) + 1.5 M (program) CAN bus, UART, HDLC
	Internal communication satellite	
MIMU	Accelerometer bias	$< \pm 7$ mg
	Accelerometer bias stability	30 μg (3 months)
	Accelerometer scale factor	1.25 $\pm$ 0.10 mA/g
	Accelerometer scale factor stability	80 ppm (3 months)
	Accelerometer measurement range	$\pm$ 25 g
	Acceleration resolution	$<10^{-5}$ g
	Gyro linearity	<0.05% range
	Gyroscope scale factor	50 mV/degree/s
	Gyro resolution	<0.004 degree/s
	Gyro output noise	<0.02 degree/s/$\sqrt{\text{Hz}}$
	Gyroscope measurement range	$\pm$100 degree/s
Control and communication	Remote control	
	Number of receivers	Dual hot backup
	Carrier frequency	2094.417 MHz
	Modulation	PCM/CPFSK
	Code rate	9.6 kbps
	BER	1×10^{-6}

Continued

Table 1.12 Overall Technical Indicators *Continued*

	Receiver sensitivity demodulation instruction	−108 dBm
	Telemetry	PCM/DPSK
	Modulation	Dual cold backup
	Number of transmitters	2274.480 MHz
	Carrier frequency	Better than $\pm 1 \times 10^{-5}$
	Frequency stability	Convolution coding (2, 1, 7), RS (255, 223)
	Error correction coding	
	Code rate	19.2 kbps
	Transmit power	No less than 320 mW
	Antenna	2 anntenas for send and 2 anntenas for receive
	Antenna type	≥−1.5 dB
	Gain	
	Directivity	Omnidirectional
	Polarization	Circular polarization
	Impedance	50 Ω
	VSWR	Better than 1:1.5
	Network (including diplexers, filters, four-port network and feeder)	Insertion loss: ≤5 dB Impedance: 50 Ω SWR better than 1:1.5
Structure	The basic way bearing	Wear contact module case
	Solar cell substrate	Aluminum honeycomb core aluminum panels
	Module case	Aluminum
Thermal control	Approach	The combination of passive and active
	Temperature range (cabin)	−25°C to +55°C
	Battery	0°C to +40°C
	Extraterrestrial components	−80°C to +80°C

Dynamic input condition analogue power supply and demand situation is as follows:

- Orbital altitude: 600 km;
- Descending node place local time: 11:00;
- Side solar array output power (tested data): 14.16 W;
- Top surface of the solar array output power (tested data): 7.08 W.

As the satellite separates from the carrier, the stars triaxial angular rate is not greater than 4 degrees/s, with the attitude angle being uncertain. Accordingly the simulation initial value is determined as follows: three-axes angular rate is 4 degrees/s or −4 degrees/s; three-axes attitude angle between −180 and 180 degrees any value. Calculation results show that when the satellite is in a

Table 1.13 Satellite Physical Quality Measurement Statistics

Component Name	Mass (g)	Number	Total Quantity (g)	Notes
Payload bay	7052	1	7052	Including: a payload bay module compartment, propulsion systems, MIMU module, battery pack, the momentum wheel, camera, RF antenna
BCR module case	1408	1	1408	Including: BCR module boxes, BCR circuit board module compartment floor, Z-axis magnetorquer
PCM/PDM module case	1458	1	1458	Including: PCM/PDM module box structure, PCM/PDM board
RF module assembly	4902	1	4902	Including: RF0 modules and module RF1
GPS module case	724	1	724	704
OBC module case	878	1	878	Including: OBC module box structure, OBC board
TTC module case	1052	1	1052	Including: TTC module box structure, TTC board
ADCS module case	1016	1	1016	Including: ADCS module box structure, ADCS board
Module cover plate assembly	968	1	968	Comprising: propulsion system wiring boxes, modules cover plate, windsurfing fixed side pod
Top windsurfing assembly	994	1	994	Comprising: a top windsurfing, magnetometer, GPS antenna, RF antenna
Side windsurfing (with magnetorquer)	610	3	1830	
Side windsurfing (without magnetorquer)	418	3	1254	
Cable network	930	1 set	930	Comprising: a bus cable networks and promote the cable
Other structural connector		Several	438	Comprising: a screw coupling sleeve, pin
Total quality satellite missions			24,904	
Star Arrow adapter	4266	1	4266	
Total quality satellite carrier			29,170	

free rollover state, for all of the above cases, poor power supply for the initial attitude (40 degrees, −40 degrees, −40 degrees), the initial angular rate (4 degrees/s, 4 degrees/s, 4 degrees/s).

Under this light district average power can provide initial conditions for windsurfing at 21.8 W and full track average power of 13.9 W. Light district average power windsurfing in the three axes stabilized to provide for 23.8 W and full track average power of 15.2 W (Tables 1.14 and 1.15).

Table 1.14 Satellite Power Distribution (Unit: W)

Subsystems or Instrument	Long-Term Consumption	Short-Term Power	
	Assign a Value (W)	Power (W)	Working Time (min)
Power supply	< 2.5		
TTC			
RX (dual hot backup)	4		
	2.6		After separation and each over the top
OBC	<1.5		
Attitude control	<2.9		
Camera		0.41 (photographic) 0.65 (transmission)	< 10 min/time
MIMU		<2.4	<2 min/time
Advance		2.65	<5 min/time
GPS receiver		2.5	<5 min/time

Table 1.15 Satellite Energy Margin (Unit: W)

Working Condition	Power Requirements Equivalent Time	Windsurfing Output Requirements	Energy Margin	Explanation
Smallest system	<9.1	10.71	3.19	Orbit mode
Transmitter normally open	<10.9	12.82	2.38	Platform testing and attitude acquisition mode

Table 1.16 Nanosatellite Reliability Allocation

Subsystems or Instrument	Flight Reliability
Structure	0.995
Thermal control	0.995
Power supply	0.97
Monitoring and control communication	0.97
Attitude control	0.95
Advance	0.99
Number of tubes	0.97
Camera	0.98
GPS receiver	0.98
MIMU	0.99
The entire star	>0.65 (smallest system)

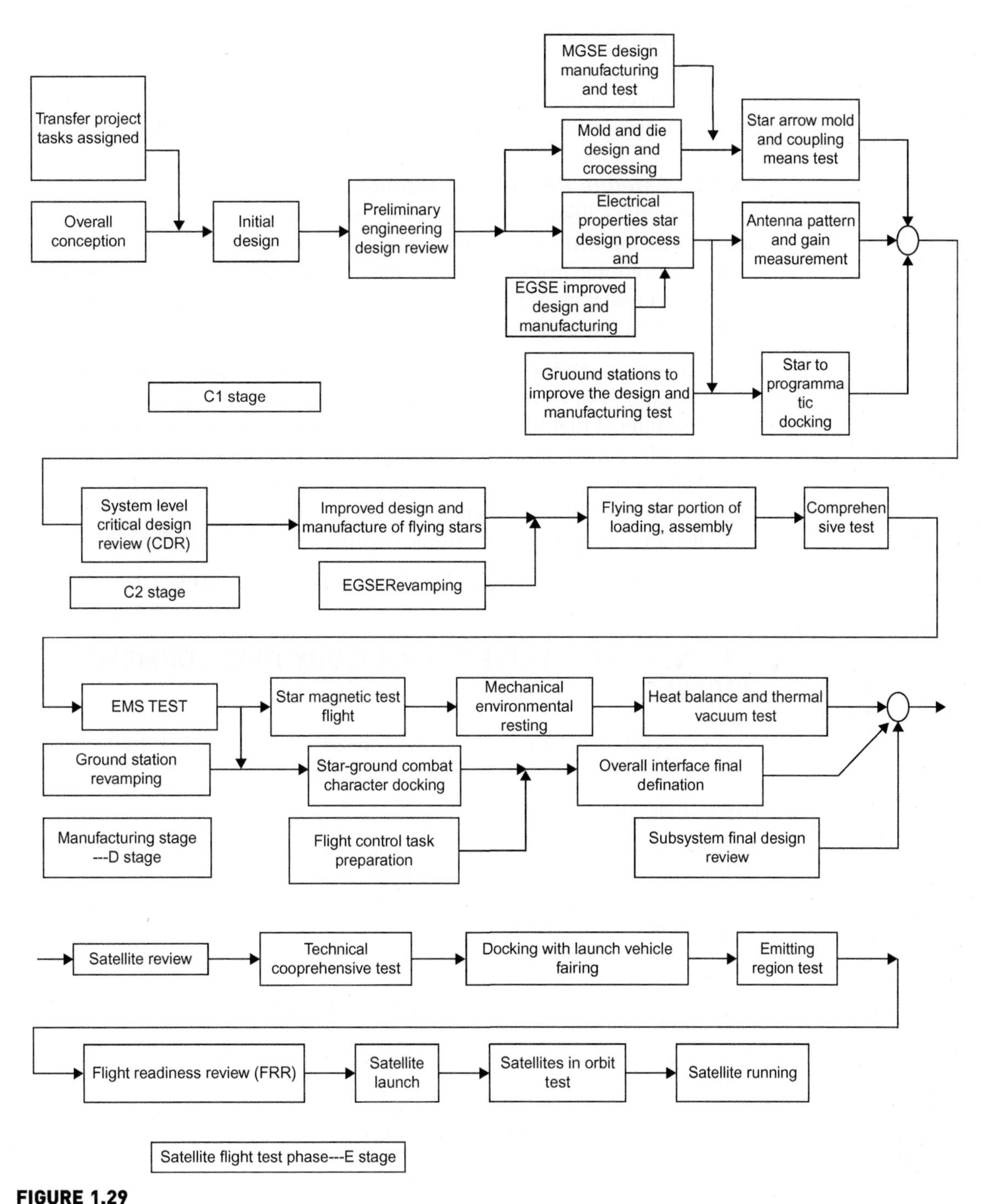

FIGURE 1.29

NS-1's development of technological processes.

1.6.2.3 Propellant Distribution

Nano-satellite propulsion system is a kind of test load, mainly for the use of small satellites network technology tests, and can be used in the satellite orbit correction, track maintenance, and other tasks; the propellant is 0.3 kg.

1.6.2.4 Satellite Reliability Allocation

A nanosatellite's lifetime and reliability are influenced by the level of technology, development, and progress of funding constraints. Therefore it is necessary to determine both a certain nature, and reliability through all possible efforts. Nano satellite is a small satellite platform with very high density technology, which must use reliable design. The design principles are as follows:

1. Strengthen the integrated design of electronic circuits and enhance the integration of electronic systems;
2. Use filtered highly integrated highly reliability electronic components, to reduce their number;
3. Make full use of ground and flight-proven design with strict environmental assessment;
4. Redundant design of electronic circuits.

The subsystem reliability index distribution is shown in Table 1.16.

1.7 NS-1 SATELLITE TECHNOLOGY DEVELOPMENT PROCESS

NS-1's development of technological processes is shown in Fig. 1.29.

References

[1] W. Xiji, L. Dayao, Satellite Design, Shanghai Scientific & Technical Publishers, Shanghai, December 1997.

[2] J.R. Wertz, W.J. Larson, Space Mission Analysis and Design, third ed, Kluwer Academic Publishers, Dordrecht, The Netherlands, 1999.

[3] Y. Houman, F. Hanlin, Achievements and prospect of spacecraft system design technology, Spacecr. Eng. 17 (4) (2008) 1–24.

[4] P. Chengrong, Spacecraft System Design, Science and Technology of China Press, Beijing, November 2010.

[5] Y. Zheng, L. Tao, Application of CMOS image sensor in space technology, Opt. Tech. 28 (1) (January 2002) 42–60.

[6] M.N. Sweeting, C. Fang-Yun, Network of low cost small satellites for monitoring and mitigation of natural disasters, in: 47th International Astronautical Congress, paper number IAF-96-C.1.09, Beijing, China, October 7–11, 1996.

[7] G. Yuksel, C. Ozkaptan, U. Orlu, A. Bradford, L.M. Gomes, M.N. Sweeting, BILSAT-1: a low-cost, agile, Earth observation microsatellite for Turkey, IAF-01, IAA, 11.4.01, in: 53rd International Astronautical Congress, Houston, TX, USA, October 2002.

[8] M.N. Sweeting, Micro satellite and mini satellite programmes at the University of Surrey for Effective Technology Transfer and Training in satellite engineering, in: Proceedings of the International Symposium on Satellite Communications and Remote Sensing, Xi'an, China, September 20–22, 1995.

[9] C. Underwood, G. Richardson, J. Savignol, SNAP-1: a low cost modular COTS-based nano-satellite—design, construction, launch and early operations phase, in: 15th Annual Small Satellite Conference, Logan, UT, USA, August 2001.

[10] J. Xun, X. Jianping, Y. Zheng, Error correcting system of satellite data memory with variable mode realized by FPGA, Appl. Electron. Tech. 28 (8) (2002).

[11] D. Yanwei, F. Junming, Y. Zheng, Design and simulation for thermal control system of nanosatellite, J. Syst. Simul. 18 (1) (January 2006) 89–97.

[12] X. Jianping, Y. Zheng, L. Jian-Hua, J. Ward, W. Sun, CAN network application as a backbone bus in micro-satellite, Spacecr. Eng. 9 (1) (2000) 107–112.

[13] Y. Zheng, Y. Shijie, L. Yang, Remote sensing system for nanosatellite with CMOS imaging sensor, J. Tsinghua Univ. 44 (8) (2004) 1047–1050.

CHAPTER 2

Multidisciplinary Design Optimization of a Micro/Nano Satellite System

2.1 OVERVIEW

Currently, satellite technology has shown a main trend towards being less heavy and much smaller, accompanied with the development of advanced techniques and ideas, causing the naissance of the micro/nano satellite. With the help of integration techniques and flexible design methods, the micro/nano satellite is becoming much more intensively integrated and multidisciplinarily coupled. As a consequence, its design has inevitably become a procedure of iteration, as with the traditional satellite.

How to fully assemble all subsystems to obtain an integrated optimum satellite, reduce researching cost effectively, and enhance the holistic performance, has been a major consideration for satellite developers and end users. For medium-sized and large satellites, in most cases, the subsystems are separated, and the satellite is separated from the whole system, which means having to construct with an over-reliance on expert experience. However, when it comes to micro/nano satellites, their fast upgrading and low-cost requirements have made the development significantly different from traditional ones.

To achieve a faster, cheaper and lower-cost process of micro/nano satellite design, researchers should not only explore novel technologies, but also improve the existing traditional way to build satellites, by studying and adopting a new designing method.

In general, the desired functions and environmental adaptability are two basic considerations for micro/nano satellite design. A specific function can be achieved by several solutions, and an effective solution may conflict with the environmental requirements. Therefore, the design and optimization of a satellite have to be adjusted to adapt to the particular working and natural environment, which brings many constraints for spacecraft design.

The design and optimization of micro/nano satellites is a process starting with application requirements and physical restrictions, optimizing and selecting

Space Microsystems and Micro/Nano Satellites. DOI: http://dx.doi.org/10.1016/B978-0-12-812672-1.00002-3

solutions with systematic engineering methods, studying characteristics of each constituting element and subsystem, determining interfaces and interrelationships between subsystems, and adjusting the technical parameters and operation indicators of the entire system. These can be concluded as a method of systematic construct and optimization, with various specifications [1], which is known as Multidisciplinary design optimization (MDO) today.

The MDO method of micro/nano satellites studied and discussed in this chapter aims at building a man–machine cooperative computing system to assist effectively spacecraft design and shorten the design period significantly. The related thesis and method in this chapter can be further developed and adopted for other satellite systems.

2.1.1 Introduction of Complex Systems Modeling

A scientific methodology of complex system analyzing and design is crucial for satellite design, analysis, optimization, construction, integration, and operation. As the problems appearing in satellite R&D are becoming more complicated and much larger in scope, a series of new methodologies has been developed, such as the IDEF series methods, architectural framework technology, and object-oriented development method (OODM) [2], etc.

2.1.1.1 IDEF Series Modeling Methods

IDEF modeling methods [3], which were proposed by the US Air Force in the late 1970s, are series methods for system analysis and design, and have been widely used in manufacturing and aerospace systems. The widely used IDEF0–IDEF4 are described in Table 2.1.

Of all these methods, the IDEF0 method describes the functional activities and interrelationships between different complex systems. Its principal ideas came from structured-analysis theory, systematology, and information theory.

The IDEF0 method describes the relationship between the system and its parts. The IDEF0 chart is the model built with IDEF0. In an IDEF0 chart, system

Table 2.1 IDEF Modeling Methods

Name	Main Functions
IDEF0	System structure and function modeling
IDEF1	Information modeling
IDEF2	Simulation model design
IDEF3	System workflow and process description capture
IDEF4	Object-oriented design

structures and all complicated interrelationships between each subsystem are top-down decomposed and aredescribed clearly with standard figures.

2.1.1.2 Architectural Framework Technology (AFT)

"Architecture," which refers to the satellite constitution structures and their interrelationship, is a basic principle for system design and development. "Architecture framework" is the guidance to regulate architecture design.

The United States summarized its successful experience on complex system development, which is represented by C^4ISR, and published the *C4ISR Architecture Framework* [4] in 1996. Later, in 2003, the US presented the *DoD Architecture Framework and Software Architecture Workshop Report* [5] in which it proposed the epoch-making theory of systematic engineering. So far, the AFT has become an important weapon design method for the American military and was soon implemented and followed by many countries.

At present, the United States and some other countries have built up an entire set of practical system engineering techniques. Many supporting software programs and complementary tools such as DOORS, System Architect, TAU, G2, and DOCExpress etc., have been developed.

2.1.2 Domestic Studies on Satellite System Design and Optimization

Space technology has had a history of more than 50 years in China. A series of practice-proven system design methods, standards, and regulations, such as satellite system design specification, general designing standard for communication satellite, design guidelines for satellite orbit, spacecraft system design evaluation data packet guideline, and satellite reliability design guideline, etc. have been developed. In domestic satellite systems, the basic system engineering process of each research stage includes: project analysis, function analysis and distribution, comprehensive design, system verification, and relative management activities. The comprehensive design is to design and optimize at the system level, which means working out a systematic physical structure according to the output of the procedure function analysis and allocation. In China, the system is verified by analysis (modeling and stimulation), demonstrative experimentation, and testing (on the ground and on board).

Domestically, there are seven stages in the development of a new spacecraft model, including: project demonstration, feasibility demonstration, project design, prototype development, spacecraft development, application experiment, and product improvement.

Here we take nano satellite NS-1 of Tsinghua University as an example to introduce the typical design process of micro/nano satellites.

NS-1 nano satellite was launched on April 18, 2004. Its research flow is shown in Fig. 2.1. Firstly, a top-level design program is settled according to the research requirements, then the nano satellite platform, effective load, reliability, and launching trajectory scheme are designed in parallel. Herein, satellite platform and effective load are determined by each subsystem and element, restrained by the reliability requirements. Finally, a proper trajectory and launch tool are chosen for system test and satellite launching.

As shown in Fig. 2.2, a micro/nano satellite system contains many subsystems and involves complex multidisciplinary calculations. To optimize our

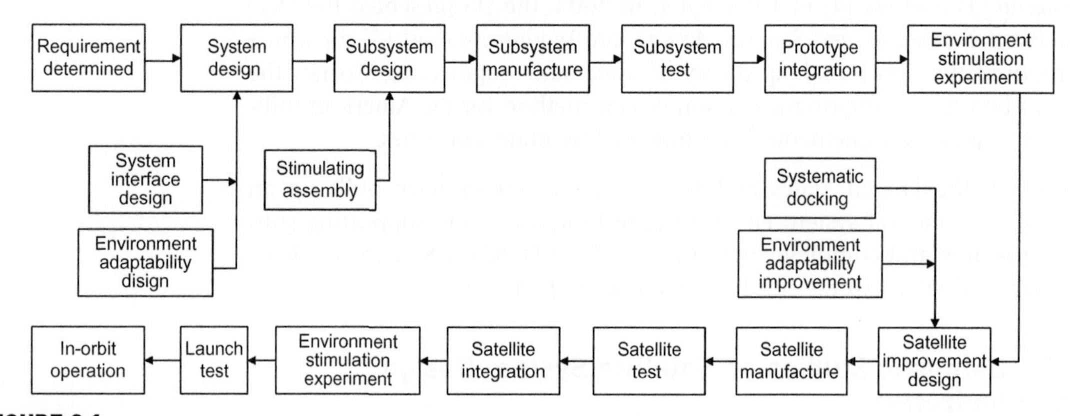

FIGURE 2.1
Micro/nano satellite research flow.

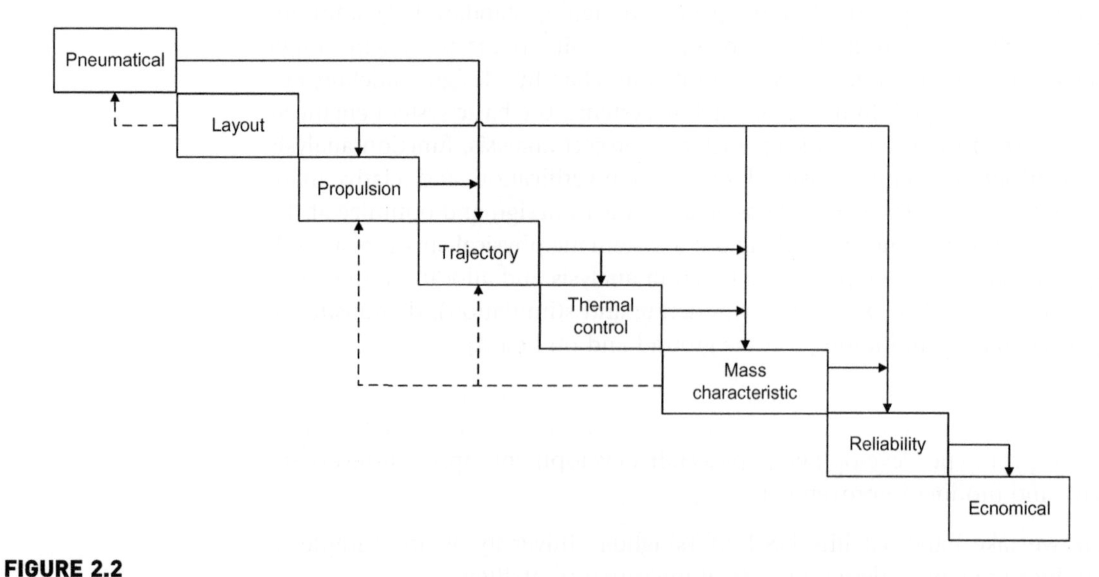

FIGURE 2.2
System architecture matrix of a nano satellite.

design, the coupling relationship between each subsystem and every discipline must be seriously considered. The coupling relationship above can be demonstrated in the system architecture matrix.

Accordingly, micro/nano satellite overall design and optimization is a multidisciplinary design process, which requires researchers to figure out the interrelationship between the satellite system and each subsystem multidisciplinary parameter, and to optimize the overall system under restrictions of guidelines on satellite performance and function. In recent years, the MDO has been developing rapidly, featuring its great advantage in multitarget optimization of an intricately coupled system, and is particularly fitting for micro/nano satellite overall design.

2.2 MICRO/NANO SATELLITE MDO

2.2.1 Overview

Micro/nano satellite MDO is a new study area which was gradually formed in the late 1980s. Since the first *AIAA/ISSMO Multidisciplinary Analysis and Optimization Conference* [6] in 1986, which is now held twice a year, the world has witnessed an upsurge in MDO research, where many universities, research institutions, and companies are involved. So far, rich productions have been obtained and theoretical study is being transformed into engineering application. A number of institutions, such as AIAA, NASA, USAF, and ISSMO, etc., have successfully held many international conferences on MDO research, such as the biennial *World Congress of Structural and Multidisciplinary Optimization.*

In 1991, the AIAA MDO Technical Committee (MDO TC) published the *AIAA White Paper,* where they introduced MDO in detail, and pointed out the orientation for future MDO studies. In the white paper, MDO is defined as a methodology for the optimal design of complex engineering systems and subsystems that coherently exploits the synergism of mutually interacting phenomena using high-fidelity analysis with formal optimization. The main ideology of MDO is: $\Delta_{Design} = \left(\sum_i \Delta_{Discipline,i}\right) + \Delta_{MDO}$, i.e., the whole is greater than the sum of the parts [7].

In general, MDO in micro/nano satellite design can be described mathematically as:

$$\begin{aligned} &\max/\min F(X) \\ &\text{s.t.} \quad a_i \le g_i(X) \le b_i \end{aligned}$$

where $F(X)$ is objective function, $g_i(X)$ is constraint condition, and X is design variable.

The MDO conceptual components were first brought up in 1995 by Sobieski [8], a Polish-American scientist. Later, in 2001, the MDO TC revised the description by adding a new category of "management and culture practice" into MDO conceptual components and highly summarized each aspect of MDO research and its industrial applications [9]. As can be seen in Table 2.2, Chen from the National University of Defense Technology summed up the specific content and requirements of each conceptual component [10].

Fig. 2.3 demonstrates an evolutionary process of satellite design, which is depicted as phases from conceptual to preliminary to detail design and then manufacturing and production [11]. As this process evolves, design freedom

Table 2.2 MDO Conceptual Components

Information Management and Processing	Analysis Ability and Approximation	Design Formulation and Solution	Management and Culture Practice
MDO framework and systematic structure	Analysis and sensitivity	Objectives	Organization structure
Calculation requirements	Parameterized geometric model	Problem decomposition and organization	MDO practice
Visualization of designing space	Approximation and modification	Optimization and assumption	Acceptation, confirmation, cost and benefit training
	Broadness and deepness requirements		
	High fidelity and analysis and experiment integration		

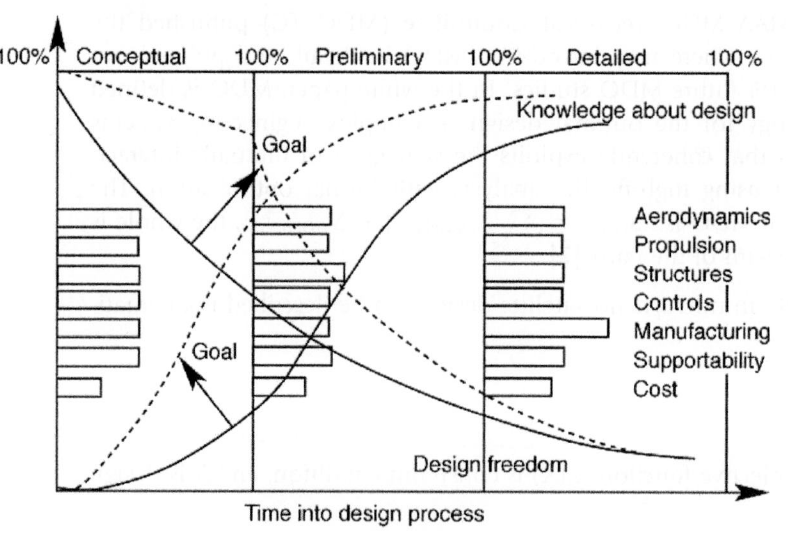

FIGURE 2.3
MDO in micro/nano satellite design.

decays rapidly, while knowledge about the object of design increases. The dashed line projection from the "Knowledge about design" curve reflects the requirement that more knowledge will have to be brought forward to the conceptual and preliminary design phases. The dashed line projection from the "Design freedom" curve reflects the need to retain experimentation, and human reasoning.

2.2.2 MDO Development

MDO technology for micro/nano satellites optimizes a complex spacecraft systematically by exploiting and taking advantage of the synergetic mechanisms between each subsystem (or disciplinary). The core idea of the technology can be expressed as: a complex system design process which organizes and manages the design process with proper optimization strategy, decomposing the complex system into several modern engineering organization-consistent subsystems, taking advantage of existent industrial multidisciplinary analyzing tools to unite the abundant knowledge and experience of subsystems by a distributed computer network, to shorten the design period, lower development costs, and improve product quality and competiveness [12].

In general, MDO research can be summarized into three aspects: integration of design-oriented multidisciplinary analyzing software, concurrent multidisciplinary design for system optimization, and supporting environment of distributed MDO computing.

As an important research area, study of the MDO developing environment has made great progress while a series of MDO-based program developing systems are already in operation, some of which had been commercialized. Some overseas software enterprises have provided a number of world-famous MDO integration tools such as iSIGHT, Modelcenter, Optimus, BossQuaur, and so on. Some well-known MDO softwares and their developers are listed in Table 2.3.

Table 2.3 Some Computing Software for MDO

General Computing Tools	Embedded Optimization Tools	Process Integration Tools
Excel (Microsoft)	SolidWorks-Cosmos	ModelCenter, CenterLink (Phoenix Int.)
MATLAB (Mathworks)	GENESIS (VRD)	iSIGHT, FIPER (Engineous)
Mathematica (Wolfram)		model FRONTIER (Esteco)
		Optimus
		BossQuaur

2.2.3 Domestic MDO Research and Applications

Compared with overseas, MDO research in China began late. However, in recent years, the country has seen a flourishing development of the aerospace and shipbuilding industries, making MDO a new favorite of the Chinese government and companies. So far, Tsinghua University, National University of Defense Technology, and many other institutions have done much fundamental research into MDO.

At present, the study and application of DMO in China are mainly in complex systems such as aircraft and missiles. In terms of satellite system designing, especially with micro/nano satellite systems, MDO is worth significant study. In the government's 11th Five-Year Spacecraft planning, optimization technology of satellite systems was listed as one of the key technologies. The application of MDO in satellite system design can be concluded into the following three aspects:

1. Modeling of satellite system parameters;
2. Digital stimulation platform of satellite;
3. Specific mission target-based optimization algorithm.

2.2.4 Characteristics of MDO for Micro/Nano Satellite

The MDO process of micro/nano satellites has its own characteristics:

1. Different from regular satellites, micro/nano satellites usually have similar payloads. Therefore, optimization of micro/nano satellite systems requires taking account of the entire effective payload system into the computing process, rather than considering only the satellite platform.
2. Micro/nano satellites usually perform as a stellar or a formation, thus they should be designed at a system level, which will unavoidably increase the analyzing complexity.
3. Micro/nano satellites are relatively very light and are usually launched together with other spacecraft, making the cost on launching negligible.
4. Micro/nano satellites have a short design period and fast updating rate, which requires mature optimization and a design platform to accelerate production.

2.3 MDO ALGORITHM FOR MICRO/NANO SATELLITES

The optimization algorithm is basically a searching process/rule following some theory or mechanism to find solutions to customer requirements.

Multidisciplinary design optimization (MDO) aims at building a reasonable optimization system by modeling some particular problems, and working out a proper optimization strategy for minimal calculation and communication burden, thus structurally reducing the calculation complexity and management complexity of MDO. The main subject of MDO is the description of the problem, which includes decomposition and coordination of the problem, transmission method of designing information, etc. When the MDO method is settled, it can be combined with an optimizer and algorithm of the method to solve the MDO problem. In conclusion, the MDO algorithm is the core of MDO, and the most important and active subject in the MDO area. This section explains the collaborative optimization method and its improvement.

2.3.1 Overview

Collaborative optimization (CO) is an MDO method originally suggested by Professor Kroo and others from Stanford University in the mid-1990s [13–15]. To some extent, the appearance of CO can be attributed to constraints of the consistency optimization, which had been taken advantage of in CO. However, there are distinct differences between CO and consistency-constrained optimization: the former is a multistage process, where each subsystem will both be analyzed and optimized, while the latter is a single-stage process without subsystem optimization.

2.3.1.1 Basic Theory of CO Methodology

When considering subsystem design optimization, the effect of other subsystems can be temporarily ignored. That is, the purpose of subsystem MDO is to minimize the difference between its real optimization solution and systematically settled target solution. Then, the inconsistencies between each subsystem or discipline are collaborated by systematically optimization, in which systematically and disciplinary optimization are iterated multiple times to finally reach a proper solution which meets the subsystem consistency requirements.

2.3.1.2 Basic Framework of CO Methodology

According to its fundamental theory, the basic framework of CO is demonstrated in Fig. 2.4.

As can be seen from the diagram, the original MDO problem is decomposed into N multidisciplinary units, each of which becomes an independent subproblem. Each subproblem consists of two parts: a subsystem optimizer and an analysis submodule. Accordingly, the design parameters of the task are grouped into N subsets: $[X_1, X_2, \ldots, X_N]$. To be more specific, each of these subsets consists of two kinds of parameters: single disciplinary design

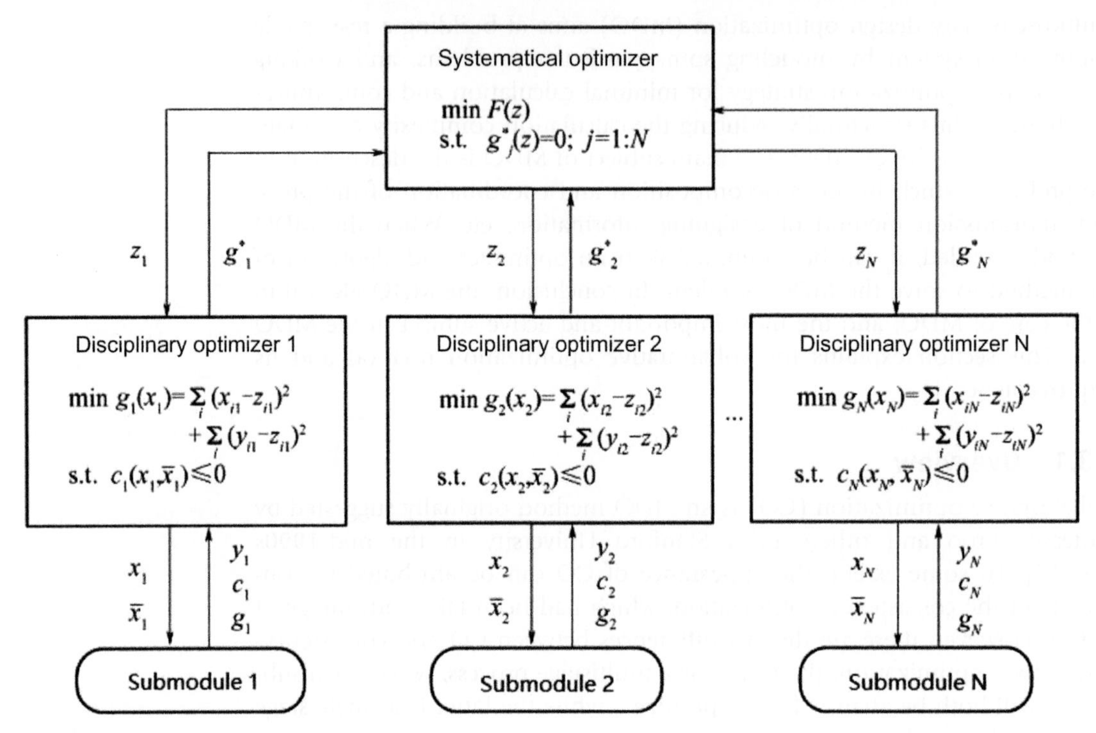

FIGURE 2.4
Basic framework of CO methodology.

parameters $\overline{x_i}$ and multidisciplinary design parameters x_i. Likewise, the original constraint sets c are also grouped into N nonintersecting subsets, named local constraints of MDO subproblems. Meanwhile, the analysis output of submodules provides multidisciplinary effected coupling parameters y_i. To designate target values passed to the subspace optimizer, the CO methodology adds a system-level optimizer; this optimizer specifies the target values of the design parameters and passes them to each of the subproblems. The system-level optimizer's objective is to adjust the parameter values so that the objective function is minimized while the system-level constraints are satisfied.

2.3.1.3 Solution of CO Methodology

To solve a CO problem, assume that the system-level optimizer chooses an initial set of design parameters z. These parameters are grouped and passed on to subsystems as fixed input z_i. Then their corresponding analysis module accepts the optimizers' output $\overline{x_i}$ and x_i as input to generate parameter y_i, local constraints c_i, and the difference between local variable values and the target values is passed to the subsystem optimizer g_i. The objective of disciplinary optimization is to minimize $g_i(x_i)$ while the local constraints $c_i(x_i, \overline{x_i}) \geq 0$ are satisfied. After that, the solution of disciplinary optimization g_i^* will be submitted to the system-level optimizer, which will reevaluate the optimization strategy and update the system design parameter z. Then a new iteration starts with new design parameters passed to each subsystem optimizer. The process stops when a system-level solution which satisfies all consistency restraints is obtained.

2.3.2 Disadvantages of CO Methodology

Even though the CO methodology is especially suitable for satellite MDO problems, several of its shortcomings have been perceived through comparison of large numbers of numerical examples and analysis of mathematical principles of the CO algorithm. Some of these are:

1. The CO algorithm optimizes each subsystem toward an objective value independent from the system value to reduce systematical analysis, therefore, the optimization objective value will probably oscillate significantly as the system value approaches an optimal number, resulting in an increase in the iteration number.
2. The CO algorithm uses systematic equality constraints to ensure consistency between subjects, and optimizes the problem with lower-level optimization to minimize disciplinary inconsistency. However, since highly nonlinear consistent equality constraints are introduced into systematical optimization, the Lagrange multiplier may not exist, or the Jacobi matrix of consistency constraints may be discontinuous at the optimized solution, which means the systematic optimization fails to meet the requirements of the Kuhn–Tucker condition. As a result, the CO algorithm often fails to provide a global optimal solution, or even diverges sometimes.
3. The CO algorithm provides a practical optimal solution only when all the equality constraints are satisfied, rather than output a better solution after each iteration.
4. The CO algorithm solves optimization problems with continuous parameters only, however, there are large numbers of discrete parameters (decision variables) in spacecraft designs, which cannot be solved by the CO algorithm.

2.3.3 Improvement of CO Methodology

According to Section 2.3.2, an improvement process of the CO methodology is inevitably required before it is applied in satellite MDO problems.

At present, there are two main approaches: reformulate optimization and apply a modern optimization algorithm into the optimizers.

2.3.3.1 Reformulate Optimization

As mentioned above, the system-level nonlinear consistency constraints of the CO process often result in local optimization or convergence failure, meanwhile, as an inherent shortage of the special formulae of CO, the relative optimizing independence between subsystems and the whole system will lead to excessive analysis. Therefore, the CO method has to be reformulated.

Reformulate the original system-level CO problem:

$$\begin{aligned}&\min F(z)\\ &\text{s.t.}\quad J_i^*(z) = \sum (x_j^* - z_j)^2 + \sum (y_j^* - z_j)^2 = 0\end{aligned}$$

where $F(z)$ is the system target function, and $J_i^*(z)$ is the subsystem optimization objective.

Apparently, the most effective way to eliminate the effect of system nonlinear consistency constraints and relate the optimization target of the subsystem to that of the entire system is to include $J_i^*(z)$ in the expression. A penalty function is added and the MDO problem reformulated as follows:

$$\min F'(z) = F(z) + \sum \omega_i J_i^*(z)$$

The new expression is a constraints free nonlinear optimization formula, which is much easier to solve.

2.3.3.2 Application of a Modern Optimization Algorithm

Generally, standard CO methodology uses traditional numerical optimization method, such as sequential quadratic programming (SQP). However, these algorithms have stringent constraints on problem structure, therefore they unable to offer solutions for a discrete problem, and so the advanced modern optimization algorithm was introduced in Ref. [16].

The genetic algorithm (GA) [17–22] is a general optimization algorithm which is in high parallel, random and self-adapted. It adopts biogenetic points of view, and mimics mechanisms of natural selection, heredity, and variation to increase individual fitness. This process embodies Darwin evolutionism: "survival of the fittest."

The GA describes problems as the most fit survival process of the "chromosome." At first, it is necessary to select an initial population that may contain the potential solution, which is composed of a certain amount of genetic coded individuals. Each individual is a unit with the "chromosome." The population will be generated by operations including copying, crossing, and variation, making better genes for individuals, until the fittest unit is determined and brought up.

2.4 RESEARCH OF THE MICRO/NANO SATELLITE MDO FRAMEWORK

The design of a satellite is usually divided and carried out by several geographically separated teams of different disciplines, using their own analyses and design software to perform the optimization. To achieve the coordination of various disciplines and maintain the autonomy of each team at the same time, which will utilize the full experience of experts in various disciplines, MDO must organize the design optimization effectively by full use of a modern computer network, achieving optimum solutions with concurrent computation. Therefore, it is necessary to study a better framework of MDO, and to develop the distributed MDO platform.

2.4.1 The Ideal Development Environment of MDO

Some characteristics of satellite design, including difficulty to find the solution, complexity of computation and distribution organization, put forward a significant request to the MDO development environment, including the framework, structure of problems, selection of optimization methods, execution of the framework, design information acquirement, as Table 2.4 shows.

Table 2.4 Features of Ideal MDO Development Environment

Features	Requirement
Requirement of developing environment	Good GUI
	Object-oriented technology
	Extendibility of structure
	Large-scale data handling
	Distributed collaborative design
	Standardization of architectural elements
Setting up principles	Avoiding the bottom driver programming
	Easy to reconfigure
	Code reusable
	Avoiding fully integrated unchangeable
	Debugging tools
Selection of optimization methods	Adapt to the mathematical model
	Support multiple optimization methods
	Tolerate expensive cost of function estimates
	Robustness
	The ability to characterize the design space
	Provide sensitive data
Execution of the framework	Batch process and optimize automatically
	Parallel optimization
	Heterogeneous distributed network
	High-performance computing
	Interactive process monitor and control
Design data acquirement	Database management
	Visualization of result
	Execution status monitoring
	Fault-tolerant mechanism

2.4.2 The Framework of MDO

There are three main kinds of MDO framework to select. The general MDO framework includes FIDO, iSIGHT, LMS, Optimus, etc. Frameworks with specific techniques include Fiper, CSD, Caffe, etc. Frameworks for specific optimization problems include AMESim, CJOopt, etc.

1. iSIGHT
   iSIGHT is an important MDO tool that can automate the process and optimize for several objects. It provides a visual and smart process for the user to set up the optimization platform. At the same time, it also provides a software interface for most mainstream CAE analysis tools. On the platform, users can easily and quickly drag and drop elements to create the complex process flow, to set and modify the design parameters and the solution objective, and to automatically carry out the analysis for multiple cycles. iSIGHT also provides software packages to carry out experiment design, optimization method selection, approximation model selection, and six-sigma design, which can help users understand the product system and make up the interface of design variables and optimization objectives.
2. Fiper
   Fiper is an enterprise tool which was extended from iSIGHT, providing multidisciplinary simulation and optimization system for commercial use. Its framework and functions are shown in Fig. 2.5. By the web form, users can modify the simulation parameters, execute the analysis process, and manage and analyze the entire workflow, component library and running environment, etc. Fiper also allows staff to share their workflow and establish a complex workflow collaboratively. Different groups also can work together by information exchange at the enterprise level. Fiper has advanced features to integrate product life-cycle management (PLM) systems and simulation life-cycle management (SLM) systems, to acquire CAD models from PLM and SLM, manage the simulation data, and balance and evaluate the results from different schemes.
3. AMESim
   AMESim was released in 1995 by IMAGINE, to provide professional solutions for hydraulic/mechanical systems modeling, simulation, and dynamic analysis. Based on the expertise of IMAGINE, AMESim has good performance in engineering design. Especially for hydrodynamic (fluid and gas), mechanical, fluid thermal, and control systems, AMESim provides a superior simulation environment and flexible total solution, allowing users to set up an engineering design platform to finish multidisciplinary modeling and developed analysis.

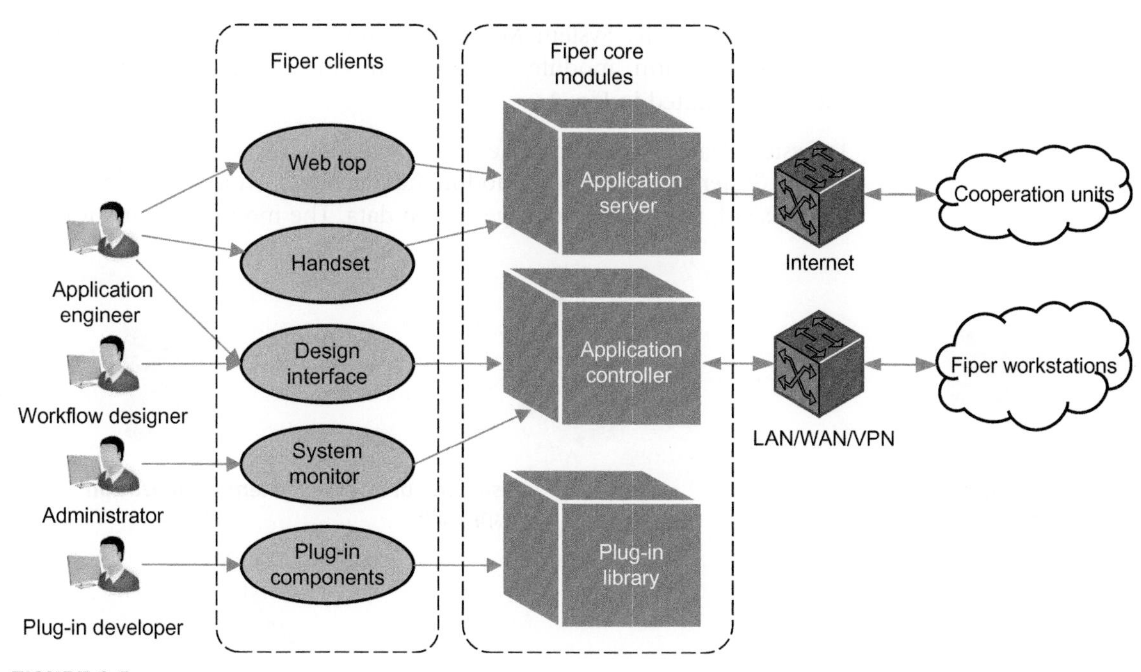

FIGURE 2.5
Architecture and functions of Fiper.

2.5 THE MDO PLATFORM FOR MICRO/NANO SATELLITES

The MDO platform for micro/nano satellites discussed here was established by Tsinghua University, and named SDIDE (Spacecraft Distributed Integrated Design Environment) [23], which is based on a cooperative method and distributed technology. SDIDE is set up on iSIGHT, integrating analysis tools from various disciplines, including STK, MATLAB, UG NX, I-DEAS, ANSYS, etc. With the help of iSIGHT-Net module, distributed modeling and computation can be done easily. SDIDE can help shorten satellite development time, and improve design work. On the SDIDE platform, a collaborative optimization algorithm is improved with a genetic algorithm and local optimization technique. For the launched satellites "Hangtian Tsinghua 1" and "NS-1," related models and data result, with better field temperature and the weight set as the objectives, obtaining a good solution after optimization on SDIDE.

2.5.1 The Basic Framework of SDIDE v1.0

The first version of SDIDE is v1.0. In SDIDE v1.0, the basic framework is set up. The platform contains seven modules: System Data Manage module,

Task Control module, System Monitoring module, Configuration module, System Optimization module, Database module, and Discipline/Design Subnet, as illustrated in Fig. 2.6.

1. System Data Manage module
   The System Data Manage module handles data on a system level, and can send, receive and manage system data. The module automatically

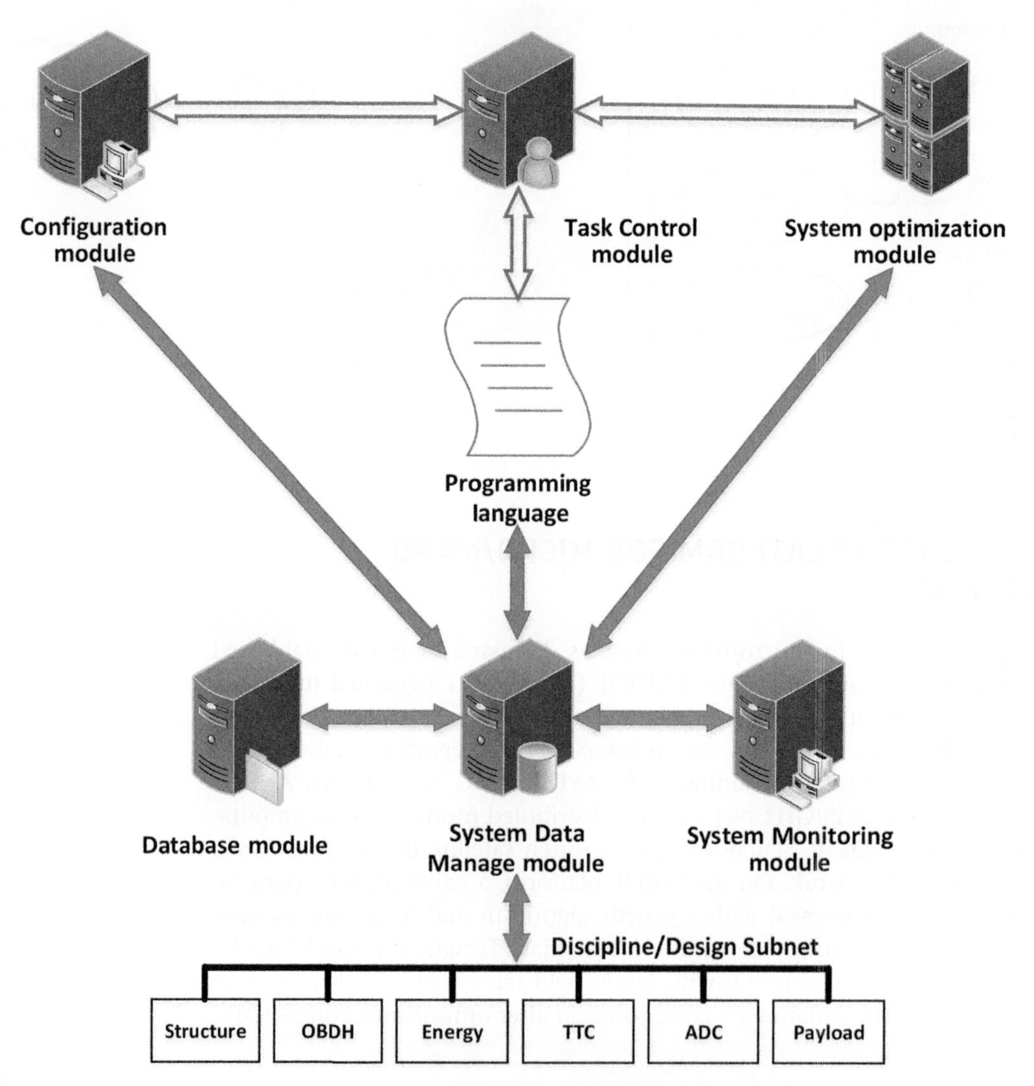

FIGURE 2.6
Framework of SDIDE v1.0.

receives data from every discipline, updates and saves it to the Database module, to real-time update the system data. At the same time, the System Data Manage module records the previous system data and status, and sends data renew instructions to all modules to start an update.

2. Task Control module
   The Task Control module manages, monitors, and controls the whole task procedure, acquires the task status, and starts or stops services and the Discipline/Design Subnet when needed.

   Interactive GUI is an important part of the Task Control module. By the GUI, users can start the task when the MDO problem and objective are configured. Then the module operates as indicated in the task flow, starting the Discipline/Design Subnets in sequence or in parallel. The status of each Discipline/Design Subnet and subtask are shown on the GUI, by which the user can operate and handle events.

   While the optimization task is working, users can pause the system task or Subnet, and the previous system data and status will be saved and sent to the System Data Manage module and Discipline/Design Subnets, allowing the computation to hibernate until needed to continue to run or stop.
3. System Monitoring module
   The System Monitoring module is set up to acquire and monitor the design information during the optimization, and to show the process data and the final result.

   For satellite design and optimization, the process of optimization iteration is very important as the final optimum solution. Therefore, the System Monitor module must support users to get necessary information and data at any moment in time, including design variables, constraints, object functions, and any other values that may be required. The module also provides analysis function for users to discover features and evaluate feasibility of the design.
4. Configuration module
   The first action is carried out in the Configuration module to begin a MDO task. In this module, users can define and configure design variables, initial values, object functions, constraints, and optimization strategy. The process flow and information interactions are also defined.

   When users finish their configuration work, many elements, including blocks and figures and values, are shown in the GUI. Then the Configuration module translates and compiles those elements to ASCII format files. This kind of ASCII file is "executable" for software inside SDIDE.

   As mentioned above, when optimization starts, users can pause the process and configure parameters as needed.

5. System Optimization module
   The System Optimization module is the key and energy source of the system that can be optimized. It is a library of MDO methods and optimization algorithms, providing basic strategies to construct the MDO task, and providing functions to the optimizer.
   For users, the System Optimization module is a kind of black box, providing services to upper modules. The MDO methods include single-level optimization, concurrent subspace optimization (CSSO) and collaboration optimization (CO), etc. The optimization algorithms include numerical algorithms such as the feasible direction method, steepest descent method, sequence linear programming (SLP) method, sequential quadratic programming method, etc.; and search algorithms such as random search algorithm, genetic algorithm, simulated annealing algorithm, etc. More numerical algorithms and search algorithms can easily link up and integrate in the System Optimization module.
6. Database module
   The Database module stores all data and information during satellite design and optimization, playing the role of data center in the SDIDE framework.
   The Database module is operated by instructions from the System Data Manage module, including actions such as reading, storing, searching, etc. In some cases, the System Data Manage module can transfer file pointers to other module, allowing the specific module to operate on the Database module. Most modules are separated in a distributed framework, but the Database module must be set up to be a kind of network database.
   Massive amounts of data are produced during satellite design and optimization. For example, the finite element method is widely used in mechanical analysis and thermal analysis, containing hundreds and even tens of thousands of nodes in the finite element model. Huge matrices are necessary to handle the data of these nodes as a whole. To ensure continual read and write operations, the Database module must have strong store and search ability and perfect performance.
7. Discipline/Design Subnet
   The Discipline/Design Subnet of SDIDE is a kind of distributed computing network, consisting of several separated computers for different disciplines. Each computer is in charge of one kind local discipline design and analyze, following instructions from the System Data Manage module, using data from other modules and local input. All the process data and final data are sent to the System Data Manage module.

All needed disciplines, corresponding professional database, design and analyze software, computers for satellite design and optimization are included in the Discipline/Design Subnet. The network is star-based, and there are no direct connections or channels between different disciplines. To exchange design data and other information between each other, a discipline must send the request to the System Data Manage module.

2.5.2 Improvement of SDIDE

After SDIDE v1.0 was launched, a great deal of progress was made in the MDO framework, optimization method, etc., from which SDIDE v2.0 was developed. The architecture of SDIDE v2.0 is shown in Fig. 2.7.

Compared with SDIDE v1.0, the SDIDE v2.0 has some new features:

1. SDIDE v2.0 was developed from Fiper, containing existing optimization algorithms which are provided by Fiper and MATLAB. Some more algorithms were introduced and improved upon, including CO, CSSO, MOPCSSO, which will expand the facilities of SDIDE.
2. SDIDE v2.0 is enhanced for conceptual design, and satellite overall design scheme selection, comparison, and optimization are realized. This version can manage multitarget models to satisfy a variety of optimization needs.
3. Professional toolkits are developed based on Fiper, including STK components, UG components, ANSYS components, and MATLAB components, which is of benefit for CAD/CAE analysis.
4. A new web module was added, providing better service for distributed cooperative design and optimization. It also has some new features, including people management, knowledge management, project management, etc., and the GUI is now more user-friendly.
5. SDIDE v2.0 has two types of new service: local resource service and distributed resource service, providing various documents, reports, and videos for developers.

2.5.3 Features and Function of SDIDE 2.0 System

2.5.3.1 Main Features

1. Technical features: developed with C# language on Microsoft. NET framework, introduced in WCF technology of service-oriented architecture (SOA) to issue and invoke local and distributed resource services.
2. Functional features: providing more key functions including requirement management, people management, project management, solution building, local resource service, distributed resource service, etc., by which users can acquire various information/data.
3. Easy to integrate: by using modular approaches, modules and their functions are defined clearly, making them easy to rebuild.

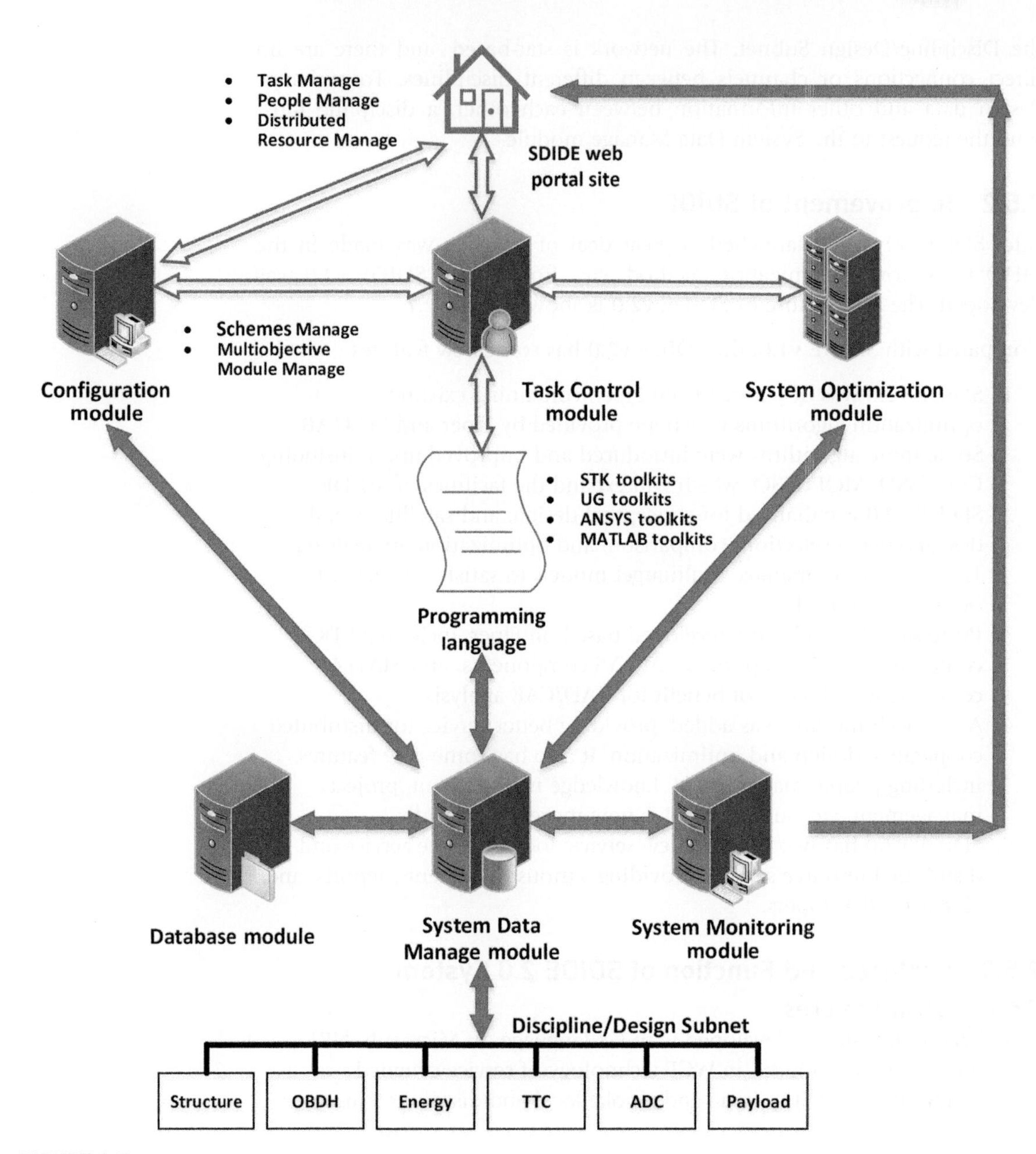

FIGURE 2.7
Framework of SDIDE v2.0.

4. Componentization: componentization of resource services of each discipline provides good features for resource discovery, release, and integration.
5. Externalization of tacit knowledge: allowing users to acquire tacit knowledge which is presented as solutions and maintenance by the system in SDIDE v2.0.

2.5.3.2 *Main Functions*

1. Function of requirement modeling and analysis during satellite design. Qualitative and quantitative methods are both provided to transform between application requirements and technical requirements.
2. Function of team building and knowledge acquirement. This function allow users to build up team and necessary knowledge as the project proceeds.
3. Function of resource knowledge discovery, release, and usage. By this function, the resource knowledge provider and consumer come into being and exchange resource knowledge naturally.
4. Function of knowledge integration. Different users can acquire integrated knowledge from local and distributed resource environments.

2.5.4 Integration of CAD/CAE and Further Development

In SDIDE v2.0, professional toolkits are provided for CAD/CAE analysis, such as UG, ANSYS, STK, MATLAB, etc. By these toolkits, the design and optimization have the ability for total parametric modeling, command stream control, and process automation. Fig. 2.8 shows how these work.

These professional toolkits are developed by the following methods.

1. Toolkits for UG
   The toolkits for UG are developed and integrated by UG/Open API module, which is also called the User Function. This is an assembly that can be accessed by programs and change UG models. It contains nearly 2000 library functions, by which the corresponding UG functions can be operated in C and C++ programs, such as GUI operation, file operation, and database operation.
2. Toolkits for ANSYS
   The toolkits for ANSYS arranges ANSYS commands by APDL (ANSYS Parametric Design Language), providing parametric programs to users. This kind of program can actualize the whole process of ANSYS optimization, including CAD parameter modeling, grid division and control, material definition loading, boundary conditions, parametric control and solve.

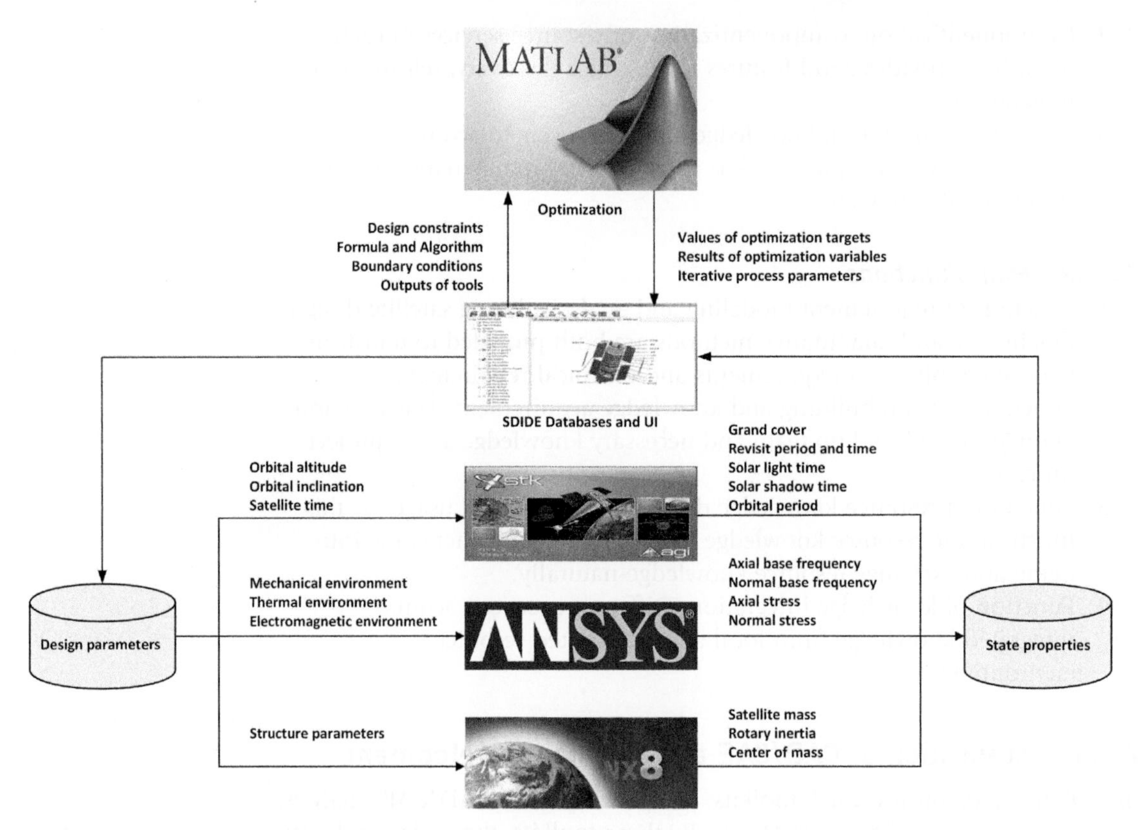

FIGURE 2.8
Workflow between SDIDE and the toolkits.

In these toolkits, the system can perform finite element analysis of force, heat, electromagnetism, and their couplings.

3. Toolkits for STK
   The toolkits for STK are developed by STK/Connect module in Visual C++ environment. The STK/Connect module is a set of API functions that can embed in VC programs to operate and respond to STK commands.
   The toolkits can calculate and display on-orbit operation status, cover ratio, and sunshine hours of satellites.
4. Toolkits for MATLAB
   The toolkits for MATLAB are developed with MATLAB Web Services technology, which allows MATLAB to run as a pure calculation engine on the server. The client calls the Web Services in SDIDE, transferring parameters and commands to the server, and getting the results after calculation.

References

[1] J.R. Wertz, W.J. Larson, Space Mission Analysis and Design, Microcosm Press, El Segundo, CA, 1999.

[2] W. Heijenga, View definition in OODBS without queries: a concept to support schema-like views, in: Proceedings of 2nd International Conference on Databases and Information Systems, 1996.

[3] R.J. Mayer, M.K. Painter, P.S. Dewitte, IDEF Family of Methods for Concurrent Engineering and Business Re-engineering Applications, Knowledge Based Systems Inc, College Station, TX, 1994.

[4] C4ISR Architecture Working Group, C4ISR Architecture Framework Version 2.0, Carnegie Mellon University, Pittsburgh, PA, 1997.

[5] W.G. Wood, M. Barbacci, P. Clements, et al., DoD architecture framework and software architecture workshop report [EB/OL], March 2003 (February 2008). <http://www.ichnet.org/DODAF%20SEI%20report.pdf>.

[6] OL, <http://www.aiaa.org/content.cfm?pageid = 230&lumeetingid = 1384>.

[7] AIAA Multidisciplinary Design Optimization Technical Committee, Current state of the art on multidisciplinary design optimization (MDO). An AIAA White Paper, 1991.

[8] J. Sobieszczanski-Sobieski, T. Haftka, Multidisciplinary aerospace design optimization: survey of recent developments, Struct. Optim. 14 (1) (1997) 1–23.

[9] S. Kodiyalam, Multidisciplinary aerospace systems optimization-computational aerosciences (CAS) project. NASA/CR-2001-211053, 2001.

[10] C. Qifeng, Distributed coevolutionary multidisciplinary design optimization methods for flying vehicles. National University of Defense Technology, April 2003.

[11] Z. Wang, X. Chen, W. Luo, M. Design, Optimization Theory and Applications, National Defense Industry Press, Beijing, 2006.

[12] E.R. Taylor, Evaluation of multidisciplinary design optimization techniques as applied to spacecraft design, in: Proceedings of Aerospace Conference, IEEE, Oakland, CA, 2000, pp. 371–384.

[13] R.D. Braun, A.A. Moore, I.M. Kroo, Collaborative approach to launch vehicle design, J. Spacecr. Rockets 34 (4) (1997) 478–486.

[14] R.D. Braun, Collaborative optimization: an architecture for large-scale distributed design. PhD thesis, Department of Aeronautics and Astronautics, Stanford University, May 1996.

[15] R.D. Braun, I.M. Kroo, Development and application of the collaborative architecture in a multidisciplinary design environment. Virginia, NASA Langley Technical Report Seroer, 1995.

[16] W. Ling, Intelligent Optimization Algorithms with Applications, Tsinghua University Press, Beijing, 2004.

[17] L. Yong, K. Lishan, C. Yuping, Nonnumerical Parallel Algorithm (Volume 2): Genetic Algorithm, Science Press, Beijing, 2000.

[18] M.A. Stelmack, N. Nakashima, S.M. Batill, Genetic algorithms for mixed discrete/continuous optimization in multidisciplinary design, in: AIAA: Symposium on Multidisciplinary Analysis and Optimization, 1998:984771.

[19] R.A.E. Makinen, J. Periaux, J. Toivanen, Multidisciplinary shape optimization in aerodynamics and electromagnetics using genetic algorithms, Int. J. Numer. Methods Fluids 30 (1999) 149–159.

[20] G. Rudolph, Convergence analysis of canonical genetic algorithms, IEEE Trans. Neural Netw. 5 (1) (1994) 96–101.

[21] T. Back, H.P. Schwefel, Evolution strategies I: variants and their computational implementation, Genetic Algorthms in Engineering and Computer Science, John Wiley & Sons Ltd, Chichester, 1995.

[22] J. Morrison, F. Oppacher, A general model of co-evolution for genetic algorithms, Artificial Neural Nets and Genetic Algorithms, Springer Vienna, Wien, 1999, pp. 262–268.

[23] F. Junming, Research on Applying Collaborative Optimization to the Multidisciplinary Design Optimization of Satellites Design, Tsinghua University, Beijing, 2006.

CHAPTER 3

Attitude Determination and Control System of the Micro/Nano Satellite

The high-precision and high-performance attitude determination and control system (ADCS) of the micro/nano satellite are the basic conditions for a satellite to run efficiently as the accomplishment of the mission of satellites relies on the performance of this instrument as well as being determined by the precision of the attitude control. From the international development trend, ADCS are approximately 40% of total development costs and this is the critical technology in the development of the micro/nano satellite.

To enhance the performance of ADCS, we should take advantage of new technology to develop lightweight and high-performance attitude-sensing and measurement devices and actuators [1–5], because of the limitation of power consumption and volume, while optimizing this comprehensive system by excavating the capacity of sensors when designing the system [6–10] to enhance the density of functionality of the satellite. Researching new data fusion and control methods to decrease the demand on hardware through software compensation is an effective and significant measure.

In this chapter, a comprehensive satellite was modeled, including environmental disturbance, kinematics, dynamics, attitude sensors, actuators, etc., according to the design of the NS-2 satellite. The attitude and control methods were researched on the basis of models used to fulfill the requirements of this task and to enhance the performance of the attitude control system. Reasonable and effective control schemes were proposed. The context in the chapter is necessary guidance for and demonstration of this appliance in engineering.

3.1 SPACE ENVIRONMENT OF THE MICRO/NANO SATELLITE

Designers of ADCS should understand the dynamics of the controlled object comprehensively, as well as the space environment and the mathematics of disturbance control. Understanding the environment [11] is very important

Space Microsystems and Micro/Nano Satellites. DOI: http://dx.doi.org/10.1016/B978-0-12-812672-1.00003-5

for choosing control laws and actuators to take full advantage of the space environment for passive and semiactive attitude control.

Space environment torque consist of the solar radiation pressure, gravity gradient, magnetic fields, and aerodynamics. The magnitudes of these depends on the orbit height, mass distribution, shape, surface, solar activity, atmosphere density, magnets on the satellite, and attitude motion.

3.1.1 Gravity Environment

Bodies are subjected by gradient moment if they have asymmetric mass distribution and are in a gravitational field that is inversely proportional to the square of the distance that was researched by D'Alembert and Euler in 1749. Lagrange explained why there is always a dark side of the moon in 1780. After that Laplace and others illustrated the equation to express the gravity gradient moment when researching the libration of the moon. It is relatively complex to calculate gravity and gravity moment since they depend not only on the mass distribution of the satellite and the mass distribution of the attractive body but also the location and the attitude of the satellite. To obtain a practical and precise enough equation in engineering, we assume the following. There is a single attractive object, the Earth. The distribution of the attractive object is one of spherical symmetry. The size of the satellite is far less than the size of the attractive object. The satellite itself is a single object. The equation of gravity gradient moment is as follows:

$$\boldsymbol{N}_{d,g} = \frac{3\mu}{r^3}\boldsymbol{R} \times \boldsymbol{I}\boldsymbol{R} \tag{3.1}$$

where μ is the gravitational constant of the Earth $\mu = 398{,}600.44\ \mathrm{km}^3/\mathrm{s}^2$, r is the distance from the center of the Earth to the centroid of the satellite, $\boldsymbol{R}$ is a unit vector that points from the center of the Earth to the centroid of the satellite, and $\boldsymbol{I}$ is the moment of the inertia matrix of the satellite.

When the attitude angles are infinitesimal, the two order and high-order infinitesimal are neglected, we obtain:

$$\boldsymbol{N}_{d,g} = 3\omega_0^2 \begin{bmatrix} (I_z - I_y)\varphi + I_{yz} - I_{xz}\theta \\ (I_z - I_x)\theta - I_{xz} - I_{xy}\varphi \\ I_{yz}\theta + I_{xz}\varphi \end{bmatrix} \tag{3.2}$$

When $\boldsymbol{I}$ is a diagonal matrix, Eq. (3.2) can be reduced to Eq. (3.3).

$$\boldsymbol{N}_{d,g} = 3\omega_0^2 \begin{bmatrix} (I_z - I_y)\varphi \\ (I_z - I_x)\theta \\ 0 \end{bmatrix} \tag{3.3}$$

3.1.2 Atmosphere Environment

The gas movement beyond 120 km above sea level can be treated as free molecule flow, that is, the mean free path of molecules is far larger than the characteristic size of the satellite. Aerodynamic forces and torque rely on the atmospheric density, the size and the shape of the satellite, surface characteristics, the velocity of the gas flow, and the attitude of the satellite, etc. In physics, aerodynamic forces and torque are caused by collisions between the satellite and gas molecules. Gas molecules that hit the satellite are generally emitted again. However, there is no interaction between the secondary molecule and the incidence molecule under the conditions of free molecule flow. Therefore, the total momentum can be calculated by adding the independent calculation of the incident flow and the secondary flow. Meanwhile, the free molecule flow method can simplify the calculation of satellites, of which the aerodynamic shape is complex, by dividing it into several simple parts for calculation.

To obtain a practical equation with the aerodynamic torque calculation, we assume the following. Gas molecules reaching the surface transfer all their momentum to the satellite. Thermal motion average speed calculated by the maximum Maxwell speed in probability is 1 km/s approximately, which is less than the speed of the satellite. The impulse caused by molecules leaving the surface is negligible. The simplified equation is as follows:

$$N_{d,a} = \frac{\rho V_R^2}{2} C_D A_p \cdot \mathbf{c}_p \times \boldsymbol{v} \tag{3.4}$$

$\frac{\rho V_R^2}{2}$ denotes the dynamic head, C_D is the drag coefficient within 2.2–2.6. A_p is the cross-sectional area, $\mathbf{c}_p$ is the vector from the centroid to the center of pressure, and $\boldsymbol{v}$ is the unit vector of the air flow.

The following is an example:

Altitude 630 km: $\rho = 4.31 \times 10^{-13}$ kg/m^3

Velocity of the satellite: $v = \sqrt{\frac{\mu}{r}} = \sqrt{\frac{3.986 \times 10^{14}}{7008.137 \times 10^3}} = 7.5417 \times 10^3 \ m/s$

Assume the drag coefficient as: $C_D = 2.5$

Magnitude of vector from the centroid to the center of pressure:

$c_p = 0.2$ m

Then, $N_a = 1.5934 \times 10^{-5}$ Nm.

3.1.3 Electromagnetic Environment

Magnetic moment is caused by the interaction between the effective dipole magnetic moment and the local geomagnetic field. The dipole magnetic moment of the satellite is caused by the loop current as well as by residual and permanent magnetism in the magnetic material. The geomagnetic field at the location of the satellite has a relationship with the orbit height, longitude, latitude, attitude, and solar activity. The magnetic torque can be used to dampen the nutation, eliminate the rotation, control the precession (though it is a disturbance torque), and can have a positive effect on attitude.

Understanding and modeling the geomagnetic field are necessary to estimate the geomagnetic torque. Geomagnetic fields are classified into inner-originated fields and outer-originated fields according to the origin of the field. Inner-originated fields are caused by the inner structure of the Earth, while outer-originated fields originate from the current system around the Earth, such as the ionospheric current, the ring current, magnetic top layer current, etc. These current systems are changing constantly. They are influenced by solar activity, magnetic storms, etc. Inner-originated fields contain a basic magnetic field and an inducted magnetic field. The basic magnetic field is very stable as it is caused by the core of the Earth. The magnetic pole on the surface of the Earth moves very slowly despite being caused by a change of current. The primary component in the harmonic function model of the geomagnetic field or the simplified di-inducted magnetic field is expressed by a dipole model.

Residual magnetic torque is caused by the interaction of the residual magnet of the satellite and the geomagnetic field. The basic geomagnetic field that is a potential field in the space outer the Earth satisfies the Laplace equation. It is expressed as spheric harmonic function, as in Eq. (3.5).

$$\phi = R_e \sum_{n=1}^{\infty} \sum_{m=0}^{\infty} \left(\frac{R_e}{r}\right)^{n+1} (g_n^m \cos m\lambda + h_n^m \sin\lambda) P_n^m(\cos\theta) \tag{3.5}$$

$R_e = 6378.14$ km is the radius of the Earth at the equator, g_n^m and h_n^m are the harmonic coefficients, λ is the geographic longitude, and θ is the geocentric colatitude. $P_n^m(x)$ is the associated Legendre function as:

$$P_n^m(x) = (-1)^m \frac{(1-x^2)^{\frac{m}{2}}}{2^n n!} \frac{d^{n+m}}{dx^{n+m}} (x^2-1)^n \tag{3.6}$$

In the geographic coordinate system, magnetic induction of the geomagnetic field $\boldsymbol{B}_\lambda$ is defined as:

$$\boldsymbol{B}_\lambda = -\nabla\phi = \left[\frac{1}{r} \cdot \frac{\partial\phi}{\partial\theta} \quad -\frac{1}{r\sin\theta} \cdot \frac{\partial\phi}{\partial\lambda} \frac{\partial\phi}{\partial r}\right]^{\mathrm{T}} \tag{3.7}$$

From the relationship of the geographic coordinate, the inertial coordinate, and the body coordinate, we obtain $\boldsymbol{B}_b = \boldsymbol{A}_{bi}\boldsymbol{A}_{i\lambda}\boldsymbol{B}_\lambda$. Assuming the residual magnetic moment is $\boldsymbol{M}_b$ (Am^2), then the residual magnetic torque is:

$$\boldsymbol{N}_{d,m} = \boldsymbol{M}_{residual} \times \boldsymbol{B}_b \tag{3.8}$$

Estimating the residual magnetic torque as follows:

Orbit radius: $r = 7.008137 \times 10^8$ cm
Geomagnetic torque: $M = 8 \times 10^{25}$ Gcm^3
Residual magnetic moment of the satellite: $D = 0.15$ Am^2
Then: $N_m = 6.973 \times 10^{-6}$ Nm

3.2 ATTITUDE DYNAMICS OF THE MICRO/NANO SATELLITE

The movement of the attitude of satellites can be expressed in a series of differential equations. These differential equations express the process of attitude movement about time. There are two groups of equations: kinetics equations and dynamics equations. Kinetics equations describe the movement of the satellite, while dynamics equations show the relationship between the attitude movement and torques on the satellite. A description of the movement of satellites and coordinate systems, as well as modeling kinetics and dynamics equations of satellites, is given in this section.

3.2.1 Coordinate System

According to the requirement of tasks and modes of the ADCS [12–17], the following coordinate systems would be used to describe the attitude movement.

Earth centered inertial coordinate system $O_eX_iY_iZ_i$ (ECI)

The origin of ECI is at the center of mass of the Earth. The O_eX_i axis points to the mean equinox, O_eY_i points to the 90 degrees angle of right ascension. O_eZ_i points to the North Pole.

Body coordinate system $OX_bY_bZ_b$ (body system)

The origin of the body system is located at the center of the mass of satellites. OX_b, OY_b, and OZ_b are permanently fixed on the satellite body. OZ_b points to the observation orientation. OX_b points to the solar panel, the heading direction. OY_b points to the surface without the solar panel that constructs a right-headed coordinate system with OX_b and OZ_b.

Flight coordinate system (body inertial system) $OX_fY_fZ_f$ (flight system)

The origin of the flight coordinate system is located at the center of the mass of satellites. OZ_f points to the object to be observed. OX_f, OY_f, and OZ_f construct an inertial coordinate system and the flight moves in constant velocity in the ECI and the transform matrix is a constant matrix. The flight system is the reference system of satellites, while the achievement of the ADCS is to coincide the flight system and body

system when satellites run in a three-axes stable inertial orientation mode.
Earth orbit coordinate system $OX_oY_oZ_o$ (orbit system)
The origin of the orbit system is located at the centroid of the mass of satellites. OZ_o points to the centroid of the Earth. OY_o is perpendicular to the plane of the orbit points to the negative direction of the angular speed. OX_o is perpendicular to and right-handed to OY_o and OZ_o. For circular orbit, OX_o points to the direction of the velocity of satellites. The orbit system would be the reference system when the satellite works in an assistant orientation mode.

3.2.2 Representation of Attitude

Mathematic equations to express attitude have been used and developed for a long time. Many methods were introduced successively, such as direction cosine matrix, Euler angle, quaternion, and modified Rodrigues parameters (MRPs). These methods have their advantages, disadvantages, and suitable occasions for use. Quaternion is used to describe the attitude in this book, while direction cosine matrix, Euler angle, and MRPs are used in some situation when necessary.

1. Direction cosine matrix
   Considering two right-handed orthogonal coordinate systems as follows:

$$\{\hat{\mathbf{r}}\} \equiv \begin{Bmatrix} \hat{r}_1 \\ \hat{r}_2 \\ \hat{r}_3 \end{Bmatrix}, \{\hat{\boldsymbol{b}}\} \equiv \begin{Bmatrix} \hat{b}_1 \\ \hat{b}_2 \\ \hat{b}_3 \end{Bmatrix}, \tag{3.9}$$

   where $\{\hat{\boldsymbol{r}}\}$ is the reference coordinate system, $\{\hat{\boldsymbol{b}}\}$ is the body coordinate system, $\hat{\boldsymbol{r}}_i$ and $\hat{\boldsymbol{b}}_i$ $(i = 1, 2, 3)$ are three unit vectors along their axis direction, respectively.
   Assume the transformation matrix from reference system to body system is $\boldsymbol{A}_{br} = \{A_{br,ij}\}_{3\times 3}$, then:

$$A_{br,ij} = \cos\alpha_{ij} = \langle \hat{\boldsymbol{b}}_i, \hat{\boldsymbol{r}}_j \rangle, \tag{3.10}$$

   where $\alpha_{ij} \in [-\pi, \pi)$ is the angle between $\hat{\boldsymbol{b}}_i$ and $\hat{\boldsymbol{r}}_j$, $i, j = 1, 2, 3$.
   $\boldsymbol{A}_{br}$ is the direction cosine matrix and is also called the attitude matrix. $\boldsymbol{A}_{br}$, of which the determinate equals 1, is an orthogonal matrix that $\boldsymbol{A}_{br}$ has to satisfy to the following conditions:

$$\boldsymbol{A}_{br}^{\mathrm{T}} \boldsymbol{A}_{br} = \boldsymbol{A}_{br} \boldsymbol{A}_{br}^{\mathrm{T}} = \boldsymbol{I}_3 \tag{3.11}$$

$$\det \boldsymbol{A}_{br} \equiv +1 \tag{3.12}$$

Direction cosine matrix corresponds to physics attitudes. Its advantage is that there is no singular solution and it is not necessary to solve a triangular equation but six redundancy parameters.

2. Axis-angle parameters
According to Euler's rotation theorem, in three-dimensional space, any displacement of a rigid body, such that a point on the rigid body remains fixed, is equivalent to a single rotation with angle Φ about some axis that runs through the fixed point. The axis is called the Euler axis and is expressed in unit vector $\boldsymbol{e} = [e_1 \quad e_2 \quad e_3]^{\mathrm{T}}$ and Φ is defined as the rotation angle. The attitude of the body system in the reference system can be expressed as rotation from the reference system to the body system. The axis-angle parameters are defined as:

$$\Theta = \boldsymbol{e}\Phi \tag{3.13}$$

The most outstanding advantage of axis-angle parameters is their direct physical meaning since the attitude is expressed as an angle Φ with direction and without redundancy. It is convenient to describe attitude in single parameters, therefore it is often used to express the error vector of the attitude. Assuming the angles between the corresponding axis of the body system and the reference system are $\phi_x, \phi_y,$ and ϕ_z, the total error of attitude is as in Eq. (3.14) for small attitude angles:

$$\Phi = \sqrt{\frac{\phi_x^2 + \phi_y^2 + \phi_z^2}{2}} \tag{3.14}$$

However, axis-angle parameters correspond to physical attitude. It can be expressed as a globe attitude.

3. Euler angles
According to the Euler's rotation theorem, the rotation of a rigid body about a fixed point can be divided into several types of rotation. The body system can be obtained by rotating the reference system three times, with each rotation about the axis of the system that would be rotated. The angle of each rotation is the Euler angle. The Euler angles of three rotations are ψ, φ, θ. According to the sequence of rotation, there are 12 types of sequence to rotate. In general, a 3−1−2 sequence is often used in describing the attitude of three-axes stable satellites. In this situation, ψ, φ, θ express the yaw, the roll, and the pitch, respectively.
Form the definition, the direction cosine matrix from the body system to the reference system is as follows:

$$A_{br} = A_{312}(\psi, \varphi, \theta) = L_2(\theta)L_1(\varphi)L_3(\psi)$$

$$= \begin{bmatrix} \cos\theta & 0 & -\sin\theta \\ 0 & 1 & 0 \\ \sin\theta & 0 & \cos\theta \end{bmatrix} \begin{bmatrix} 1 & 0 & 0 \\ 0 & \cos\varphi & \sin\varphi \\ 0 & -\sin\varphi & \cos\varphi \end{bmatrix} \begin{bmatrix} \cos\psi & \sin\psi & 0 \\ -\sin\psi & \cos\psi & 0 \\ 0 & 0 & 1 \end{bmatrix}$$

$$= \begin{bmatrix} \cos\theta\cos\psi - \sin\theta\sin\varphi\sin\psi & \cos\theta\sin\psi + \sin\theta\sin\varphi\cos\psi & -\sin\theta\cos\varphi \\ -\cos\varphi\sin\psi & \cos\varphi\cos\psi & \sin\varphi \\ \sin\theta\cos\psi + \cos\theta\sin\varphi\sin\psi & \sin\theta\sin\psi - \cos\theta\sin\varphi\cos\psi & \cos\theta\cos\varphi \end{bmatrix} \tag{3.15}$$

The known direction cosine matrix $A_{br} = \{A_{br,ij}\}_{3\times 3}$, and the Euler angles (3−1−2 sequence) are:

$$\varphi = \arcsin(A_{br,23}), \theta = \arctan\left(-\frac{A_{br,13}}{A_{br,33}}\right), \psi = \arctan\left(-\frac{A_{br,21}}{A_{br,22}}\right) \tag{3.16}$$

where $\psi, \varphi, \theta \in [-\pi/2, \pi/2]$, when $\varphi = \pi/2$, ψ, θ are singular values. When ψ, φ, θ are small angles, Eq. (3.15) can be expressed as:

$$A_{br} = \begin{bmatrix} 1 & \psi & -\theta \\ -\psi & 1 & \varphi \\ \theta & -\varphi & 1 \end{bmatrix} \tag{3.17}$$

The most serious problem with Euler angles is their singular value. When $\varphi = 90$ degrees, the direction of the first rotation and the third rotation coincide. In a kinetic equation, zero occurs in the denominator of fractions. On the other head, there is a large amount of triangle function calculations to get the attitude matrix from Euler angles and this brings a heavy computational load. However, it has intuitive physical meaning without redundancy and is suitable for expressions such as plots in this book.

4. Quaternion

According to the Euler's rotation theorem, any displacement of a rigid body is equivalent to rotation about an axis. Assuming the vector describing the Euler axis is $\boldsymbol{e} = [e_1 \quad e_2 \quad e_3]^{\mathrm{T}}$ and the rotation angle is Φ, the quaternion is defined as:

$$\bar{q}_{br} = \begin{bmatrix} q_{br} \\ q_4 \end{bmatrix} = \begin{bmatrix} q_1 \\ q_2 \\ q_3 \\ q_4 \end{bmatrix} = \begin{bmatrix} e_1 \sin\frac{\Phi}{2} \\ e_2 \sin\frac{\Phi}{2} \\ e_3 \sin\frac{\Phi}{2} \\ \cos\frac{\Phi}{2} \end{bmatrix} = \begin{bmatrix} \boldsymbol{e}\sin\frac{\Phi}{2} \\ \cos\frac{\Phi}{2} \end{bmatrix} \tag{3.18}$$

$\boldsymbol{q}_{br}$ is the vector part of $\overline{\boldsymbol{q}}_{br}$, q_4 is the scalar part of $\overline{\boldsymbol{q}}_{br}$. From the definition, $\overline{\boldsymbol{q}}_{br}$ and $-\overline{\boldsymbol{q}}_{br}$ correspond to the same physical attitude. To avoid ambiguity, all $q_4 > 0$ in this book.

Quaternion must be normalized because only three components can express whole attitudes and are not independent:

$$\left\|\overline{\boldsymbol{q}}_{br}\right\| = \sqrt{\overline{\boldsymbol{q}}_{br}^{\mathrm{T}}\overline{\boldsymbol{q}}_{br}} = \sqrt{\left\|\boldsymbol{q}\right\|^2 + q_4^2} \equiv 1 \tag{3.19}$$

This limitation is called the constant norm constraint. Calculation need to satisfy strictly this constraint.

According to the rule of quaternion, $\overline{\boldsymbol{q}}_{br}$ is the negative inverse calculation. Assuming the attitude matrix is $\boldsymbol{A}_{br} = \{A_{br,ij}\}_{3\times 3}$, then the quaternion is:

$$\overline{\boldsymbol{q}}_{br}^{-1} = \begin{bmatrix} -\boldsymbol{q}_{br} \\ q_4 \end{bmatrix} = \begin{bmatrix} -q_1 \\ -q_2 \\ -q_3 \\ q_4 \end{bmatrix} \tag{3.20}$$

The attitude matrix expressed by the quaternion is:

$$\boldsymbol{A}_{br} = \begin{bmatrix} q_1^2 - q_2^2 - q_3^2 + q_4^2 & 2(q_1q_2 + q_3q_4) & 2(q_1q_3 - q_2q_4) \\ 2(q_1q_2 - q_3q_4) & -q_1^2 + q_2^2 - q_3^2 + q_4^2 & 2(q_2q_3 + q_1q_4) \\ 2(q_1q_3 + q_2q_4) & 2(q_2q_3 - q_1q_4) & -q_1^2 - q_2^2 + q_3^2 + q_4^2 \end{bmatrix} \tag{3.21}$$

$$\begin{cases} q_4 = \pm\dfrac{1}{2}\sqrt{A_{br,11} + A_{br,22} + A_{br,33}} \\ q_1 = \dfrac{A_{br,23} - A_{br,32}}{4q_4} \\ q_2 = \dfrac{A_{br,31} - A_{br,13}}{4q_4} \\ q_3 = \dfrac{A_{br,12} - A_{br,21}}{4q_4} \end{cases} \tag{3.22}$$

From Eq. (3.22), the quaternion calculated from the attitude matrix $\boldsymbol{A}_{br} = \{A_{br,ij}\}_{3\times 3}$ has two solutions. To avoid this, we define $q_4 \geq 0$ as $q_4 = (1/2)\sqrt{A_{br,11} + A_{br,22} + A_{br,33}}$.

When Φ is small enough, we obtain:

$$\boldsymbol{A}_{br} = \boldsymbol{I}_3 - 2[\boldsymbol{q}_{br}\times] \tag{3.23}$$

In Eq. (3.23), $\boldsymbol{I}_3$ is a 3×3 identity matrix and $[\boldsymbol{q}_{br}\times]$ is the skew-symmetric matrix of $\boldsymbol{q}_{br}$, which is also called the product matrix such that:

$$[\boldsymbol{q}_{br}\times] = \begin{bmatrix} 0 & -q_3 & q_2 \\ q_3 & 0 & -q_1 \\ -q_2 & q_1 & 0 \end{bmatrix} \tag{3.24}$$

When Φ is small enough, the vector part has a relationship with the Euler angles as follows:

$$q_{br} \approx \frac{1}{2}\begin{bmatrix} \varphi \\ \theta \\ \psi \end{bmatrix} \tag{3.25}$$

According to the equation of the direction cosine matrix and the Euler angle, the equation to calculate Euler angles from quaternion can be obtained as:

$$\begin{cases} \varphi = \arcsin\left[2(q_1q_2 + q_3q_4)\right] \\ \theta = \arctan\left[2(q_1q_3 - q_2q_4)/(-q_1^2 - q_2^2 + q_3^2 + q_4^2)\right] \\ \psi = \arctan\left[2(q_1q_2 - q_3q_4)/(-q_1^2 + q_2^2 - q_3^2 + q_4^2)\right] \end{cases} \tag{3.26}$$

Once the Euler angles are known, the quaternion can be calculate using quaternion multiplication. The quaternions of three times rotations (Eq. 3.27) rotate from reference system to body system in the sequence 3–1–2.

$$\begin{aligned} \overline{q}_z &= \begin{bmatrix} 0 & 0 & \sin\frac{\psi}{2} & \cos\frac{\psi}{2} \end{bmatrix}^{\mathrm{T}}, \\ \overline{q}_x &= \begin{bmatrix} \sin\frac{\varphi}{2} & 0 & 0 & \cos\frac{\varphi}{2} \end{bmatrix}^{\mathrm{T}}, \\ \overline{q}_y &= \begin{bmatrix} 0 & \sin\frac{\theta}{2} & 0 & \cos\frac{\theta}{2} \end{bmatrix}^{\mathrm{T}} \end{aligned} \tag{3.27}$$

From the definition and rules of quaternion multiplication, we obtain:

$$\overline{q}_{br} = \overline{q}_z \otimes \overline{q}_x \otimes \overline{q}_y \tag{3.28}$$

Thus:

$$\begin{cases} q_1 = \cos\frac{\psi}{2}\sin\frac{\varphi}{2}\cos\frac{\theta}{2} - \sin\frac{\psi}{2}\cos\frac{\varphi}{2}\sin\frac{\theta}{2} \\ q_2 = \cos\frac{\psi}{2}\cos\frac{\varphi}{2}\sin\frac{\theta}{2} + \sin\frac{\psi}{2}\sin\frac{\varphi}{2}\cos\frac{\theta}{2} \\ q_3 = \sin\frac{\psi}{2}\cos\frac{\varphi}{2}\cos\frac{\theta}{2} + \cos\frac{\psi}{2}\sin\frac{\varphi}{2}\sin\frac{\theta}{2} \\ q_4 = \cos\frac{\psi}{2}\cos\frac{\varphi}{2}\cos\frac{\theta}{2} - \sin\frac{\psi}{2}\sin\frac{\varphi}{2}\sin\frac{\theta}{2} \end{cases} \tag{3.29}$$

5. Modified Rodrigues parameters
   The normalization of quaternions is limited in its application on some occasions. To avoid this, a parameter which is without redundancy and

that corresponds to the physical attitude is necessary. Modified Rodrigues parameters (MRPs) satisfy the above constraints. This is a dimensionality reduction transform for axis-angle parameter which is defined as:

$$p_{br} = e\left[\sin\frac{\Phi}{4} \Big/ \left(1 + \cos\frac{\Phi}{4}\right)\right] = e\tan\frac{\Phi}{4} \tag{3.30}$$

The conversion between quaternions and MRPs is as follow:

$$p_{br} = \frac{q_{br}}{1 + q_4} \tag{3.31}$$

$$\bar{q}_{br} = \frac{1}{1 + \|p_{br}\|^2}\begin{bmatrix} 2p_{br} \\ 1 - \|p_{br}\|^2 \end{bmatrix} \tag{3.32}$$

That is:

$$q_4 = \frac{1 - \|p_{br}\|^2}{1 + \|p_{br}\|^2}, q_{br} = (1 + q_4)p_{br} \tag{3.33}$$

3.3 MICRO/NANO SATELLITE ATTITUDE CONTROL SYSTEM

The satellite attitude control system is a core part of micro/nano satellites; its technology development played a key role in improving the development level of micro/nano satellites. The hardware of the satellite attitude control system [12–17] includes three parts: attitude sensors, control computer, and actuators, and its software includes the attitude determination algorithm and the calculation of control law. Currently, micro/nano satellites tend to use integrated design, an attitude control system, and other subsystems connected as closely as possible to achieve structural reuse, with the introduction of new technologies and new devices to achieve high accuracy, low power consumption, and miniaturization.

This section contains detailed design and simulation verification of attitude control systems, in accordance with the flight mission and design indicators of the NS-2 satellite.

3.3.1 Task Analysis of the NS-2 Satellite's ADCS Subsystem

The main task of the attitude determination and control system (ADCS) is to ensure that antenna pointing meets the requirements of satellite communications and the successful completion of the satellite payload flight mission.

The main flight mission of the satellite payloads is as follows:

- Intersatellite communication (one of the formation flying test missions);
- Ground imaging;
- Intersatellite laser ranging;
- MEMS devices flight test: MIMU, magnetometer, and microthruster;
- In-orbit test of new attitude control sensor: star sensor, analogue sun sensor, digital sun sensor.

The communication between satellites has no special requirements in regard to attitude control, therefore it will be mainly the design of the ADCS subsystem in ground imaging, intersatellite laser ranging, and MEMS devices flight test, to ensure that the control accuracy of the ADCS subsystem meets the normal requirements of the CMOS camera and the laser rangefinder, and also to ensure that the measurement accuracy meets the normal job requirements of the microthruster.

Taking into account that the exposure time of the CMOS camera is generally less than 200 ms during normal working, the satellite's attitude control accuracy and stability must be at least 2 degrees, 0.02 deg/s. At the same time, the laser-ranging missions require the nano satellite to have higher attitude determination accuracy and have a higher attitude maneuver ability to follow its target satellite MEMSat, meanwhile microthruster tests require the satellite to have a higher angular rate measurement accuracy, which is why the star sensor and MIMU is needed.

In addition, since the satellite is launched on board a rocket, the angular rate of the satellite after separation from the carrier will be relatively large, and may reach 5 deg/s.

Based on the above analysis, we can determine that the main task of the nano satellite's ADCS subsystem is:

- Elimination of rotation, namely eliminating the angular rate of the satellite after separation from the carrier, and establishing a basic controlled attitude;
- Establishing and maintaining an Earth-directional three-axes stabilized attitude, to ensure normal communication between the satellite and Earth, and successful completion of the task of the CMOS camera;
- Having the ability to track the target satellite MEMSat, and ensuring the tasks of laser ranging are competed.

3.3.2 Technical Specifications of the NS-2 Satellite's ADCS Subsystem

According to the design task and flight test requirements of the NS-2 satellite, the accuracy specifications of its attitude determination and control system are set at two levels. The first is the basic specifications of the platform, which must be maintained. Second, it is the test specifications that a closed loop test can achieve under the premise of all attitude sensors working properly in open-loop conditions (see Tables 3.1 and 3.2).

3.3.3 Attitude Determination and Control System Design

3.3.3.1 Functional Modules

Based on the requirements analysis of the control software for the NS-2 satellite's ADCS subsystem, we divided the control software into four functional modules:

1. Command processing module. Completing reception, interpretation, judgment, and execution of control instructions from the ground station and the OBC.
2. Attitude determination module. Gathering information from the attitude measurement section, and making corresponding filter calculations to determine the satellite's current attitude and angular rate, providing a basis for attitude control.
3. Attitude control module. According to control law this generates a control signal to drive the corresponding gesture actuator, and then controls the attitude of the satellite.

Table 3.1 The Basic Precision Specifications of a Platform

Content	Technical Specifications
Control method	Earth-directional three-axes stabilized attitude
Attitude determination accuracy	1 degree
Attitude control accuracy	2 degrees
Attitude stability	0.02 deg/s

Table 3.2 Test Accuracy Specifications

Content	Technical Specifications
Control method	Earth-directional three-axes stabilized attitude
Attitude determination accuracy	0.02 degree
Attitude control accuracy	Better than 1 degree
Attitude stability	Better than 0.01 deg/s

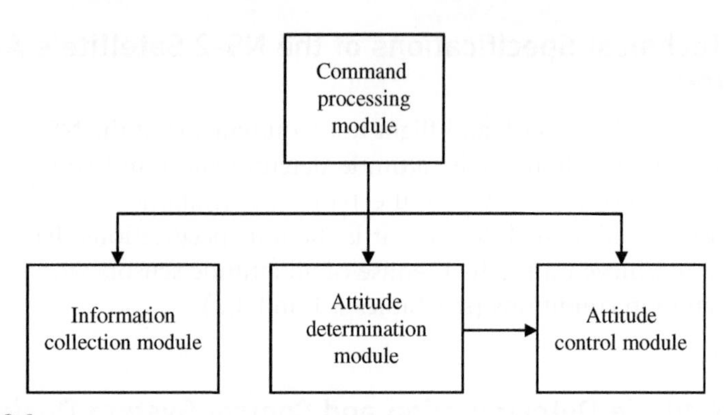

FIGURE 3.1
Control software of the NS-2 satellite's ADCS subsystem.

4. Information collection module. Gathering information about the system hardware and software, and sent to ground along with software information, which will be used to determine the working status of the ADCS subsystem.

The relationship between the four functional modules shown in Fig. 3.1.

3.3.3.2 Operating Modes

The NS-2 satellite's ADCS control software has three operating modes: operating mode 1, operating mode 2, and operating mode 3. The specific meanings of these three operating modes are as follows:

- Operating mode 1: Attitude determination mode and control mode of the ADCS control software are remotely selected by the ground station;
- Operating mode 2: The ADCS control software independently selects the attitude determination mode and the control mode in accordance with certain conditions;
- Operating mode 3: Attitude determination mode of the ADCS control software is remotely selected by the ground station, the control law does not run, and the instructions for the momentum wheel speed and magnetorquer's duty cycle are remotely determined by the ground station.

3.3.3.3 Task Stages

Since the NS-2 satellite needs to complete in-orbit testing tasks of the star sensor, sun sensor, and MIMU, this simulation system was designed in two task stages. The first stage lets the satellite achieve Earth-directional three-axes stabilized attitude, complete in-orbit tests of payloads. In the second stage, according to the results of in-orbit tests, the control circuit has access to payloads which work properly, to achieve Earth-directional three-axes stabilized attitude or observation target directional attitude.

3.4 SOFTWARE DESIGN OF THE ATTITUDE DETERMINATION MODULE AND ATTITUDE CONTROL MODULE

3.4.1 Software Design of the Attitude Determination Module

The basic attitude component of the NS-2 satellite is a magnetometer. The function of the attitude determination module is to collect the magnetometer's measurement information and make corresponding filter calculations to determine the satellite's current attitude and angular rates, providing the basis for attitude control. Different control modes will use different filters to determine the satellite attitude.

The software process of the attitude determination module is shown in Fig. 3.2.

3.4.1.1 Rate Filter

The rate filter is used to estimate the angular rate of the satellite. It uses the magnetometer's measurement data (strength of magnetic field in the satellite body coordinate system: Bm) as input and has the characteristics of stable operation. As the accuracy requirement for the rate filter is not very high, it is not necessary to introduce the disturbing moment in the state equation. The rate filter is used to estimate the satellite's three-axes angular rate (projection of inertial angular rate in the satellite body coordinate system). The state vector (inertial angular rate) and the kinetic equation of the satellite are:

$$\boldsymbol{X} = \boldsymbol{\omega} = \begin{bmatrix} \omega_x & \omega_y & \omega_z \end{bmatrix}^T \tag{3.34}$$

$$\boldsymbol{I}\dot{\boldsymbol{\omega}}_{bi} + [\boldsymbol{\omega}_{bi} \times](\boldsymbol{I}\boldsymbol{\omega}_{bi} + \boldsymbol{h}) = \boldsymbol{T} - \dot{\boldsymbol{h}} \tag{3.35}$$

Using the spindle coordinate system, the satellite kinetic equation can be expanded as follows (only control torque of MT and MW are considered, the disturbance torque is negligible):

$$\begin{cases} \dot{\omega}_{bi1} = \dfrac{I_y - I_z}{I_x}\omega_{bi2}\omega_{bi3} + \dfrac{T_x}{I_x} + \dfrac{h_y}{I_x}\omega_{bi3} \\[2ex] \dot{\omega}_{bi2} = \dfrac{I_z - I_x}{I_y}\omega_{bi3}\omega_{bi1} + \dfrac{T_y}{I_y} - \dfrac{\dot{h}_y}{I_y} \\[2ex] \dot{\omega}_{bi3} = \dfrac{I_x - I_y}{I_z}\omega_{bi1}\omega_{bi2} + \dfrac{T_z}{I_z} - \dfrac{h_y}{I_z}\omega_{bi1} \end{cases} \tag{3.36}$$

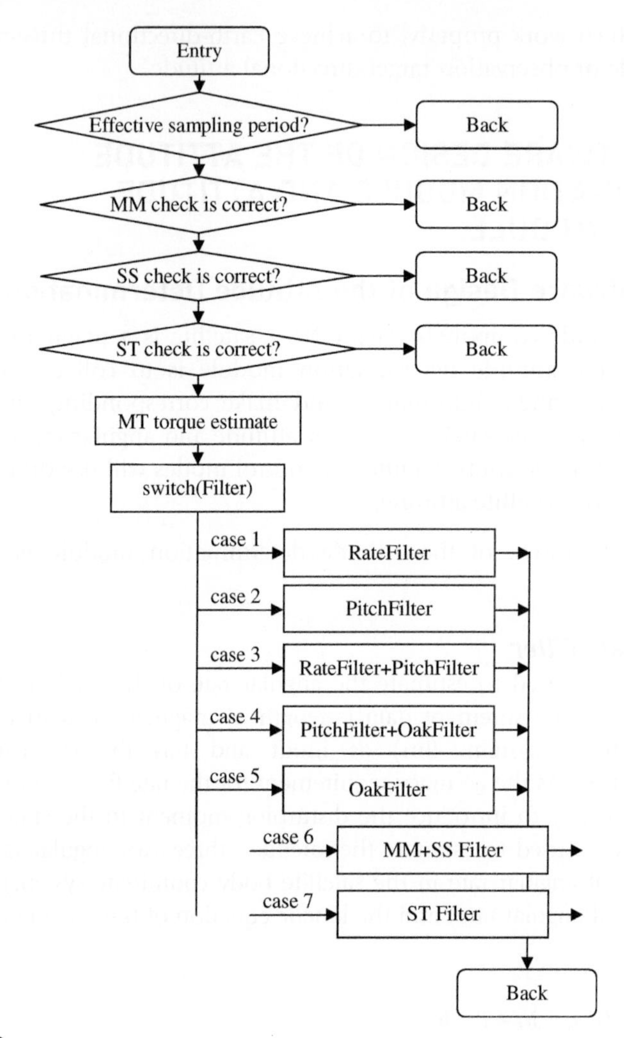

FIGURE 3.2
Software process of the attitude determination module.

The observation vector's value of the rate filter is the difference between the two adjacent magnetometer measurement values. The current magnetometer measurement value and the previous magnetometer measurement value are denoted by $\boldsymbol{B}$, $\boldsymbol{B}_{prev}$, the observation vector is:

$$Z = \boldsymbol{B} - \boldsymbol{B}_{prev} = \begin{bmatrix} B_x - B_{prev\,x} \\ B_y - B_{prev\,y} \\ B_z - B_{prev\,z} \end{bmatrix} \tag{3.37}$$

The observation equation is the conversion equation of the geomagnetic vector B between the two coordinate systems (the orbital coordinate system and the satellite body coordinate system). During the sampling period T, the satellite rotates from the original attitude to the current posture. In the period T, the angular rate of the satellite is denoted by $\boldsymbol{\omega} = [\omega_x \quad \omega_y \quad \omega_z]$, rotation angle is denoted by $\boldsymbol{\alpha} = [\alpha_x \quad \alpha_y \quad \alpha_z]$, and the transformation matrix between two coordinate systems rotating before and after is denoted by A. There following is the transforming relationship:

$$\begin{aligned} B - B_{prev} &= (A - I)B_{prev} \\ &= \begin{bmatrix} 0 & \omega_z \Delta T & -\omega_y \Delta T \\ -\omega_z \Delta T & 0 & \omega_x \Delta T \\ \omega_y \Delta T & -\omega_x \Delta T & 0 \end{bmatrix} \begin{bmatrix} B_{px} \\ B_{py} \\ B_{pz} \end{bmatrix} \\ &= \begin{bmatrix} 0 & -B_{pz} \Delta T & B_{py} \Delta T \\ B_{pz} \Delta T & 0 & -B_{px} \Delta T \\ -B_{py} \Delta T & B_{px} \Delta T & 0 \end{bmatrix} \begin{bmatrix} \omega_x \\ \omega_y \\ \omega_z \end{bmatrix} \\ &= HX \end{aligned} \tag{3.38}$$

Eq. (3.38) is the observation equation of the system.

By linearizing the state equation, we obtain transfer matrix of the system:

$$\Phi = \begin{bmatrix} 1 & \left(\frac{I_y - I_z}{I_x} \omega_z \right) \Delta T & \left(\frac{I_y - I_z}{I_x} \omega_y + \frac{h_y}{I_x} \right) \Delta T \\ \left(\frac{I_z - I_x}{I_y} \omega_z \right) \Delta T & 1 & \left(\frac{I_z - I_x}{I_y} \omega_x \right) \Delta T \\ \left(\frac{I_x - I_y}{I_z} \omega_y - \frac{h_y}{I_z} \right) \Delta T & \left(\frac{I_x - I_y}{I_z} \omega_x \right) \Delta T & 1 \end{bmatrix} \tag{3.39}$$

After obtaining the above state space of the system, the filter calculation can be made in accordance with the generic algorithm of the extended Kalman filter (Fig. 3.3).

3.4.1.2 Pitch Filter

Pitch filter is used to estimate the satellite's pitch angle and pitch angular rate. It uses three parameters as input: the magnetometer's measurement data (in the satellite body coordinate system), reference magnetic field strength value (in the orbital coordinate system) calculated from the IGRF model according to current orbital parameters, and control torque in the pitch axis direction from MT and MW.

In a rotating object, its moment of inertia for rotation of a rod about this axis is denoted by I, the torque of the object received is denoted by T, the

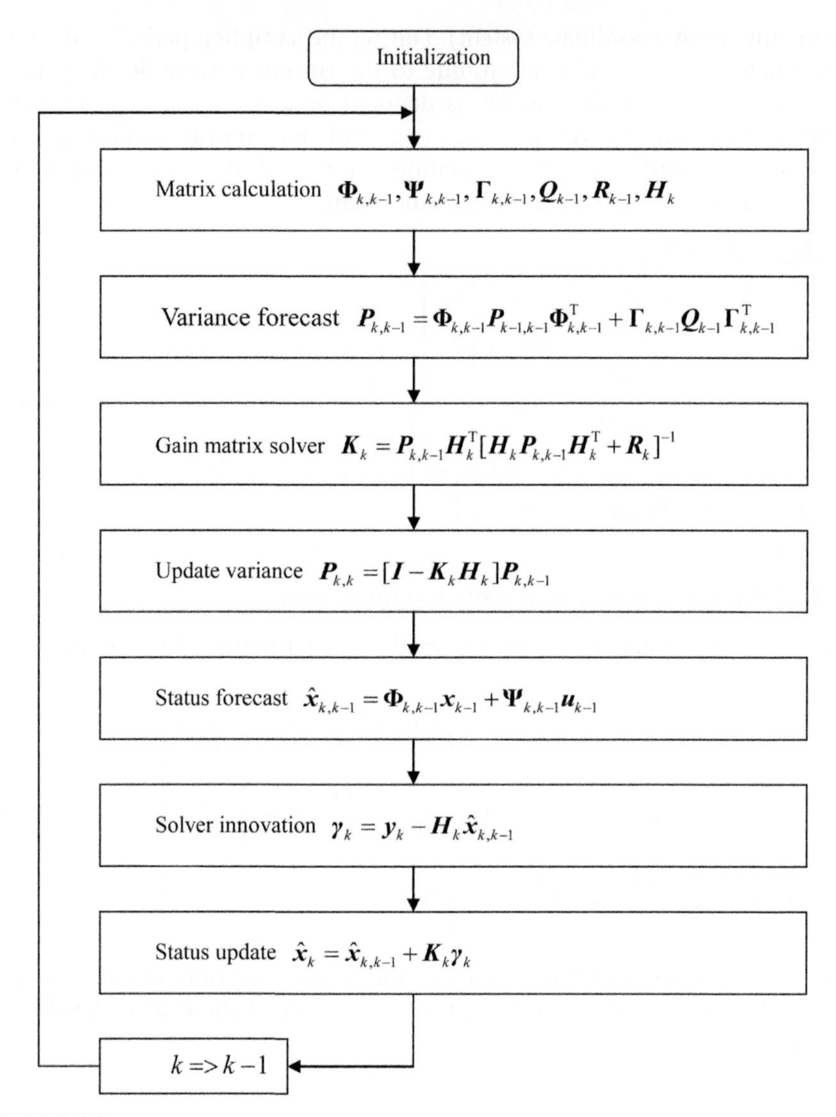

FIGURE 3.3

Algorithmic process of the MM attitude filter.

object's initial angular rate about the axis is denoted by ω_0, and angular rate at time t is denoted by ω_t. Therefore:

$$\omega_t = \omega_0 + \frac{T}{I}(t - t_0) \tag{3.40}$$

The object's average angular rate is denoted by ω, the initial angle about the axis is denoted by θ_0, and angle at time t is denoted by θ_t, therefore:

$$\theta_t = \theta_0 + \omega(t - t_0) \tag{3.41}$$

According to Eq. (3.41), the pitch angular rate ω_y and the pitch angle θ of the satellite can be calculated as follows:

$$\begin{cases} \omega_y(k) = \omega_y(k-1) + \dfrac{T_m + T_w}{I_y} T \\ \theta(k) = \theta(k-1) + \dfrac{1}{2}\left[\omega_y(k) + \omega_y(k-1)\right]T \end{cases} \tag{3.42}$$

The magnetic field vector in the orbital coordinate system and the satellite body coordinate system are denoted by B_o and B_b, respectively, giving the following relationship:

$$\boldsymbol{B}_b = \boldsymbol{A}_{bo}\boldsymbol{B}_o \tag{3.43}$$

where $\boldsymbol{A}_{bo}$ is the attitude matrix of the satellite.

$$\boldsymbol{A}_{bo} = \begin{bmatrix} \cos\theta & -\sin\theta \\ \sin\theta & \cos\theta \end{bmatrix} \tag{3.44}$$

After expanding and transposing the formula above, we obtain

$$\theta = \arctan\left(\frac{B_{oz}B_{bx} - B_{ox}B_{bz}}{B_{ox}B_{bx} + B_{oz}B_{bz}}\right) \tag{3.45}$$

From the analysis above, the algorithmic process of the pitch filter is as follows:

Estimating pitch angular rate

$$\omega_y(k) = \omega_y(k-1) + \frac{T_m + T_w}{I_y} T \tag{3.46}$$

Estimating pitch angle

$$\theta(k) = \theta(k-1) + \frac{1}{2}\left[\omega_y(k) + \omega_y(k-1)\right]T \tag{3.47}$$

Calculating correction value

$$\Delta_{\text{correct}}(k) = \arctan\left(\frac{B_{oz}(k)B_{bx}(k) - B_{ox}(k)B_{bz}(k)}{B_{ox}(k)B_{bx}(k) + B_{oz}(k)B_{bz}(k)}\right) - \theta(k) \tag{3.48}$$

Regulating the correction value

$$\Delta_{\text{correct}}(k) = \arctan\left(\frac{\sin\Delta_{\text{correct}}(k)}{\cos\Delta_{\text{correct}}(k)}\right) \tag{3.49}$$

Correcting estimated value

$$\begin{cases} \omega_y(k) = \omega_y(k) + k_1\Delta_{\text{correct}} \\ \theta(k) = \theta(k) + k_2\Delta_{\text{correct}} \end{cases} \tag{3.50}$$

Regulating the pitch angle

$$\theta(k) = \arctan\left(\frac{\sin\theta(k)}{\cos\theta(k)}\right) \tag{3.51}$$

3.4.1.3 MM Attitude Filter

The MM attitude filter can simultaneously estimate the attitude angles and angular rates of the satellite. Taking into account the advantage of quaternion in numerical calculation, we choose to use quaternion to represent the satellite's attitude. Therefore, the state vector of the system is composed of the attitude quaternion and the attitude angular rates:

$$X = \begin{bmatrix} q_{bo} & \omega_{bi} \end{bmatrix}^T = \begin{bmatrix} q_{bo1} & q_{bo2} & q_{bo3} & q_{bo4} & \omega_{bi1} & \omega_{bi2} & \omega_{bi3} \end{bmatrix}^T \tag{3.52}$$

The kinetic equation of the satellite is:

$$\boldsymbol{I}\dot{\omega}_{bi} + [\omega_{bi}\times](\boldsymbol{I}\omega_{bi} + \boldsymbol{h}) = \boldsymbol{T} - \dot{\boldsymbol{h}} \tag{3.53}$$

The kinematic equation is:

$$\dot{q}_{bo} = \frac{1}{2}\Omega(\omega_{bo})q_{bo} \tag{3.54}$$

According to Eqs. (3.53) and (3.54) and the defined state vector of the system, we obtain the state equation of the system:

$$\begin{cases} \dot{q}_{bo} = 0.5\,\Omega(\omega_{bo})q_{bo} \\ \dot{\omega}_{bi} = \boldsymbol{I}^{-1}[\boldsymbol{T} - \boldsymbol{h}[\omega_{bi}\times](\boldsymbol{I}\omega_{bi} + \boldsymbol{h}_w)] \end{cases} \tag{3.55}$$

wherein, $\hat{\omega}_{bo}$ can be obtained by computing $\hat{\omega}_{bo} = \hat{\omega}_{bi} + A(\hat{q}_{bo})\omega_{oi}^{o}$, and we can assign the magnetometer measurement value B_b to the observation vector of the system. Then the observation equation of the system is a conversion equation of the geomagnetic vector between the two coordinate systems (the orbital coordinate system and the satellite body coordinate system).

$$\boldsymbol{B}_b = \boldsymbol{A}_{bo}(\boldsymbol{q}_{bo})\boldsymbol{B}_o = h[X, t] \tag{3.56}$$

Here, the matrix H is derived as follows:

$$\begin{aligned}
&\begin{cases} \hat{V}^b = A(\hat{\overline{q}})V^r \\ V^b_{meas} = A(\hat{\overline{q}} + \delta\overline{q})V^r + \nu \end{cases} \Rightarrow \\
&z = V^b_{meas} - \hat{V}^b = \left[\frac{\partial A(\overline{q})}{\partial q_1}V^r, \frac{\partial A(\overline{q})}{\partial q_2}V^r, \frac{\partial A(\overline{q})}{\partial q_3}V^r, \frac{\partial A(\overline{q})}{\partial q_4}V^r\right]\delta\overline{q} + v \Rightarrow \\
&z = H\delta x + v, H = \left[\frac{\partial A(\overline{q})}{\partial q_1}V^r, \frac{\partial A(\overline{q})}{\partial q_2}V^r, \frac{\partial A(\overline{q})}{\partial q_3}V^r, \frac{\partial A(\overline{q})}{\partial q_4}V^r, 0_{3\times 3}\right] \\
&A(\hat{\overline{q}} + \delta\overline{q}) \approx A(\hat{\overline{q}}) + \left[\frac{\partial A(\overline{q})}{\partial q_1}, \frac{\partial A(\overline{q})}{\partial q_2}, \frac{\partial A(\overline{q})}{\partial q_3}, \frac{\partial A(\overline{q})}{\partial q_4}\right]\delta\overline{q}
\end{aligned} \tag{3.57}$$

According to the system state model identified above, we get the whole algorithmic process of MM attitude filter as shown in Fig. 3.3.

3.4.1.4 MM + SS Attitude Filter

The MM + SS attitude filter can simultaneously estimate the attitude angles and angular rates of the satellite. Here we use the same state variable and the same kinetic model as in the MM attitude filter. Assigning the magnetometer measurement value B_b and the sun sensor's measurement value D_b to the observation vector of the system, the observation equation of the system is a conversion equation of the geomagnetic vector and sun vector between the two coordinate systems (the orbital coordinate system and the satellite body coordinate system).

Theoretical derivation of the magnetometer's H matrix can be seen in Section 3.4.1.3, the derivation of sun sensor's H matrix is similar to the magnetometer's H matrix.

3.4.1.5 ST Attitude Filter

The status update section of the ST attitude filter is similar to the MM attitude filter. For the measurement update, the measurement equation is:

$$Z_{bo_s tt} = \delta q_{bo} + v \tag{3.58}$$

where, $\overline{Z}_{bo_s tt} = \hat{\overline{q}}_{bo}^{\,-1} \otimes \hat{\overline{q}}_{oi}^{\,-1} \otimes \overline{q}_{meas}$.

On the other hand, the best estimate of the difference between the measurements does not need to be cut off, which is $H = eye(4)$.

3.4.2 Attitude Control Module

The satellite attitude control system hardware is composed of an attitude sensor, control computer, and actuator, while the software includes an attitude determination algorithm and control law calculation. As the core part of a Nanosatellite, its technological development plays a key role in improving microsatellite developing level. In order to enhance structure reusability, Nanosatellites tend to adopt an integrated design at present, with close connection between the attitude control system and other subsystems. Simultaneously, new technologies and devices are introduced to achieve high accuracy, low power consumption, and miniaturization.

This section gives detailed design and simulation of the attitude control system [18–21] based on the NS-2 Nanosat's flight mission and design indices.

3.4.2.1 Magnetorquer Control Module

1. Nutation damping

   If the triaxial magnetorquer of the satellite is produced according to the following formula

$$M = -K\dot{B} \tag{3.59}$$

then the function of nutation damping can be achieved. "K" in this formula is a constant greater than zero. Generally, K takes large values because magnetorquers usually work in a saturated state in the rate-damp mode. Each axis component of the geomagnetic field in the satellite system can be obtained from a direct difference method:

$$\dot{B}_i = \frac{B_i(k) - B_i(k-1)}{T} \tag{3.60}$$

The damping control algorithm in the above representation has a low amount of calculation required and fast convergence speed, which applies to the despun stage after satellite–launcher separation. However, this algorithm is directly affected by measurement noise from each axis magnetometer, which leads to low control accuracy. To enhance control accuracy, the angle between each magnetic component and the magnetic field can be determined at first

$$\alpha_i = \arccos\frac{B_i}{\sqrt{B_x^2 + B_y^2 + B_z^2}} \qquad (i = x, y, z). \tag{3.61}$$

The derivative of α_i is

$$\dot{\alpha}_i(k) = \frac{\alpha_i(k) - \alpha_i(k-1)}{T} \tag{3.62}$$

And the control law is given by

$$M = -K\dot{\alpha}_i \tag{3.63}$$

where K is a constant greater than zero. This can reduce the effects of magnetometer noise and improve control accuracy.

   Remark: B dot is different from B product. B product needs information about angular velocity while the control objective of B dot is to keep triaxial angular velocity at 0.

2. Pitch axis vortex

   If the reference angular rate of the satellite's pitch axis is ω_y and the actual angular rate is ω_y', the angular rate error is defined by

$$\Delta\omega_y = \omega_y' - \omega_y \tag{3.64}$$

The angular rate of the satellite's pitch axis can be controlled by the following formula

$$M_z = Sign(B_x)K\Delta\omega_y \tag{3.65}$$

or

$$M_x = Sign(B_z)K\Delta\omega_y. \tag{3.66}$$

Magnetorquer control in the momentum wheel spinning process is defined as

$$\mathbf{e} = \begin{bmatrix} e_x \\ e_y \\ e_z \end{bmatrix} = \begin{bmatrix} 0 \\ k_{dy}(\Omega - \Omega') \\ 0 \end{bmatrix} \tag{3.67}$$

And the output of the magnetic momentum is

$$\mathbf{M} = \mathbf{b}_b \times \mathbf{e} \tag{3.68}$$

After the three-axes stabilized y bias momentum wheel starts spinning (which means unloading saturation in the y direction), the control error of the three-axes magnetorquer is defined as

$$\mathbf{e} = \begin{bmatrix} e_x \\ e_y \\ e_z \end{bmatrix} = \begin{bmatrix} k_{dx}\omega_x + k_{pz}\psi \\ k_{dy}(\Omega - \Omega') \\ k_{dz}\omega_z + k_{px}\phi \end{bmatrix} \tag{3.69}$$

where k represents the control coefficient, Ω, Ω' are actual and rated speed of the momentum wheel, ψ is yaw, θ is pitch, and ϕ is roll. The normalized vector of the geomagnetic field under the satellite's coordinate system is $\mathbf{b}_b$ and the control law of three-axes stabilization is

$$\mathbf{M} = \mathbf{b}_b \times \mathbf{e} \tag{3.70}$$

Remark: for the control mode of MT, the torque needs to be calculated at first (which is supposed to be in a negative direction). Then, the magnetic momentum is given by $\mathbf{M} = \mathbf{b}_b \times \mathbf{e}$ where y wheel deceleration takes the same control as in the preceding part and Ω' is set differently.

Zero-momentum controls yields (three-axes magnetorquer unloads three-axes wheel):

$$\mathbf{e} = \begin{bmatrix} e_x \\ e_y \\ e_z \end{bmatrix} = \begin{bmatrix} h_x \\ h_y \\ h_z \end{bmatrix} = \begin{bmatrix} J_{hx}\omega_{hx} \\ J_{hy}\omega_{hy} \\ J_{hz}\omega_{hz} \end{bmatrix} \tag{3.71}$$

Remark: unloading the momentum wheel means controlling the predetermined angular momentum (angular velocity) of the momentum wheel (Fig. 3.4).

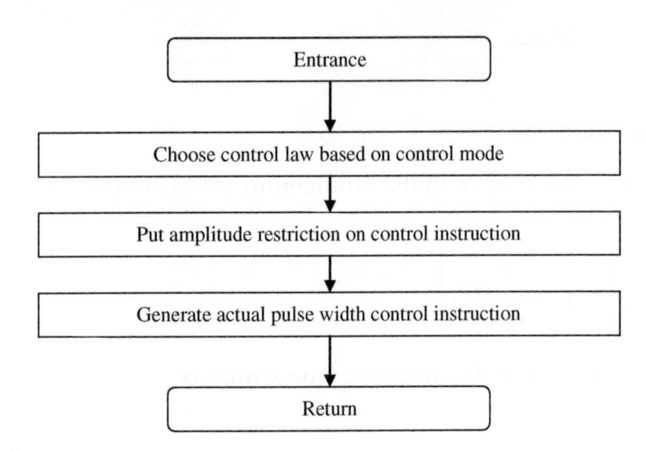

FIGURE 3.4
Flow chart of MT control module.

3.4.2.2 Control Module of the Momentum Wheel

1. Bias momentum state

   The momentum wheel is used for controlling the attitude and angular rate of the satellite's pitch plane. Letting pitch angular rate and pitch angle of the body be ω_y and θ, the demanded control momentum is calculated as

$$M = k_p\theta + k_d\omega_y \tag{3.72}$$

   where k_p, k_d are control coefficients.

   The momentum wheel of the NS-2 Nanosatellite is under speed control mode. The demanded increment of rotational speed can be derived from Eq. (3.72)

$$\Delta\Omega = \frac{M\cdot\Delta T}{J} = \frac{(k_p\theta + k_d\omega_y)\cdot\Delta T}{J} \tag{3.73}$$

   where ΔT is the sampling period. The control instruction of the momentum wheel is

$$\Omega = \Omega_{prev} + \Delta\Omega \tag{3.74}$$

   where Ω_{prev} is the previous control instruction of rotational speed.

   The flow chart of the momentum wheel control module is illustrated in Fig. 3.5

2. Zero-momentum state

   Let the Euler angular rate and angle of the body be $\omega_x, \omega_y, \omega_z$ and $\theta_x, \theta_y, \theta_z$. The demanded control momentum is calculated as

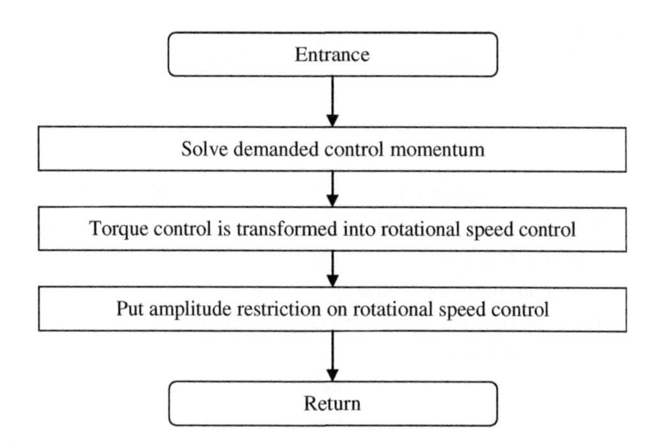

FIGURE 3.5
Flow chart of the MW control module.

$$\mathbf{e} = \begin{bmatrix} e_x \\ e_y \\ e_z \end{bmatrix} = \begin{bmatrix} k_{dx}\omega_x + k_{px}\theta_x \\ k_{dy}\omega_y + k_{py}\theta_y \\ k_{dz}\omega_z + k_{pz}\theta_z \end{bmatrix} \tag{3.75}$$

3.5 NS-2 NANO SATELLITE ADCS SUBSYSTEM SIMULATION

After the ADCS subsystem design is complete, mathematical simulation is required to verify its correctness, control accuracy, control stability, and stability of the main control software. The environmental torque acting on the satellite in simulation is almost the same as the actual environmental torque where the body is located.

The relevant parameters in simulation are shown in Table 3.3.

3.5.1 Control Mode 1

After the separation of the satellite and the carrier (or without attitude control for a long time), three axes have a certain angular rate and attitude will also be in a random state. According to the carrier information provided by the satellite system, here it is assumed that the stars triaxial angular rates are -4 deg/s and attitude angles are 40 degrees when the attitude control starts. According to flight procedures of the NS-2 satellite ADCS subsystem, the ADCS subsystem should first enter control mode 1. Mode 1 is used to control the angular rate of the satellite X-axis and Z-axis by damping, so that the angular rates of the two axes are small enough to enter the next control mode. Control mode

Table 3.3 Relevant Parameters in Simulation of the NS-2 Nano Satellite

Orbit altitude		600 km
Atmospheric density(600 km)		2.94×10^{-13} kg/m^3
Aerodynamic drag coefficient		2.5
Properties of the satellite	Inertia of the satellite I	[0.696 0.696 0.551] kg/m^2
	Maximum moment of MT	0.7 Am^2
	Inertia of MW J	1.34×10^{-4} kg/m^2
	Maximum moment of MW	6.7×10^{-5} Nm
	Distance between pressure center and centroid	0.04 m
	Remanent magnetism	0.005 Am^2
	Reflection coefficient on surface	$\beta = 0.6$
Environment torque	Atmospheric resistance torque	Fixed value $[0.1\ 1.6\ 1.0] \times 10^{-7}$ Nm
	Solar pressure torque	Fixed value $[5.6\ 5.6\ 5.6] \times 10^{-8}$ Nm
	Geomagnetic torque	Actual simulation
	Gravity gradient torque	Actual simulation
	Total torque after stabilizing (3σ)	X-axis $(-2 \sim 3) \times 10^{-7}$ Nm
		Y-axis $(-1 \sim 5) \times 10^{-7}$ Nm
		Z-axis $(0 \sim 3) \times 10^{-7}$ Nm
Earth's magnetic field	Space environment simulation	Earth's main magnetic field takes 9×9 order (ignoring external magnetic field)
	Solving ADCS magnetic field	Earth's main magnetic field takes 6×6 order (ignoring external magnetic field)
Initial conditions	Initial angular rate (deg/s)	$[-4, -4, -4]$
	Initial attitude (degree)	[40, 40, 40]
MM noise	Constant bias	0.1 uT
	Random noise (3σ)	0.4 uT

1 may last the time period from 0 to 15,000 seconds. The curves of the satellite orbital angular rate and three-axes magnetic torque output during this time are shown in Figs. 3.6 and 3.7, respectively.

The control cycle of this mode is 1 second.

As shown in the satellite angular rate curve (Fig. 3.6), control mode 1 can effectively dampen the angular rates of satellite X-axis and Z-axis, so that the rates are reduced to near zero, to create a good state for entering the next control mode. Therefore, control mode 1 is correct and effective.

Control mode 1 only dampens angular rates of the X-axis and, Z-axis, while the Y-axis angular rate is uncontrolled. When the angular rates of the X-axis

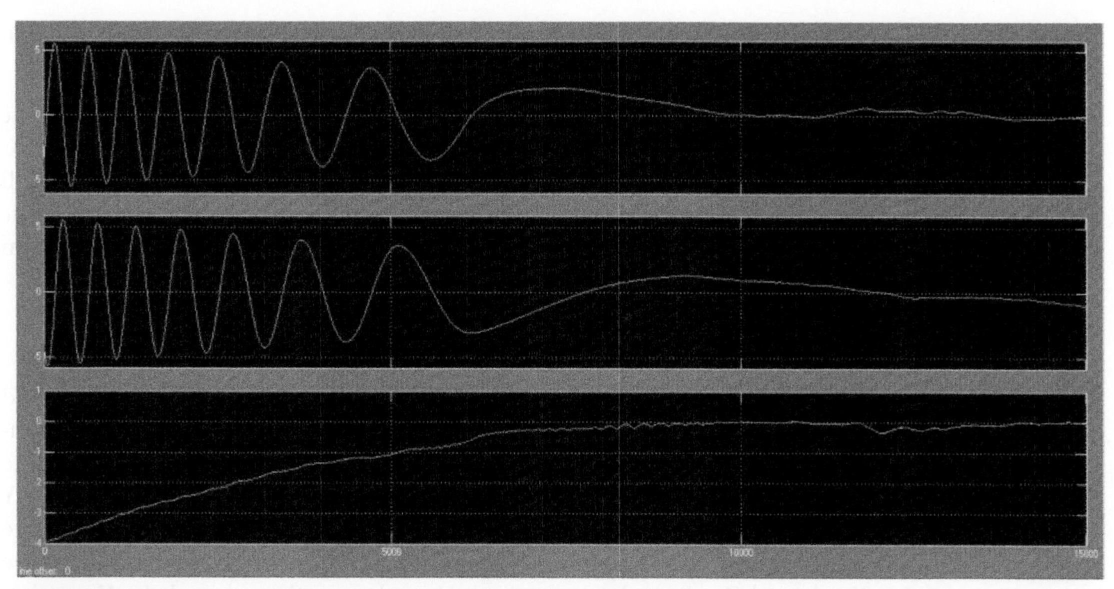

FIGURE 3.6
Satellite orbit angular rate of control mode 1 (deg/s).

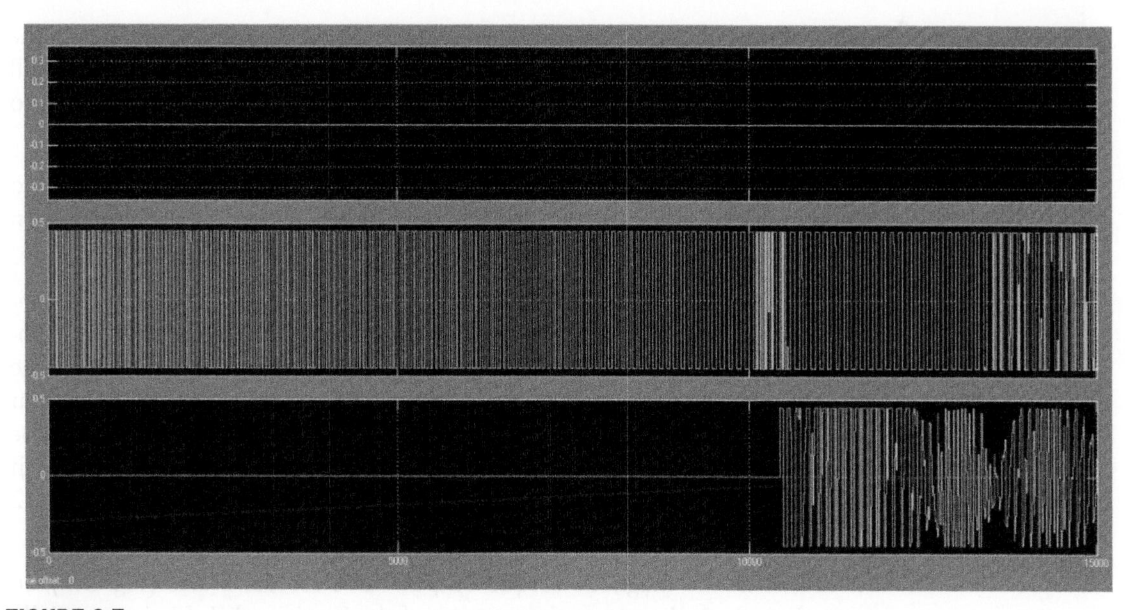

FIGURE 3.7
Three-axes magnetic torque output of control mode 1 (Am^2).

and Z-axis dampen to near zero, the Y-axis angular rate will be maintained at a substantially constant value (this may also be near a value of zero).

As shown in the Y-axis magnetic torque output curve (Fig. 3.7), during the damping process in control mode 1 for damping of the angular rates of X-axis and Z-axis, the Y-axis magnetic moment is basically working at full capacity, demanding a large power supply. In this mode, the X-axis magnetic moment does not work and the magnetorquer of the Z-axis starts from 10,500 seconds later.

3.5.2 Control Mode 2

After the process of control mode 1, the angular velocity of the X-axis and Z-axis has already been damped to near zero. But at this moment, the angular rate and direction of the Y-axis will be in a random state, so the satellite attitude is still not captured. Control mode 2 is used to establish a rough Y Thomson gesture for the satellite. The control process is to make the satellite rotate around the Y-axis while continuing to dampen the angular rate of the X-axis and Z-axis at the same time. Here the rotation around the Y-axis has two purposes: First, to process the astral Y-axis to a position which is perpendicular to the orbital plane; second, to reserve angular momentum for absorbing disturbing moment when the momentum wheel starts. The cycle of this mode control is 1 second.

After control mode 2, the angular velocity of the Y-axis should be in the vicinity of a predetermined value, angular rates of X-axis and Z-axis are still about zero and the Y-axis should be in an approximately vertical posture to the track surface.

The period of control mode 2 in the simulation lasts from 15,000 to 19,000 seconds. The satellite angular rate curve for this mode is shown in Fig. 3.8. The triaxial magnetic torque output curve is shown in Fig. 3.9.

As shown in the satellite angular velocity curve (Fig. 3.8), as long as the control mode 1 establishment can achieve the expected control objectives, control mode 2 will be able to successfully complete the task: to make the satellite rotate around the axes in a predetermined speed; to dampen angular rates of X-axis and Z-axis to remain near zero; to make the satellite Y-axis be approximately perpendicular to the orbital plane. These preparations make it ready for the momentum wheel to start working. Therefore, control mode 2 is correct and effective.

As shown in the magnetic torque output curve for the Y-axis (Fig. 3.9), in control mode 2, the magnetic moment of the Y-axis and Z-axis is essentially at full capacity, demanding great electrical power. In this mode, the X-axis magnetic moment does not work.

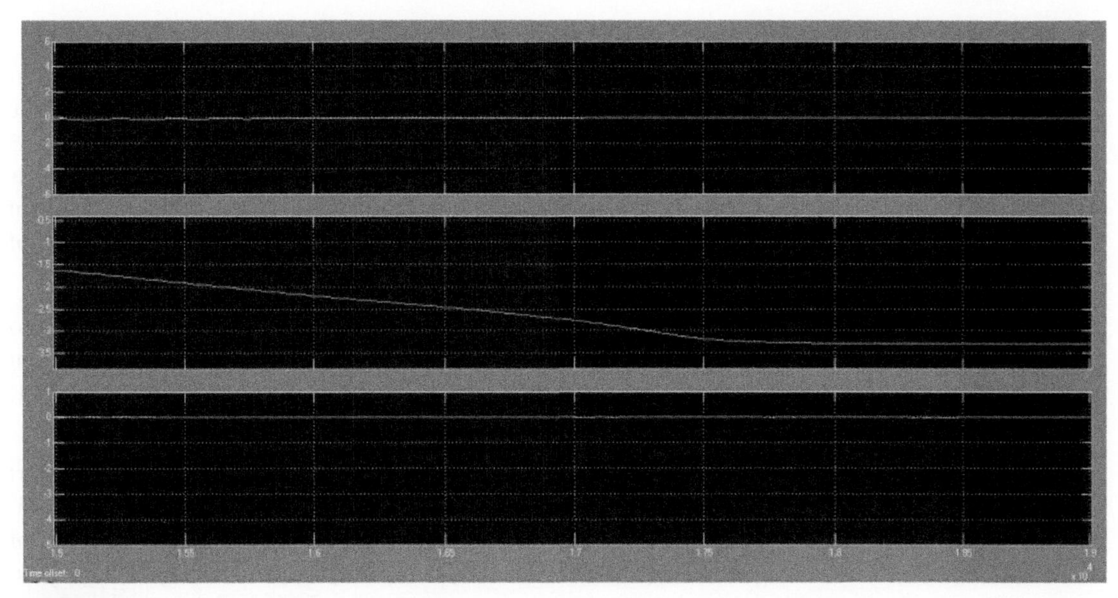

FIGURE 3.8
Satellite angular velocity curve of control mode 2 (deg/s).

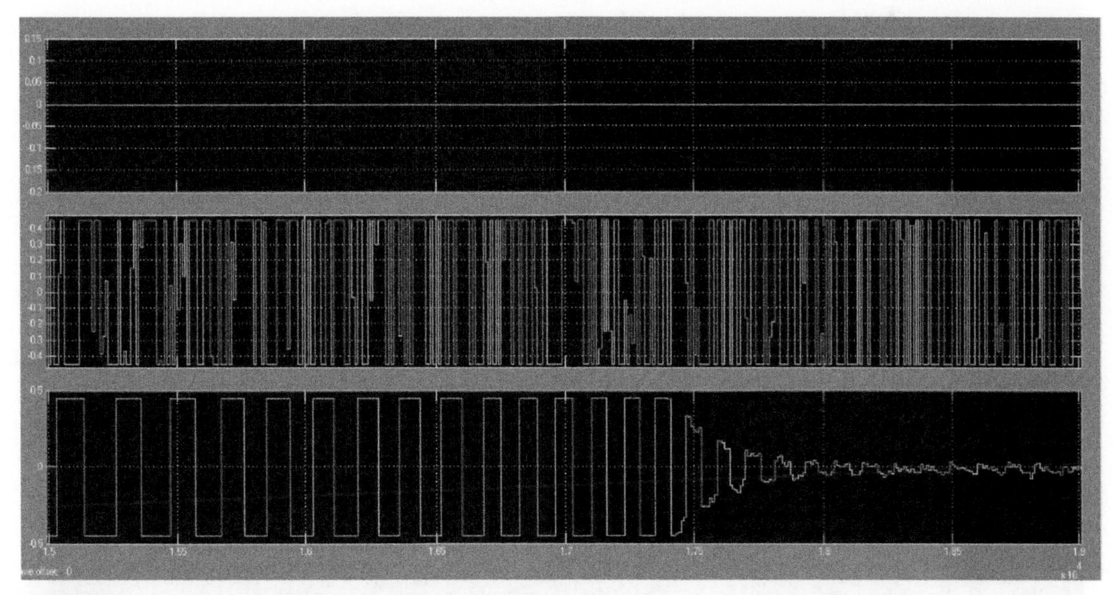

FIGURE 3.9
Three-axes magnetic torque output curve of control mode 2 (Am^2).

3.5.3 Control Mode 3

In control mode 2, the satellite Y-axis is already approximately perpendicular to the orbital plane and the angular rate of the Y-axis also remains in the vicinity of a predetermined value. Then the satellite can enter control mode 3 and the momentum wheel starts to work. When the momentum wheel is maintained near the rated speed (2800 rpm) and the triaxial angular rate of the satellite remains near zero, control mode 3 is completed. The cycle of this mode control is 1 second.

The period of control mode 3 in the simulation lasts from 19,000 to 20,000 seconds. The satellite angular rate curve of this mode is shown in Fig. 3.10; the three-axes magnetic torque output curve is shown in Fig. 3.11; and the speed curve of the momentum wheel is shown in Fig. 3.12.

As shown in the satellite angular velocity curve (Fig. 3.10), during the start process of the momentum wheel, the nutation of the Y-axis increases significantly. However, due to the short time of this process and the damping of Y-axis magnetic torquer, the Y-axis can still keep approximately vertical posture with the track surface.

As shown in the speed curve of the momentum wheel (Fig. 3.11), the momentum wheel can increase the speed to the rating speed stably and remain in the vicinity of the rating speed.

Therefore control mode 3 is correct and effective.

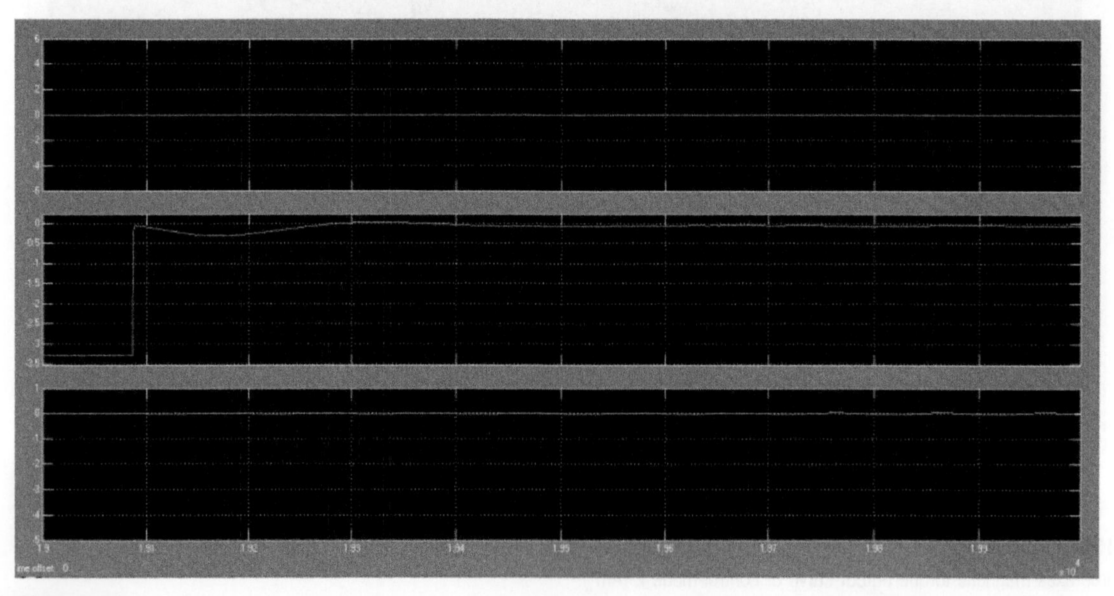

FIGURE 3.10

Satellite angular rate curve of control mode 3 (deg/s).

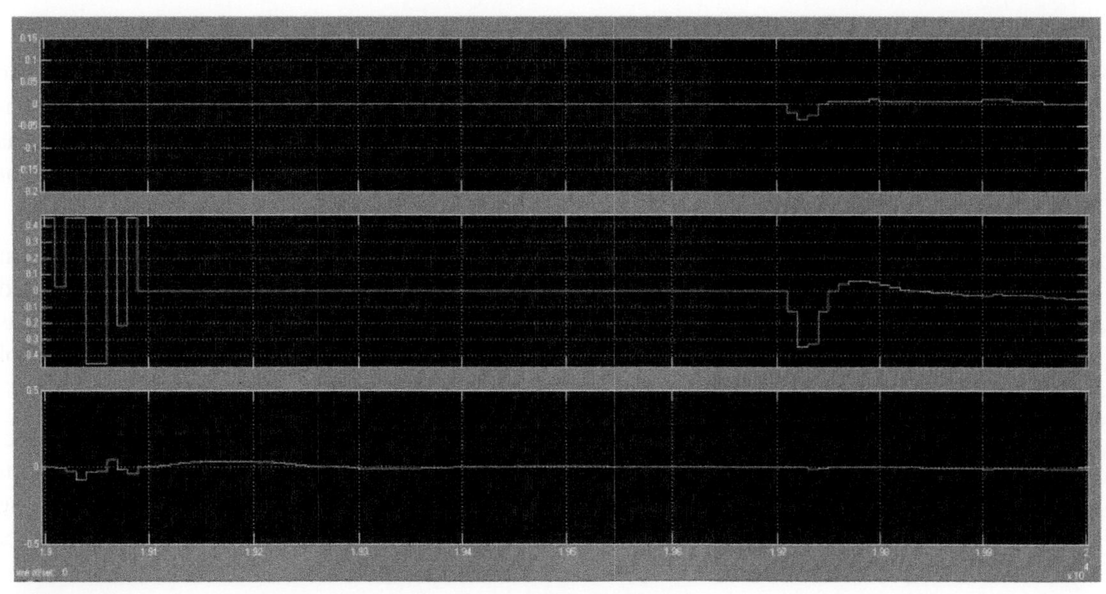

FIGURE 3.11

Three-axes magnetic torque output curve of control mode 3 (Am^2).

FIGURE 3.12

Speed curve of the momentum wheel of control mode 3 (rpm).

3.5.4 Control Mode 4

After control mode 3, the satellite three-axes angular rates are about zero, the Y-axis is approximately perpendicular to the orbital plane, and the momentum wheel has been maintained in the vicinity of the rated speed. At this moment, the satellite can enter control mode 4—the stabilized mode for bias momentum state of Y wheel to the oriented three axes.

In control mode 4, the Y wheel works under the polarization momentum state to control the attitude of the pitch axis; Y-axis magnetic torque is used to control the attitude of the X-axis and Z-axis; magnetorquers of the X-axis and Z-axis are used to unload the momentum wheel, preventing lost control due to saturation. The control cycle of this mode is 10 seconds.

The period of control mode 4 in the simulation lasts from 20,000 to 47,000 seconds. The satellite three-axes attitude angle curve of this mode is shown in Fig. 3.13; the three-axes orbital angular velocity curve is shown in Fig. 3.14; the speed curve of the momentum wheel is shown in Fig. 3.15; and the three-axes magnetic torque output curve is shown in Fig. 3.16.

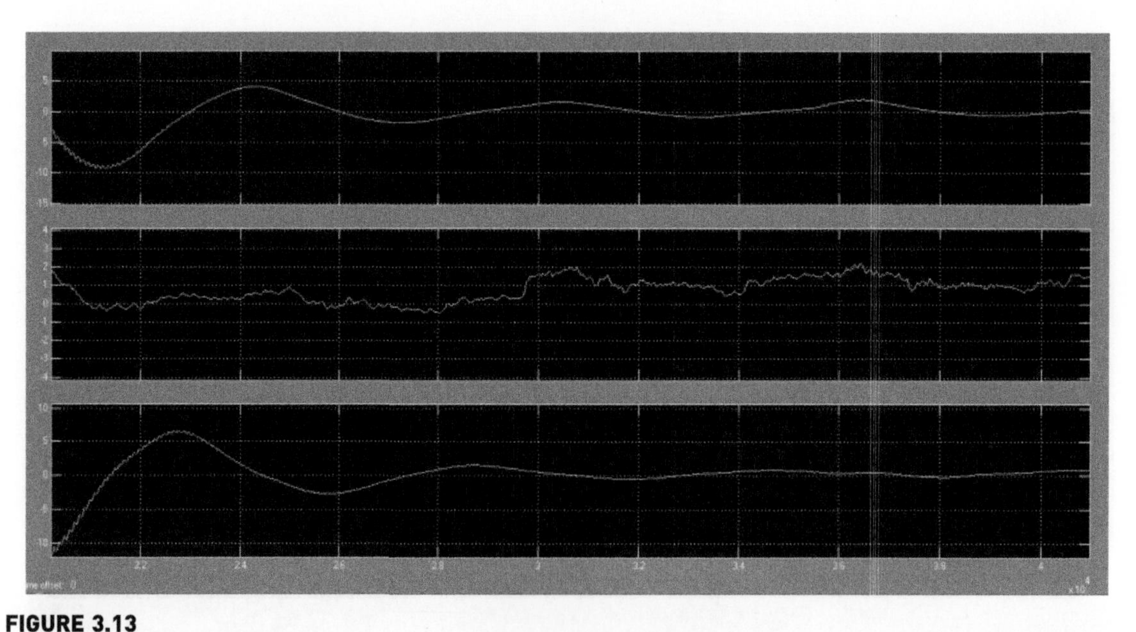

FIGURE 3.13
Three-axes attitude angle curve in the control mode of the Y wheel bias momentum (degree).

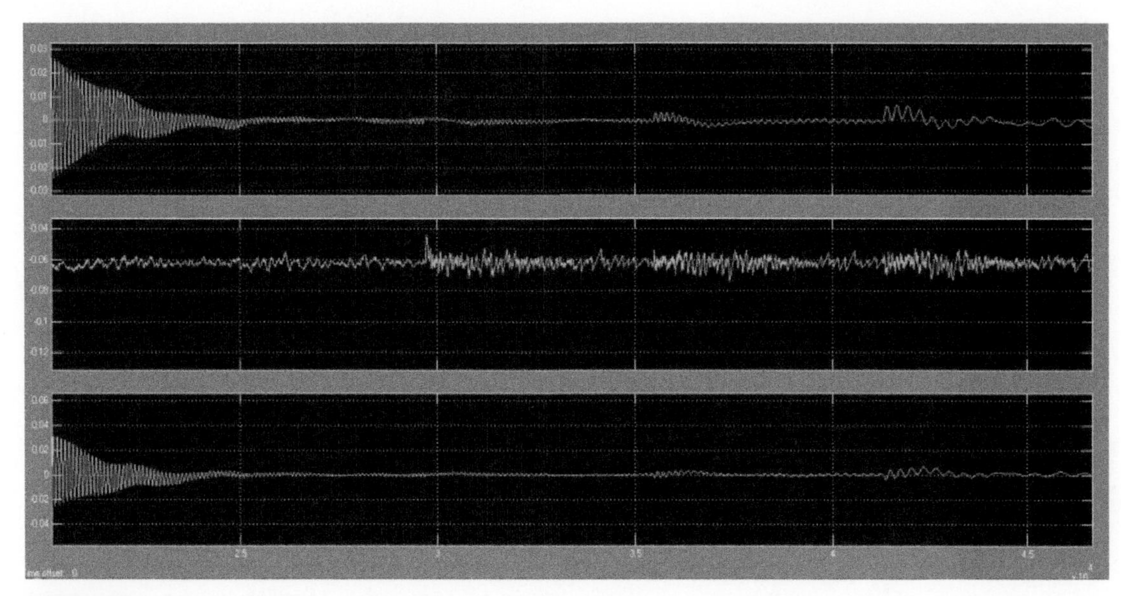

FIGURE 3.14
Three-axes orbital angular velocity curve in the control mode of the Y wheel bias momentum (deg/s).

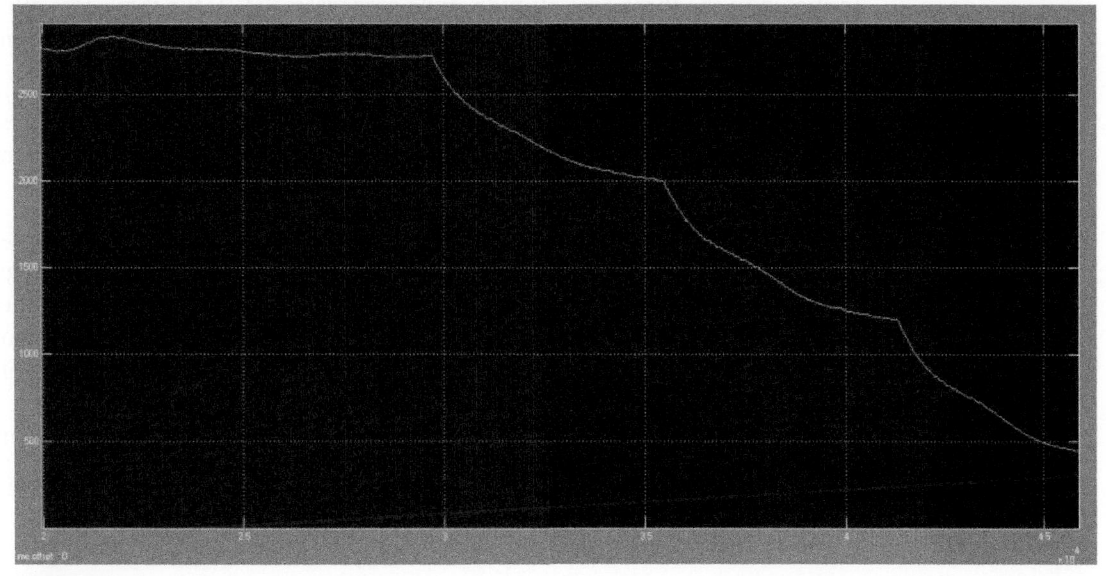

FIGURE 3.15
Speed curve of the Y wheel in control mode 4 (rpm).

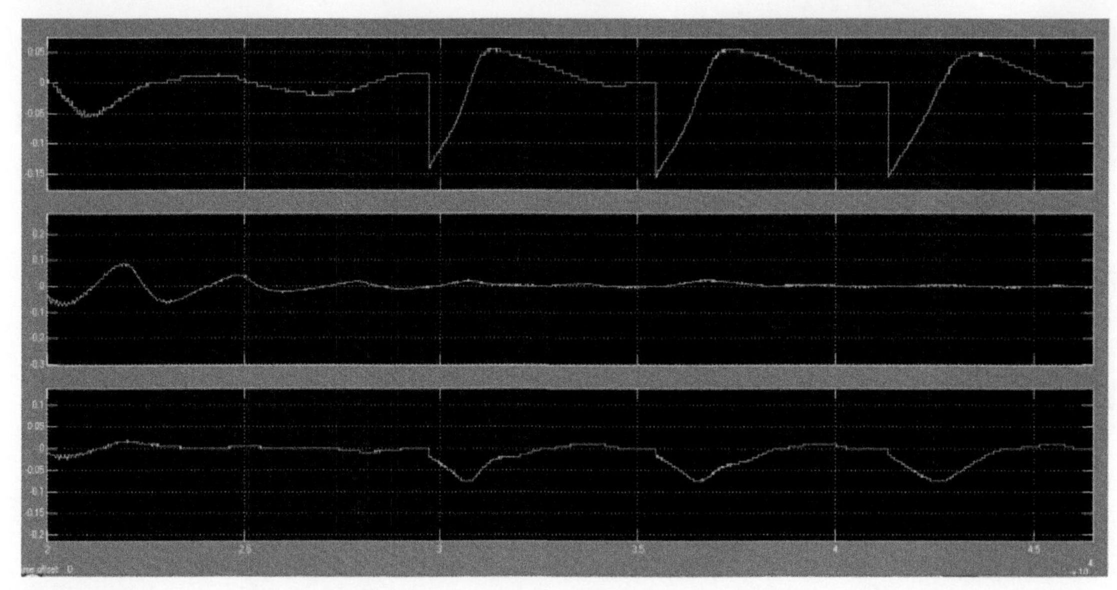

FIGURE 3.16
Three-axes magnetic torque output curve in control mode 4 (Am^2).

As shown in the three-axes attitude angle curve (Fig. 3.13), the three-axes attitude control accuracy of the NS-2 satellite in this control mode is better than 3 degrees.

As shown in the three-axes attitude angle rate curve (Fig. 3.14), the three-axes attitude angular rate of the NS-2 satellite in this control mode is less than 0.01 deg/s.

As shown in the speed of the Y wheel (Fig. 3.15), under the influence of disturbance torque, the speed of the momentum wheel is always changing; however the magnetorquer keeps unloading the momentum wheel, so the rotational speed of the Y wheel decreases steadily, so that the satellite can complete a smooth transition from the bias momentum operating state of the wheel Y to reaction wheel state of the Y wheel while the ground three-axes state remains at a relatively stable situation, providing the conditions for the X-axis and Z-axis reaction wheel to start.

3.5.5 Control Mode 5

After control mode 4, the satellite has been working in the three-axes stabilized mode and the stability of three axes has been able to achieve the general technical requirements, however the attitude determination and control accuracy of three-axes stabilization do not meet the general requirements, while the speed of the Y wheel has been reduced to 500 rpm, providing sufficient preparations for the X, Y, Z wheel to work round the zero-momentum control state.

The period of control mode 5 in the simulation lasts from 47,000 to 100,000 seconds. The satellite three-axes attitude angle curve of this control mode is shown in Fig. 3.17; the estimation error curve of the three-axes attitude is shown in Fig. 3.18; the angular velocity curve of the three-axes orbital is shown in Fig. 3.19; the speed curve of the momentum wheel is shown in Fig. 3.20; and the three-axes magnetic torque output curve is shown in Fig. 3.21.

As shown in the three-axes attitude angle curve (Fig. 3.17), in this control mode, three-axes attitude control accuracy of the NS-2 satellites is better than 1 degree, meeting the overall performance requirements.

As shown in the measurement error curve of three-axes attitude (Fig. 3.18), in this control mode, measurement error of the NS-2 satellite-axis attitude is less than ± 0.003 (3σ), meeting the overall requirement index.

As shown in the angle rate curve of the three-axes attitude (Fig. 3.19), in this control mode, the three-axes attitude angular rate of NS-2 satellites is less than 0.01 deg/s, meeting the overall requirement index.

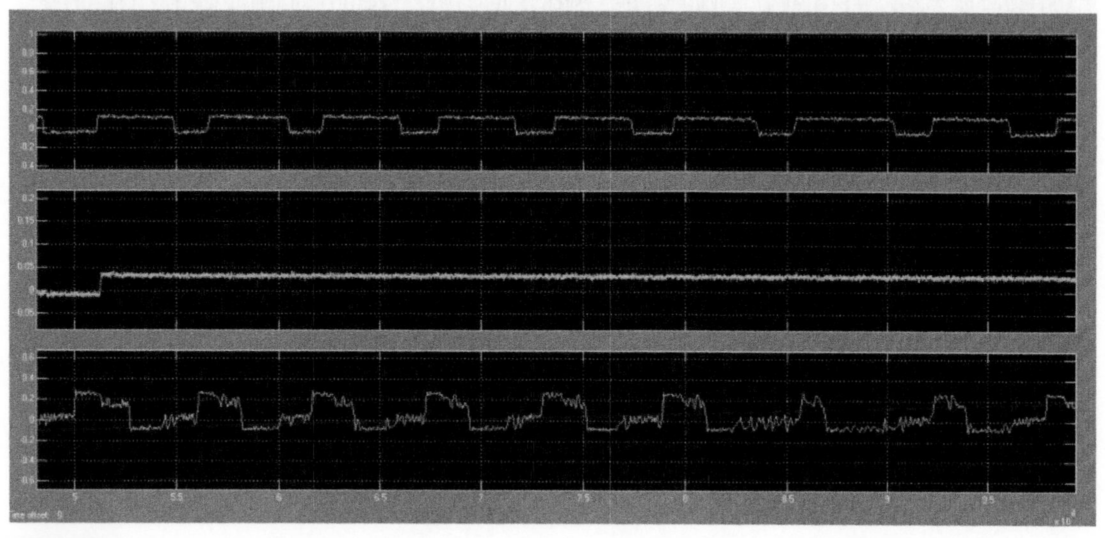

FIGURE 3.17

Three-axes attitude angle curve in control mode 5 (degree).

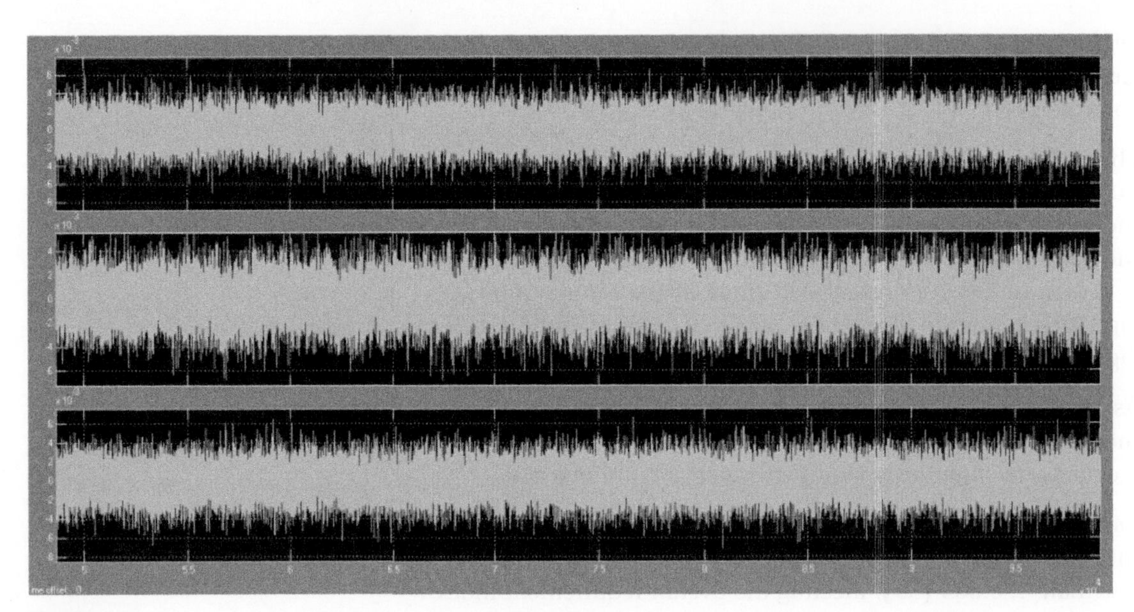

FIGURE 3.18
Estimation error curve of the three-axes attitude in control mode 5 (degree).

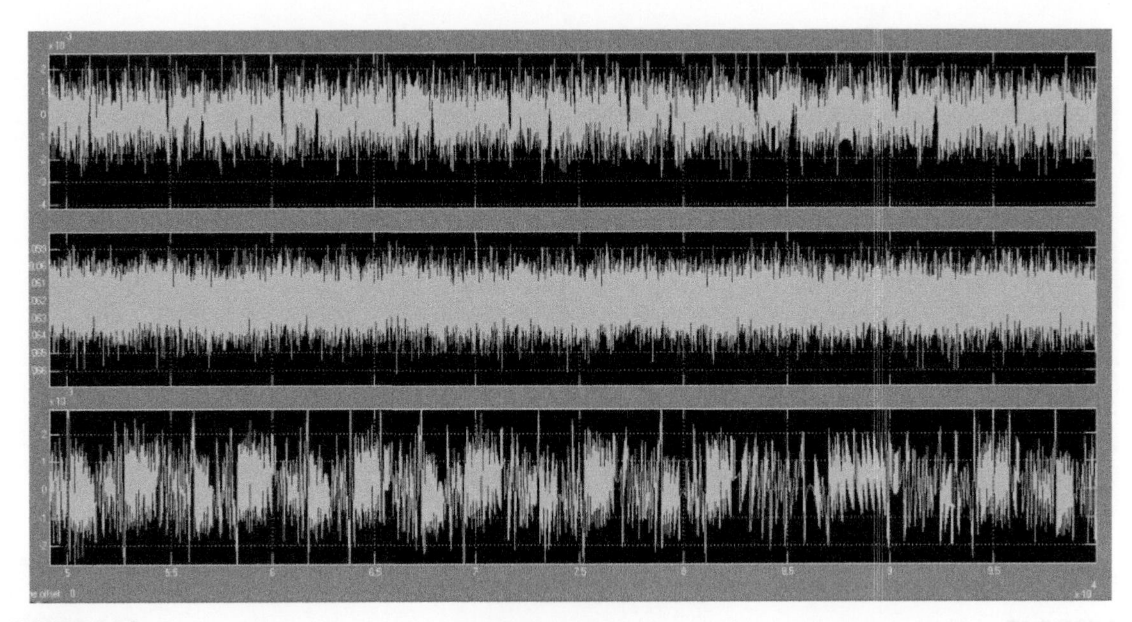

FIGURE 3.19
Angular velocity curve of the three-axes orbital in control mode 5 (deg/s).

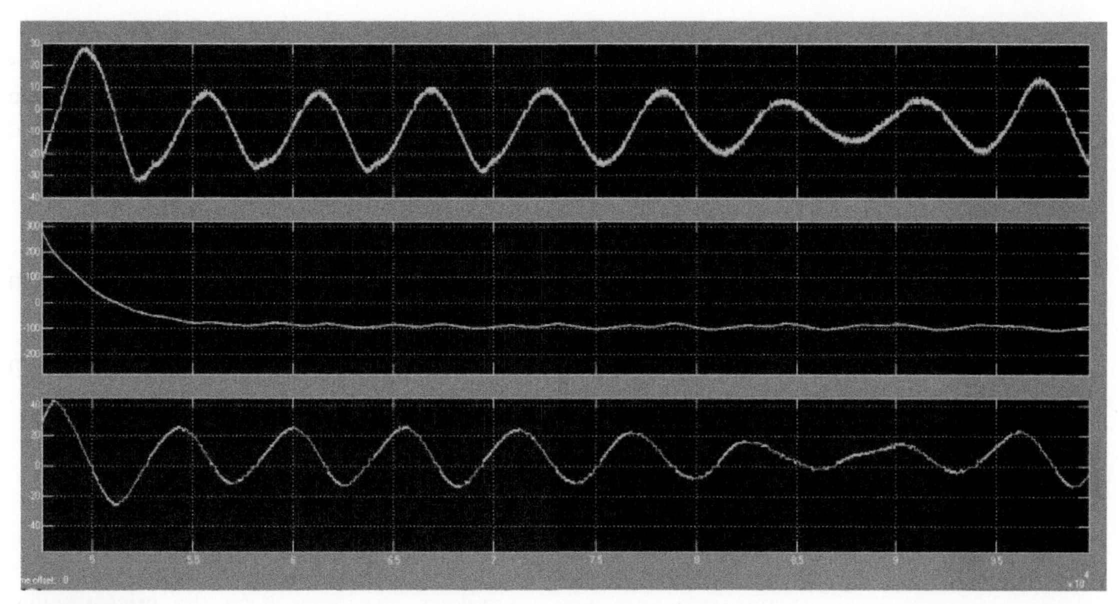

FIGURE 3.20

Speed curve of the X, Y, Z wheel in control mode 5 (rpm).

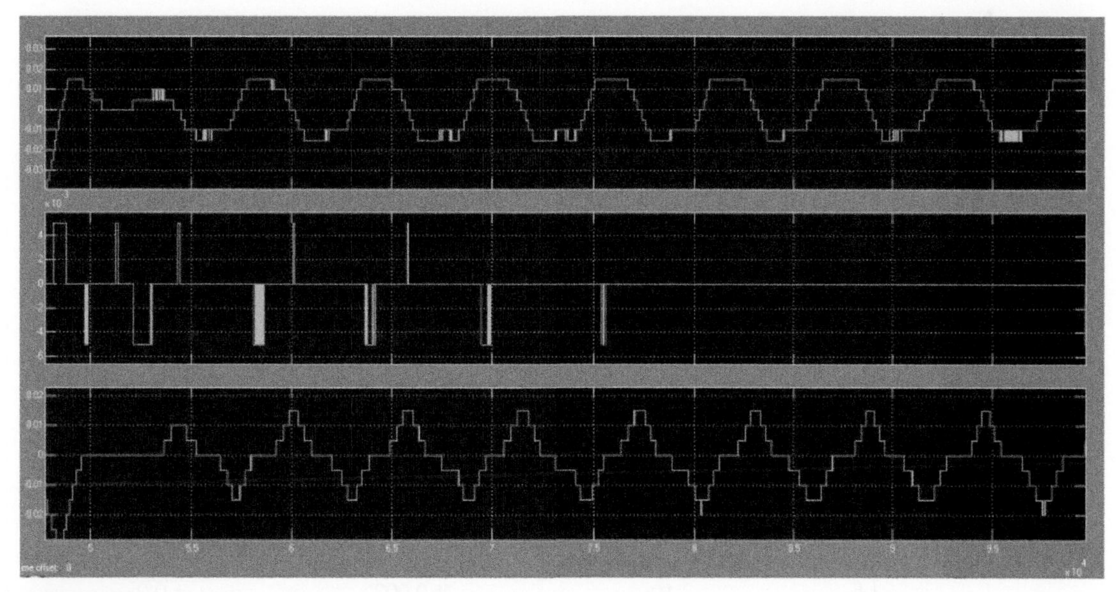

FIGURE 3.21

Magnetic torque output curve in control mode 5 (Am^2).

As shown in the speed of momentum wheel (Fig. 3.20), due to the effects of disturbance torque, the rotational speed of the momentum wheel is always changing; however the magnetorquer keeps unloading the momentum wheel, therefore the speed of the momentum wheel is always maintained near the rated speed without saturation.

3.5.6 Simulation Conclusions

The simulation results of the various control modes are summarized in Tables 3.4 and 3.5.

The following conclusions can be obtained from the above simulation results:

- The magnetorquer of the NS-2 satellite mission is capable of various control modes;
- The momentum wheel of the NS-2 satellite is capable of carrying out the attitude control task. This mainly refers to two aspects: First, to provide sufficient bias momentum to enable fixed axis character of the satellite pitch axis; the second is when the magnetic torque is unloaded, the angular momentum is sufficient to absorb the disturbance of the pitch channel of the satellite, in order to ensure the control accuracy of the pitch channel;

Table 3.4 Torque Simulation Results of the NS-2 Satellite Under Normal Circumstances

		Control Results (3σ)	
Control Mode	**Control Cycle (s)**	**Angular Rate (deg/s)**	**Attitude (degree)**
1	1	± 0.1	
		± 0.1	
2	1	± 0.1	± 3
		-6 ± 0.1	
		± 0.1	± 3
3	1	± 0.05	± 3
		± 0.02	± 1
		± 0.05	± 3
4	10	± 0.02	± 1
		-0.05 to -0.07	$0 \sim 2$
		± 0.01	$0 \sim 2$
5	10	± 0.003	± 0.2
		-0.006 to -0.0065	$0 \sim 0.05$
		± 0.0025	$-0.1 \sim 0.4$

Table 3.5 Stabilized Triaxial Measurement Accuracy of the NS-2 Satellite in a Normal Environment Torque

Accuracy of attitude measurement (degree) (3σ)	± 0.003
	± 0.003
	± 0.003

- The control accuracy and stability of the ADCS subsystem for the NS-2 satellite are fully able to meet the design specifications and general requirements;
- Each control mode of the ADCS master control software for the NS2 satellite can perform its function and achieve the intended purpose, providing favorable conditions for the next control mode;
- The ADCS master software of the NS-2 satellite can run for a long time and has good robust stability.

References

[1] T. Li, Study on Microsatellite Magnetic Attitude Control and Three Axis Passivity Stabilization, National University of Defense Technology, Changsha, 2002.

[2] Y. Guo, Study on Bias Momentum Micro-Satellite Attitude Control System, Nanjing University of Astronautics, Nanjing, 2008.

[3] Y. Xie, Study on Micro-Satellite Attitude Control System, Nanjing University of Astronautics, Nanjing, 2007.

[4] Y. Yang, Magnetic Control System on Small Satellite Design Simulation, National Space Science Center, CAS, Beijing, 2006.

[5] L. Bo, Study on the Attitude Determination and Control System of Three Axis Stabilized Satellite, Northwestern Polytechnical University, Xi'an, 2001.

[6] Z. Wang, The Attitude Control System Design and Simulation of Microsatellite, Nanjing University of Astronautics, Beijing, 2007.

[7] L. Liu, Simulation Technology of Control System of Satellite, China Astronautic Publishing House, Beijing, 2003.

[8] S. Tu, Satellite Attitude Dynamics and Control, China Astronautic Publishing House, Beijing, 2006.

[9] J. Yu, Technology and Application of Modern Small Satellite. 2004, Shanghai Science Popularity Press, Shanghai.

[10] R. Zhang, Dynamics and Control of Satellite Orbit and Attitude, Beihang University Press, Beijing, 1998.

[11] H. Du, Z. Ye, Handbook for Space Environment of Low Earth Orbit Spacecraft, National Defense Industry Press, Beijing, 1996.

[12] X. Xu, Development of the Simulation Software Platform for Satellite Orbit and Attitude Dynamics, Tsinghua University, Beijing, 2002.

[13] J. Zhou, Spacecraft Principle, Northwestern Polytechnical University Press, Xi'an, 2001.

[14] K. Zhao, Hangtian-Tsinghua 1 Satellite and Its Application, China Astronautic Publishing House, Beijing, 2002.

[15] Y. Qian, Studies on the Attitude Determination and Control System of High-Precision Three-Axis Stabilized Satellite, Northwestern Polytechnical University, Xi'an, 2002.

[16] X. Zhang, L. Lin, X. Suo, Satellite control system simulation technics, Comput. Simul. 17 (2) (2000) 57–59.

[17] Y.-G. Jian, Z.-Q. Wang, B. Zhang, Simulation research of three-axis satellite attitude control period, Comput. Eng. Des. 29 (03) (2008) 716–718.

[18] Y.-H. Geng, H.-T. Cui, Attitude control system design for small satellite earth acquisition with active magnetic control, Acta Aeronaut. Astronaut. Sin. 21 (2) (2000) 142–145.

[19] M. Chen, S.-J. Zhang, Y.-J. Xing, Y.-C. Zhang, Combined attitude control method of small satellite in the case of reaction wheel failed, J. Harbin Inst. Technol. 39 (5) (2007) 811–816.

[20] W. Wang, B. Zhang, J. Yang, Three-axis stabilization technology of NS-1 satellite, in: Proceedings of 12th National Conference on Space and Motion Body Control Technology, 2006.

[21] C. Duan, Combined Attitude Control of Small Satellite by Using Reaction Wheels and Magnetorquers, National University of Defense Technology, Changsha, 2007.

CHAPTER 4

Micro/Nano Satellite Integrated Electronic System

4.1 OUTLINE

In recent years, space missions have become more and more diverse and complex. This raises a higher requirement for the development of micro/nano satellites, especially on on-board electronic systems. The rapid development of modern electronic technology, computing technology, control technology, and information engineering provides a new implementation approach and implementation possibilities to meet these requirements. In this context, the new field of micro/nano satellite integrated electronic technology was born.

A micro/nano satellite integrated electronic system refers to an on-board computer (OBC) for the management of the core, and for the information link between various types of electronic equipment and payloads to provide information management services data processing and transmission. Integrated electronic systems include the system layer component, the energy, and the information core satellite.

From the traditional satellite design point of view, the satellite consists of subsystems, such as construction, power, thermal control, attitude and orbit monitoring and control, data management, etc. In accordance with the requirements of the satellite, subsystems have their own characteristics for the design, processing, debugging and then manufacture of a satellite. As satellite peripherals are managed by multiple independent crew members, so the on-board computer is always managed by indirect peripherals. The disadvantage of this approach is the management circuit complexity, as resources can not be shared.

4.1.1 Integrated Electronic Design Ideas

The idea of an integrated electronic system design is to share resources and information, beyond those of the various subsystems, hardware, and software. Integrated electronic systems are not like a traditional satellite design

Space Microsystems and Micro/Nano Satellites. DOI: http://dx.doi.org/10.1016/B978-0-12-812672-1.00004-7

with the composition of a number of subsystems, but with the onboard computer as the core, each unit on the satellite is considered as a whole, connected through a smart bus before the hardware functions are implemented by the software. By adopting this integrated design to achieve the functionality of the software, not only the types but also the quantities of electronic components could be greatly reduced, so that both can meet the functional requirements of the satellite and minimize the types and quantities of electronic components that system required, simplifying the hardware complexity and improving system reliability.

An integrated electronic system aims to be "universal, series, modular," organically combined to achieve information sharing and integration of satellites, so as to improve performance, reduce costs, accelerate the speed of development of space missions, and meet rapid response requirements [1].

4.2 INTEGRATED ELECTRONIC SYSTEM MICRO/NANO SATELLITES

According to the integrated electronic system of Tsinghua University NS-1 satellites, the advantages of an integrated electronic system can be summed up by the following five aspects (Sections 4.2.1–4.2.5).

4.2.1 High Level of Integration

Using multiple integration technology, large-scale integrated circuit design and manufacturing methods of system integration are used. This enables the mechanical components and the sensors, actuators, microprocessors, and other electronic and optical systems to be integrated in a very small space such as a micro/nano satellite or subsystems. This design also improves the reliability of the satellite.

4.2.2 High Processing Performance

The on-board computer is responsible for information collection, processing, distribution and storage of platform, and for the satellite information and functional integration. With increase in requirements of the on-board computer performance in data processing capability, so the development of a high-performance on-board computer with a reconfigurable design becomes very important. The main objective is to develop a reasonable selection of onboard computer architecture, without a significant increase in hardware premised by bus arbitration techniques and providing alternative resources by strict fault detection, improving system reliability, and the processing and fault tolerance to meet micro, nano satellite high processing capacity, high reliability, and long life demands.

4.2.3 Highly Modular

The modular technology for electronic systems and even micro/nano satellites, enhances the flexibility and scalability of the system, so that in a short period of time the design, production, assembly and upgrading of the satellite system becomes possible, so that the development significantly shortens the cycle.

4.2.4 Highly Intelligent

Tsinghua University NS-1 satellites can customize the stable operation of the main remote intervention, supplemented by ground control. When a failure occurs, the system has independent switching capacity to maintain normal operation; this minimizes the dependence on ground stations, improving independent operational capabilities.

4.2.5 Relatively High Degree of Reliability

Micro/nano satellites, due to their smaller size, low power consumption, in addition to the necessary hardware redundancy, but also by the means of software, appear to be effective in fault diagnosis and treatment to improve system reliability. Microsatellite-integrated electronic system has implemented a high degree of reliability [2].

4.3 MICRO/NANO ELECTRONIC SATELLITE INTEGRATED ELECTRONIC SYSTEM ARCHITECTURE

Micro/nano satellite structure is basically hierarchical distributed by bus to connect the subsystems.

Tsinghua University NS-1 satellites, launched in 2004, are typical micro/nano satellites. Fig. 4.1 shows an electronic system architecture diagram for a "NS-1" satellite. The OBC module, telemetry, and telecontrol (TTC) module are core modules of the integrated electronic system. These modules communicate with other subsystems via CAN bus and asynchronous serial interfaces. The onboard computer hardware resources of the micro/nano satellite integrated electronic system, the core management of the entire system, have the basic function of providing time management and storage management, as the key subsystems to achieve satellite platform and payload capabilities. Onboard computer requirements for the hardware system include high reliability and flexibility. The onboard computer typically uses embedded computer systems, consisting of an embedded microprocessor, whose performance determines the overall performance of the onboard computer. Telemetry and telecontrol (TTC) is a fundamental part of the micro/nano satellite integrated electronic system, and enables the ground station and the

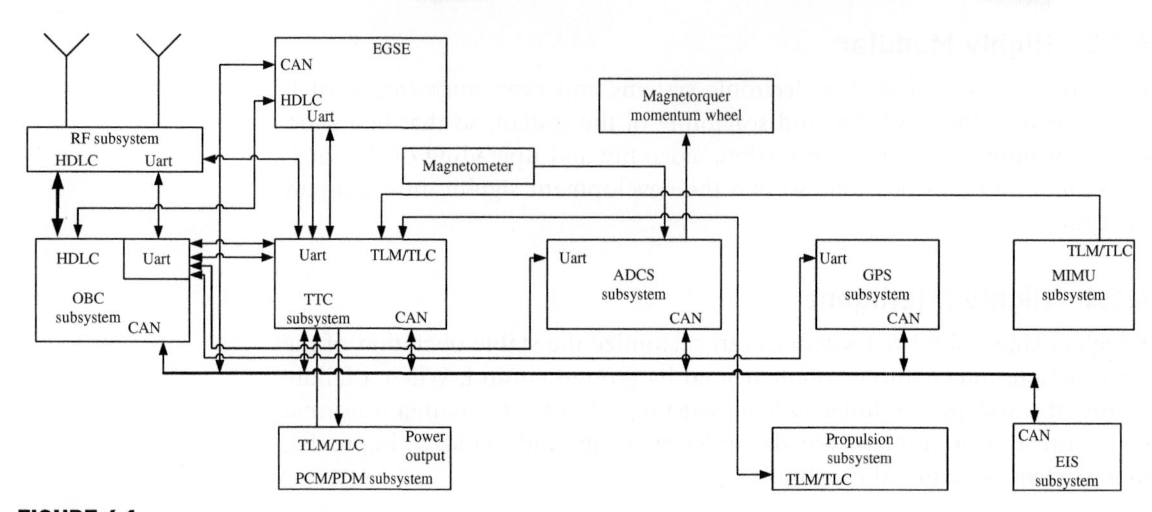

FIGURE 4.1
NS-1 satellite electronic system architecture.

onboard computer control to manage the various operating modes. The main task of the telemetry and telecontrol subsystem is to measure the satellite operating parameters, verify that the satellite is operating normally, according to the requirements of the satellite's operating mode switching and system restructuring, and also transfer data payloads for ground use.

According to the actual situation and the different mission requirements of different stages, integrated electronic system controls the controllers of each function module by the system bus and makes functional reorganization, which enables satellites to achieve different modes of operation after completing self-working mode adjustment. The integrated electronic operating mode can be divided into two types: direct and indirect modes of operation.

1. Direct application of several operating modes:
   a. In the initial state, with the integrated electronic system under the direct mode, the onboard computer is turned off, telemetry module is active and will periodically transmit satellite platform information to ground stations, is controlled by the ground station and transmits control results back to the ground station.
   b. Without much work or continuous data needs, the satellite will be switched to the energy-saving mode; the integrated electronic system operates in direct mode, with only telemetry and telecontrol module default monitoring and control functions.

2. Then the application of several operating modes:
   a. In the satellite's attitude initial adjustment phase: the integrated electronic system will switch from the direct mode to working under the indirect mode; the telemetry module is active, with on-board computer in working condition. The onboard computer replaces telemetry and telecontrol module function, starts attitude control software implementation of satellite attitude control, and ensures payload capabilities.
   b. In the payload experimental stage: After the completion of the satellite attitude control functions, the satellite will be equipped with the functions of load performance and functional verification experiment. The onboard computer performs according to the specific requirements of the load test, by control of telemetry and telecontrol module restructuring satellite.

4.4 TECHNICAL SPECIFICATIONS

Specifications of the on board computer	
CPU	Intel SA-1110
EPROM:	512 kB Bootloarder 512 kB Program Memory
Program memory space:	1.5 MB (TMR)
Data memory space:	6 MB
CAN node:	666 kbps
Communications controller:	HDLC Uplink 9.6 kbps Downlink 19.2 kbps
Power consumption:	Less than 1.5 W (average)
Telemetry/remote indicator	
Dual-mode redundancy telemetry encoder	
Telemetry channels:	48 analogue channels (AN 0–AN 47), 16 digital input channels (DIN 0–DIN 15), remote commander feedback channel 32
ADC resolution:	12 bits
Analogue normal operating input voltage range:	0 to +5 V
Analogue maximum allowable input voltage range:	−0.5 V to +10 V
Digital inputs:	Low voltage 0–0.8 V, high voltage 3.5–5 V
TMR remote decoder	
Design of the remote command channel 128, with actual use of 32	
Communication port:	CAN bus interface 666 kbps
Asynchronous serial port:	9.6 kbps
Remote sensing and control module power consumption:	Power consumption less than 0.8 W

4.5 SELECT COMPUTER ARCHITECTURE

Commercial off-the-shelf computer technology has been used in many applications, but due to the microsatellite with its own characteristics, design with an onboard computer has created many differences.

These differences include:

1. Since the implementation of microsatellite space missions, as failure is costly, the addition of an OBC is required with high processing performance and high reliability requirements, which is the main aim of high commercial computer processing performance.
2. In order to achieve attitude control, the OBC needs to perform mathematical operations, task scheduling, and task management, which means that the service is not intensively computational, but has intensive management (or is control intensive).
3. As the service management tasks (such as data access) are mostly concurrent, to ensure the high real-time tasks running, the service management software is generally designed in the form of multitasking.

The main functions of the onboard computer main are "greater system reliability" and "higher management efficiency." These two points are the principles behind the selection of the OBC architecture [2].

4.6 THE ON-BOARD COMPUTER DESIGN

4.6.1 The Subsystem Block Diagram of an On-Board Computer

The schematic block diagram of an OBC subsystem is shown in Fig. 4.2.

4.6.2 The Bus Architecture

Two different data buses (i.e., the unbuffered local bus and buffered bus) are designed in an OBC system.

4.6.2.1 The Unbuffered Local Bus (USA, USD, UnOE, UnWE, URD_nWR)

The unbuffered bus has not been driven. It is connected to the Boot PROM. This guarantees that the initializations of SA1110 and other peripherals can be completed smoothly after powering on, as well as memory test, peripheral monitoring, etc.

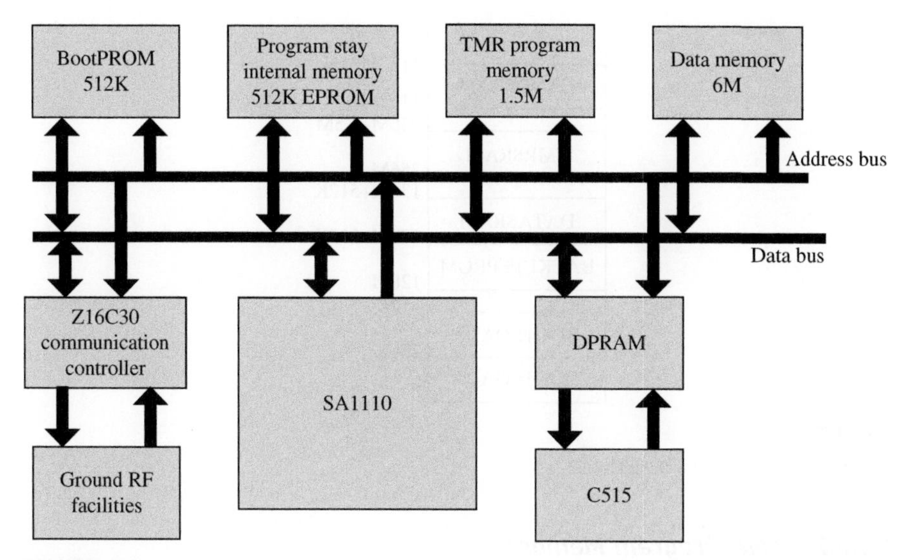

FIGURE 4.2
The block diagram of an OBC system.

4.6.2.2 The Buffered Bus (SA, SD, nOE, nWE, RD_nWR)

General peripherals are connected to the buffered data bus, including EPROM, triple modular redundancy program memory, data memory, HDLC communication controller and DPRAM, etc.

4.6.2.3 The CPU Clock and Reset Circuit Design

The CPU externally connects two crystals at 32.768 kHz (Y101) and 3.686 MHz (Y102). The Y101 is used for the real-time clock, OS clock, and peripherals after being internally driven. Y102 is used for the core of SA1110 after frequency multiplication through a phase-locked loop.

There are three types of system resets, namely the power on reset, direct remote reset from TTC, and indirect remote reset from the CAN node. The three signals are connected by an "AND" gate, connected to the reset of SA1110.

4.6.3 The Memory System Design

The memory system of the OBC consists of EPROM, TMR program memory, RAMDISK, and data memory.

4.6.3.1 EPROM

Two pieces of EPROM respectively store the boot program and operating system with their applications. Their sizes are both 512 kB. The operating system can be loaded from EPROM directly or from the ground.

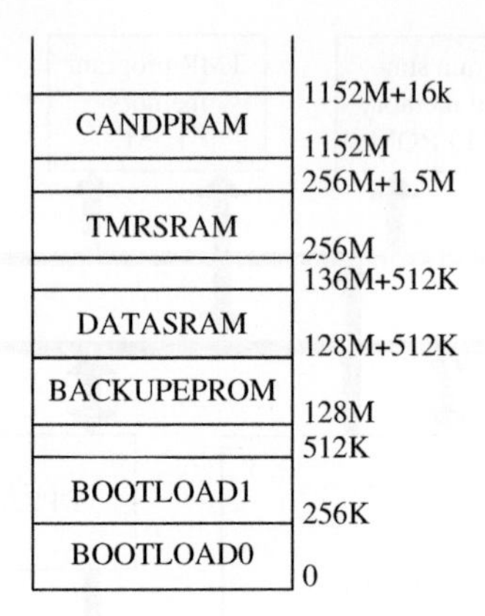

FIGURE 4.3
The OBC memory space allocation map.

4.6.3.2 TMR Program Memory

The system has a 1.5 MB TMR program memory. The SRAM uses Cypress company's CY62146V (256K × 16 bits).

4.6.3.3 Data Memory

The system design uses a 6 MB data memory, which is combined with 12 pieces of CY62146. The data width is 16 bits. The R/W signal of each chip is generated by BCU, and the chip select signal is generated by the nCSI and address decoder of SA1110.

The data memory does not have a hardware check. The system memory map is shown in Fig. 4.3.

4.6.4 The Communication Controller

The synchronous communication of OBC adopts the HDLC data link control protocol. The HDLC control chip uses the universal serial communication controller Z16C30, produced by Zilog. This chip uses 16 bits data bus, data R/T modules, and CPU data buffer asynchronous operation. It has advanced architecture and is convenient to use. This simplifies the connection with the CPU to the greatest extent possible.

The design adopts a data address nonmultiplexing method corresponding to the bus. The 16 bits data lines are connected to the system bus.

When accessing registers, accessing the channel command/address register (CCAR) in order to implement indirect addressing is first needed. Therefore, each access of one register is composed of two read/write (R/W) access cycles, first with the write CCAR for addressing, second by completing the content access of target register. When the implementation of data FIFO access, according to the access mode is read or write, automatically address to the data receive or transmit FIFO.

The bus access cycle of Z16C30 (up to 10 MHz) is slower than the system bus (66 MHz). Hence it is necessary to insert waiting cycles when accessing Z16C30.

The CPU accesses the HDLC controller by interrupt. The two interrupts INT A and INT B are generated by two channels entering the CPU through the GPIO. The interrupt vector is directly determined by the GPIO number.

4.6.5 CAN Node

The system adopts a piece of C515 to control the CAN node. Fig. 4.4 shows its modular block diagram.

The data exchange between CAN and SA1110 is implemented through one piece of DPRAM. The left side of DPRAM is controlled by C515, while its right side is controlled by SA1110. The DPRAM adopts IDT70V06 of IDT. The DPRAM provides MailBox on hardware, guaranteeing data safety. The highest bit of DPRAM is the MailBox for CPU on the right side. When C515 writes the MailBox, INTR will be valid and an interrupt is generated. SA1110 can read the MailBox and operate the DPRAM. In the same way,

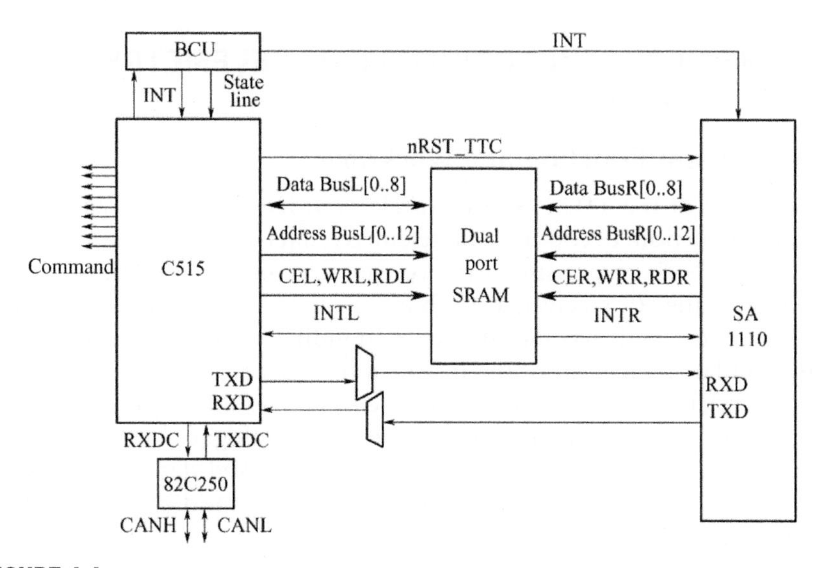

FIGURE 4.4
The OBC CAN module design.

the next-highest bit is the MailBox for CPU at the left side, which SA1110 can use to communicate with C515.

The RXDC and TXDC output of C515 are converted by 82C250 into CAN differential signal, which is then transferred to the CAN network.

4.6.6 The Power Supply Unit

The voltage of power supplied to the OBC is +5 V. SA1110 needs two working voltages: +1.5 and +3.3 V. The FPGA (Field–Programmable Gate Array) on the orbiting satellite should be ACTEL A54SX16A, which also needs two working voltages: 3.3 and 2.5 V. Therefore, the design adopts two pieces of MAXIM's MAX1755EEE, one of which provides +5 to +1.5 V and +3.3 V conversion, while the other provides 2.5 V power. The MAX1755 is a two-output highly efficient DC-DC converter. The output conversion efficiency of +3.3 V is up to 95%, and the load current is 2 A; the output conversion efficiency of +1.5 V is up to 92%, and the load current is 1.5 A. Its quiescent current is 170 μA, and the operating temperature is −40 to 85°C. In order to reduce power consumption, MAX1755's SHDNC pin is connected to SA1110's power enable pin. When SA1110 is in sleep mode, it uses an SA_PWR_EN pin to shut down the MX1755 +1.5 V voltage conversion, so as to reduce OBC power consumption.

4.7 THE OPERATING PRINCIPLE OF TELEMETRY AND TELECONTROL

4.7.1 System Function and Workflow

The structure of the TTC is shown in Fig. 4.5. It mainly consists of telecontrol instruction decoding and telemetry data acquisition. A remote control decoder is responsible for receiving telecontrol data frames transmitted from the RF receiver, CAN bus, or an asynchronous serial interface, decoding and writing the corresponding control bits into the instruction latch. Each bit of the instruction latch corresponds to a switch state. The telemetry data acquisition system receives and packs the digital information on the satellite into frames. For analogue data, it applies A/D conversion first and then packs them. The packed data are provided to the RF downlink and then transmitted to the ground station or OBC for further processing.

Both telemetry and telecontrol of TTC have two operating modes: direct mode and indirect mode.

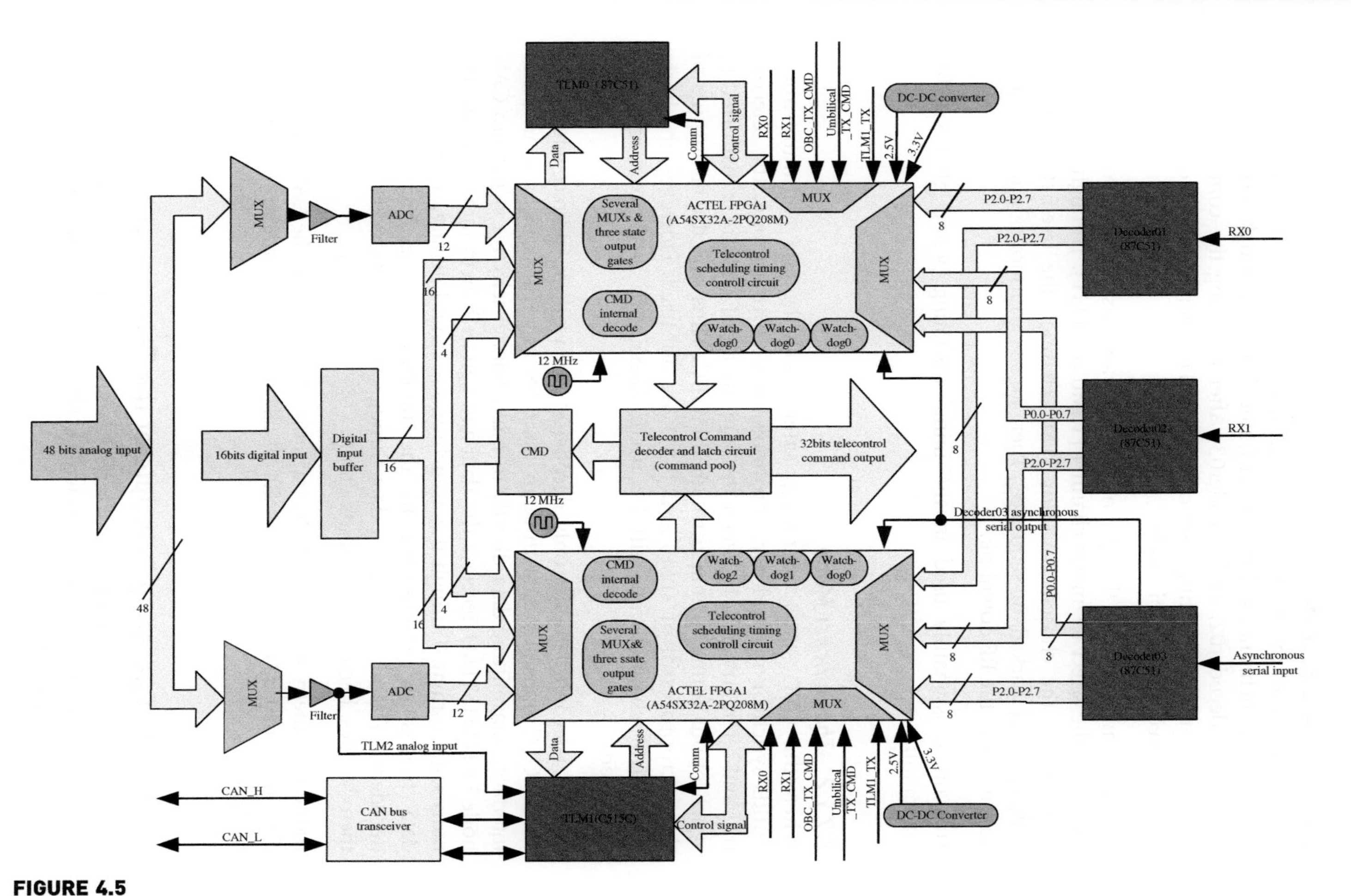

FIGURE 4.5
TTC structure diagram.

4.7.1.1 The Direct Mode

Direct telecontrol mode: the ground station transmits basic telecontrol command (in asynchronous simple data frame format) through RX0 or RX1 to decoder01, decoder02, and decoder03. After decoding the control word is output to latch, for latching and outputting. Decoder01 and decoder02 only work in direct telecontrol mode.

Direct telemetry mode: the ground station transmits basic telecontrol commands to the select telemetry channel (TLM0 or TLM1). Through the microprocessor of the chosen TLM0 or TLM1 channel, the telemetry data are collected and packed in sequence. Then an asynchronous link is used to transmit it to the RF downlink.

This mode is mainly used in the initial operating period after satellite separation.

4.7.1.2 The Indirect Mode

When the OBC starts operating, satellite–Earth data communication builds synchronous and asynchronous links. In this mode, the telecontrol and telemetry data between the satellite and Earth are transmitted in synchronous frame format through the OBC. Thus, the direct outputs of the microprocessor of TLM0 and TLM1 are not used. However, the basic telecontrol channel can still be used. These two situations are now discussed.

Indirect telecontrol mode: only decoder03 can work at indirect telecontrol mode. OBC transmits data through the CAN bus (for asynchronous serial interface, transmitting data directly to multichannel selective switch) to CAN node in TLM1. TLM1 receives the data and converts them into an asynchronous serial format. Through the asynchronous serial interface, asynchronous data frames are transmitted to the multichannel selective switch of decoder03. Decoder03 receives data through the asynchronous serial input, then decodes, latches and outputs them.

Indirect telemetry mode: the OBC transmits data through the CAN bus to the CAN node in TLM1. TLM1 transmits all telemetry data through the CAN bus to the OBC.

4.7.2 The Operating Principle of the Telecontrol Unit

The whole system frame structure mainly consists of three decoders (decoder01, decoder02, and decoder03), FPGA (FPGA0 and FPGA1) chips, and output interface.

4.7.2.1 Decoder01 and Decoder02

The decoder01 and decoder02 work in direct telecontrol mode. Their circuit structures and functions are the same. They are implemented by two 8 bit microprocessors 87C52, two clocks, and two FPGA internal gate arrays. The specific design is shown in Fig. 4.6. In order to achieve better redundancy, the instruction decoded by SCM is output through the two ports of SCM to the corresponding FPGA for processing. P0 and P2 ports are used for 8 bit command data bus output, while P1 port is used for control line output.

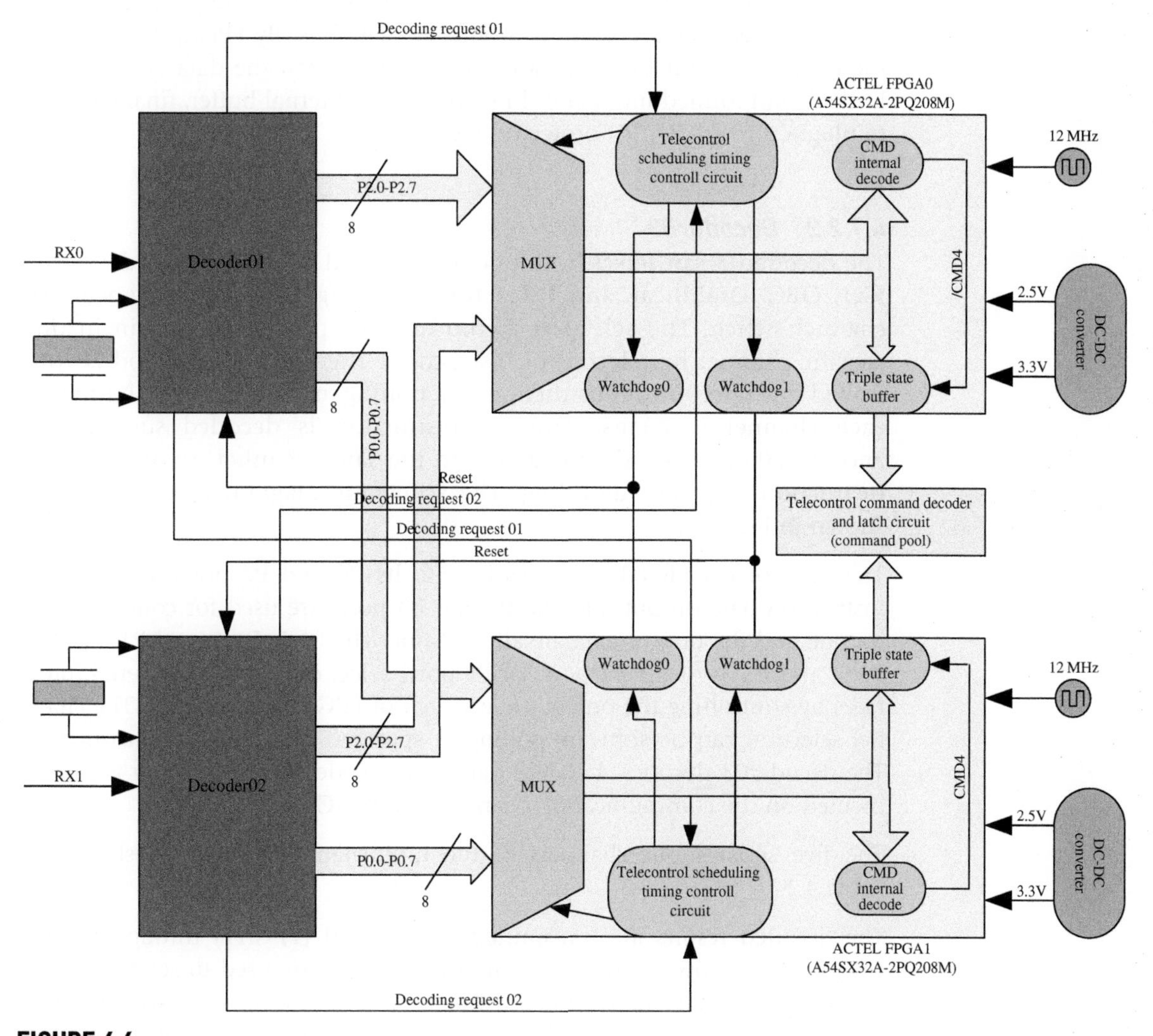

FIGURE 4.6

Decoder01 and decoder02.

The processing procedure: the telecontrol instruction from RX0 (or RX1) enters the SCM asynchronous serial interface of decoder01 (or decoder02) through an asynchronous serial interface. The data processed by SCM are output through the P0 and P2 port. To implement the complete redundancy from decoder SCM to the FPGA circuit, the related control signals of the P0 and P1 ports are connected to FPGA0, the related control signal of the P2 and P1 ports are connected to FPGA1.

The inner sequential logic of FPGA0 (or FPGA1) makes the three telecontrol decoders obtain the control of the bus circularly. The decoder which has the control power puts the valid data onto the data bus. By FPGA0 (or FPGA1), the data are output to the data decoding latch. Then the data are decoded, latched, and formed into control instruction by internal buffer, finally output to plugin through 4.7 kΩ protective resistance.

4.7.2.2 Decoder03

The decoder03 can select from one of five serial interface input data RX0, RX1, OBC, Umbilical, and TLM1 for processing, by controlling a multichannel switch. The selection method can be polling or remain in one specific channel by telecontrol instruction. The selection method is controlled by telecontrol instruction. In polling mode, the dwell time of each channel is 70 ms. After an instruction is decoded successfully, another 70 ms dwell time is added in the same channel, so as to satisfy instruction sequence decoding. The schematic diagram of decoder03 is shown in Fig. 4.7.

The SCM of decoder03 also uses 87C52. Its P0 and P2 ports are used for instruction data output bus, the P1 and P3 ports are used for control signal output, and the output data are all the same. The P1.5, P1.6, and P1.7 (P3.5, P3.6, and P3.7) of the P1 port (or P3 port) select one of the five serial interfaces by controlling the one-out-five switch of FPGA0 (or FPGA1). The channel selection can be software polling or specified by telecontrol instruction. The decoder03 decodes the valid data. For decoder03, redundancy is implemented on the communication channel (except SCM).

The five serial input channels of the multichannel switch are shown in Table 4.1.

The decoded results are transmitted into FPGA0 (FPGA1) through the P0 (P2) port. The lower four bits of the P1 (P3) port are used to send a decoding application signal. Together with the decoded result of decoder01 and decoder02, wait for FPGA0 (FPGA1) to process. The following processing procedure is the same with decoder01 and decoder02.

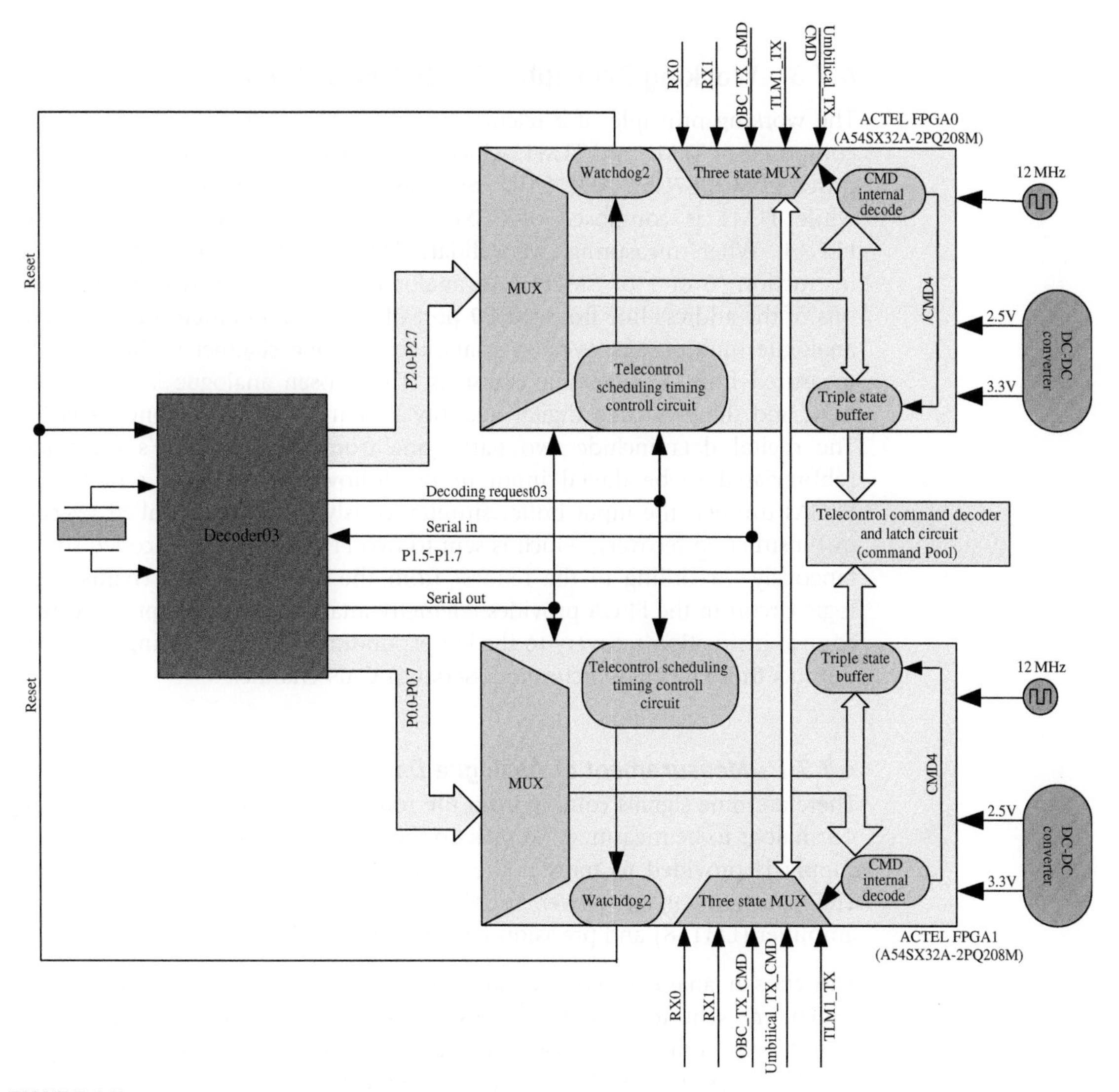

FIGURE 4.7
The schematic design of decoder03.

Table 4.1 Serial Input Channels List

Channel Number	Connection	Speed (bps)
0	RX0	9600
1	RX1	9600
2	OBC	9600
3	Umbilical	9600
4	TLM1(CAN)	9600

4.7.3 Working Principle of Telemetry Units

The working principle of a telemetry unit is shown in Fig. 4.8. This unit is composed of TLM0 and TLM1, which are mutually redundant. TLM0 is composed of SCM87C52, ADC (A/D conversion), and relevant circuits of FPGA0; while TLM1 is composed of SCMC515C, ADC, and relevant circuits of FPGA1. When measuring, we validate TLM0 or TLM1 through telemetry instruction (0 or 1 of CMD4). An analogue channel is chosen by six lower bits of the address line from the P2 port of the SCM and then control several analogue multiplexers. We can guarantee that one channel is chosen to be measured from 48 analogue channels. The chosen analogue is sent to the ADC and the generated digital quantity is sent to the corresponding FPGA. The digital data include two parts: one from other satellite's subsystem (abbreviated to be digital input in the following), which is sent to two FPGAs through the input buffer simultaneously; the other is called telecontrol instruction recovery, which is sent to two FPGAs after being coded simultaneously. According to the request from the valid telemetry channel, the logic circuit in the FPGA provides telemetry analogue, digital input, and telecontrol instruction recovery to the corresponding SCM for framing, and then outputs those by an asynchronous-serial or CAN channel.

4.7.3.1 Measurement of Analogue Data

There are nine signals coming from the resistance temperature sensor among 48 analogs to be measured. To pick up these signals a constant-voltage power supply is provided to transfer the resistance signal into the voltage signal. The constant-voltage power supply is mainly composed by operational amplifier (LM158) and precision voltage regulator (LM129).

The chosen analogue voltages are divided into two lines for redundancy, which are sent to two ADC lines, respectively. SCM87C52 and C515C, respectively, control two ADCs from the processor. The conversion clocks of two AD7880s are provided by the corresponding FPGA.

The inner ADC (10 bits) of SCMC515C, forming the third telemetry channel, works as the supplement of the TLM1. It can replace the TLM1 by switching the instruction when AD7780 of the TLM1 is invalidated.

The result after A/D conversion is sent to FPGA0 (or FPGA1). The SCM chooses one from the analog conversion results, data input and the recovery data of telecontrol instruction by controlling multiway switch. Then, the chose data is processed in microprocessor.

The result can be either output to OBC, RF, or Umbilical by an asynchronous serial port, or to the OBC or Umbilical by the CAN bus.

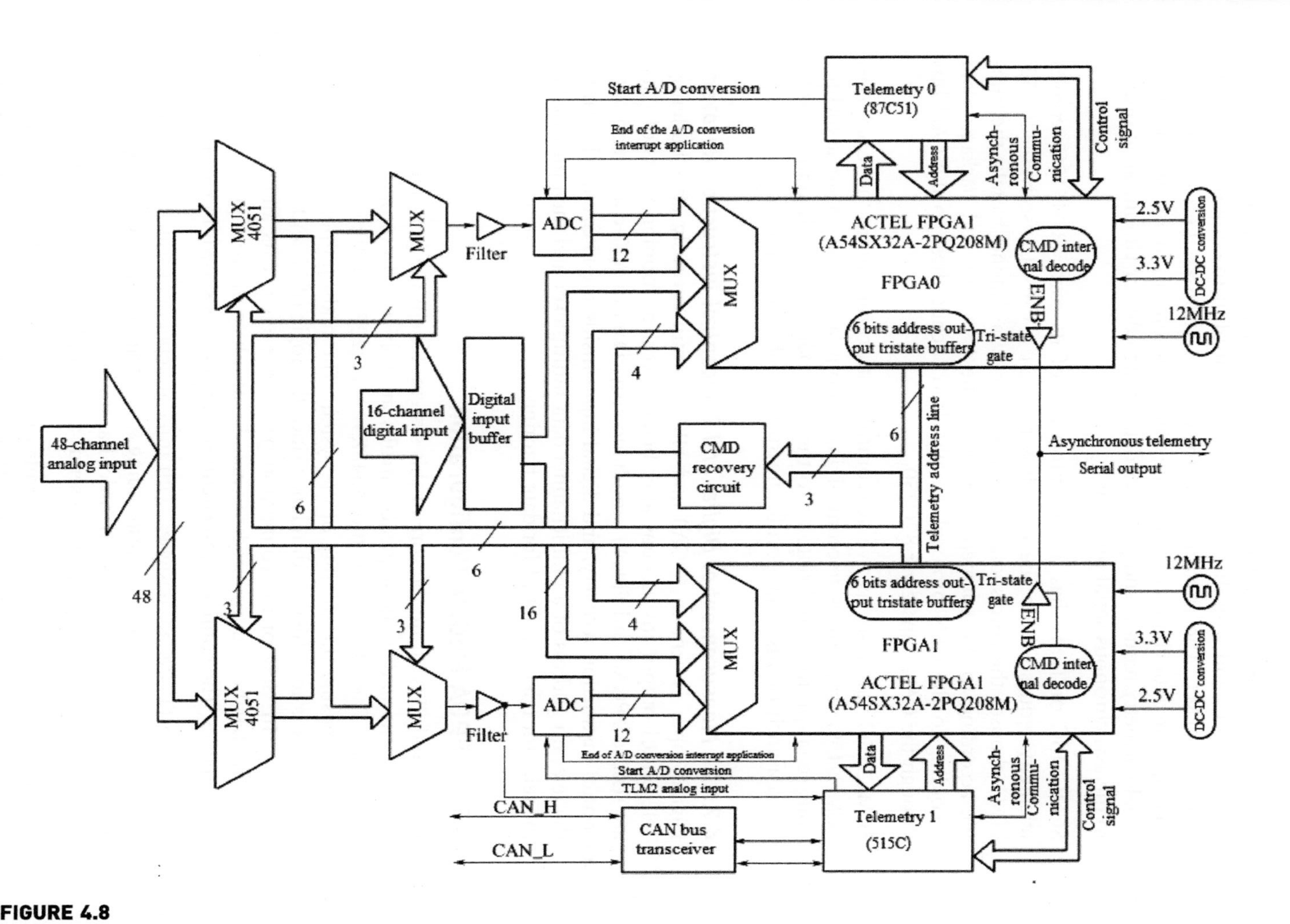

FIGURE 4.8
Working principle diagram of a telemetry unit.

Both of these processors can receive asynchronous serial telemetry requests from the OBC and Umbilical. However, telecontrol instruction decides to use TLM0 or TLM1. The chosen telemetry channel chooses the corresponding analogue or data channel to take the sample according to an asynchronous serial telemetry request and packages data to the OBC or Umbilical according to the asynchronous serial protocol.

SCMC515C of TLM1 has a CAN processor of its own. SCM515C receives telemetry requests from the CAN bus through a CAN transceiver 82C250. TLM1 chooses the corresponding analogue or data channel to take the sample according to the telemetry request of the CAN bus and packages data according to the CAN protocol. Finally, TLM1 sends these to the CAN bus through the CAN transceiver 82C250.

SCMC515C can also receive telecontrol instructions from the CAN bus. SCMC515C reorganizes telecontrol instructions according to an asynchronous serial protocol and outputs to the 03decoder of the telecontrol system by an asynchronous serial port for instructions after decoding.

4.7.3.2 Measurement of Digital and Telecontrol Instruction Recovery

Digital input is sent to FPGA0 and FPGA1 simultaneously after buffering. Telecontrol output instructions after being decoded, forming 8 bits of data, are sent to FPGA0 and FPGA1 simultaneously. Data of A/D conversion, telecontrol instruction recovery, and digital input inside the FPGA are chosen by the relevant address of a multichannel data multiplexer and then provided to the relevant telemetry SCM for data framing.

4.7.4 FPGA Module Configuration

FPGA0 and FPGA1 are the core of this system. Except the port, all the digital logic circuits and work sequence circuits are achieved by programming on the FPGA. The FPGA uses the A54SX32A2PQ208M antifuse device from ACTEL to ensure the stability of the system. The main purpose of designing two FPGAs is to achieve redundancy since the logic of FPGA0 and FPGA1 is the same. In view of the connection parts between FPGAs and telecontrol, telemetry has been introduced; we mainly introduce the configuration circuit of the FPGA. The configuration circuit of the FPGA is shown in Fig. 4.9.

Both of the FPGAs are designed with three independent watchdog circuits to reset the three decoders. Corresponding watchdog outputs are put together by a tristate gate to the SCM. If a decoder has no instruction output in the setting time, the corresponding watchdog sends a reset pulse to reset the decoder.

The switch of the two FPGAs depends on the instruction to the tristate output gate. When the instruction is 0, the output of FPGA0 is valid while FPGA1 is in a high-impedance state. When the instruction is 1, the situation

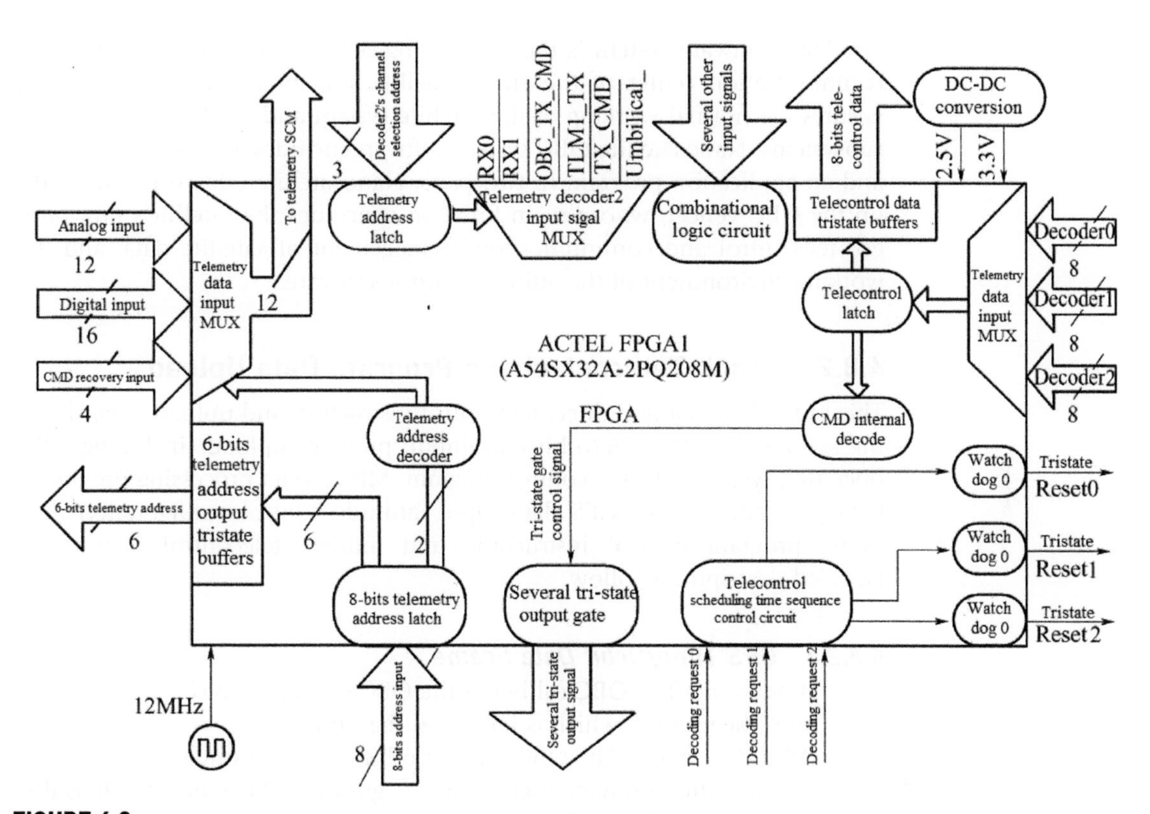

FIGURE 4.9
Configuration diagram of the FPGA.

is contrary to the above. In order to reduce the influence of any one failure to another, we concatenate 4.7 kΩ resistance on connected pins of two FPGAs before connection.

4.7.5 Voltage Conversion Module

The power supply voltage of the core inside FPGA is 2.5 V, while the power subsystem provides 5 V. As a result, we design two lines of the same voltage conversion circuit from 2.5 to 5 V. The two FPGAs supply power separately, which is also designed with double redundancy.

4.8 OBC SOFTWARE REQUIREMENTS ANALYSIS

4.8.1 Requirement Specification

The OBC software system is the center which coordinates each subsystem on the star and the bond which effectively connects the ground and the satellite.

The OBC software system is the dispatching center, management center, and communication center of the entire satellite system. Its efficient operation ensures the normal working of the satellite system and the function of other subsystems. Nanosatellite OBC system software includes a bootstrap program and an application program in two parts. Nanosatellite OBC software works on the OBC veneer, whose main function concludes the satellite's task project, telecontrol and communication, management of satellite data, and the working environment of the attitude control software.

4.8.2 Uplink Telecontrol and Program Data Upload

This can uplink many indirect telecontrol instruction and upload data, showing as follows: GPS time/orbit data input; program upload (including OBC operating system, attitude control program, MIMU data processing program); GPS program upload; GPS data input; attitude control configuration data input; program control instruction; and indirect telecontrol instruction. Detailed descriptions follow.

4.8.2.1 GPS Time/Orbit Data Frame

1. Time frame: The OBC calibrates the OBC clock after receiving GPS time from the ground, which is known as the initial time.
2. Orbit data frame: The OBC transfer orbit data to attitude control software after receiving them from the ground, which are known as the initial values of GPS data.

4.8.2.2 Program Upload

The operating system and applications can be uploaded to prevent the failure of application and guiding from the OBC EPROM, as well as to add new functional modules to the application.

4.8.2.3 GPS Program Upload

The OBC receives the GPS program and sends it to the GPS receiver by the CAN bus.

4.8.2.4 GPS Receiver Data Input

The OBC receives GPS data input from the ground and sends it to the GPS receiver by the CAN bus.

4.8.2.5 Attitude Control System Data Input

The OBC receives attitude control system data from the ground and sends it to attitude control software.

4.8.2.6 Program Control Instruction

The program control instruction specifies an absolute GPS time and executes some switch instructions. These instructions are sent to the TTC's telecontrol equipment by the OBC's CAN bus or operated by the OBC.

4.8.2.7 Indirect Telecontrol Instruction

The OBC receives indirect telecontrol instructions from the ground and sends it to subsystem.

4.8.2.8 Request Packet Downlink

The downlink request packet includes:

1. GPS data request;
2. Camera data request;
3. Attitude control configuration data request;
4. WOD data request;
5. MIMU data request;
6. Current position/velocity vector request;
7. Attitude control log file;
8. OBC log file.

4.8.2.9 Data Flow Diagram

The data flow diagram shown in Fig. 4.10 is after OBC software runs normally.

4.9 SOFTWARE SYSTEM DESIGN

The OBC software is divided into boot program and application program parts, which are described in the following.

4.9.1 Bootloader Design

The bootloader mainly works in initializing OBC hardware, receiving boot commands from the ground, and making the operating system and application of the OBC run well. The bootloader is solidified on the PROM of the satellite and is important for OBC applications. Consequently, the bootloader uses the procedure-oriented basic programming method rather than the operating system.

4.9.2 OBC Power-On Boot Process

The OBC starts to execute the initialization program from PROM, which includes initializing the register of the CPU and so on after OBC power on. THe OBC moves the bootloader from PROM to RAM and starts to run the

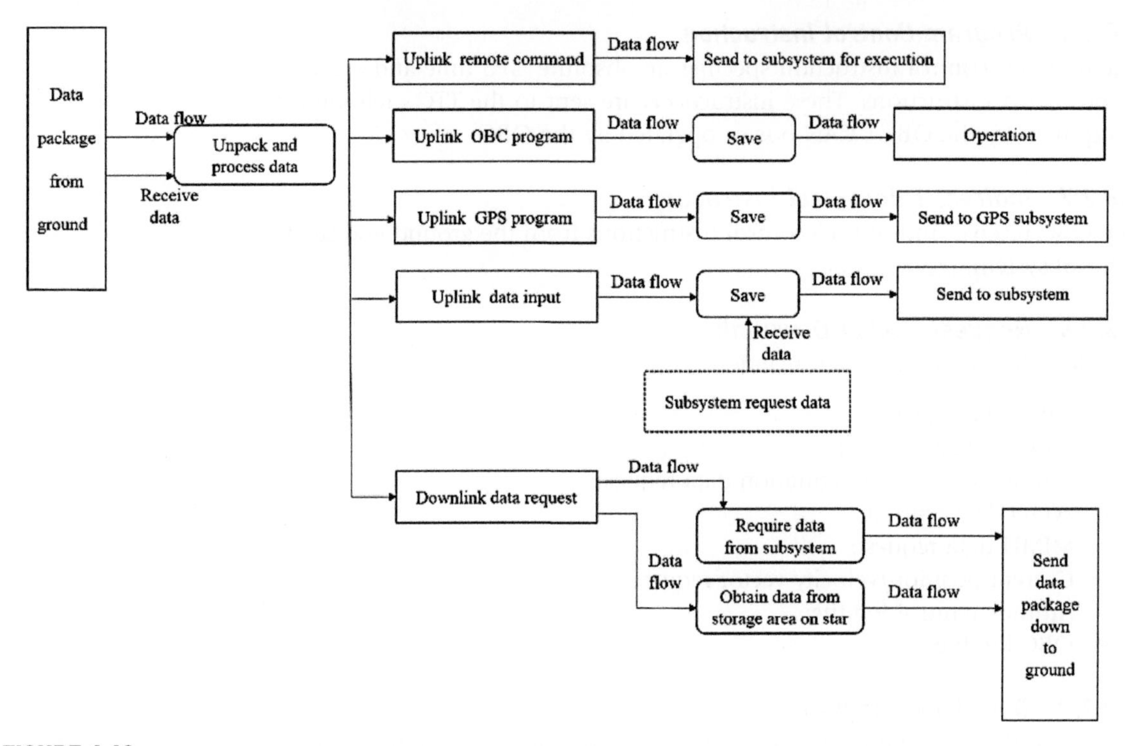

FIGURE 4.10
The data flow diagram on the star.

bootloader after initializing hardware. The bootloader sends a self-check state to the ground constantly, then, the ground sends instructions to the OBC by a synchronous serial port informing the OBC to boot the operating system and application from PROM or to upload from the ground.

If the OBC boots from PROM on star, the OBC will copy the operating system and applications from PROM to RAM and then execute in the RAM.

If the OBC uploads a program from the ground, it will run synchronous communication protocol and start to receive the operating system and applications from the ground. Then, it will put the program into RAM and execute. Because the overhead time of the satellite is limited, we will divide the program into several data blocks and upload by block if the overhead time is too short to upload all of the program. Finally, we integrate on star.

The space allocation of PROM and the program storage area are shown in Fig. 4.11.

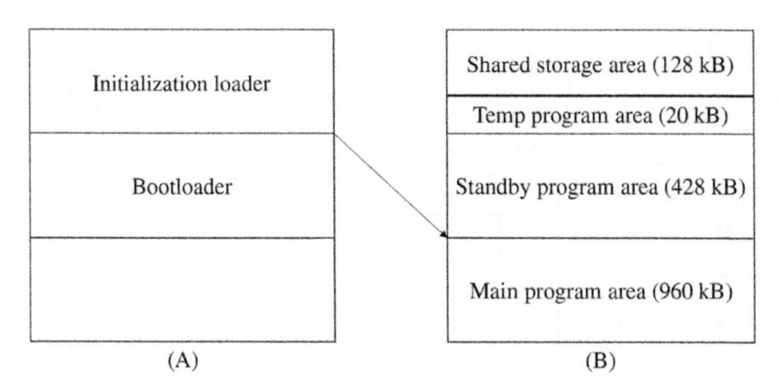

FIGURE 4.11
Allocation of storage space. (A) EPROM space allocation, (B) program storage area space allocation.

4.9.3 Bootloader Flow

The bootloader flow is shown in Fig. 4.12.

4.9.4 Application Design

The application of OBC software develops on the platform of a Micro-Controller Operating System (uc/os) embedded real-time multitask operating system. The main function process shows as Fig. 4.13.

4.9.5 Task Partitioning

After the software runs, it creates nine tasks:

1. Time disposal task;
2. Synchronous data distributing task;
3. Attitude controlling task;
4. Synchronal receiving task;
5. Synchronously transmitting mission;
6. Camera transmission mission;
7. GPS data injection mission;
8. WOD data storage mission;
9. Real-time telemetry mission.

Their priority is reduced in turn [3].

4.9.6 Mission Description

The mission function is described in Table 4.2.

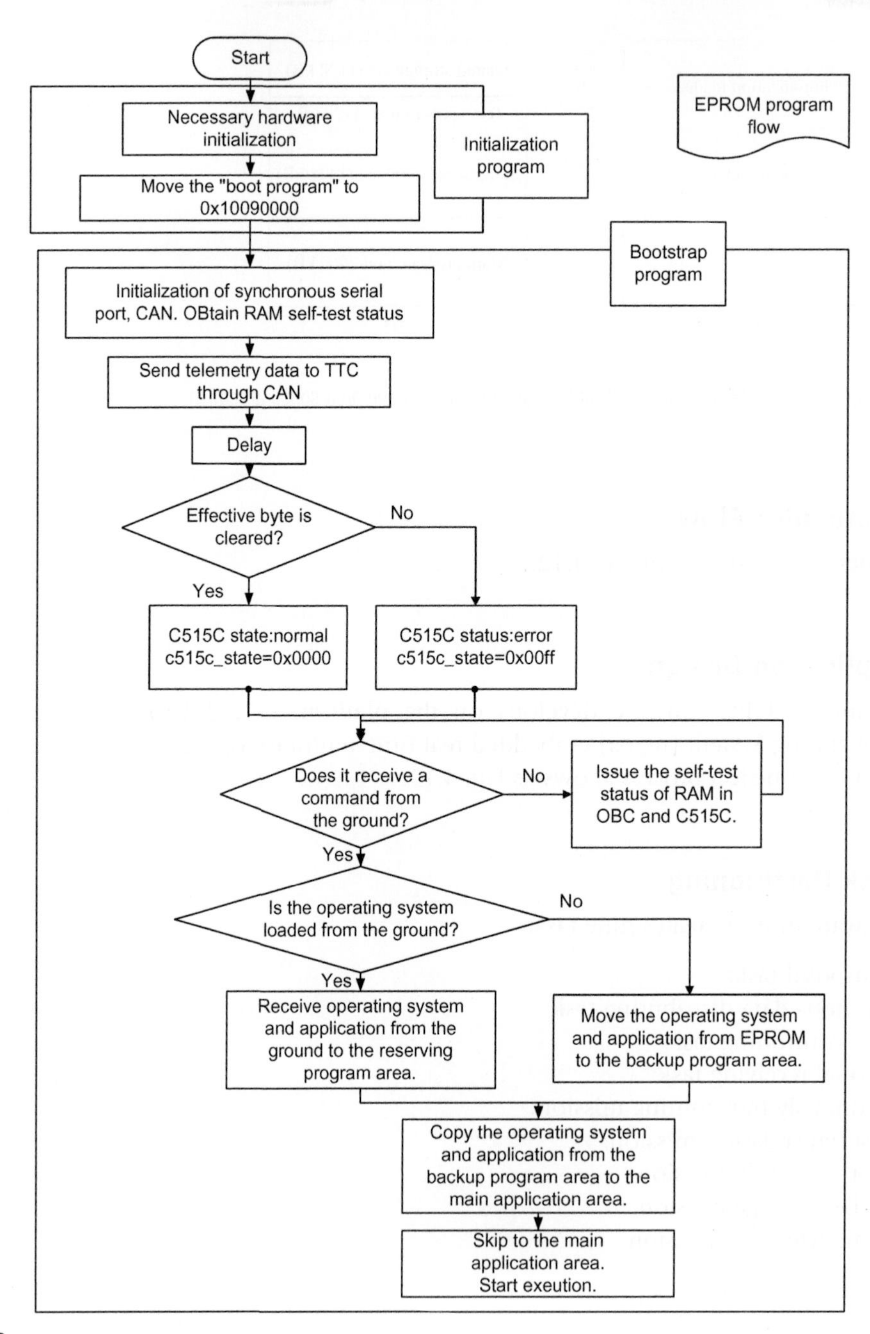

FIGURE 4.12
Flow chart of the bootstrap program.

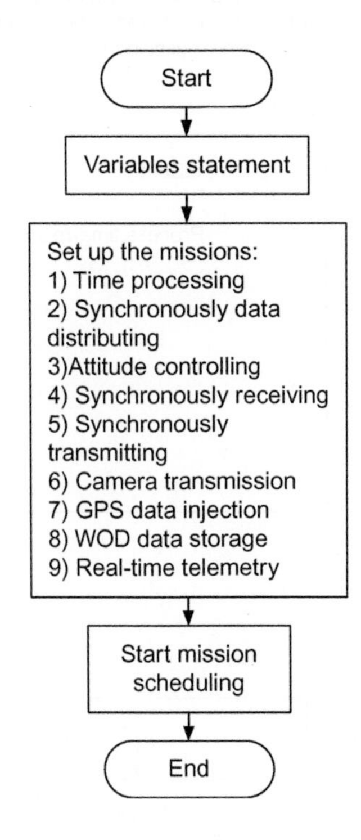

FIGURE 4.13
Flow chart of the main function.

Table 4.2 Mission Functional Description

Mission Name	Operating Conditions	Functional Description
Time processing	To one second	Perform all time-related operations: check if there is a timeout of the transmitted data or check if there is a programmed instruction execution time
Synchronously data distributing	Synchronous serial data received	Send different types of packets to different message queues
Attitude controlling	Attitude controlling operation period (1–10 s)	Calculate the attitude controlling
Synchronously receiving	Receive data packets from the ground	Receive uploaded data and unpack the data to process, as the uploaded data are a command or the injected program data
Synchronously transmitting	Send data to the ground	Send data to the ground by the synchronously communicating protocol
Camera transmission	The camera is finished	Save photos from the camera to the memory of the OBC
GPS data injection	GPS requests OBC for data injection	When GPS is on or reseted, it requests the OBC for data injection of GPS time and the number of tracks. Then the task listens to the GPS request. If there is a request, it injects data through the CAN into the GPS
WOD (World Orbit Database)	To cycle time of WOD storage	According to the storage cycle timing acquisition of WOD, the collected data, including TTC indirect telemetry data and telemetry data for attitude control, will be stored in the WOD storage area (1 M); when a WOD is requested on the ground, the data packet in the request time range will be sent to the ground
Real-time telemetry		Issue real-time indirect telemetry data and telemetry data for attitude control of TTC to the ground

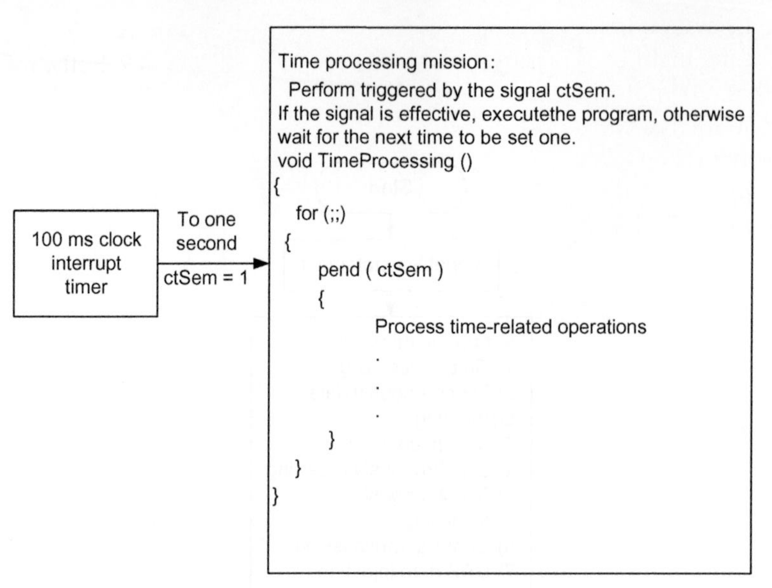

FIGURE 4.14

Static structure diagram of the time processing mission.

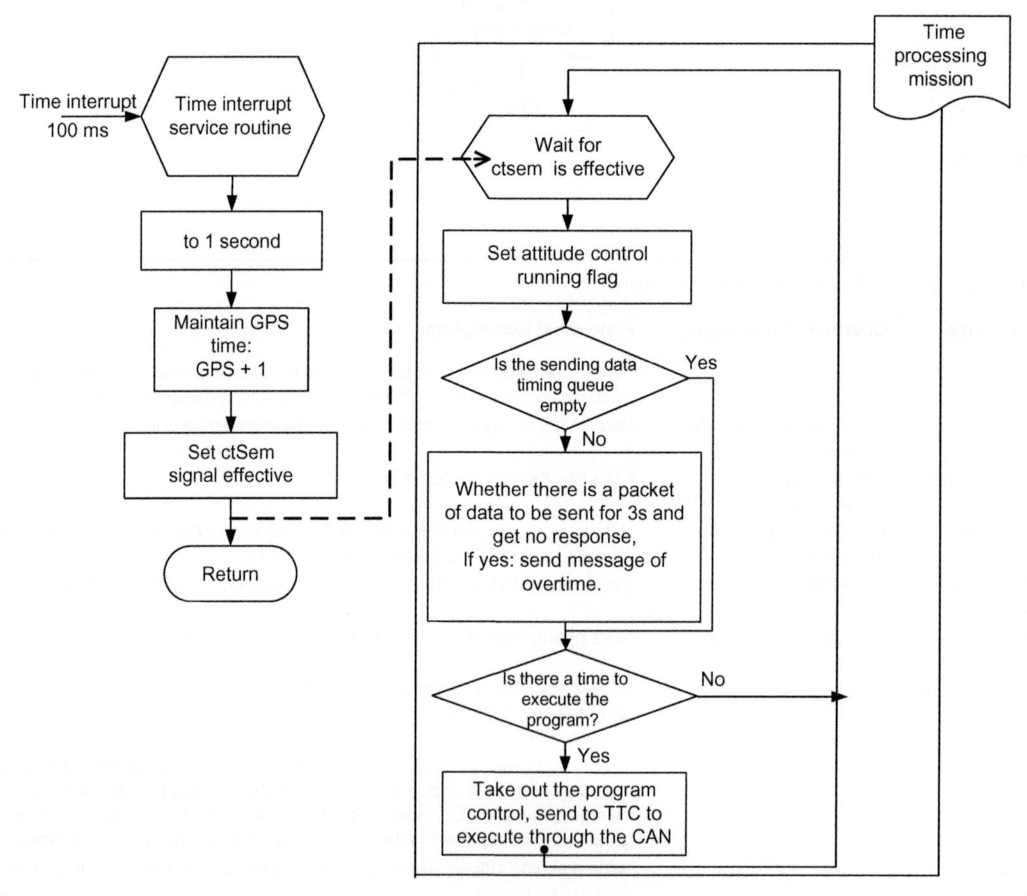

FIGURE 4.15

Flow chart of the time processing mission.

Table 4.3 Time Processing Mission

Mission Name	Timer_isr
Functional description	Set flag running of attitude controlling; check if synchronization is sent overtime; check if there is a program controlling instruction to be executed
Operating conditions	To one second.

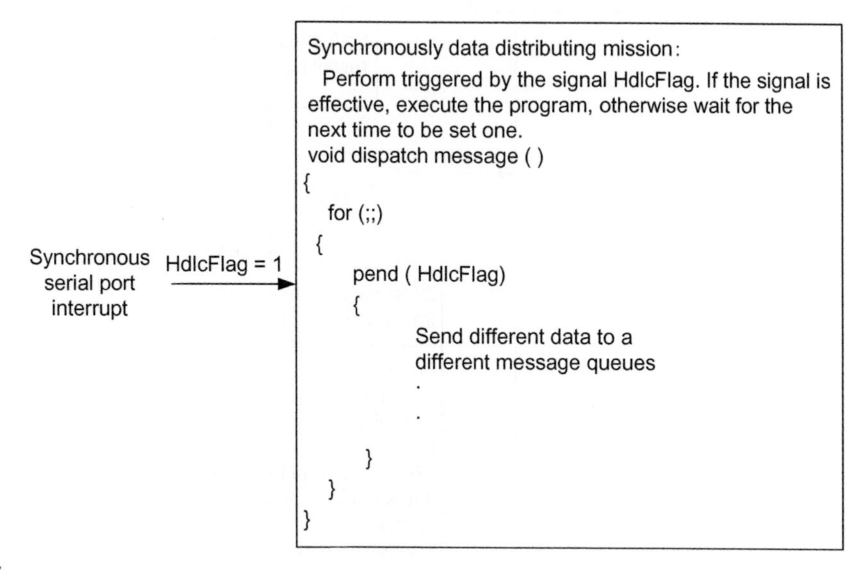

FIGURE 4.16
Static structure diagram of a synchronous data distributing mission.

4.9.7 Mission Design

1. Time processing mission (Figs. 4.14 and 4.15, Table 4.3)
2. Synchronous data distributing mission (Figs. 4.16 and 4.17, Table 4.4)
3. Attitude controlling mission (Figs. 4.18 and 4.19, Table 4.5)
4. Synchronously receiving mission (Figs. 4.20 and 4.21, Table 4.6)

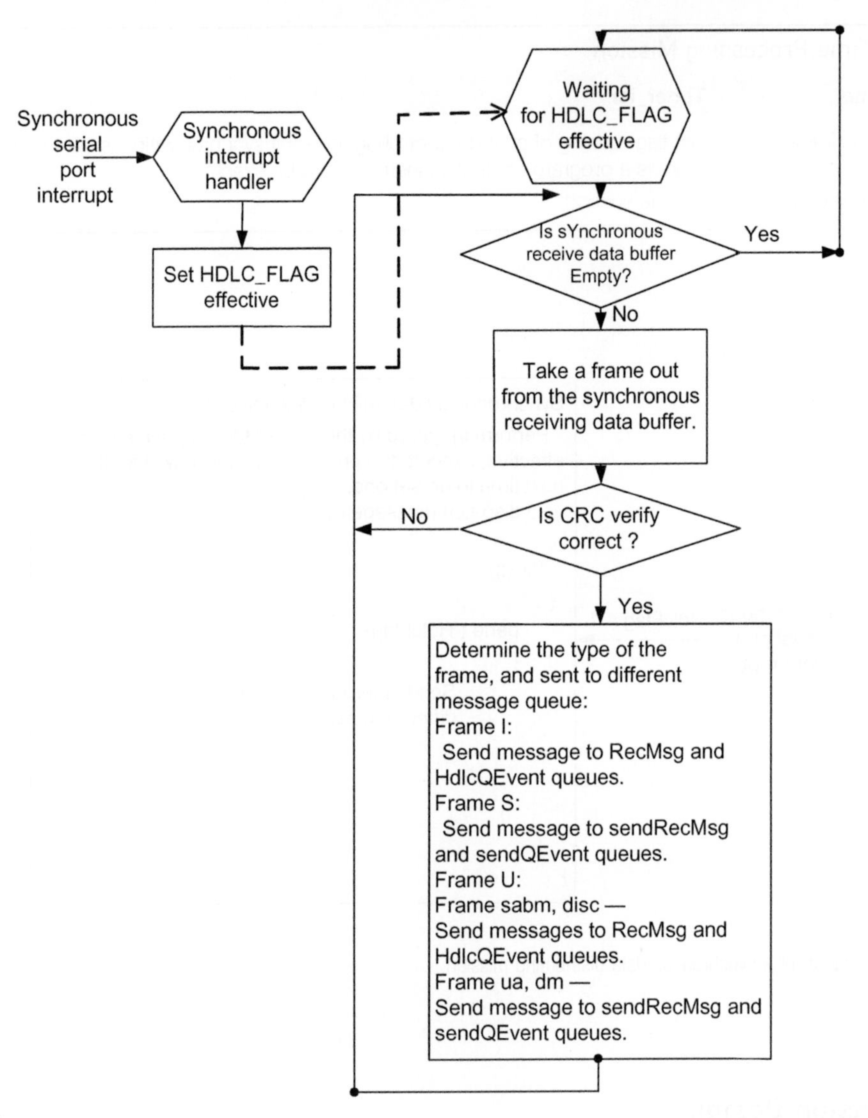

FIGURE 4.17
Flow chart of a synchronous data distributing mission.

Table 4.4 Synchronous Data Distributing Mission

Mission Name	Dispatch_message
Functional description	Analyze synchronous data frame, send different message queues, based on a data frame, supervisory frame, or unnumbered frame
Operating conditions	There are data in synchronous serial port.

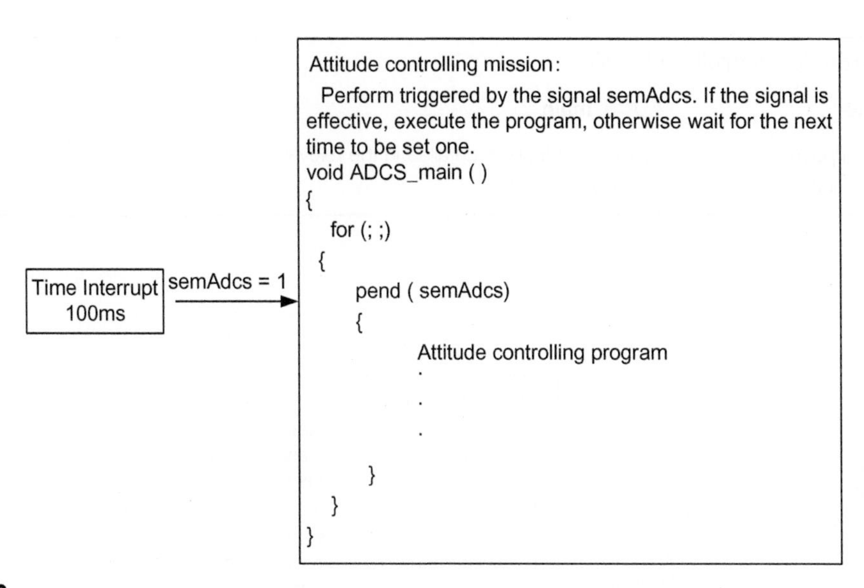

FIGURE 4.18

Static structure diagram of an attitude controlling mission.

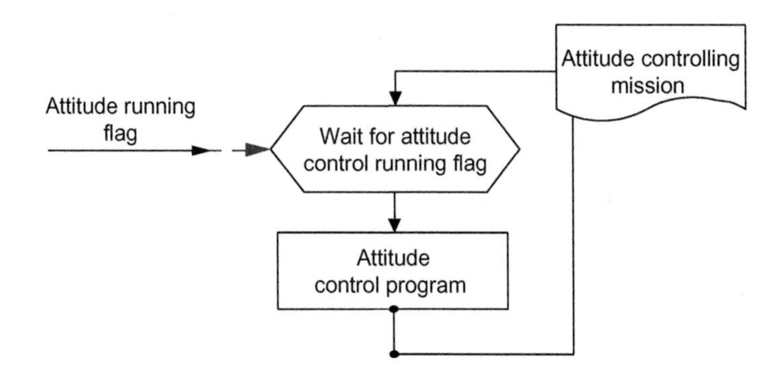

FIGURE 4.19

Flow chart of an attitude controlling mission.

Table 4.5 Attitude Controlling Mission

Mission Name	ADCS_main
Functional description	Attitude controlling, detailed in the relevant attitude control software documentation
Operating conditions	To a control cycle

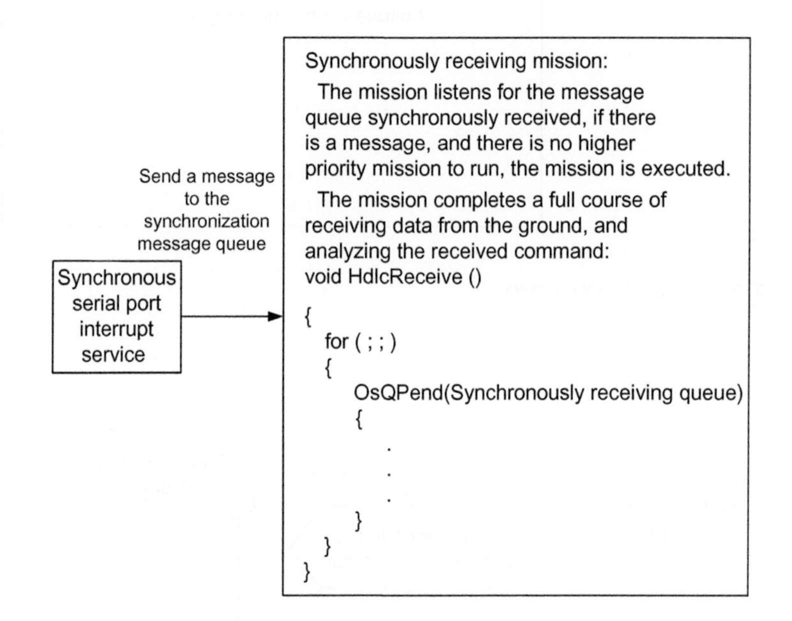

FIGURE 4.20
Static structure diagram of a synchronously receiving mission.

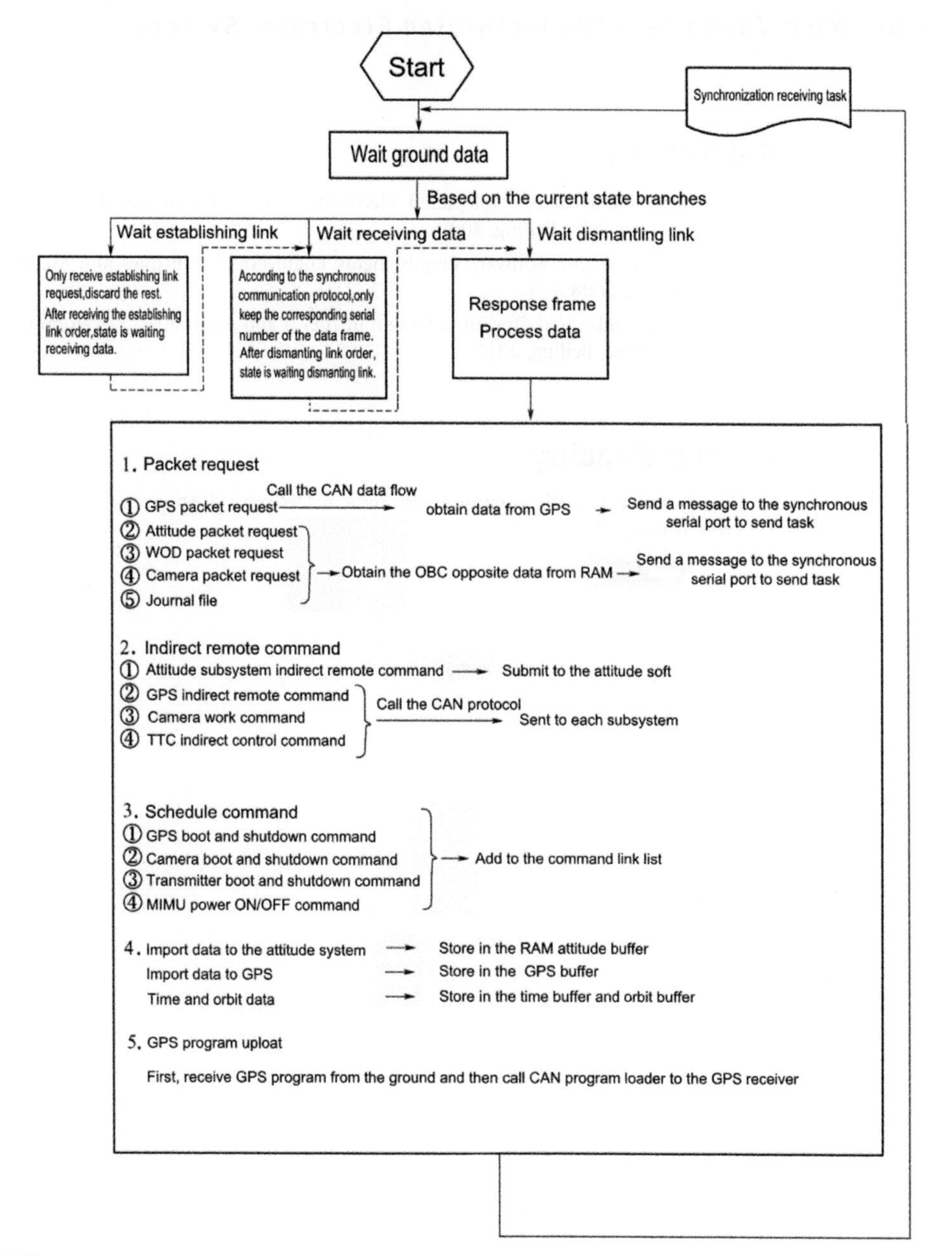

FIGURE 4.21

Flow chart of a synchronously receiving mission.

Table 4.6 Synchronously Receiving Mission

Mission Name	HdlcReceive
Functional description	Run a synchronous communication protocol, receive a complete data stream, and analyze the data
Operating conditions	There are messages in the synchronously receiving message queue

References

[1] H.X. Tian, Small Satellite Integrated Electronic System and Its Key Technology Research, Tsinghua University, BeiJing, 2009.

[2] Z. Jin, L. Liu, Y. Jin, Software Requirements Engineering: Principles and Methods, Science Press, BeiJing, 2008.

[3] X.W. Wang, Embedded Operating System μC/OS-II and Application Development, Tsinghua University Press, BeiJing, 2012.

Further Reading

J.M. He, Embedded 32-Bit Microprocessor System Design and Application, Electronic Industry Press, BeiJing, 2006.

CHAPTER 5

Ground Tests of Micro/Nano Satellites

After each subsystem and component of the micro/nano satellite has successfully passed the subsystem-level test, the joint test has been qualified, and been assembled in the satellite body, the performance and functionality of the satellite will be done for the full test, including an electrical interface test between each subsystem and an electromagnetic compatibility test. This is in general called a ground electrical comprehensive test. The purposes of the ground electrical comprehensive test are as follows:

1. Test if the main performance and functionality of all satellite subsystems meet the design requirements;
2. Correctness and compatibility test on the interfaces between all satellite subsystems;
3. Correctness test on the interface and communication protocols between EGSE (electrical ground support equipment) and subsystems of the satellite;
4. Correctness test on the EGSE software;
5. Test if the satellite meets the overall technical indicators, especially whether the technical indicators remain unchanged or not after various virtual environment tests.

The ground electrical comprehensive test is an important and indispensable step in research and manufacture on satellites. Through it, researchers can discover: technical indicators which may not meet the design requirements, imperfect functions, electrical interfaces which do not match requirements, design defects, and so on. And this work can help designers to improve the design accordingly. Finally, based on these test data, designers can effectively improve match, compatibility, and reliability, eliminate hidden troubles, and ensure the flying success of satellite [1].

Space Microsystems and Micro/Nano Satellites. DOI: http://dx.doi.org/10.1016/B978-0-12-812672-1.00005-9

5.1 TESTING PHASES

According to the development procedure for satellites, the ground electrical comprehensive test is divided into nine stages—from desktop testing to testing in the launch site area. The stages of testing are shown in Fig. 5.1.

5.1.1 Desktop Testing

The purposes of desktop testing are as follows: checking if the main performance and functionality of all satellite subsystems meet design requirements; checking if correctness and matching of interfaces meet the requirements of each subsystem; checking the correctness and rationality of inspection and certification systems of satellites and ground test equipment interface design; finding design defects and process defects of the satellite and testing system; verification of a correct and rational design process.

Desktop joint testing includes: testing on the subsystem, matching tests between subsystems and the EGSE, testing on minimum systems of the satellite, and system-level testing.

5.1.2 Testing in Environment Experiments [2,3]

5.1.2.1 Mechanical Environmental Testing

The purposes of mechanical environmental testing are as follows: verification of the adaptability of all kinds of equipment and instruments which may work properly in the mechanical environmental, and obtaining relevant parameters in this environment. Because satellites do not work and power on in the course of the flight of the rocket, the satellite can autonomously power on after the separation. And the satellite do not power on in the mechanical environmental phase. In the course of mechanical environmental testing, the main work of the electrical test is: status determination before the testing, not working in testing, and review status after testing. There are two aspects to note:

1. Power on before the environmental testing and confirm that the satellite is in normal condition;
2. Power on after the environmental testing and confirm that the satellite is in normal condition.

5.1.2.2 Thermal Cycle Testing and Thermal Vacuum Testing

The purposes of thermal cycle testing and thermal vacuum testing are as follows: verification of the adaptability of all kinds of equipment and instruments which may work properly in thermal environmental, obtaining the temperature data and changing rules for key equipment if needed. The main tasks are as follows:

1. Power on before testing and confirm that the satellite is in normal condition;

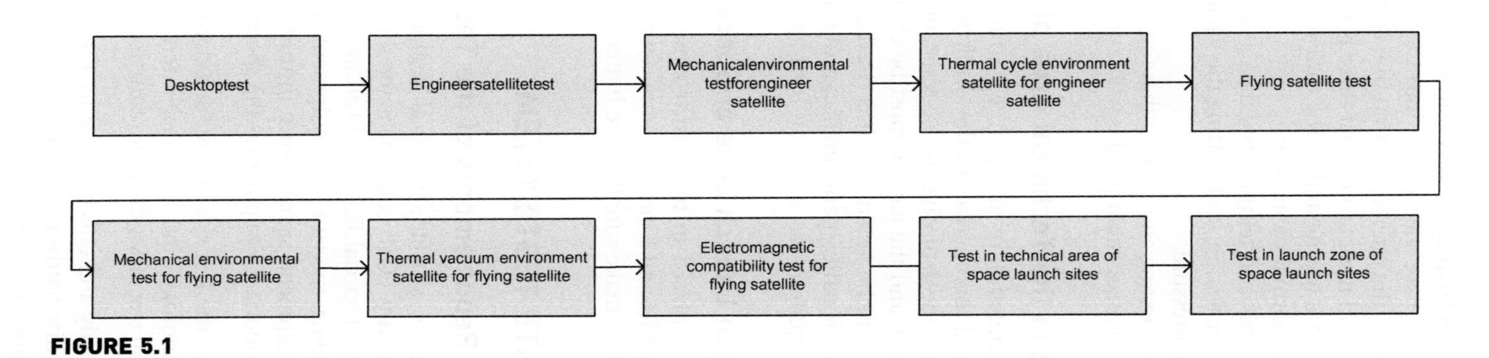

FIGURE 5.1
Procedure of the ground electrical comprehensive test.

2. Set up the high-frequency cable and normal cable through the bulkhead, and place the satellite into the testing jar;
3. To confirm that the satellite is in normal condition, the vacuum canister is evacuated, and it is ensured that there is no vacuum discharging;
4. Power on before testing and in the canister and confirm that the satellite is in normal condition;
5. Recording status and temperature data at all stages in testing;
6. Power off after testing, and putting the vacuum canister back to normal atmospheric pressure;
7. Power on after testing and confirm that the satellite is in normal condition;
8. Summary and assessment for the whole test.

5.1.3 Testing in the Technical Area of Space Launch Sites

1. After the completion of transport, each function performance parameter and the interface of the whole satellite must be rechecked and reviewed; the explosive device resistance and switch must be acknowledged; and the battery must be charged;
2. The satellite will be tested after it is fixed on to the rocket;
3. Confirm functionality, performance, and the status of compliance with the design requirements, and wait for transition.

5.1.4 Testing in Launch Zone of Space Launch Sites [4]

When the rocket is placed in the launch zone of the space launch site, the researchers must review all functions, parameters, and interfaces of the entire satellite, and, finally, make sure that the battery is fully charged.

5.2 SATELLITE TEST SYSTEM

5.2.1 Design Requirements of the Testing System [5]

In order to complete satellite's electrical testing tasks, we must design and configure our necessary EGSE equipment for the satellite. Generally, the ESGE equipment consists of special EGSE and various general purpose test equipment components, including: special EGSE equipment, satellite simulator, sun simulator, general test equipment, signal source instrument, C spectrum analyzer, and so on. The most important equipment is the special EGSE equipment.

The design rules of the test system are "standardization, universal, compact," sophisticated technology, and universal interface. According to the requirements of the overall test tasks, the main functions of the EGSE are as follows:

1. It can correctly and effectively verify the correctness of the subsystems on the satellite hardware and software electrical function;
2. It can provide power to each subsystem and control it;

3. It can verify compatibility and correctness of interfaces between the various subsystems, send telecontrol commands and inject data, and verify the results of commands in the satellite;
4. It can receive and demodulate telemetry signals, analyze telemetry data and display it in real-time, record continuously working data, playback data offline, analyze data and back it up, and print data information;
5. It can self-calibrate and self-test;
6. It is capable of manual operation, with timely charge and power;
7. It can test the whole satellite at every stage;
8. It has a flight simulator and operate flight test program;
9. It is capable of error detection;
10. It has modular design, and it is easy to carry out maintenance.

5.2.2 The Composition of the Satellite Test System

The ground electrical comprehensive test equipment is made up of special EGSE equipment and various general purpose test equipment components. During the AIT stage of satellite development, the power of each subsystem or component is supplied by the EGSE equipment, which is generated by the sun simulator and assigned by the EGSE switchbox, and the battery is charged by sun simulator. The EGSE control-computer send telecontrol commands, and the EGSE telemetry-computer receives the telemetry data and resolves them. All computers can communicate by TCP/IP protocol.

The components of ground electrical comprehensive test equipment are shown in Fig. 5.2 and Table 5.1. In Fig. 5.2, the internal block is EGSE equipment. EGSE has two test loops, one is by cable, and the other is RF. The next section will introduce these in detail.

5.2.3 Loop Selection for a Ground Electrical Performance Test

For the satellite electrical performance test system, an uplink is used to control and excite the measured object and a downlink is used to collect the corresponding situation and to gather and process the output data of instruments and equipment. For the ground integrated test, the calibration procedure of the satellite system is carried out step by step through a minimum testing path.

Uplinking and downlinking with different properties consist of different testing paths, having different characteristics and application scenes. In the following, we will introduce several typical testing loops.

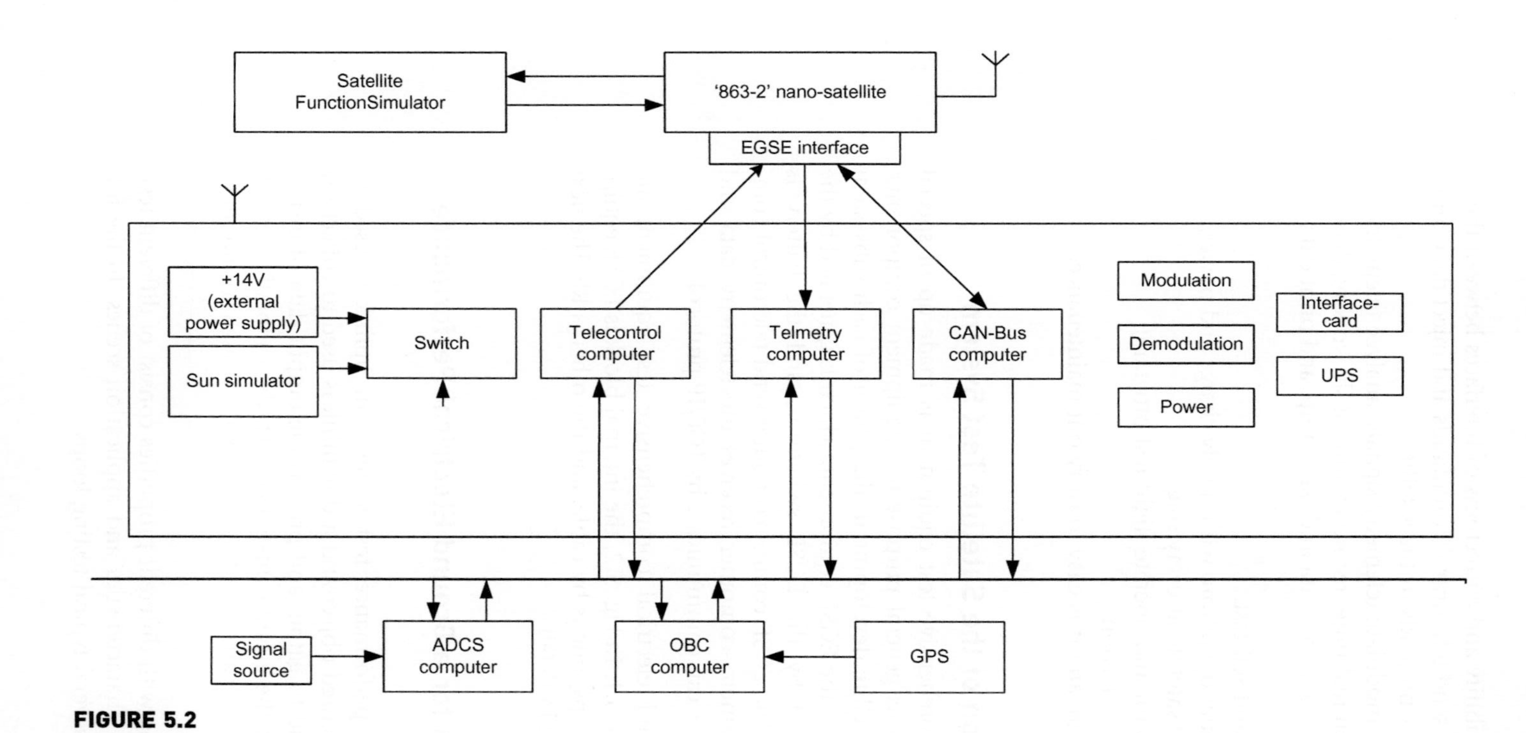

FIGURE 5.2
Main configuration diagram of the EGSE equipment.

Table 5.1 The Components of EGSE

No.	Name	Function
1	Baseband section	• (255.223) error correction code, R–S code, (1/2,7) convolutional codes • Function of forwarding data
2	Transceiver and receiver	• Receive RF signal from 2200 to 2300 MHz, and track it. PCM/CPPSK modulation, receiver noise coefficient is better than 1.5 dB • Receiver input dynamic range: better than 60 dB • Transceiver RF signal from 2025 to 2125 MHz. PCM/DPSK/PM modulation, maximum output power better than 0 dBm
3	Multifunction box	• Modulation (232-PCM-BPSK), demodulation (PCM-232), TLC, TLM, switch, pseudo-random code generation, error coder test, TLM signal source
4	Charge box	• Power supply, constant current output • Charging timing
5	Switchbox	• Switch for power for satellite and charging of battery • Display of charge voltage
6	Main cabinet	• Carrier for whole EGSE

5.2.3.1 TM/TC Testing Loop

The TM/TC testing loop is the main loop, which is used by each subsystem to the maximum extent. Since this testing loop is quite similar to the orbital state of the satellite, the TM/TC computer of the EGSE is mainly responsible for processing the data of this loop.

The TM/TC testing loop consists of measurement and control equipment of the ground test as well as the TTC and RF subsystems. The uplink implements the direct transmission and verification of EGSE instructions such as telecontrol commands, time/orbital data, attitude control configuration, equipment state parameters, etc. The downlink EGSE implements satellite telemetry data receiving, real-time processing, and measurement record.

To ensure the normal operation of this loop, two distinct approaches are required to check the smoothness of the uplink and downlink communications.

1. Carrier signals sent from the satellite-borne transmitter are demodulated into video telemetry PCM code signals via the EGSE wireless measurement and control subsystem, and sent to EGSE telemetry front end, processed and displayed by the telemetry computer. In the mean time, the binary phase shift keying (BPSK) signals sent from the EGSE telecontrol front end are modulated into carrier PM and then sent to the satellite. The procedure mentioned above is the main loop measurement between the satellite and ground (RF + TTC). Applications of this testing loop include:

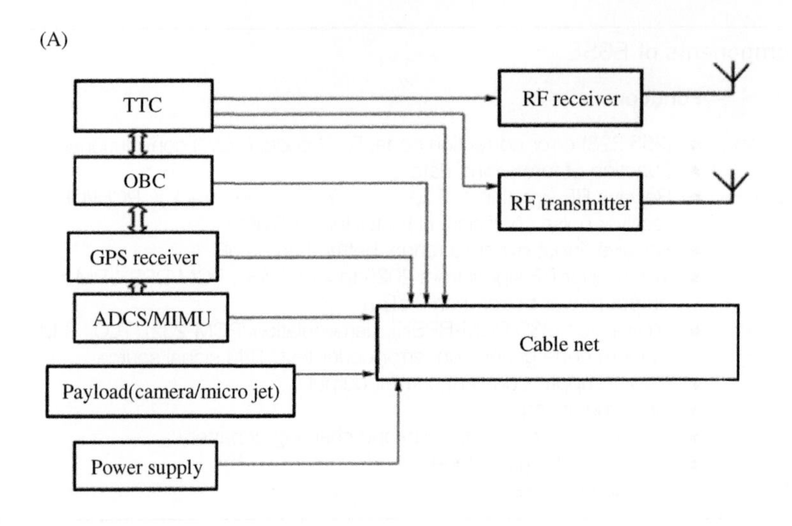

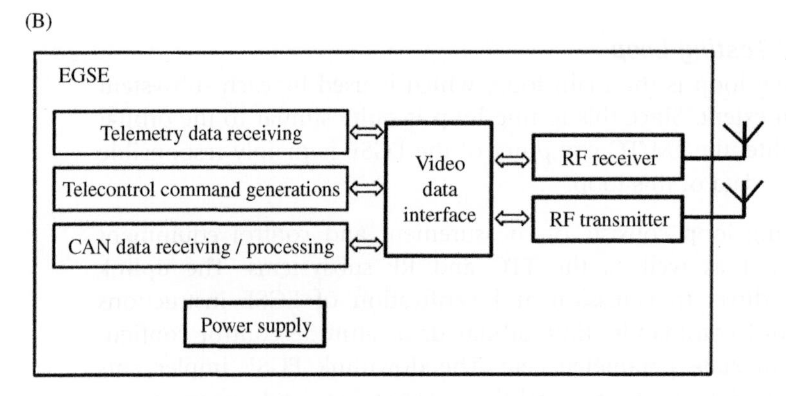

FIGURE 5.3

Principle block diagram of an RF wireless test. (A) Satellite internal block diagram; (B) EGSE block diagram (wireless signal transmission).

receiving and processing of telemetry data, generation and transmission of telecontrol commands, testing of satellite-borne telemetry and telecontrol devices, monitoring and testing of each subsystem's telemetry data and working conditions, transmitting control commands to each subsystem via onboard telecontrol devices and obtaining the situation of onboard devices by loop comparison. Furthermore, the control strategy and logic of OBC and ADCS subsystems are inspected by program uploading and comparison. Fig. 5.3 shows the RF wireless testing principle.

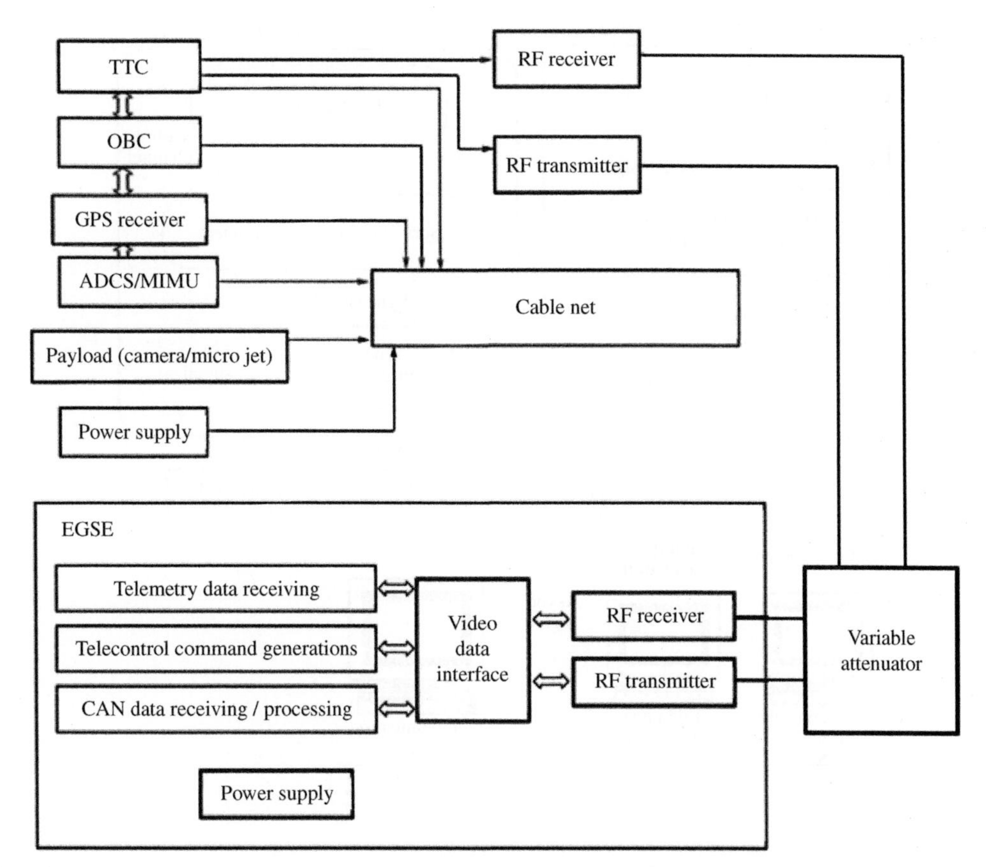

FIGURE 5.4
Principle block diagram of an RF system test.

2. In the case of interference in the wireless channel, the EGSE sends the telemetry BPSK signals directly from the satellite to the telemetry front end through the signal attenuator and cables, and meanwhile uses cables to send the telecontrol BPSK signals directly from the telecontrol front end to the satellite. This part is mainly used to test the high-frequency performance before the RF wireless entrance, and the performance of the antenna itself is tested under specific conditions (required in a subsystem test). In Fig. 5.4, the upper part is a satellite internal block diagram, and the lower part is an EGSE block diagram.

The testing of the TM/TC loop needs to be carried out on the satellite through wireless channels, and its internal structure is shown in Fig. 5.5.

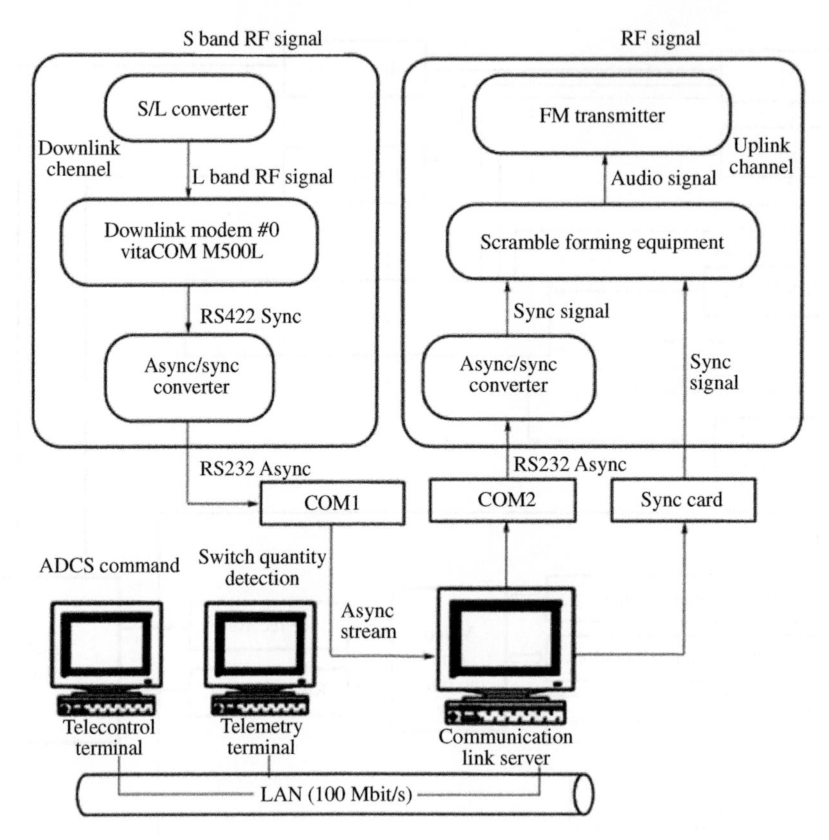

FIGURE 5.5
Principle block diagram of wireless testing system (TM/TC testing loop).

5.2.3.2 Uplink Channel Composition

The main function of the uplink channel is to provide the transmission channel for the wireless test system, including the sync card, async/sync converter (Fig. 5.6), scramble forming equipment, and FM transmitter.

1. Sync card. This is connected to the communication link server through a PCI slot, transmitting the OBC uplink synchronization signal to the scramble-forming equipment.
2. Async/sync converter. This converts the RS232 asynchronous TTC uplink signal (output from the computer serial port) to the synchronous digital signal, and then transmits it to the scramble-forming equipment.
3. Scramble-forming equipment. This equipment is ground test equipment of the satellite ground measurement and control subsystem

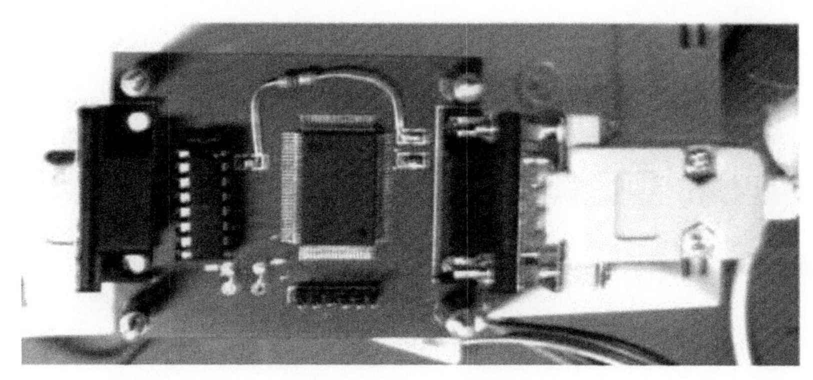

FIGURE 5.6
Async/sync converter.

FIGURE 5.7
Ground test equipment of the satellite ground measurement and control subsystem.

(shown in Fig. 5.7). Its main usage is scrambling and shaping filtering of the uplink baseband data coming from the sync card or async card, and sending the processed audio baseband signal to the FM transmitter. Related parameters are as follows.

- Output code rate: 9.6 kbit/s;
- Output voltage: 1 ± 0.5 V adjustable (shaped sine wave);
- Output impedance: ≤ 1 kΩ;
- Output scrambling polynomial: $1 + X12 + X17$;
- Working temperature: 0–40°C;
- Interfaces: Lotus head output audio signal (hole), external data input CD-9 (hole).

4. FM transmitter (in the same cabinet with downlink S/L convertor) (shown in Fig. 5.8). This is responsible for FM modulation of baseband signals. Transmit frequency point: 2094.417 MHz. Interfaces: input audio signal BNC, output FM signal BNC.

FIGURE 5.8
Nanosatellite transmitter cabinet (including FM transmitter and downlink covertor).

FIGURE 5.9
Modem Vitacom M5500L.

5.2.3.3 Downlink Channel Composition

The main function of the downlink channel composition is to provide a receiving channel for the wireless measurement system. It consists of an S/L band frequency convertor, satellite BPSK modem, and async/sync interface converter.

1. S/L band frequency convertor (integrated in the transmitter cabinet of the nanosatellite, as shown in Fig. 5.8). Due to the requirements of the modem band, signals need to be transformed by the convertor from S band to L band. The local oscillator frequency of the convertor is 3490 MHz, the center frequency of the input S band signal is 2274.48 MHz, and the center frequency of the output L band signal is $3490 - 2274.48 = 1215.52$ MHz.

   Interfaces: Input S band downlink signal BNC, output L band downlink signal BNC.
2. Satellite BPSK modem. For LEO satellites, the downlink demodulator should have fairly good anti-Doppler frequency shift ability. Since this modem continues to be used in the ground station of satellite A, BER performance should be tested under the Doppler frequency shift condition. Optional modems are CDM570L, Vitacom M5000L, and Vitacom 5500L (as shown in Fig. 5.9).

   The selected modem is M5500L from Vitacom, with main functions as below: frequency changing of L band signal, intermediate frequency amplification, demodulation, differential solution, perturbation, and

FIGURE 5.10
Synchronous/asynchronous interface converter.

FIGURE 5.11
Measurement and control terminal and network subsystem.

Viterbi decoding. The type of demodulation is BPSK, the perturbation method is IESS308.

3. Async/sync interface converter (shown in Fig. 5.10). The async/sync interface converter implements the interface conversion from the synchronous RS422 interface of downlink modem to the asynchronous interface RS232 of computer.

   Interfaces: Conversion from RS422 (25 pin) to RS232 (9 pin).

5.2.3.4 Measurement and Control Terminal and Network Subsystem (Fig. 5.11)

In the test process, the requirement on the continuity of telemetry and remote control data is not strict, and the laboratory resources are limited.

Therefore, the measurement and control terminal and the network subsystem are composed of two computers and routers for the practical system. One computer is responsible for realizing both the functions of the communication link server and the remote control terminal. The other computer is used as a telemetry terminal. The data exchange between these two computers is supported by the local area network and the serial port.

5.2.3.5 Cable Test Loop

This is a loop formed by the umbilical cord (Umbilical) of the satellite test, including the satellite power distribution control and measurement, and CAN bus monitoring and measurement. The nanosatellite completes the satellite power supply (solar simulator) control, bus voltage measurement, battery voltage measurement, and provides the TTC, OBC, and CAN bus test interface through the test umbilical cable and special test equipment interface on the satellite (Fig. 5.12).

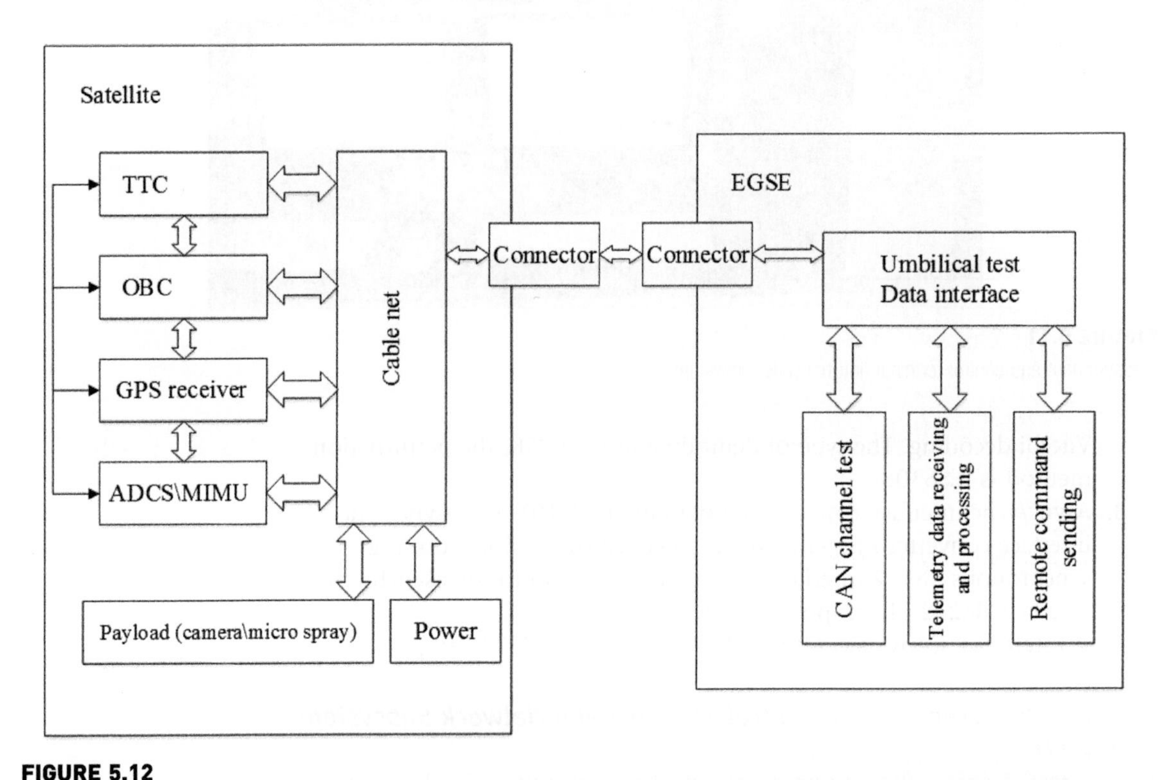

FIGURE 5.12
The principle block diagram of the system.

5.3 GROUND TESTING SCHEME

Before the test, a comprehensive test program combining the satellite overall design scheme and the data flow on the satellite should be developed (Fig. 5.13).

The main points of the design of satellite ground test program are as follows:

1. The satellite ground integrated test should focus on the function of the whole satellite and the performance test of the main part. In the test scheme the test loop and the test point should be designed.
2. Electrical performance tests include remote uplink, telemetry downlink (radio/wired) test loop, and cord cable test loop; interface matching is mainly based on an umbilical cord test in the early stages of testing; the satellite function test is mainly based on RF uplink/downlink in the late stages of testing.

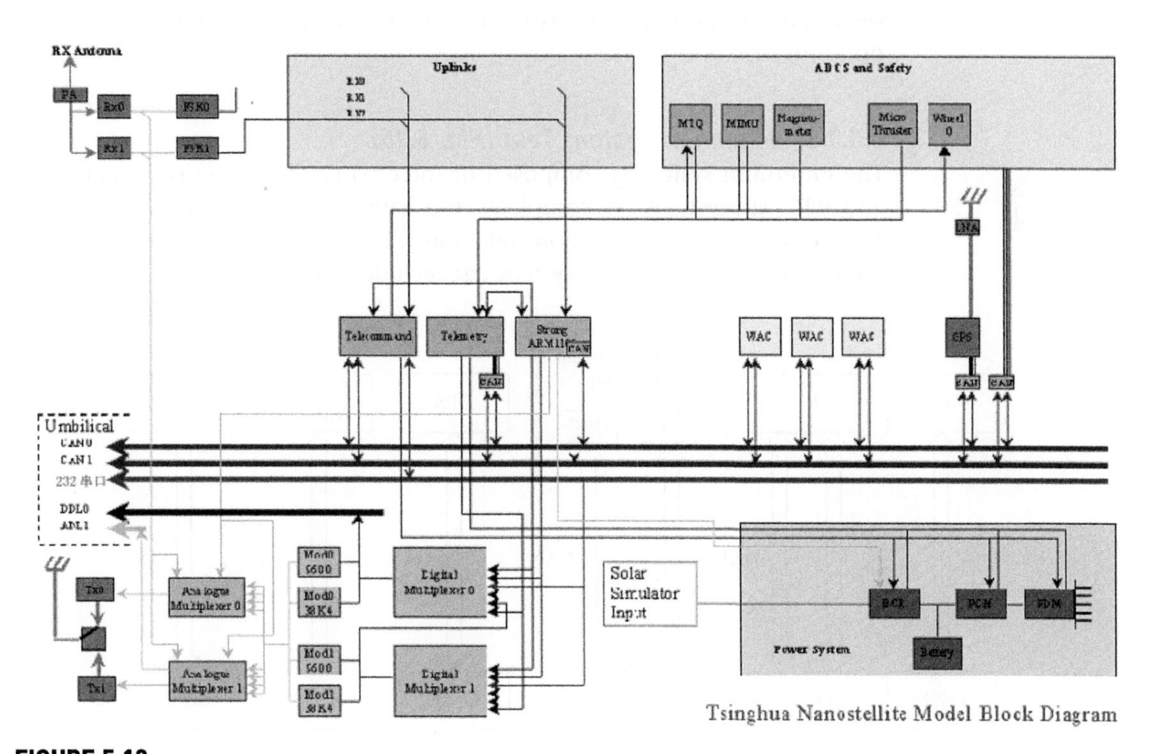

FIGURE 5.13

Satellite data flow diagram.

3. Test equipment should reflect the principle of "standardization, generalization, small metallurgical"; the test software is developed according to the software engineering method; test procedures should have the ability to distinguish faults.
4. The ability to ensure the safety of the test operator and on-board equipment.

5.3.1 Ground Integrated Test Procedure [6,7]

5.3.1.1 Acceptance Test of the Subsystem

The acceptance requirements of the power subsystem, RF subsystem, data management subsystem, GPS subsystem, attitude control subsystem, effective load (camera/micro jet) are in accordance with subsystem acceptance requirements and subsystem acceptance rules.

5.3.1.2 Matching Test for the Subsystem and EGSE (Fig. 5.14)

The matching test of each subsystem and EGSE includes the power check, the signal path inspection, the inspection of the interface, and the communication protocol.

5.3.1.3 Minimum System Test (Fig. 5.15)

The minimum system is composed of three parts: the power subsystem, RF, and TTC subsystems. It completes the satellite power supply, telemetry/remote control communication link, and it is the minimum system for satellite survival. The minimum system test includes the following:

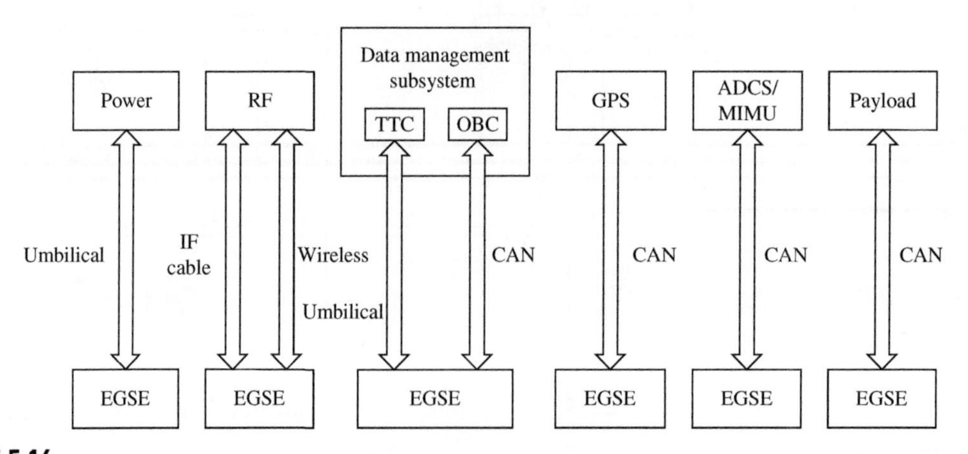

FIGURE 5.14

Principle block diagram of the matching of the system and the EGSE of the umbilical cord test.

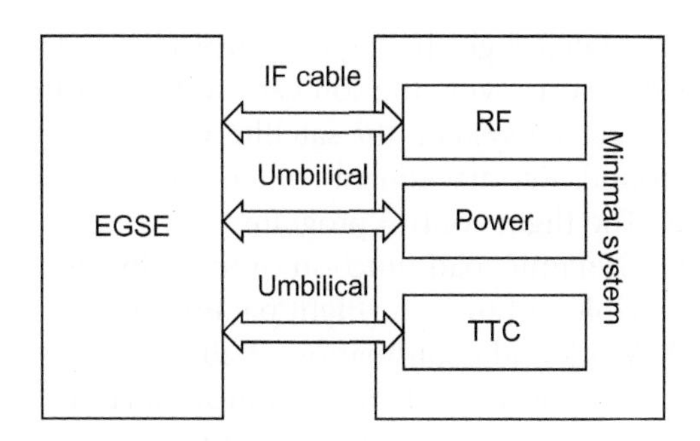

FIGURE 5.15
EGSE and the minimum system matching test block diagram.

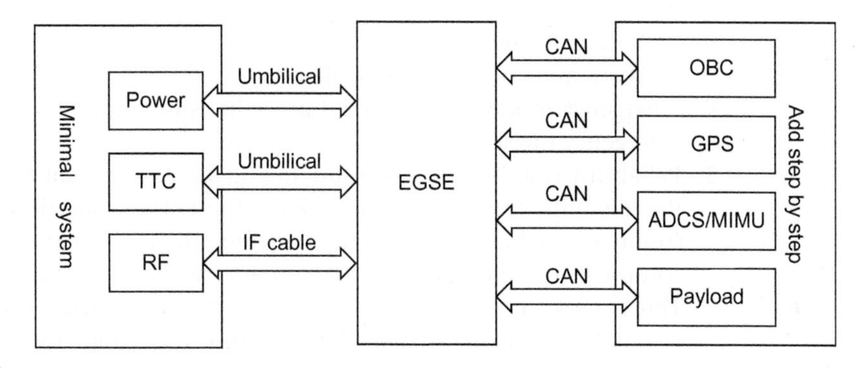

FIGURE 5.16
Block diagram of a satellite ground test.

1. Power supply of the satellite system and minimum system power consumption.
2. The performance of uplink and downlink.
3. System works and coordination check.
4. Coordination and verification of the format and content of telemetry data.
5. Receiving and implementation of remote control instructions.

5.3.1.4 System-Level Test of a Satellite

1. CAN bus test (OBC, GPS, ADCS effective load subsystem test). CAN controller sends up the instruction, OBC receives the CAN data stream; when the TM/TC loop is not ready or fails, we can use the computer RS232 serial port to monitor the telemetry data, and test OBC, GPS, ADCS/MIMU, and the payload. It also verifies the reliability of the TM/TC (Fig. 5.16).

2. Simulated flight II (flight control status). Flight control status, OBC works to provide the software and hardware platform of the attitude control subsystem, the satellite system sets up three-axes stability to the ground. Key assessment contents are as follows:
   a. The flight control program;
   b. Dynamic load function of the software module;
   c. Upload function (flight parameter binding);
   d. Work and coordination of all systems;
   e. Telemetry and remote control function;
   f. Satellite system power consumption.
3. Simulated flight III (autonomous flight), focus on the assessment of the following:
   a. Coordination, validation of satellite flight procedures;
   b. Upload function;
   c. Packet telemetry function;
   d. Work and coordination of all systems.
4. Emergency mode. Satellite flight procedures in failure mode or emergency mode.
5. Reliability test. In order to ensure the quality of the joint test, increase the reliability test content after the joint test. The inspection of power supply of the battery pack is implemented three times after the completion of the joint test in accordance with the simulated flight III. Three checks are qualified to pass. The battery-powered peak voltage test and the battery discharge lower limit test are implemented. If there is a failure, a supplementary test or long-term test should be implemented according to the specific circumstances.
6. Primary power supply test in simulated light period.
7. Combined power supply for analogue peak load in different operating modes.

5.3.2 Comprehensive Test of Ground Electrical Performance

5.3.2.1 Content of Comprehensive Testing of Electrical Properties

1. To test the correctness, rationality and matching of the overall electrical design of the satellite and the correctness of grounding system.
2. To test whether the electrical performance and parameters of each subsystem meet the requirements of the general.
3. Power supply inspection.
4. To test whether the system can complete the required functions in the system level conditions or not. Test order: power supply subsystem, RF subsystem, data management subsystem, attitude control subsystem (including MIMU), GPS subsystem, payload (including camera and microinjection).

Table 5.2 Test Schedule for Each Subsystem of the Satellite

No.	Subsystem	Test Content
1	Satellite cable network	• Contact check • Contact insulation inspection • Check interface table
2	Power subsystem	• Remote command check • Power supply inspection for each subsystem • Telemetry parameter check • Subsystem and system power consumption • Simulated peak load combined power supply test
3	Pipe system TTC	• Check the matching performance of telemetry/remote communication protocol with frame structure • Check the transmission performance of the telemetry data and the remote control instructions, etc.
4	RF subsystem	• Check the index matching performance of the RF communication link
5	ADCS subsystem	• Power supply inspection of components • The electric interface and communication interface protocol of the sun sensor • Electrical interface and communication interface protocol of the magnetometer • Electric interface and communication interface protocol of the magnetic torque converter • Electric interface of the reaction wheel, speed • The polarity of the sun sensor
6	MIMU subsystem	• Electric interface and communication interface protocol of gyroscope
7	GPS subsystem	• Electric interface and communication protocol of GPS • Check the performance of GPS data transmission
8	Camera subsystem	• According to the instructions of the TTC node controller, the storage function of the ground image and image data are completed under the control of the FPGA, and the image data are output at the request of the TTC controller • TTC node controller: complete the power switch control of the camera system and the temperature and current detection, communicate with OBC, complete the function of the camera control and data transmission • Power switch control circuit and voltage conversion circuit • Temperature and current monitoring circuit
9	Microinjection system	• A / D acquisition: Achieve the acquisition and processing of the pressure and temperature using the A/D of 515 • CAN communication: upload pressure and temperature data to TTC through CAN, get the control command of the TTC or OBC to open the engine • Engine control: Check the control instruction, execute the instruction which has passed the check

5.3.2.2 Main Test Content

1. To check the matching of the interface between the subsystems, including the correctness of the electromechanical, photoelectric and thermoelectric interfaces.
2. Inspection instruction, the reliability and accuracy of the information channel transfer (uplink and downlink parameter format definition).

3. Main function and performance test of the satellite in all working modes.
4. Simulate flight procedures of the satellite in space, test the feasibility and correctness of the flight procedures.
5. In the first phase of the entire satellite electrical performance test, the compatibility of the satellite to Earth interface is also required. Check the correctness of the control channel, check the correctness of the test software, the telemetry/remote control parameters and the switch state; check the correctness of the test outline and the rules.
6. Control and monitor the working status of instruments and equipments on the satellite.
7. Simulate a faultly working condition or check the main function and performance of the satellite under different excitation signals.
8. Record the working process and test data in real time and in a continuous manner.

Test contents of satellite subsystems are listed in Table 5.2.

References

[1] H. Guowei, Reliability Test Technology/Reliability, Maintainability and Supportability, National Defense Industry Press, Beijing, 1995.

[2] X. Jianqiang, Rocket and Satellite Products Test, China Aerospace Publishing House, Beijing, 2012.

[3] K. ShouQuan, Satellite Environmental Engineering and Simulation Test, China Aerospace Publishing House, Beijing, 1993.

[4] GJB 152A, Requirements for electromagnetic emission and sensitivity of military equipment and subsystems, 1997.

[5] QJ 2266, Electromagnetic compatibility requirements for aerospace systems, 1992.

[6] L. Yanxiao, Reliability Design and Analysis, National Defense Industry Press, Beijing, 1995.

[7] QJ 1408A, Reliability assurance requirements for aerospace products, 1998.

CHAPTER 6

Advanced Space Optical Attitude Sensor

6.1 INTRODUCTION TO THE ADVANCED SPACE OPTICAL ATTITUDE SENSOR

The advanced space optical attitude sensor refers to the sensor series that determines attitude through sensing a celestial body. Sun sensor and star sensor are two representative advanced space optical attitude sensors. The development of microsatellites and microsatellite formation flying technology requires a high-accuracy, rapid measurement and control system of satellite attitude, such as the Grace project [1], Techsat-21 project [2], LISA [3], or ST-5 [4]. A high-accuracy and high-reliability attitude sensory system is the optimal choice for these missions. With superiority of high accuracy, being drift free and having a long service life, the advanced space optical attitude sensor not only is the key basic component for the survival and performance improvement of spacecraft, but it also has important strategic significance in various astronautic applications (e.g., ground remote sensing, deep-space exploration, and scientific experimentation). The sun and fixed stars are two major astronomical sensing objects. Therefore, sun sensors and star sensors are the most widely applied systems to determine the attitude of satellites and have become the basic and core component of satellite attitude.

6.1.1 Introduction to the Sun Sensor and Star Sensor

The sun is the most important energy source of the satellite system. Satellites that fail to locate the sun or whose solar panels fail to get accurate sun orientation are regarded as unserviceable. Meanwhile, optical devices and thermal control systems on satellites are designed strictly according to the incidence direction of the sun. As a result, the sun sensors is essential attitude sensors on satellites. The sun, a bright point light source with high sensitivity and identifiability, is convenient for sensor design and for the purposes of the attitude determination algorithm. The sun sensor has a wide range of field of view (FOV) (from minutes of arc to

Space Microsystems and Micro/Nano Satellites. DOI: http://dx.doi.org/10.1016/B978-0-12-812672-1.00006-0

128×128 degrees) and a wide resolution range (from degrees to arc-seconds), thus making it the optimal attitude-sensing component of various spacecraft. In engineering technology, a sun–Earth system composed of a sun sensor and an Earth sensor is the most convenient and popular method of attitude determination. Traditional sun sensors have poor accuracy and are susceptible to reflected light from the Earth. To address these problems, this book describes a sun sensor designed to use area array APS CMOS technology, MEMS diaphragm technology, and algorithms related with prediction, extraction, and images.

Currently, since star sensors are universally accepted as the most precise absolute spacecraft attitude measurement component [5], they are widely used for high-accuracy attitude determination in Earth orbiters, deep space probes, large space structures, and small satellites. Compared to other attitude sensors, such as sun sensor, star sensors have many advantages: (1) adequate high pointing accuracy; (2) providing three-axes absolute attitude of the spacecraft instead of one-axis or two-axes attitude of a sun sensor or magnetometer; (3) strong fixed star detection in any direction and all-round attitude information under the assistance of an appropriate FOV and sensitivity, which is incomparable to other attitude sensors.

Outside China, traditional star sensors are widely applied in the aerospace field, with bright prospects. However, most of the existing star sensors were developed for microsatellites, which are difficult to apply in microsatellites directly. As a result, many international research institutions and star sensor development organizations have shifted their attention to microsatellite-oriented new star sensors [6,7], aiming to win a place in the microminiature sensor and microsatellite field. With the development of space mission and microsatellite technologies, China's aerospace future proposes an urgent demand for high-accuracy, high-dynamic and small star sensors. Many of China's existing high-accuracy satellites use imported star sensors as the major attitude measurement unit, while domestic star sensors generally serve as backups or in tests [8,9]. All of these star sensors were developed for microsatellites without consideration to the application characteristics of microsatellites, thus they are unable to be used in the microsatellite field directly. In this book, a small low-power consumption, high-accuracy and high-dynamic star sensor was developed by using new technologies and methods according to the requirements of microsatellites on the attitude sensing system. The research results not only can improve China's microsatellite attitude measurement accuracy, but also are of important significance in improving China's independent spacecraft R&D capability and space application of the new optical system, and reduce dependency on foreign key technologies.

6.1.2 Spacecraft Attitude Sensor Overview

To have a clear understanding on the principle and method of spacecraft attitude measurement and a further understanding on the significance of star sensors in the aerospace field as well as its difference with other attitude sensors, this book provides a brief introduction to the major attitude sensors used in spacecraft.

Generally speaking, spacecraft attitude is described by orientation relative to a certain coordinate system (generally the coordinate system relative to the inertial celestial sphere). This requires orientations of one or more directional vectors in relation to the spacecraft coordination system and specific coordinate system, respectively. Moreover, these vectors shall be known relative to the specific coordinate system. Common vectors include sun vector, Earth center vector, navstar vector, and Earth magnetic field vector. When the attitude sensor can measure the orientations of these vectors in relation to the sensor coordination system or spacecraft coordination system, the spacecraft attitude at the specific coordinate system can be further calculated.

6.1.2.1 Sun Sensor

The sun sensor is a sensor that measures the vector of the sun in the spacecraft coordination system directly. As one kind of optical sensor, the sun sensor is used on almost all satellites. The sun is distinctive from other light sources (e.g., background stars) surrounding the Earth due to its shining light. Furthermore, the sun can be viewed as a point to Earth satellite due to the great distance to the Earth. This simplifies the sun sensor design significantly. Additionally, most satellites orient their solar panels dead against the sun in order to collect solar energy for normal operation. However, some components or devices have to be kept away from the sun in order to prevent stray light and heating. The sun sensor plays an important role in all of these aspects. Although sun sensors have abundant design methods and styles, they include three major types: analogue sun sensor, sun-appearing sensor, and digital sun sensor [10].

6.1.2.2 Infrared Horizon Sensor

The infrared horizon sensor is a sensor that measures the vector of the Earth in the spacecraft coordinate system directly, which is also called the Earth sensor. Since it can calculate the spacecraft attitude in relation to the Earth directly, it is widely applied in Earth-related missions, such as communication satellites (TDRSS), meteorological satellites (GOES), Earth resource satellites (LANDSAT), etc. [11]. Different from the sun, the Earth can not be viewed as a point, especially to satellites near the Earth orbit. Since the Earth occupies up to 40% of the FOV of satellites, detection of the Earth is inadequate for satellite attitude determination. Therefore, most Earth sensors adopt horizon examination. The horizon sensor, one kind of infrared sensor, detects the warm Earth surface/cold space ratio. Its difficulties lie in the sensor–Earth discrimination and selection of its surrounding sensitization threshold, which is changeable due to the effects of the Earth's atmosphere and reflected sun rays.

6.1.2.3 Magnetometer

The magnetometer is a common vector measurement sensor used on spacecraft. It is characteristic of: (1) simultaneous measurement of the direction and strength of the magnetic field; (2) low power consumption and light weight; (3) reliable measurement results within a certain orbit range; (4) wide temperature endurance range; and (5) no moving components. However, based on the Earth magnetic field model (an approximate description of the Earth's magnetic field), the magnetometer brings large measurement errors and is not qualified as a high-accuracy attitude sensor. Additionally, the Earth's magnetic intensity is inversely proportional to the cubic geocentric distance. Therefore, when the spacecraft reaches higher than 1000 km orbit, the spacecraft attitude is mainly influenced by the residual magnetic bias within the satellites, instead of the Earth's magnetic field, thus making the magnetometer lose accurate measuring ability.

6.1.2.4 Star Sensor

The star sensor determines spacecraft attitude according to the orientations of the navstar in the spacecraft coordination system. Firstly, the star sensor measures the vector of the navstar in the spacecraft coordination system. Subsequently, it gains the vector of the navstar in the corresponding inertia coordination system through star pattern recognition. Thirdly, the transformation matrix from the inertia coordination system and the spacecraft coordination system, that is, spacecraft attitude in the inertia coordination system, can be obtained by analyzing the relationship between these two vectors of the navstar.

The star sensor is superior to other sensors for its high precision (up to an arc-second). However, compared to other sensors, the traditional star sensor has many shortcomings, such as higher cost, larger size and heavier weight, higher power consumption, more complicated computer hardware and software, and greater difficulties in design and debugging. Moreover, the star sensor is more susceptible to external disturbances (e.g., stray light in space). As a result, the star sensor requires strict installation environments free of disturbances from the sun, Earth, moon, and the satellite. Although the star sensor has so many shortcomings, it is still highly appreciated in the aerospace application field for its high-accuracy absolute attitude measurement.

6.1.2.5 Inertial Sensor (Gyro)

The inertial sensor, also known as the inertial navigation system (INS), uses an accelerometer and gyro to determine spacecraft attitude in relation to the inertial system. The accelerometer is used to test the motion acceleration of the carrier, which is then used to calculate the real-time location of the carrier. The gyro achieves essential reference coordinates (navigation coordinates) for the above-mentioned navigation positioning. The gyro demands no external information and has no optical or electrical connections with the external environment. These contribute to good concealment and applications under all meteorological conditions of the gyro. However, since the gyro generally involves high-speed weak rotating parts and may produce drift, it is inferior in independent high-accuracy long-term operation and requires a star sensor for modification.

The above-mentioned five attitude sensors are compared in this book (Table 6.1).

Table 6.1 illustrates that a small high-accuracy sun sensor and a star sensor enjoy bright development prospects in satellite attitude determination.

6.1.3 Sun Sensor Overview

6.1.3.1 Principles of the Sun Sensor

The sun sensor is the most common and important attitude measurement device used on satellites. Early sun sensors mainly included sun-appearing

Table 6.1 Comparison of Common Attitude Sensors [12]

Name	Reference System	Advantages	Disadvantages
Sun sensor	Sun	Low-power consumption, light weight, clear and bright sun, and generally an essential component for solar panel and instrument protection	Invisible in some orbits and poor accuracy (about 1 degree)
Earth sensor	Earth	Applicable in near-Earth orbit, fuzzy boundary, and easy to analyze	Generally, it needs to scan the horizon line; poor accuracy (about 0.1 degree); tight orbit–attitude coupling
Magnetic field sensor	Earth magnetic field	Economic, low-power consumption, and applicable in near-Earth orbit	Poor accuracy (about 0.5 degree) and limited applications within low orbits. Satellite magnetic balancing is needed
Star sensor	Stars	High accuracy (arc-second), independent movement from orbits, and applicable in any position in space	Heavy weight, complexity, high cost, and time-consuming star recognition
Gyro	Inertia	Demands no external sensor, independent movement from orbits, time efficiency, and high accuracy	Has drift and high-speed weak rotating parts, as well as large power consumption and mass

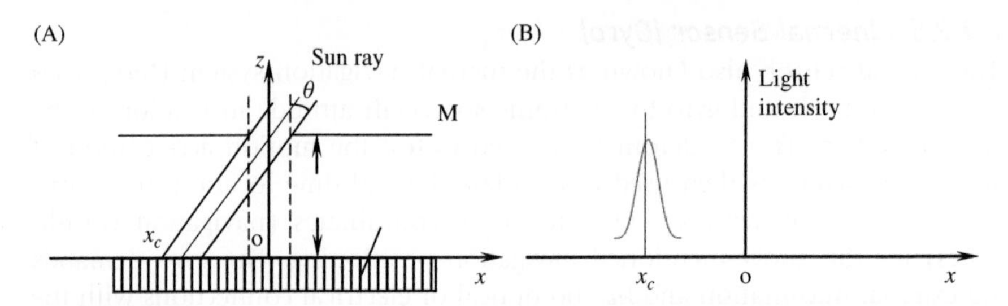

FIGURE 6.1
Homotaxial digital sun sensor. (A) Structure, (B) light intensity distribution.

sensors and analogue sun sensors [13] with poor accuracy, which were used for initial solar capture and low-accuracy determination of solar azimuth [14]. The recent rapid development of astronautical technology, especially microsatellite technology, requires sun sensors with wide FOV and high accuracy, which can not be satisfied by the traditional design plans and achievement approaches. Therefore, a homotaxial digital sun sensor [15] with significantly higher accuracy (<0.1 degree) was developed and has become the mainstream of the current sun sensor application field. Homotaxial digital sun sensor imaging is shown in Fig. 6.1.

Sun ray imaging at x_c on the charge coupled device (CCD) is through the diaphragm apertures. The corresponding light intensity distribution on CCD is shown in Fig. 6.1B. Images are read through a follow-up processing algorithm and the energy center of imaging points is calculated. On this basis, x_c can be positioned accurately. Next, the incidence angle of the sun's rays can be calculated from CCD–diaphragm height: $\tan\theta = x_c/h$.

Vertical installation of two homotaxial digital sun sensors can construct a traditional biaxial sun sensor, which is the current application mainstream of homotaxial digital sun sensors. Fig. 6.2 presents some typical products at home and abroad [16–19].

6.1.3.2 Development Trends and Major Problems of Sun Sensors

With the electronical technology development, area array photosensitive probes are widely applied in space [20], such as CCD or CMOS [21]. A sun sensor constructed from area array imaging sensors has attracted a great deal of research attention [22–24]. These sun sensors, with high accuracy and strong anti-interference, can replace star sensors on some occasions [25]. Since the diaphragm of the sun sensor has a single-aperture diaphragm (Fig. 6.3A) and array-type diaphragm (Fig. 6.3B), the sun sensor also has a single-aperture sun sensor and array-type sensor [26].

FIGURE 6.2
Some typical digital sun sensors. (A) DSS2 (China), (B) DSS2 (Japan), (C) ASS3 (China), (D) DSSS.

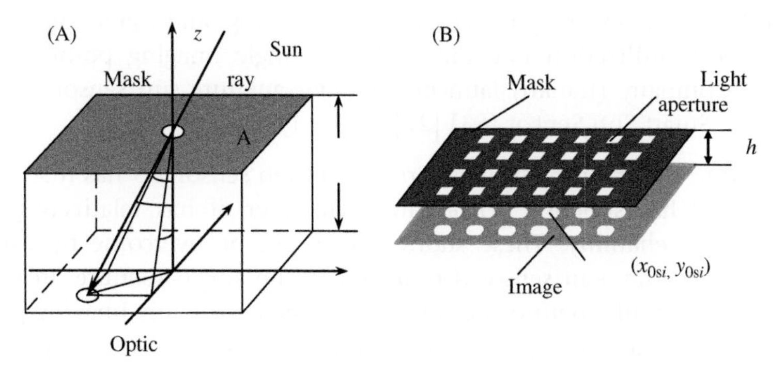

FIGURE 6.3
Area array sun sensor. (A) Single-aperture mask, (B) array-type mask.

The basic principles of single-aperture sun sensor are: sun rays imaging on the photosensitive probe through the aperture; the coordinates (x_c, y_c) of the center of sun ray imaging points are calculated through the centroid method; furthermore, the solar angles (α and β) are acquired based on the trigonometric function:

$$\alpha = tg^{-1}\left(\frac{x_c}{h}\right) \qquad \beta = tg^{-1}\left(\frac{y_c}{h}\right) \tag{6.1}$$

FIGURE 6.4
The smart sun sensor (S3).

where h is the distance between the aperture and the photosensitive probe, which is fixed according to the above analysis. The accuracy of the solar azimuth is mainly determined by the calculation accuracy of (x_c, y_c). Imaging through this sun sensor can achieve 25 pixels on its photosensitive probe, thus increasing its accuracy compared to a traditional linear array sensor. However, it is difficult to further improving the accuracy of this sun sensor due to the simple extraction, recognition, processing, and operation, as well as low image utilization rate caused by its single imaging point. Galileo Avionica Company (Italian) launched a single-aperture sun sensor recently, named the Smart Sun Sensor (S3) [27] (Fig. 6.4).

Compared to the traditional linear array CCD sun sensor, S3 has made great progress and has superior performance. However, it has relatively poorer accuracy and reliability. These shortcomings can be overcome by using a multiaperture array sun sensor. Compared to the single-aperture area array sun sensor, the multiaperture area array sun sensor has many imaging points, which can provide filtering processing, thus increasing the accuracy by one order of magnitude [28]. Meanwhile, much related information among these imaging apertures is available and can be integrated to improve the environmental adaptation of the sun sensor. This ensures the high accuracy and normal operation of the sun sensor even when some imaging points are unavailable.

The advantages and disadvantages of the single-aperture sun sensor and array sun sensor include the following:

1. Accuracy: Since the accuracy of the single-aperture sun sensor is mainly determined by the processing accuracy of the center of the sun ray

imaging points and only involves one imaging point, its image processing accuracy can only reach about 0.1 pixels. On the contrary, the processing accuracy of sun ray imaging points of the array sun sensor is proportional to the square root of the arrays. Suppose the processing accuracy of a single aperture is δ and the aperture array is $N \times N$, then the final processing accuracy of the center of total imaging points is δ/N. This plays a significant role in improving the accuracy of the sun sensor.

2. Reliability: The single-aperture sun sensor has a catastrophic threat: reliability. Complete or partial aperture blockage may affect the processing accuracy significantly and even result in system strikes. In contrast, an array sun sensor with large distribution area of apertures still can maintain normal operation and satisfy processing accuracy through reasonable algorithms even when some apertures are blocked.
3. Requirements of the imaging sensor: The single-aperture sun sensor has higher requirements of the imaging sensor because any partial bad pixels or uneven pixel response of the imaging sensor (which may be the location of the sun ray imaging point) will affect the processing accuracy and even functions of the sun sensor significantly. In contrast, the array sun sensor has $N \times N$ imaging points distributed within a wide region of the imaging sensor. Therefore, some bad pixels will not affect the system completely.
4. Image exposure and read [29]: The single-aperture sun sensor or traditional linear array CCD sun sensor generally reads images through direct overall exposure. Due to their small data size, they can follow the procedure from overall exposure, reading, processing, and calculation successively. However, the array sun sensor involves large data size and the complicated processing method requires more advanced image exposure and read methods as well as advanced data processing.
5. Algorithm complexity: The array sun sensor involves a far more complicated algorithm compared to the single-aperture sun sensor. Before accurate application, the array sun sensor has to identify each imaging point on the imaging sensor accurately. Since these sun ray imaging points are the same and every imaging point has a certain imaging error, the first imaging point may be wrongly identified or even lost (a more serious situation). Therefore, it requires an algorithm with strong robustness and high reliability.

6.1.3.3 Research Significance

Accuracy and reliability are the principal decisive factors of sun sensor quality. Therefore, multiaperture area array sun sensors represent the future of sun sensors. Currently, the development of such multiaperture area array sun sensors is mainly restricted by the high accuracy, high reliability, and high update rate of the algorithm. Based on the research achievements during the "Tenth Five-Year" and "Eleventh Five-Year," this book introduces an advanced sun sensor algorithm integrating hardware and software—Future Extraction and Image Correlation (FEIC) algorithm—to achieve breakthroughs on the key technologies of the multiaperture area array sun sensor. It takes the APS (Active Pixel Sensor) CMOS photosensitive probe with window exposure and random reading function as its imaging sensor, and takes the advantage of correlation among multiple array-arranged sun ray imaging points to read images directly through direct exposure and the read method within the effective region. To accelerate capture of the initial position of current sunlight imaging image, a binarization template image is used for binary matching with it. The position shift between the current image and the direct solar radiation image (x_c, y_c) can be calculated precisely through energy matching of the template image during direct solar radiation and the current sun ray imaging image. On this basis, a high-accuracy solar angle can be gained. Theoretically, a sun sensor involving an $N \times N$ apertures array diaphragm still can maintain high-accuracy normal operation under the absence of any N imaging points.

6.1.4 Star Sensor Overview

6.1.4.1 Navstar

Celestial bodies keep in relative balance due to gravitation and movement. Visually, all celestial bodies seem equidistant and their relationship with observers is just like the relationship between a point on a sphere and the center of a sphere. The imaging sphere centered at the observer with random radius is called a celestial sphere.

A fixed star is the basic reference point of a star sensor. According to years of astronomical observations, every fixed star occupies a relative fixed position in the celestial sphere and is generally expressed by the celestial spherical coordinates (α, δ) (Fig. 6.5). According to the relationship between rectangular coordinates and spherical coordinates, the direction vector of every fixed star under the celestial rectangular coordinates can be acquired.

$$v = \begin{bmatrix} \cos\alpha\cos\delta \\ \sin\alpha\cos\delta \\ \sin\delta \end{bmatrix} \tag{6.2}$$

Fixed stars with satisfying imaging conditions of the star sensor are selected from the star library to form the navstar catalogue. This navstar catalogue is a one-off, solidified in the memory of the star sensor on the ground.

$$v = \begin{bmatrix} \cos\alpha\cos\delta \\ \sin\alpha\cos\delta \\ \sin\delta \end{bmatrix} \quad (6\text{-}2)$$

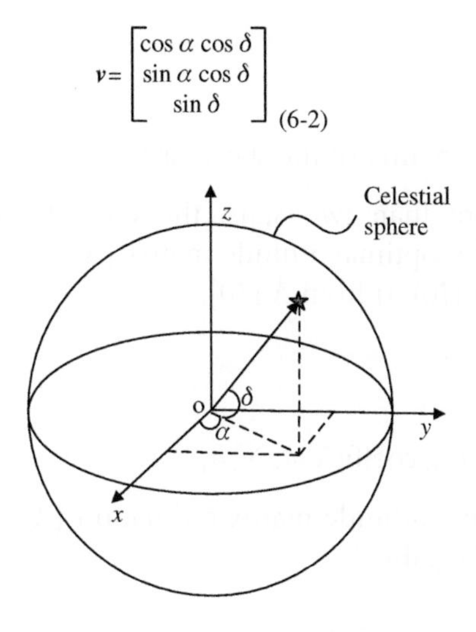

FIGURE 6.5
Descriptive relationship of navstar between celestial spherical coordinates and rectangular coordinates.

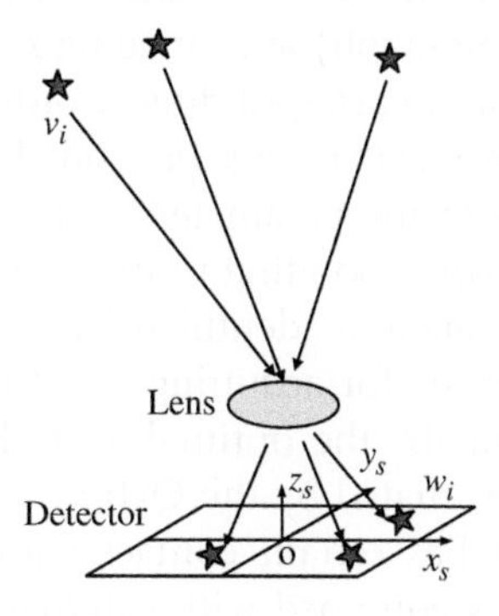

FIGURE 6.6
Imaging principle of a star sensor.

6.1.4.2 Principle of Measurement

At the attitude matrix (A) of the star sensor in the celestial coordinates, the navstar (s_i) (with a direction vector in the corresponding celestial coordinates (v_i)) and its direction vector in the star sensor coordinates (w_i) can be detected according to the aperture imaging principle of star sensor (Fig. 6.6).

(x_0, y_0) are the coordinates of the spindle center of the star center on the probe; (x_i, y_i) are the coordinates of s_i on the probe; and f is the focal length of the star sensor. Then, w_i can be expressed as:

$$w_i = \frac{1}{\sqrt{(x_i - x_0)^2 + (y_i - y_0)^2 + f^2}} \begin{bmatrix} -(x_i - x_0) \\ -(y_i - y_0) \\ f \end{bmatrix} \quad (6.3)$$

Ideally,

$$w_i = Av_i \tag{6.4}$$

where A is the attitude matrix of the star sensor.

When observing more than two stars, the QUEST method can be used directly to calculate the optimal attitude matrix (A_q), minimizing the following objective function ($J(A_q)$) from A [30].

$$J(A_q) = \frac{1}{2}\sum_{i=1}^{n} \alpha_i \left\| w_i - A_q v_i \right\|^2 \tag{6.5}$$

where α_i is the weighting coefficient, $\sum \alpha_i = 1$.

In this way, the optimal attitude matrix estimation (A_q) of the star sensor in the inertia space can be gained.

6.1.4.3 Components of the Star Sensor

The principle of measurement reveals that, firstly, the star sensor has to acquire a star pattern (photograph) and imaging s_i on the probe of star sensor. Secondly, star points and corresponding coordinates are extracted. Later, the direction vector of these star points (w_i) on the star sensor coordinates can be acquired according to the parameters of the star sensor. Thirdly, s_i in the navstar catalogue and corresponding to the current star imaging point on the star sensor coordinate plane is identified by using the star identification technology, which is the basis for acquiring the direction vector (v_i) of s_i in the inertia coordinates. Finally, the optimal attitude matrix (A_q) of the star sensor at the moment is estimated by the QUEST method. As a part used on spacecraft, the star sensor has certain contact with other parts of the star. Therefore, the star sensor is equipped with external interfaces. The workflow of the star sensor is shown in Fig. 6.7.

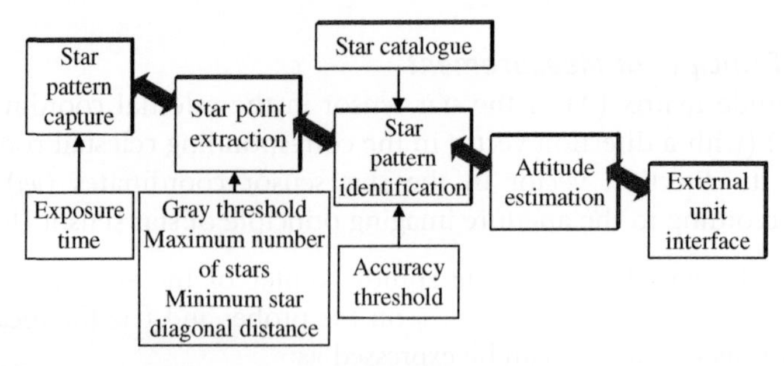

FIGURE 6.7
Workflow of the star sensor.

Current star sensors in the international market are basically subject to the principle and workflow illustrated above. To distinguish them from the early tracking star sensor, these star sensors are called peering star sensors.

6.1.4.4 Development Status of CCD Star Sensor at Home and Abroad

Various commercial CCD autonomous star sensors have been developed. Table 6.2 lists some of these that have been applied successfully on spacecraft.

Table 6.2 reveals the current basic level of commercial CCD star sensors in foreign countries: weight: 3–5 kg; power consumption: about 10 W; data updating rate: 1–10 Hz. Given a FOV of 10×10 degrees, the pointing accuracy reaches 1″ with relative lower roll accuracy (about 5″–10″). Given a FOV of 20×20 degrees, the point accuracy reaches 10″ with relative lower roll accuracy (generally 30″–50″).

Domestic CCD star sensors have also achieved some progress after more than two decades of development. The CCD probe-based star sensor (18 arc-seconds accuracy and 24 h-a-day capture) developed by 502 Institute of Aerospace Science and Technology Group accomplished the on-orbit task successfully, symbolizing the new development stage of star sensors in China. Shanghai Aerospace Control Engineering Research Institute and many other related organizations have conducted on-orbit flight tests of their star sensors. Additionally, the Institute of Optics and Electronics (Chengdu), National Astronomical Observatories of China (NAOC), Changchun Institute of Optics, Fine Mechanics and Physics, and Beijing University of Aeronautics and Astronautics (BUAA) have also made abundant research into star sensors and have developed corresponding physical star sensors. However, domestic star sensors are significantly inferior to the more well-known foreign sensors:

1. Poorer pointing accuracy: Under 10×10 degrees, foreign star sensors can achieve a pointing accuracy of 1″, while domestic star sensors can only reach about 10 arc-seconds. The point accuracy of domestic star sensors under larger fields of view is even poorer.
2. Longer initial capture time: Foreign star sensors generally cost 5s and some even cost only 1s, while domestic stars sensors generally cost about 30s.
3. Poorer anti-interference: Most domestic star sensors have abnormal operation during on-orbit service, caused by the entering of stray sun light and reflected light from the Earth into the FOV. Additionally, domestic star sensors are easily broken down in the South Atlantic region due to abnormal irradiation.

Table 6.2 Typical Commercial CCD Star Sensors

CT-633 [31] (CT-600 series) (BALL Aerospace & Technology Corp., U.S.A.)	
FOV: 20 × 20 degrees Accuracy (1σ): 10″ (pitch/yaw), 40″ (roll) Capture time: 10 s Data updating rate: 5 Hz Power consumption: 8 W Standard lens hood Weight: 2.5 kg	
SED16 [32] (latest model of SED series) (EADS Sodern Corp., France)	
FOV: 25 × 25 degrees Accuracy: 10″ (pitch/yaw) 55″ (roll) Capture time: 3 s Data updating rate: 10 Hz Power consumption: 7.5 W Standard lens hood: 35 and 25 degrees Weight: 3 kg	
CALTRAC [33] TM (EMS Technologies Inc., Canada)	
FOV: 36 × 27 degrees Accuracy: 0.005 degree (pitch/yaw), 0.02 degree (roll) Capture time: 0.5 s Data updating rate: 25 Hz Standard lens hood	

Continued

Table 6.2 Typical Commercial CCD Star Sensors *Continued*

CT-633 [31] (CT-600 series) (BALL Aerospace & Technology Corp., U.S.A.)	
Astro-15 [34] (latest model of Astro series) (Jenaoptronik, Germany)	
FOV: 13.8 × 13.8 degrees Accuracy (1σ): 1″ (pitch/yaw), 10″ (roll) Data updating rate: 4 Hz Power consumption: 10 W (max.), 9 W (norm.) Standard lens hood: 35 and 25 degrees Weight: 4.3 kg (lens hood excluded)	
ASC [35]—Technical University of Denmark (Denmark)	
FOV: 18 × 12 degrees Accuracy: (1σ): 1″ (pitch/yaw), 5″ (roll) Initial capture time: 5 s Data updating rate: 4 Hz Power consumption: 7.8 W Weight: 1 kg (lens hood excluded)	

6.1.4.5 Status of APS CMOS Star Sensor at Home and Abroad

Currently, the CCD star sensor is approaching maturity in foreign countries, but the APS CMOS star sensor is still in the research stage with few on-orbit application cases. Compared to the CCD star sensor, the CMOS star sensor has attracted more attention from international research institutions because of its unique advantages. Some research projects on the CMOS star sensor have been launched accordingly. The Miniature Star Tracker (JPL, United States) and ASTRO-APS Star Sensor (Jenaoptronik, Germany) are two typical CMOS star sensors.

The "Micro APS Star Tracker" [36] with ultra-low power consumption (70 mW), ultrahigh data updating rate (50 Hz) and ultralight weight (42 g) developed by JPL (United States) is exhibited in Fig. 6.8. Its electronical system only contains two pieces of chips: one piece of exclusive APS CMOS chip and one piece of application-specific integrated circuit (ASIC) chip that integrates with the I^2C interface, memory, and 8051 microprocessor.

FIGURE 6.8
Model of the Miniature Star Tracker.

Actually, this micro APS star sensor is just a sensing head. It reads characters and outputs initial pixel data through the APS window, but leaves all calculations to the on-board computer, which explains its small size, light weight, and low power consumption. However, it is only used to demonstrate the concept of the APS CMOS star sensor and has no practical application significance due to its small practice physiognomy, low accuracy, and uncertain reliability.

The APS CMOS star sensor (ASTRO series, Germany Jenaoptronik) [37] has achieved great progress. Based on the Alphabus/Large platform of ASTRO series, the development of this APS CMOS star sensor focuses on its economic efficiency and feasibility, aiming to lay a foundation for the follow-up large-scale development of the APS CMOS star sensor series. The technical indexes include: weight <1.5 kg; power consumption <5 W; attitude updating rate: 10 Hz; accuracy: pitch/yaw <9 arc-second (3σ), roll <45 arc-second (3σ); design life: 15 years. Although these technical indexes seem similar to those of the existing CCD star sensor, the weight and power consumption are lowered. This is mainly because: (1) the high integration of APS CMOS and simple peripheral circuit system downsized the PCB (printed circuit board) design and other auxiliary components; (2) APS CMOS integrates circuits for AD conversion within the chip, sampling and read, which not only lowers power consumption, but also improves the reliability of the star sensor. Fig. 6.9 shows the complete principled sample of the APS MCOS star sensor (ASTRO series, Jenaoptronik, Germany).

The APS CMOS star sensor also has achieved some progress in China in the past 5 years. 502 Institute of Aerospace Science and Technology Group,

FIGURE 6.9
Principled sample of the APS MCOS star sensor (ASTRO series). (A) Principle mode, (B) principled sample.

NAOC, Changchun Institute of Optics, Fine Mechanics and Physics, and BUAA participated in the research and a principle sample machine was developed accordingly. Moreover, the star sensor developed by 502 Institute of Aerospace Science and Technology Group and Tsinghua University is in on-orbit service.

Therefore, future domestic research into the star sensor will focus on accuracy and speed improvement to catch up with the international level. Meanwhile, existing star sensors developed in China are inapplicable to microsatellites because of their high power consumption and heavy weight. To address these problems, this book introduces several research results in high-accuracy microstar sensors achieved by this research team, including new technologies and new methods.

6.1.5 Framework

Based on the APS CMOS sun sensor and star sensor developed by the research team, this book introduces the research methods and ideas of high-accuracy, high-dynamic, high-reliability and full-autonomous micro sun sensors and star sensors based on the APS CMOS imaging sensor, aiming to provide minisatellites, especially microsatellite system accurate real-time attitude information, to improve the navigation performance of microsatellites and explore new ways for the development of practical new optical attitude sensors.

In this book, the developed APS CMOS micro sun sensor and star sensor, as well as the key technologies, are introduced in two parts.

6.2 TECHNICAL RESEARCH OF THE APS MICRO SUN SENSOR

According to the associated mission requirements and micro/nano satellite development idea of Tsinghua University, microminiaturization of on-board functional units is the key to micro/nano satellite technology. During the "Tenth Five-Year Plan," Tsinghua University carried out a technical study on the APS CMOS sun sensor under the support of the 973 Project and made a test flight on an XX-3 satellite under the arrangement of related national departments. These are the research basis for this book.

6.2.1 Overview

The operating principle of a single-aperture sun sensor is similar to a sundial (Fig. 6.10A). A sun spot is developed from the sun ray projection on to the optical probe through the aperture on the diaphragm. The geometric center of the projected sun spot can be used to determine the solar angle.

The working principle of a multiaperture sun sensor is similar to that of a multislit sun sensor. A sun spot array will be developed from the sun ray projection on the APS probe through apertures on the diaphragm. The geometric center of each sun spot can be used to determine the solar angle of each sun spot. The mean solar angle of all sun spots is viewed as the sensor output. Compared to the single-aperture sun sensor, the multiaperture sun sensor has higher measurement accuracy and wider FOV, which are related to the aperture array on the diaphragm (Fig. 6.10B).

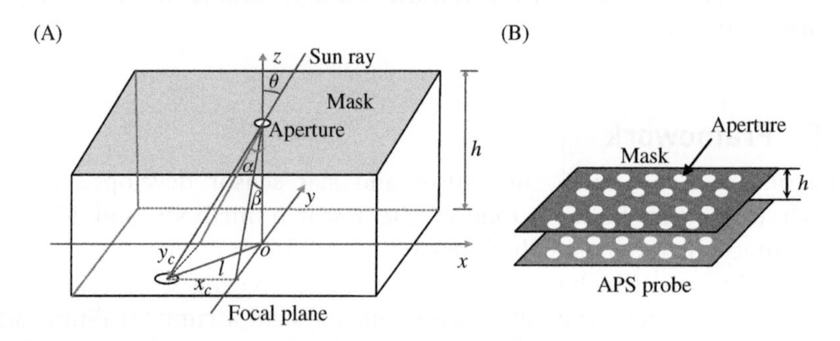

FIGURE 6.10

Working principle of a sun sensor. (A) Single-aperture sun sensor, (B) multi-aperture sun sensor.

6.2.2 Components

6.2.2.1 Structure Design

The appearance of a APS micro sun sensor is exhibited in Fig. 6.11. There is a rectangular window on the window plate. The optical system and electronical system of the sensor are installed inside the circuit box.

The internal structure of an APS micro sun sensor is shown in Fig. 6.12, including the imaging sensor and its peripheral circuit, diaphragm, image acquisition control circuit, image processing and communication module,

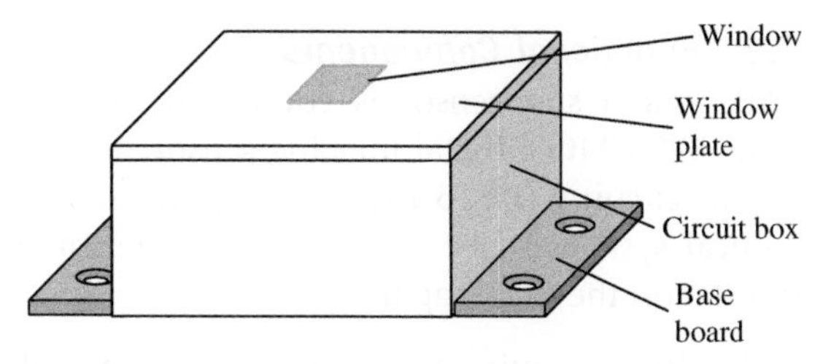

FIGURE 6.11
Appearance of an APS micro sun sensor.

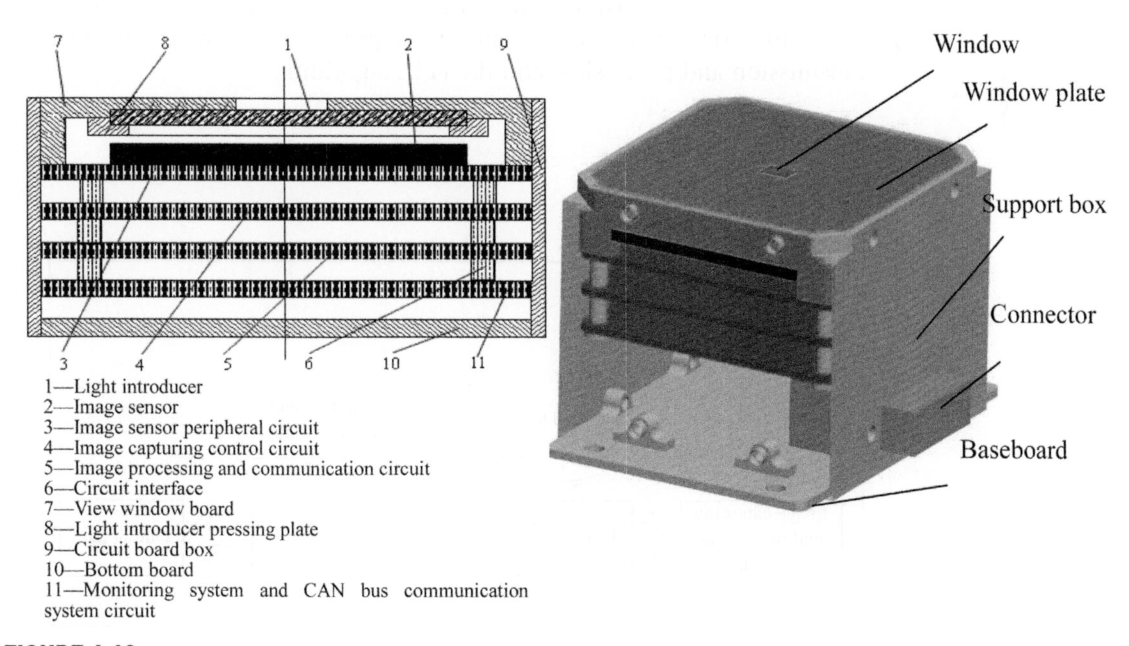

FIGURE 6.12
Internal structure and general assembly of an APS micro sun sensor.

circuit interface, window plate, circuit box, and baseboard. Parallel incident sun rays are split and projected into the sun spot array on the APS probe through apertures on the diaphragm. The image acquisition circuit composed of an imaging sensor as well as its accessory circuit and image acquisition control circuit is responsible for the acquisition of sun spot array images. The image processing and communication circuit connected with the on-board computer through the communication interface is for reading and processing images containing solar angles collected by the image acquisition module, and outputting the calculated solar angle.

6.2.2.2 Functional Components

The APS micro sun sensor is composed of an MEMS multiaperture diaphragm, APS CMOS digital imaging sensor, as well as image acquisition and processing circuitry (Fig. 6.13). The diaphragm and imaging sensor constitute the optical system of the APS micro sun sensor. Square aperture arrays are distributed on the diaphragm.

Basic algorithm: acquire images through the digital imaging sensor; use FEIC image processing to estimate barycentric coordinates of sun spots; calculate the solar angles of two axials according to the rating data and error modification data; the image acquisition and processing circuitry, the command center of the APS micro sun sensor, is responsible for image acquisition, transmission and processing, and the FEIC algorithm.

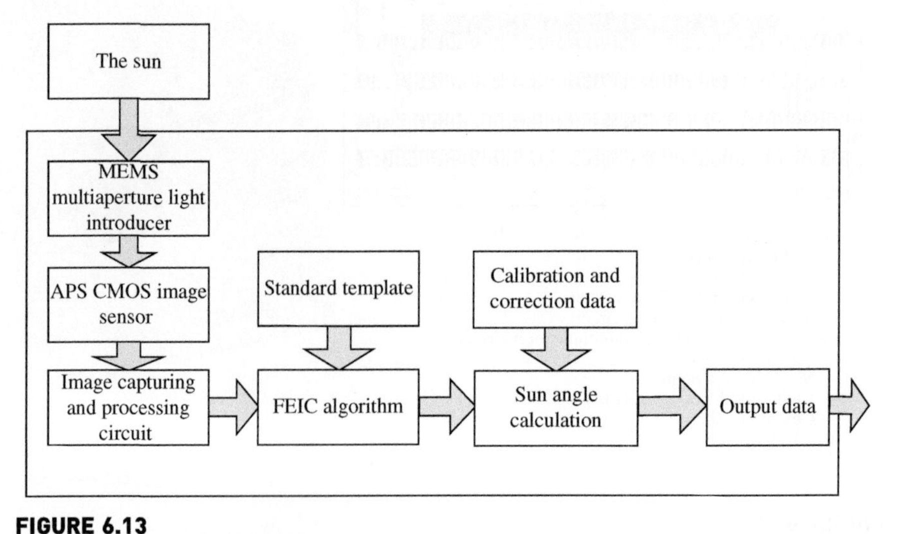

FIGURE 6.13
Workflow of the APS micro sun sensor.

6.2.3 Optical System Design

The optical system of the APS micro sun sensor mainly consists of the MEMS diaphragm and imaging sensor. Firstly, the dynamic photosensitivity range is determined according to the sensitivity and quantum efficiency of the APS CMOS Star1000 probe, which is used to design the structural parameters of the diaphragm.

6.2.4 Selection of APS Imaging Sensor

In considering the aerospace applications, technical features (radiation resistance and high temperature resistance) and technical requirements (e.g., aerospace experiences), Star1000 with high radiation resistance is selected as the photosensitive probe (Fig. 6.14).

The technical indexes of this photosensitive probe are listed in Table 6.3.

FIGURE 6.14
Star1000 APS CMOS imaging sensor.

Table 6.3 Performance Indexes of APS CMOS Imaging Sensor

Performances	Indexes
Array dimension	1024 × 1024
Pixel size	15 μm × 15 μm
Power consumption	100–350 mW
ADC output	10 parallel-by-bit
Photosensitive spectrum	400–1000 nm
Pixel filrate × photovoltaic conversion rate/FF × QE	20%

The Star1000 imaging sensor is strongly recommended by the European Space Agency (ESA) and is widely used on spacecraft. Jenaoptronik (Germany), Sodern (France), and Galileo reported successful space trials of the Star1000 imaging sensor. In China, 502 Institute and 812 Institute of Aerospace Science and Technology Group, Tsinghua University and BUAA have tested its space environmental adaptability (e.g., thermal vacuum and high/low temperatures).

6.2.4.1 Diaphragm Design

The diaphragm structure is designed specifically according to the characteristics of the Star1000 imaging sensor. The distance (h) between the diaphragm and the APS probe is determined by parameters of the diaphragm, including aperture size, aperture interval, number of aperture arrays, and exposure time.

1. Design indexes

   The optical system design is the key to the APS micro sun sensor design, which can affect its FOV and measurement accuracy directly. The optical system design is based on the design indexes of the APS micro sun sensor.
   The main design indexes of the APS micro sun sensor include:

   FOV: $\pm 64 \times \pm 64$ degrees
   Accuracy: 0.02

   Firstly, the output accuracy of each solar angle is determined according to the design indexes of the APS micro sun sensor. Then, the final design indexes can be gained according to the mean output accuracy of all solar angles. The design accuracy of a single aperture is ± 0.05 degree.

2. Design principle

   According to the diffraction law of aperture and aperture array, in the single-aperture sun sensor, parallel sun rays will pass through the aperture and project a sun spot on the receiving surface under satisfying conditions. Vertical-incident sun rays will project a round sun spot, while oblique-incident sun rays (incidence angle >0 degree) will project approximate oval sun spots and their ovality is proportional to the incidence angle. In the multiaperture sun sensor, parallel sun rays will pass through the aperture arrays and project sun spot arrays on the receiving surface. In other words, each aperture will project a sun spot on the receiving surface. Larger incidence angle, further distance between the incident plane and the receiving surface, and the smaller aperture will cause greater effect of interval between

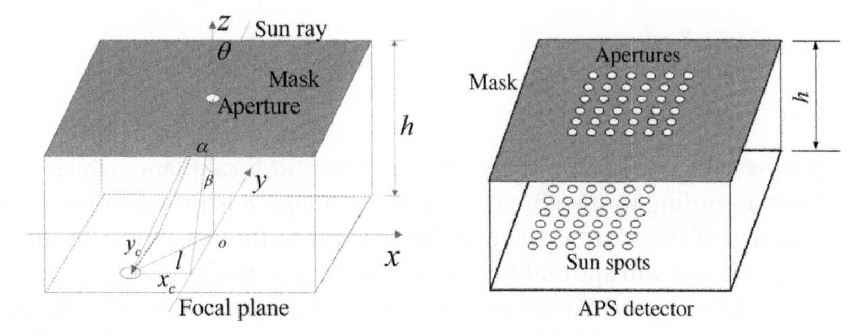

FIGURE 6.15
Image-forming principle of single-aperture and multiaperture sun sensors.

two neighboring apertures on the light intensity distribution. Sometimes, a bright strip will be projected on the receiving surface.

The basic parameter design of the optical system follows the principle that each aperture has one centroid of sun spot on the APS probe during image identification. To guarantee the centroid estimation accuracy of the FEIC algorithm, 4 × 4 pixels is the lower limit of light energy distribution. However, excessive pixels may bring heavy computations. The diffraction pattern of single aperture in the aperture arrays shall be distinguishable; otherwise, bright and dark strips will be projected on the APS probe, which will result in the failures of projection center identification and solar angle measurement of the sun sensor. Furthermore, adequate arrays are needed. According to the principle of statistics, the more arrays involved, the smaller the random error will be.

3. Imaging system modeling

Image-forming principles of single-aperture and multiaperture sun sensors are shown in Fig. 6.15.

Take the image-forming principle of single-aperture sun sensor for example. A sun ray passes through the aperture and projects a sun spot on the focal plane. In the left side image of Fig. 6.15, (x_c, y_c) are the coordinates of the projected sun spot center; l is the distance between the sun spot and the origin; h is the distance between the focal plane and diaphragm; θ is the incidence angle of sun ray; and α, β are the angle of pitch and angle of lead of sun ray (solar angle of two axials), respectively, in relation to the coordinates of the sun sensor. The relationship between these parameters can be expressed as:

$$\theta = \arctan\left(\frac{l}{h}\right), \alpha = \arctan\left(\frac{x_c}{h}\right), \beta = \arctan\left(\frac{y_c}{h}\right) \tag{6.6}$$

$$l = \sqrt{x_c^2 + y_c^2} \tag{6.7}$$

$$\tan\theta = \sqrt{(\tan\alpha)^2 + (\tan\beta)^2} \tag{6.8}$$

h represents the distance between aperture and focal plane, which is fixed according to the above analysis. Although h is proportional to the accuracy of the solar angle, it is limited within the range that ensures the projected sun spot falling in the focal plane under 128×128 degrees. Therefore, h is determined by the area array of the imaging sensor and field angle of the sun sensor. After h is determined, the measurement accuracy of the sun sensor is mainly determined by the processing accuracy of (x_c, y_c). Multiaperture arrays that can reduce random error are a good way to improve the accuracy of (x_c, y_c). Generally speaking, the processing accuracy of a single aperture can reach the subpixel level (0.05–0.1 pixels). The measurement accuracy of the sun sensor is inversely proportional to the square root of the number of apertures. More apertures will lead to higher measurement accuracy and heavier computation. From a global perspective, we design 36 apertures on the diaphragm, which can improve the accuracy of (x_c, y_c) to 0.01–0.02 pixels. To disclose the operating principle of the sun senor, the following text describes its modeling and solutions.

The above-mentioned Eqs. (6.6)–(6.8) simplify the projected sun spot. However, the projection of sun rays is more complicated in reality due to the refraction effects of the atmosphere and glass. According to Fig. 6.16, a sun ray has passed through a vacuum, quartz glass, detector protecting glass (BK7), and nitrogen, successively, before reaching the diaphragm.

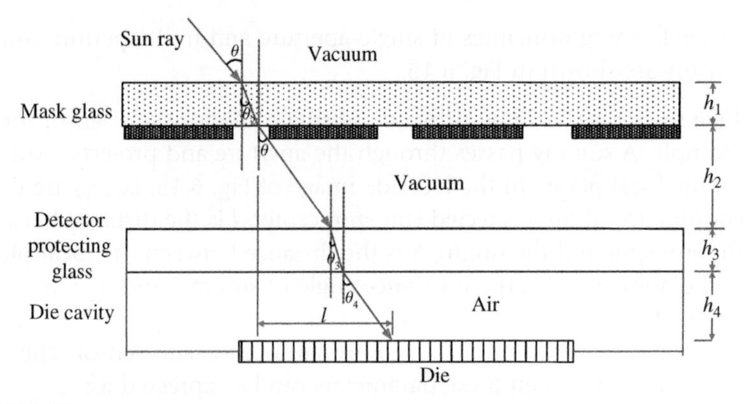

FIGURE 6.16
Sun ray tracing of a sun sensor.

Then, the function $(l \sim \theta)$ should be established as:

$$l = h_2 \tan\theta + h_3 \tan\theta_3 + h_4 \tan\theta_4 \tag{6.9}$$

$$\frac{n_{Glass}}{n_{Vacuum}} = \frac{\sin\theta}{\sin\theta_3} \tag{6.10}$$

$$\frac{n_{Air}}{n_{Glass}} = \frac{\sin\theta_3}{\sin\theta_4} \tag{6.11}$$

where $n_{Glass}, n_{Vacuum}, n_{Air}$ are the refractive indexes of BK7, vacuum and air, respectively ($n_{Vacuum} = 1$). Substituting θ_3 and θ_4 in Eqs. (6.10) and (6.11) into Eq. (6.9),

$$l = h_2\tan\theta + h_3\tan\left(\arcsin\left(\frac{\sin\theta}{n_{Glass}}\right)\right) + h_4\tan\left(\arcsin\left(\frac{\sin\theta}{n_{Air}}\right)\right) \tag{6.12}$$

Since $h_2, h_3, h_4, n_{Glass}, n_{Air}$ are constants and l is the unique function of θ, θ can be viewed as the function of l in return:

$$\theta = f(l) = f\left(\sqrt{x_c^2 + y_c^2}\right) \tag{6.13}$$

Combining Eqs. (6.1), (6.2), (6.3), and (6.8),

$$\begin{aligned} \alpha &= \arctan\frac{x_c \tan\left(f\left(\sqrt{x_c^2 + y_c^2}\right)\right)}{\sqrt{x_c^2 + y_c^2}} \\ \beta &= \arctan\frac{y_c \tan\left(f\left(\sqrt{x_c^2 + y_c^2}\right)\right)}{\sqrt{x_c^2 + y_c^2}} \end{aligned} \tag{6.14}$$

(x_c, y_c) can be gained from Eqs. (6.12) to (6.14). Then, (α, β) can be calculated.
In practical design, $h_3 = 1$ mm, $h_2 = 0.94$ mm, and $h_4 = 0.528$ mm (Fig. 6.16). When the incidence angle of the sun ray is θ_1, and supposing the refractive index of both Star1000 glass and mask glass is 1.4586, then,

$$\frac{\sin\theta_1}{\sin\theta_2} = 1.4586$$

Therefore, the relationship between the position offset of the sun spot and the incidence angle can be expressed as:

$$\begin{aligned} l &= 0.94\tan\theta_1 + 1\tan\theta_2 + 0.528\tan\theta_1 \\ &= 1\tan\theta_2 + 1.468\tan\theta_1 \qquad \text{(mm)} \end{aligned}$$

According to the refraction law,

$$n = \frac{\sin\theta_1}{\sin\theta_2}$$

Therefore, the relationship between the relative displacement and incidence angle is:

$$l = 1\tan\theta_2 + 1.468\tan\theta_1$$
$$= 1.486\tan\theta_1 + \tan\left(\arcsin\left(\frac{\sin\theta_1}{n}\right)\right) \quad (\text{mm})$$

When $0 \leq \theta_1 < 64$ degrees, the relationship between the relative displacement and $\tan\theta_1$ is shown in Fig. 6.17.

Since it is an odd function, the curve when $\theta_1 < 0$ is symmetric to the curve in Fig. 6.17.

4. Structure and process design of the diaphragm

We design 6×6 45 μm × 45 μm aperture arrays with an aperture interval of 480 μm on the diaphragm. These parameters are related to the initial capture of sun spots in Chapter 7, Miniature Inertial Measurement Unit. The structure and size of the diaphragm are shown in Fig. 6.18.

The diaphragm is stacked up by quartz glass, a Cr film, a gold film, and secondary reflection resistance layers successively from the lighting surface to the APB probe.

The glass substrate uses GJB2849-1997 qualified antiradiation quartz glass.

The mask layer is made of 2000 A gold film.

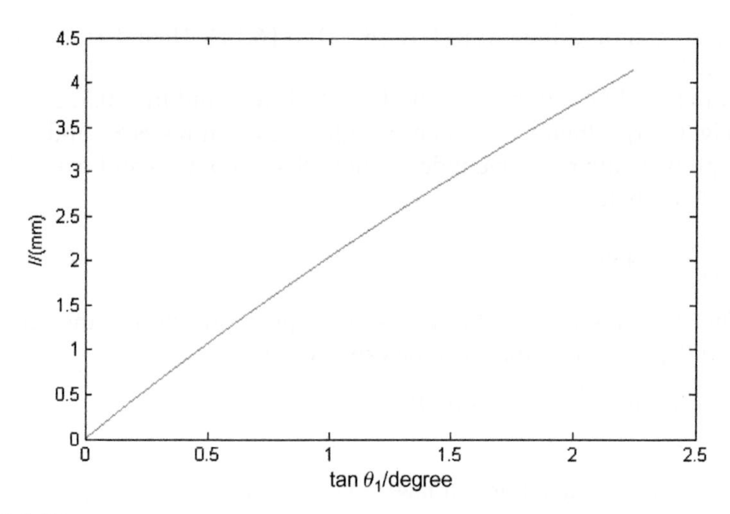

FIGURE 6.17

Relationship between relative displacement and $\tan\theta_1$.

FIGURE 6.18
Mask design.

The attenuation layer is made of 75 nm Cr film. Specific parameters are given in the following text.
The secondary reflection resistance layer is made of 60 nm Cr film.
The process flow of this diaphragm is shown in Fig. 6.19.
The mask design is drawn by using L-edit.

6.2.5 Calculation of Exposure Time

According to the parameters of the sun (spectrum: G2; apparent magnitude: -26.7; surface temperature: 5800K; energy flow: about 1.4 kW/m^2), energy of the sun is estimated at $2.51226.7 = 4.79 \times 1010$-fold of that of 0 magnitude stars. Therefore, the energy flow of 0 magnitude stars can be calculated as 2.9228×10^{-8} W/m^2.

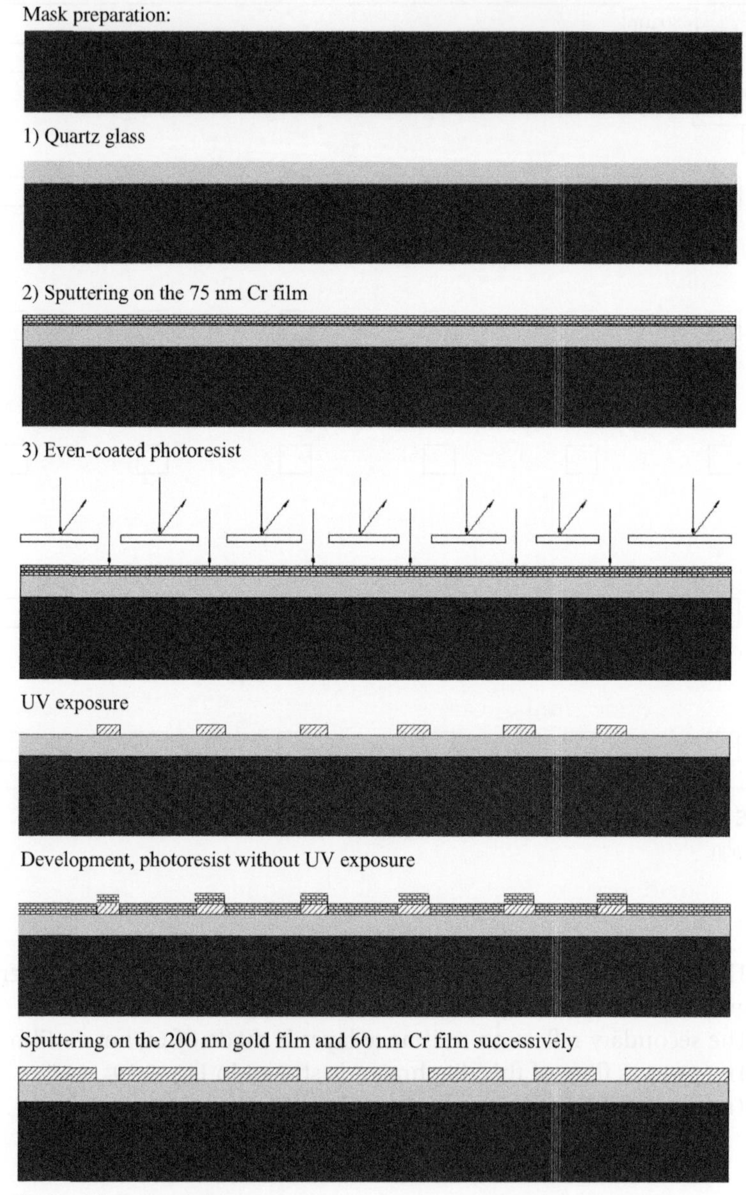

FIGURE 6.19
Process flow of the diaphragm.

Given fixed wavelength and temperature, the energy density function of black-body radiation is:

$$I(\lambda, T) = \frac{2 \cdot \pi \cdot h \cdot c^2}{\lambda^5 \cdot (e^{h \cdot c/(\lambda \cdot k_B \cdot T)} - 1)}$$

where $h = 6.626 \times 10^{-34}$ J·s, c is speed of light ($c = 2.997 \times 10^8$ m/s), k_B is the Boltzmann constant ($k_B = 1.38 \times 10^{-23}$ J/K) and, T is temperature (5800K).

Under black-body radiation, the total energies of $M_V = 0$ magnitude stars sum to 2.9228×10^{-8} W/m^2, the energy sum of all radiated wavelengths. In this case, the energy density distribution is shown in Fig. 6.20.

Star1000 spectra pass through within the range of 400–1000 nm. The corresponding energy flow is shown as follows.

According to the photon energy formula:

$$E = \frac{hc}{\lambda}$$

where E is photon energy (J), λ is photon wavelength (m), and h is Planck's constant ($h = 6.626 \times 10^{-34}$ J·s). Therefore, the energy flow of the system can be expressed by photons per second (Fig. 6.21).

Some photons convert into electrons on the photosensitive probe, which is called quantum efficiency (QE). The relation curve between the current/power of Star1000 and the wavelength is shown in Fig. 6.22.

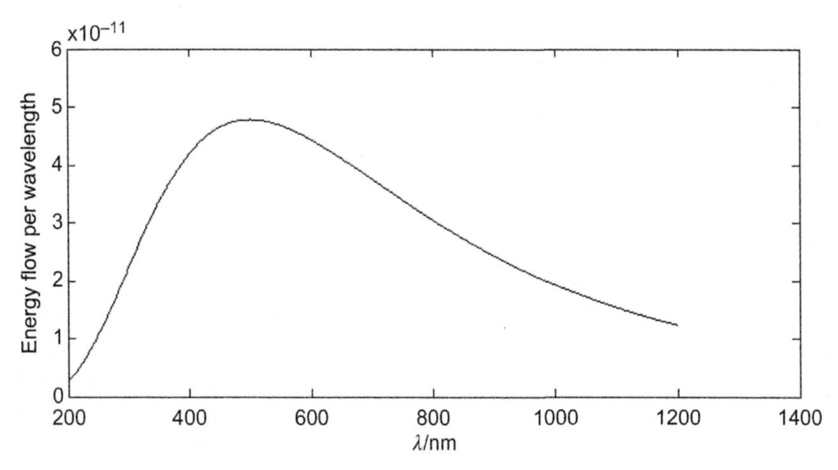

FIGURE 6.20
$M_V = 0$, $T = 5800$K, the energy flow per wavelength.

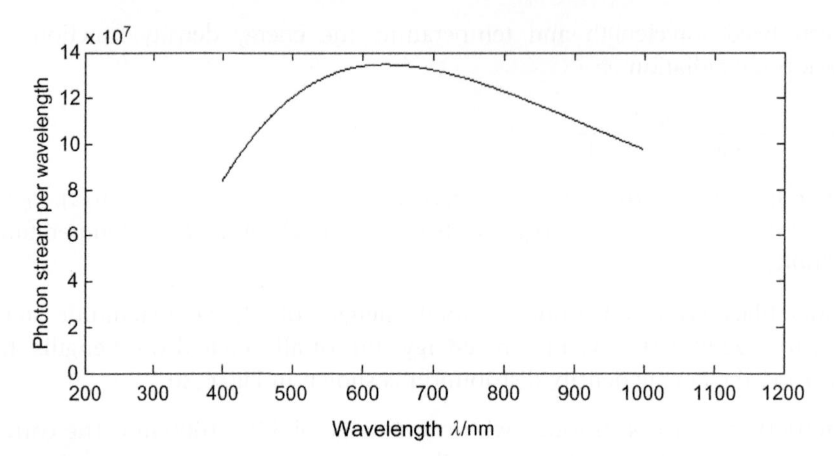

FIGURE 6.21
Photon streams distribution within 400–1000 nm wavelength ($M_V = 0$ and $T = 5800K$).

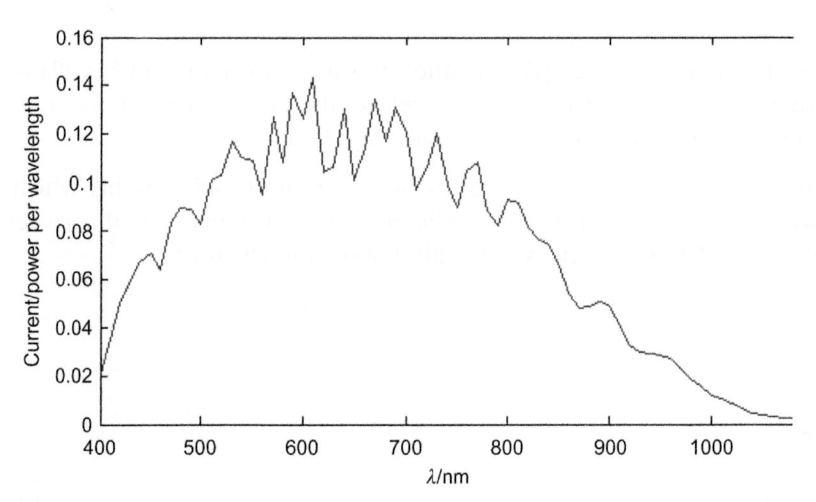

FIGURE 6.22
APS CMOS Star1000 spectral response.

According to Fig. 6.22, the practical current of the star sensor within the wavelength range of 400–1000 nm under the black-body radiation energy flow ($M_V = 0$ and $T = 5800K$) can be calculated. Since an electron carries 1.6×10^{-19} C electric quantity (1 A = 1 C/s), the relationship between the stream of electrons ($I_e(\lambda)$) and wavelength can be gained (Fig. 6.23).

After comparing Figs. 6.22 and 6.23, the QE variation of Star1000 with the wavelength can be gained (Fig. 6.24).

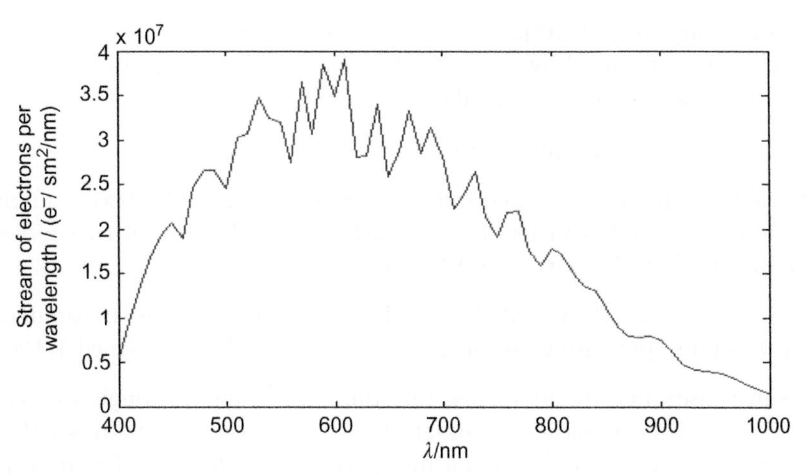

FIGURE 6.23

Relation curve between $I_e(\lambda)$ and λ ($M_V = 0$, $T = 5800K$, 1 m^2).

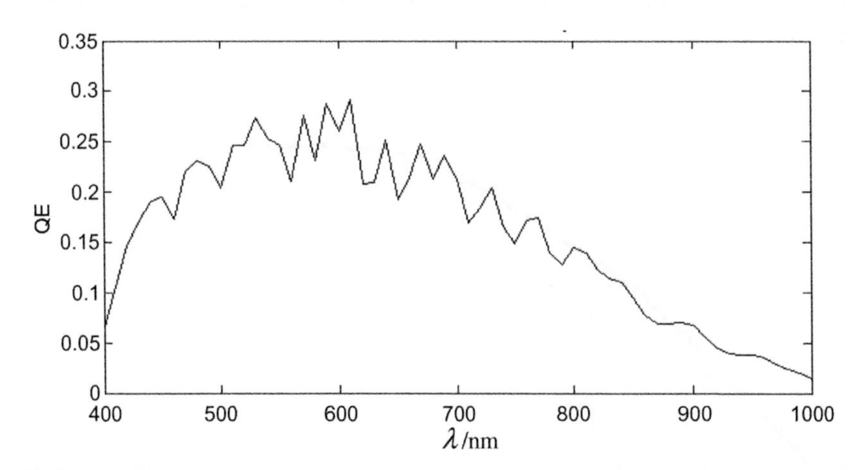

FIGURE 6.24

Wavelength-related QE variation of Star1000.

The integral of QE:

$$E_{sum} = \int_{400}^{1000} I_e(\lambda)d\lambda = 1.192 \times 10^{10} \quad (e^-/sm^2)$$

This result indicates that 1.192×10^{10} e^- can be produced on the Star1000 probe in 1 second on a 1 m^2 lens aperture by stars ($M_V = 0$, G2 spectrum).

Similarly, suppose the projection rate of the optical system is T, then the total electrons produced by the sun ($M_V = -26.8$) in the exposure time t on a 225 μm^2 pixel area can be calculated:

$$N_e = 1.192 \times 10^{10} \times 225 \times 10^{-12} \times 10^{\frac{2}{5} \times (0-(-26.8))} \times Tt \quad (e^-)$$

The relationship curve between output voltage of Star1000 sensor and electron number is shown in Fig. 6.25. To achieve excellent linearity, we determine the number of electrons as 12,000e$^-$.

If $t = 1 \times 10^{-3}$ s, $T = 0.85 \times 10^{-3}$. Considering the effects of glass transmittance and coating film processing, the attenuation of the Cr film is designed 1/1000.

The sun sensor mainly attenuates the luminous energy through Cr-coated optical glass. The Cr coat can maintain the characteristics of the original sun ray as much as possible, including the spectral characteristics. The transmittance of sun rays can be controlled by the thickness of the Cr film.

Every 7.5 nm thickness increment of the Cr film can attenuate the luminous energy by about 50%. The relationship curve between the optical power attenuation rate and thickness increment of Cr film (Δ_{Cr}) is presented in Fig. 6.26.

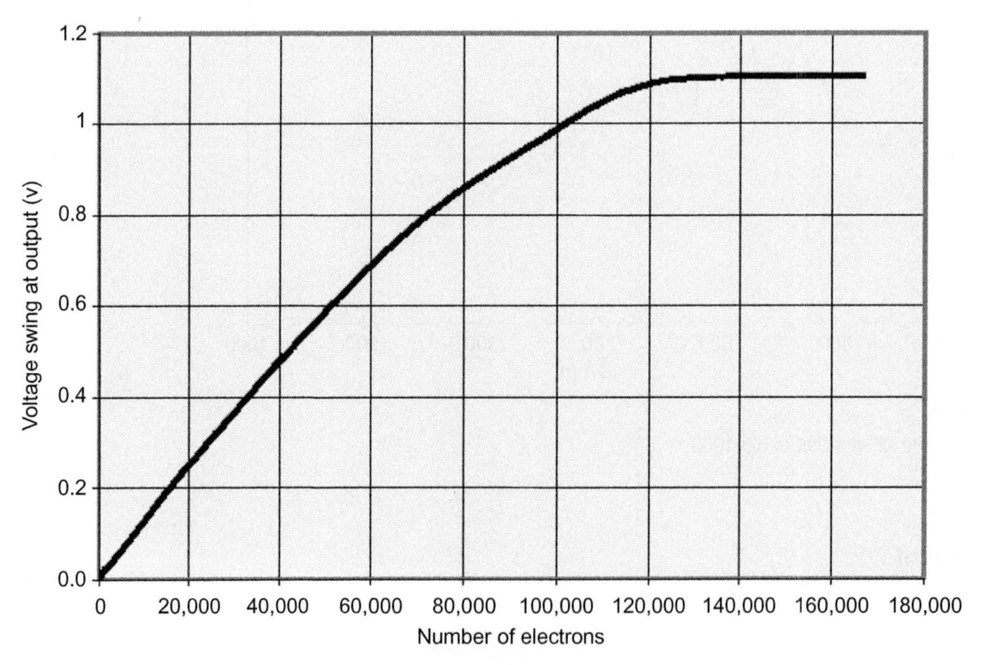

FIGURE 6.25
Electron—voltage response curve of the imaging sensor.

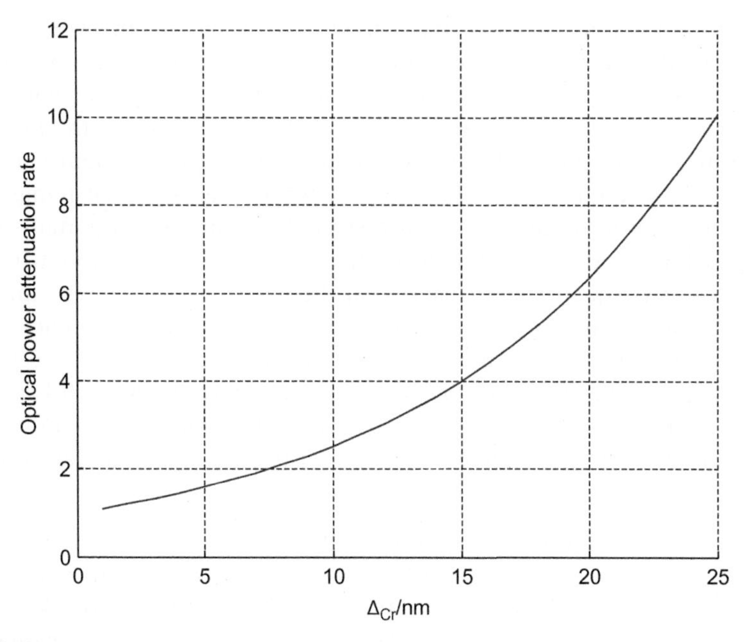

FIGURE 6.26
Relationship curve between optical power attenuation rate and thickness increment of Cr film.

In other words, if the transmittance of the optical system is T, the thickness of the Cr film (m) is:

$$m = 7.5 \log_2\left(\frac{1}{T}\right) = 75 \ (\text{nm})$$

6.2.6 FEIC Algorithm

The sun spot prediction algorithm and the subsampling technology, as well as multiwindow extraction technology of the APS imaging sensor, are called future extraction technology [38,39]. This is the basic guarantee of small size, low-power consumption, and high updating rate of the APS sun sensor. Different from the orderly reading mechanism of the CCD, the APS imaging sensor can read pixels directly according to the X and Y address registers. For a multiaperture sun sensor, pixels of sun spot arrays are far lower than the whole image array. Generally speaking, a single sun spot on the APS probe occupies about 9–25 pixels (3×3–5×5). Therefore, the 6×6 sun spot array may occupy 900 pixels as a maximum, only 1/1000th of the whole Star1000 imaging sensor.

Each sun spot is viewed as an extraction window. Therefore, the APS sun sensor has a total of 36 extraction windows, covering information of all sun spots. Under normal circumstances, small sun spot displacements exist between two successive frames of sun spots. Each sun spot in the previous frame can be mapped on the current frame according to a certain mapping rule [40]. As a result, under a fixed time interval between given frames, the mean angular velocity and sun spot displacement are two important influencing factors of the size and position of the future extraction window. We carried out a quantitative analysis on the relationship between sun spot displacement variance (Δl), sun spot displacement (l) and incidence angle (θ), and angular velocity of satellite ($\dot{\theta}$) based on two successive image frames. Suppose the time interval between two image frames is 0.1 second, $\theta = 0\text{–}64$ degrees, $\dot{\theta} = 1 \sim 3$ degrees/s, $h_2 = 0.94$ mm, $h_3 = 1.00$ mm, $h_4 = 0.528$ mm, $n_{Glass} = 1.4586$, and $n_{Air} = 1.00027$, then the relationship between l, Δl and θ, $\dot{\theta}$ is shown in Fig. 6.27.

In Fig. 6.27, the maximum sun spot displacement between two successive image frames is less than 0.06 mm (4 pixels). Except for such maximum sun spot displacement, the future extraction algorithm also has to take noise robustness as well as recursiveness and simplicity of hardware implementation into account. Therefore, our algorithm amplifies all four extraction window sides of current frame by 4 pixels, respectively, based on the image window calculated from the previous frame, forming a 13×13 pixel window (P). The center of P is (x, y), representing the current sun spot displacement compared to the sun spot at 0 degree incidence angle. According to the diaphragm

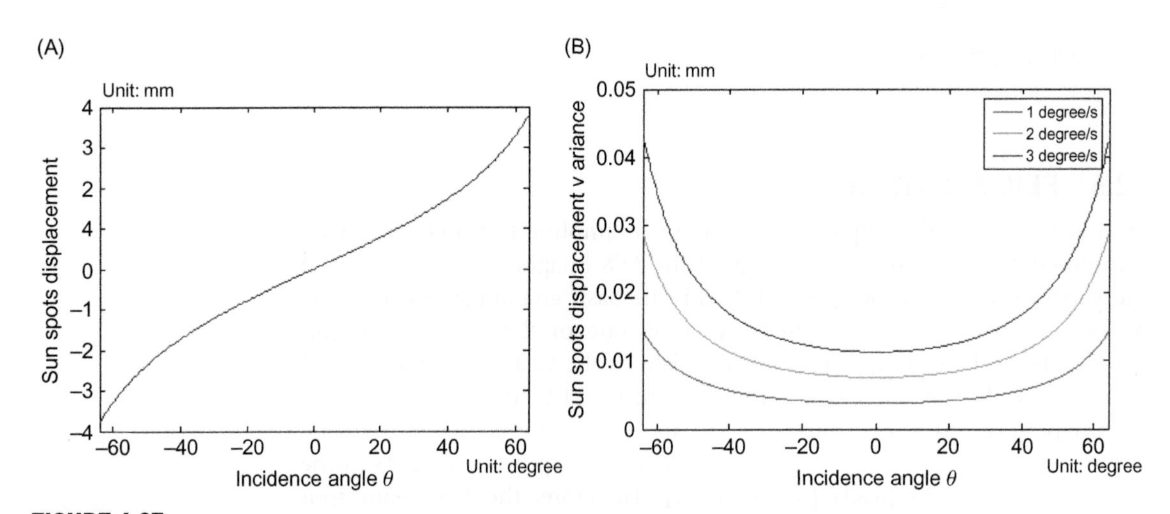

FIGURE 6.27

Relationship between l, Δl and θ, $\dot{\theta}$. (A) Relationship between sun spot displacement and θ, (B) relationship between sun spot displacement variance and θ, $\dot{\theta}$.

structure, a total of 36 windows can be extracted at a time $(P_1 \cdots P_{36})$, with corresponding center coordinates of $(x_1, y_1) \cdots (x_{36}, y_{36})$.

6.2.6.1 High-Accuracy Sun Spot Determination Based on Image Correlation Algorithm

In this book, the image correlation algorithm of the APS sun sensor is composed of a correlation algorithm based on a sun spot and centroiding algorithm based on correlation results.

1. Image correlation algorithm based on a sun spot template

   Although the image correlation algorithm plays an important role in pattern identification due to its high accuracy and reliability [41], it is seldomly used in sun sensors (especially area array imaging sensor) because of its heavy computation requirement. However, the APS CMOS sun sensor can provide fast sun image extraction and sun spot-related technologies through its future extraction mechanism.

   Window covering sun spots can be extracted directly through the future extraction algorithm. However, since a single sun spot occupies $3 \times 3 - 5 \times 5$ pixels, but the sun spot template generally is determined by 5×5 pixels, the future extracted sun spots must be amplified to 13×13 pixels in order to achieve perfect correlation (Fig. 6.28).

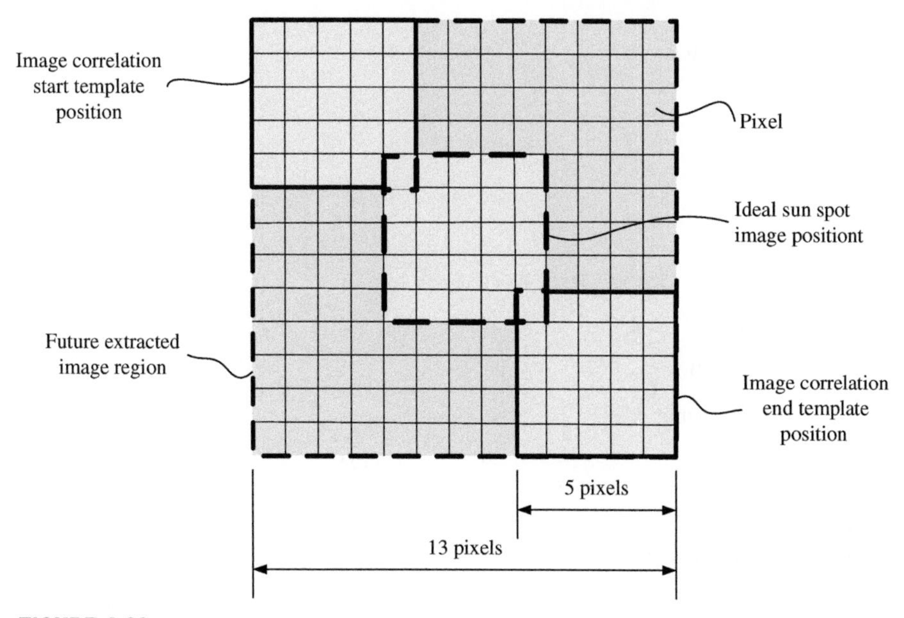

FIGURE 6.28
Future extraction and image correlation algorithm of a single sun spot.

The following text introduces the correlation algorithm of a single sun spot. Suppose the single sun spot template is T_1 (5 × 5 pixels), the future extracted image window is P_1 (13 × 13 pixels), $T_1(i,j)$ and $P_1(i,j)$ are the gray values of image (i,j) within T_1 and P_1, respectively, then the correlation algorithm can be expressed as:

$$C_1(m,n) = \sum_{i=1}^{5}\sum_{j=1}^{5} P_1(m+i, n+j)T_1(i,j) \qquad m,n = 0 \cdots 8 \tag{6.15}$$

where C_1 is the correlation matrix of a single sun spot (9 × 9 pixels). Fig. 6.29 presents a typical correlation matrix of single sun spot.

Next, we can get the correlation matrix of each sun spot (C_i). A total of 36 correlation matrixes can be gained according to the diaphragm structure, recorded as $C_1, C_2, \ldots, C_{36}$. Their corresponding templates are $T_1, T_2, \ldots, T_{36}$ and the extraction windows are $P_1, P_2, \ldots, P_{36}$.

2. Centroiding algorithm based on a sun spot correlation matrix

A centroiding algorithm is a universal algorithm for sun sensors to improve resolution during signal processing. In traditional centroiding algorithms, the centroid is calculated through the user's designated window with peak pixels [42] (Fig. 6.30).

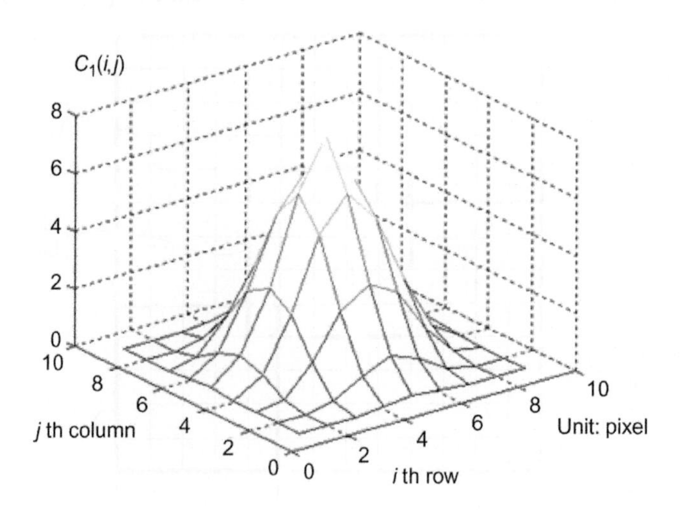

FIGURE 6.29
A typical correlation matrix C.

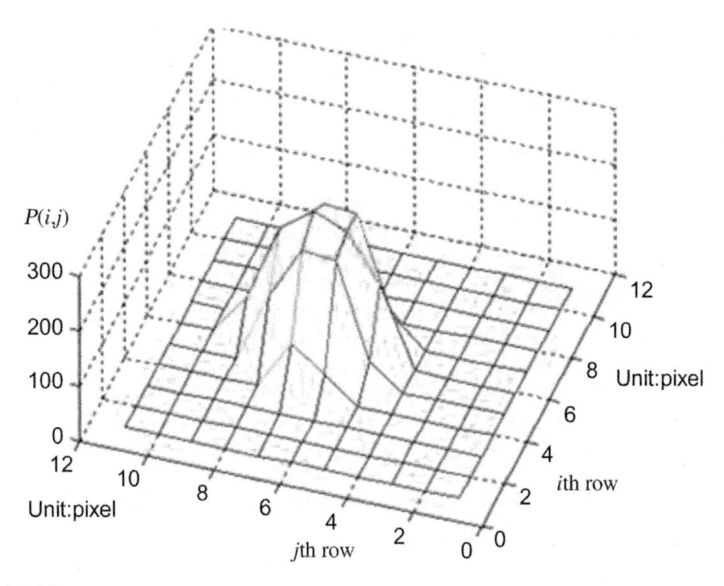

FIGURE 6.30
Traditional centroiding algorithm.

Suppose the peak pixels locate on (x_{peak}, y_{peak}) and $P(i,j)$ is the gray value of the ith row and jth column, then the window pixels used in the centroiding algorithm are:

$$x_c = \frac{\sum_{i=x_{peak}-k}^{x_{peak}+k} \sum_{j=y_{peak}-k}^{y_{peak}+k} P(i,j) \times i}{\sum_{i=x_{peak}-k}^{x_{peak}+k} \sum_{j=y_{peak}-k}^{y_{peak}+k} P(i,j)}$$
$$y_c = \frac{\sum_{i=x_{peak}-k}^{x_{peak}+k} \sum_{j=y_{peak}-k}^{y_{peak}+k} P(i,j) \times j}{\sum_{i=x_{peak}-k}^{x_{peak}+k} \sum_{j=y_{peak}-k}^{y_{peak}+k} P(i,j)} \tag{6.16}$$

In Eq. (6.16), the window size is determined by $2k + 1 \times 2k + 1$. A centroiding algorithm is susceptible to the center position. Due to effects of image noise, sun ray scattering, and intensity change [43], the center position will shift to a certain extent, especially under saturation of many pixels. Meanwhile, window size can affect the calculation accuracy to a certain extent. An over-large window will involve excessive noises into the algorithm, while an over-small window will lose useful signals. Although a centroiding algorithm has so many shortcomings, it has still proved simple and efficient in most cases.

The centroid windows in the centroiding algorithm based on the correlation matrix of sun spots are simplified into $C_1, C_2, \ldots, C_{36}$ directly. Take the first sun spot for instance; its centroid $(\tilde{x}_1, \tilde{y}_1)$ is calculated from:

$$\tilde{x}_1 = \frac{\sum_{m=0}^{8}\sum_{n=0}^{8} C_1(m,n)\times(m-4)}{\sum_{m=0}^{8}\sum_{n=0}^{8} C_1(m,n)}$$
$$\tilde{y}_1 = \frac{\sum_{m=0}^{8}\sum_{n=0}^{8} C_1(m,n)\times(n-4)}{\sum_{m=0}^{8}\sum_{n=0}^{8} C_1(m,n)} \tag{6.17}$$

Then, the displacement between the first sun spot and the template center (x_{c1}, y_{c1}) can be calculated:

$$x_{c1} = x_1 + \tilde{x}_1 - \underline{x}_1$$
$$y_{c1} = y_1 + \tilde{y}_1 - \underline{y}_1 \tag{6.18}$$

where $(\underline{x}_1, \underline{y}_1)$ is the center coordinates of the first sun spot gained in the laboratory, which will be analyzed specifically in the sun sensor test. Similarly, the center of each sun spot can be calculated from Eqs. (6.17) to (6.18). In this APS sun sensor, the displacements between 36 sun spots and the template are calculated, and recorded as $(x_{c1}, y_{c1})\cdots(x_{c36}, y_{c36})$.
Due to the image noise, the sun spot displacement under vertical sun ray incidence (x_c, y_c) can be viewed as the weighted mean of $(x_{c1}, y_{c1})\cdots(x_{c36}, y_{c36})$:

$$x_c = \sum_{i}^{36} a_i x_{ci}$$
$$y_c = \sum_{i}^{36} b_i y_{ci} \tag{6.19}$$

where a_i, b_i are weight coefficient, $\sum a_i = 1$ and $\sum b_i = 1$.

6.2.6.2 *Optimization of the Centroiding Algorithm*

According to the imaging model of the sun sensor, given a fixed attitude of the sun sensor, the displacements between all sun spots and corresponding sun spots under vertical sun ray incidence are theoretically equal:

$$x_{c1} = x_{c2} = \cdots = x_{c36}$$
$$y_{c1} = y_{c2} = \cdots = y_{c36} \tag{6.20}$$

However, this is impractical due to the imaging noise and calculation error. They differ slightly from each other (<0.2 pixel). It can be seen from Eq. (6.20) that x_{c1} includes the center pixel coordinate of the future extraction window (x_1), the centroid coordinate of template center ($\underline{x}_1$), and the center coordinate of correlation matrix $(\tilde{x}_1)\cdot\underline{x}_1$ is a constant which was stored in electronic equipment directly after being calculated in the laboratory and x_1 is an integer (pixel). We can conclude that $\Delta x_1 = \Delta x_2 = \cdots = \Delta x_{36}$ by defining $\Delta x_1 = x_1 - \underline{x}_1$, indicating that displacements between all sun spots within the future extraction window and corresponding sun spot template position are equal:

$$\begin{cases} \Delta x = \Delta x_1 = \Delta x_2 = \cdots = \Delta x_{36} \\ \Delta y = \Delta y_1 = \Delta y_2 = \cdots = \Delta y_{36} \end{cases} \tag{6.21}$$

Substituting Eqs. (6.21) and (6.20) into Eq. (6.19),

$$\begin{aligned} x_c &= \Delta x + \sum_{i=1}^{36} a_i \tilde{x}_i \\ y_c &= \Delta y + \sum_{i=1}^{36} b_i \tilde{y}_i \end{aligned} \tag{6.22}$$

This confirms that the accuracy of the sun sensor is mainly determined by the centroid in the sun spot correlation algorithm.

a_i and b_i are selected mainly though the correlation matrix:

$$a_i = b_i = \frac{\sum_{m=0}^{8}\sum_{n=0}^{8} C_i(m,n)}{\sum_{i=1}^{36}\sum_{m=0}^{8}\sum_{n=0}^{8} C_i(m,n)} \tag{6.23}$$

Substituting Eqs. (6.22) and (6.23) into Eq. (6.20),

$$\begin{aligned} x_c &= \Delta x + \frac{\sum_{i=1}^{36}\sum_{m=0}^{8}\sum_{n=0}^{8} C_i(m,n)(m-4)}{\sum_{i=1}^{36}\sum_{m=0}^{8}\sum_{n=0}^{8} C_i(m,n)} \\ y_c &= \Delta y + \frac{\sum_{i=1}^{36}\sum_{m=0}^{8}\sum_{n=0}^{8} C_i(m,n)(n-4)}{\sum_{i=1}^{36}\sum_{m=0}^{8}\sum_{n=0}^{8} C_i(m,n)} \end{aligned} \tag{6.24}$$

To reduce multiple operations, Eq. (6.24) is reordered and recombined:

$$\begin{aligned} x_c &= \Delta x + \frac{\sum_{m=0}^{8}\left(\sum_{n=0}^{8}\sum_{i=1}^{36} C_i(m,n)\right)(m-4)}{\sum_{i=1}^{36}\sum_{m=0}^{8}\sum_{n=0}^{8} C_i(m,n)} \\ y_c &= \Delta y + \frac{\sum_{n=0}^{8}\left(\sum_{m=0}^{8}\sum_{i=1}^{36} C_i(m,n)\right)(n-4)}{\sum_{i=1}^{36}\sum_{m=0}^{8}\sum_{n=0}^{8} C_i(m,n)} \end{aligned} \tag{6.25}$$

Eq. (6.25) demonstrates that the centroid of the correlation matrix is the weighted mean of all sun spots and the centroid of sum of all correlation matrixes. This chapter defines the image correlation algorithm and centroiding algorithm of sun spots as the image correlation algorithm.

When (x_c, y_c) is calculated, the future extraction point in the next frame $(\Delta x, \Delta y)$ can be viewed equal to (x_c, y_c) directly. In this way, the sun spot image position in the next frame can be known through extraction within the 13×13 pixel window centered at (x_c, y_c). On this basis, the sun spot image position in the next frame can be calculated accurately through the correlation algorithm, thus realizing the recursion of FEIC algorithm in the sun sensor.

The implementation flow of the above-mentioned future extraction and imaging correlation algorithm of APS sun sensor is: Acquire sun spot template (T_i) and center coordinate of each sun spot $((\underline{x}_i, \underline{y}_i), i = 1 \cdots 36)$ under vertical sun ray incidence in the laboratory.

Suppose $(\Delta x, \Delta y)$ is captured; Estimate the center coordinates of each sun spot in the next frame through:

$$\begin{aligned} x_i &= \text{int}[\Delta x + \underline{x}_i + 0.5] \\ y_i &= \text{int}[\Delta y + \underline{y}_i + 0.5] \end{aligned} \tag{6.26}$$

where int[x] represents maximum integers smaller than x.

Determine the extraction region of each sun spot (P_i) according to (x_i, y_i).

Calculate the correlation matrix of each sun spot (C_i) from P_i and T_i.

Calculate (x_c, y_c) and further calculate (α, β) from the correlation equation. Meanwhile, transmit (x_c, y_c) to image extraction of the next frame directly and then return to Step 3. This is the whole implementation flow of FEIC algorithm of the APS sun sensor.

6.3 TECHNICAL RESEARCH OF THE APS MICRO STAR SENSOR

6.3.1 Overview

The basic principle of the APS micro star sensor was explored during the "Tenth Five-Year" under the support of the 973 Project of "New Method and New Principle of Functional Components of Micro Spacecraft." Based on associated research results, the 863 Project of "high-performance attitude and orbital integration determination system" was further improved during the "Eleventh Five-Year," achieving many breakthroughs concerning large aperture optical design, integrated system design, and low-power consumption electronics design of the star sensor. A star sensor with excellent performances (weight: <1.5 kg; power consumption: <2 W; accuracy: 7"; initial capture time: <1 second) and higher flexibility was developed and can be applied to microsatellites as well as micro/nano satellites.

6.3.2 Development Trend of APS Technology

Compared to the sun sensor, the star sensor has higher accuracy and weaker sensitive photoelectric signals. For the APS CMOS photosensitive probe [44], the APS star sensor is inferior to the APS sun sensor in imaging and anti-interference. This complicates the APS star sensor research. However, it is still an important research direction and hot spot in the world for its low power consumption, strong radiation resistance, and simple probe operation [36]. The reading principles of a typical APS image sensor and CCD image sensor are shown in Fig. 6.31 [45].

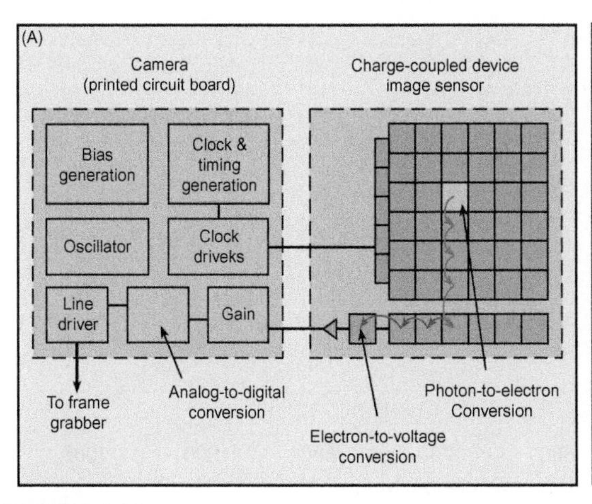

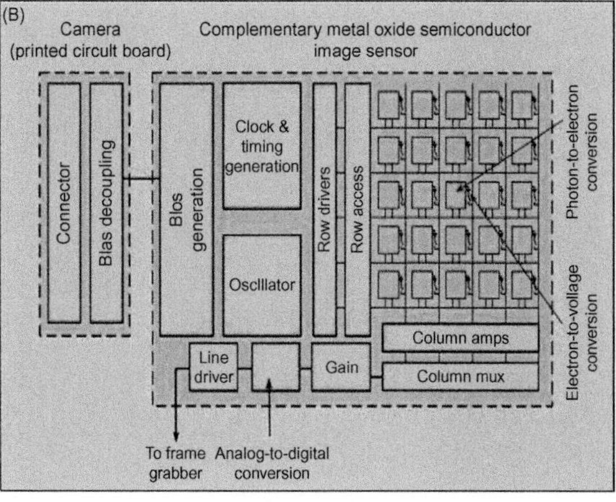

FIGURE 6.31

Reading principles of a CCD image sensor (A) and an APS image sensor (B).

Compared to the CCD image sensor, the advantages and disadvantages of an APS image sensor include the following [46].

Advantages:

1. Lower power consumption, smaller size and lighter weight;
2. Window exposure and direct access;
3. Simple application, integration convenience, single-chip digital image, and single power supply;
4. Applicable to oversized format images; compatible with other circuits;
5. Noise remains the same when data transmission is accelerated;
6. Strong radiation resistance: 10 Mrad (si),1000x CCD.

Disadvantages:

1. Lower filling factor (FF) (<50%);
2. Lower sensitivity. QE × FF generally ranges from 20% to 30%;
3. Big noise, especially the modal noise.

With the continuous development of APS CMOS, these advantages are improving gradually and great breakthroughs will be achieved in the future. Supported by ESA, CSCB have developed the HAS APS and LCMS APS sensors. Their QE and FF are improved significantly compared to that of the Star1000. Fig. 6.32 presents QE × FF of Star1000, Star250, HAS, and LCMS

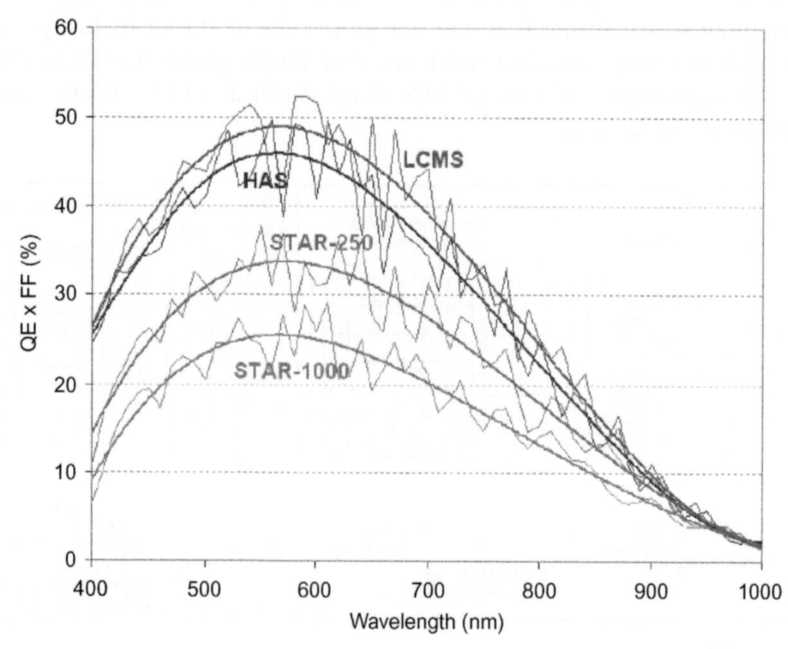

FIGURE 6.32
QE × FF of Star1000, Star250, HAS, and LCMS sensors [47].

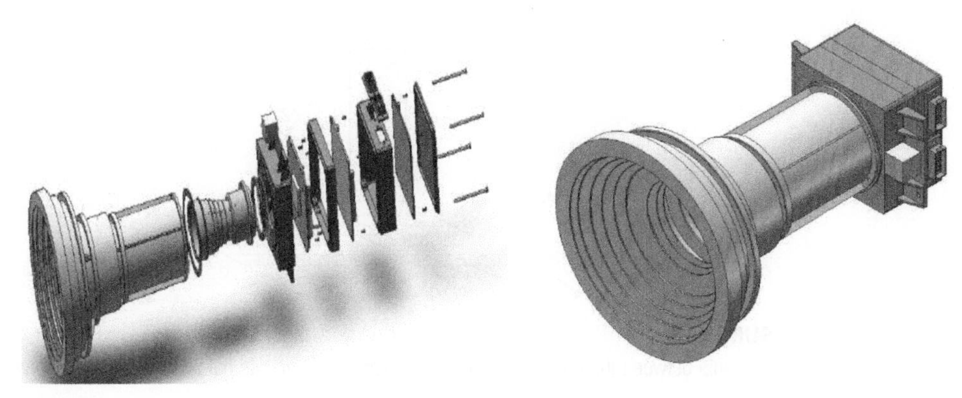

FIGURE 6.33
A star sensor.

sensors. The QE × FF of LCMS and HAS is approaching that of the CCD, laying foundations for the development of a high-accuracy APS star sensor.

The APS progress lays a solid foundation for the development of a new star sensor. The following text will introduce the core technologies as well as the basic design method and principles of the star sensor.

6.3.3 Overall Design

The star sensor is composed of a lens hood, camera lens, and electronics processing unit (Fig. 6.33). To meet the requirements of reliability and frequent modification, the electronics system uses three pieces of processing boards, namely the APS CMOS imaging system, the FPGA image acquisition and communication system, as well as the DSP star pattern identification and attitude calculation system.

6.3.3.1 Determination of FOV and Focal Distance

Generally, the pointing accuracy of a star sensor determines the FOV and focal distance. However, the FOV, focal distance, and probe size of a star sensor restrict one another (Fig. 6.34).

For example, if the designed pointing accuracy of the APS CMOS micro star sensor is 7″ (3σ), the angular resolution of each sub-pixel (1/10 pixel) can be designed 7″ according to the pixel amount of its photosensitive probe and accuracy of subpixel. Take the Star1000 imaging sensor, for instance. If $l_{pix} = 15\ \mu m$, then $\delta_{pix} = 1.5\ \mu m$ and the corresponding α can be viewed as the designed pointing accuracy (e.g., 7″). Therefore, the focal distance (f) can be calculated:

$$f \geq \frac{\delta_{pix}}{\text{atan}\,(\alpha)}$$

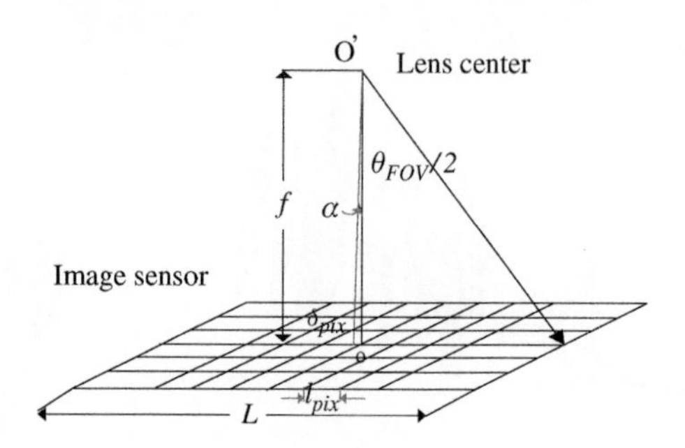

FIGURE 6.34

Relationship between the pointing accuracy of a star sensor and its parameters (e.g., FOV and focal distance).

This represents the accuracy of a single star. Theoretically, longer f leads to higher accuracy but smaller FOV. The relationship between focal distance, FOV, and probe is:

$$f \times \tan\left(\frac{\theta_{FOV}}{2}\right) = \frac{L}{2} \tag{6.27}$$

where f and θ_{FOV} are the focal distance and FOV of the star sensor, respectively, and L is the length of the photosensitive region of the probe.

To achieve the desired pointing accuracy of 7″, Star1000 (photosensitive pixel: 15 μm; photosensitive array: 1024 × 1024; photosensitive region: 15.36 mm) must have longer than 44.2 mm focal length. In this book, the focal length is determined at 50 mm in considering the industrial standard of optical lens system, processing difficulties of the lens, and constraints of FOV (17.46 degrees).

The accuracy of a single star is designed higher than 7″. Optical distortion and system installation and adjustment during the design will affect the accuracy to a certain extent. However, a star sensor generally integrates various stars during attitude calculation, thus enabling achievement of an overall accuracy of 7″.

6.3.3.2 Magnitude Determination of Navstar

According to the function provided by the SKY2000 star catalogue [48], the relationship between the average number of stars and the apparent magnitude is:

$$N(M_V) = 6.5e^{1.107\,M_V} \tag{6.28}$$

where $N(M_V)$ represents the number of navstars (magnitude $\leq M_V$) and M_V is the apparent magnitude.

Suppose stars distribute evenly on the celestial sphere, then the average number of stars within the FOV can be calculated according to the spherical degree of the FOV of the star sensor:

$$N_{FOV} = N(M_V)\frac{2\pi - 4\arccos\left[\sin^2\left(\frac{\theta_{FOV}}{2}\right)\right]}{4\pi} \tag{6.29}$$

where N_{FOV} represents the number of stars within the FOV and θ_{FOV} is the field angle.

The relationship between apparent magnitude, N_{FOV}, and θ_{FOV} can be gained through numerical simulation (Fig. 6.35).

Table 6.4 lists N_{FOV} under different magnitudes when θ_{FOV} 24.5 degrees.

The relationship between the pitch/yaw accuracy of the star sensor and the number of star is:

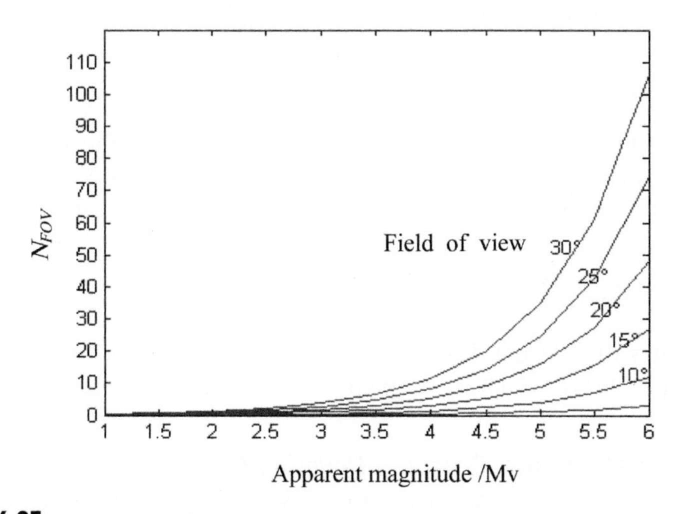

FIGURE 6.35
Relationship between apparent magnitude, N_{FOV}, and θ_{FOV} of the star sensor.

Table 6.4 Number of Stars Within 17.5 × 17.5 Degrees Under Different Magnitudes

Magnitude (M_V)	3	3.5	4	4.5	5	5.5	6
N_{FOV}	1.4	2.4	4.1	7.1	12.6	21.9	38.2

$$\sigma_{pitch,yaw} = \frac{\sigma_{star}}{\sqrt{N}} \tag{6.30}$$

where σ_{star} is the processing accuracy of a single star, N is number of stars, and $\sigma_{pitch,yaw}$ is the pitch/yaw accuracy of the star sensor.

Eq. (6.30) demonstrates that the more stars are involved, the higher the accuracy of the star sensor will be. However, this will increase system computations to tracking all navstars. The accuracy of the star sensor fails to improve significantly when it involves more than 15 stars. Considering impacts of ground experiment and model stars, the appropriate apparent magnitude of the star sensor is determined as 5.2 ($M_V = 5.2$). When the sensitive magnitude decreases, the lens diameter of the star sensor will reduce accordingly, reducing the size and weight of the star sensor and so is convenient for lens design.

Meanwhile, attention also is paid to the probability satisfying star pattern identification conditions for 24-h-a-day autonomous identification. Star pattern identification based on star diagonal distance requires at least four stars within the FOV. Abundant researches have confirmed that the number of stars within any FOV is subject to Poisson's distribution [49]:

$$P(X = k) = e^{-\lambda}\frac{\lambda^k}{k!} \tag{6.31}$$

where $P(X = k)$ is the distribution probability of star k in the FOV and λ is the average number of stars in the FOV ($\lambda = 15$ in this book). Therefore, the probability of having at least four stars in the FOV can be calculated:

$$P(X \geq 4) = 1 - \sum_{k=0}^{3} P(X = k) = 99.98\% \tag{6.32}$$

This indicates that 24-h-a-day initial attitude capture is feasible under most situations (99.98%), which is completely acceptable for a star sensor.

6.3.3.3 Initial Capture Time and Updating Rate

Different from the CCD star sensor, the APS CMOS star sensor generally has a lower QE, which requires longer exposure time and shorter processing time as much as possible to ensure the global updating rate. Generally speaking, in star sensors, the global attitude updating rate is inversely proportional to the exposure time. The attitude updating rate of a typical CCD star sensor is 5 Hz. In this book, the attitude updating rate of APS CMOS star sensor is also determined at 5 Hz.

The exposure time of the star sensor is directly related to the sensitivity of the photosensitive probe, as well as the entrance pupil diameter and transmittance of the lens. The exposure time of the APS CMOS will be introduced specifically in Chapter 7, Miniature Inertial Measurement Unit. The real-time star pattern processing method and fast 24-h-a-day autonomous star pattern identification algorithm save the initial capture time of star sensor significantly, about 1 second for an international advanced star sensor. Our designed star sensor with low magnitude, large FOV, and flow type only takes 0.5 second for the initial capture.

6.3.3.4 General Technical Indexes and Implementation Framework

Table 6.5 lists the general technical indexes of the APS CMOS micro star sensor.

Based on the comprehensive considerations on the technical indexes of a high-accuracy autonomous star sensor, the overall implementation of a star sensor will focus on high attitude measurement accuracy, real-time data processing, small size and weight, and low power consumption (Fig. 6.36).

The following text will introduce the major content of each part in this overall design and implementation plan

6.3.3.5 Optical System

As a weak target detection system, a star sensor has strict requirements on optical imaging and stray light inhibition. According to the general requirements of a star sensor, we designed the lens properties as:

1. Focal distance: 50 mm;
2. Relative aperture: 1/1.25;
3. Spectral band: 500–850 nm;
4. FOV: $2\omega = 20$ degrees, no vignetting;
5. Relative distortion: <0.1%
6. Transmittance: >72%.

According to the deep analysis results, a complicated double-Gauss optical structure is used as the optical system of the CMOS star sensor, thus promising large FOV and relative aperture. Additionally, no glued components or

Table 6.5 General Technical Indexes of APS CMOS Micro Star Sensor

FOV	17.5 × 17.5 Degrees
Sensitive magnitude	$\geq 5.2\ M_V$
Accuracy (3σ)	<7″ (pitch/yaw), 35″ (roll)
Data updating rate	≥ 5 Hz
24-h-a-day autonomous attitude capture time	≤ 1 s
Weight	≤ 1.1 kg (lens hood excluded)
Power consumption	≤ 1.5 W

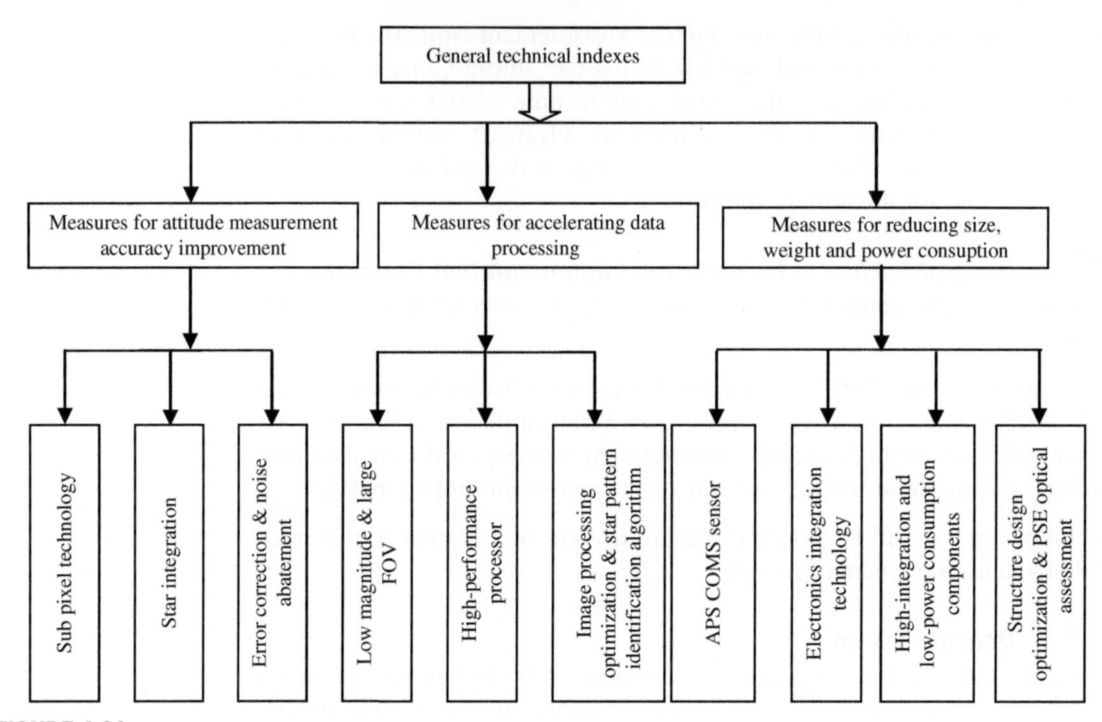

FIGURE 6.36
Implementation framework of an APS CMOS micro star sensor.

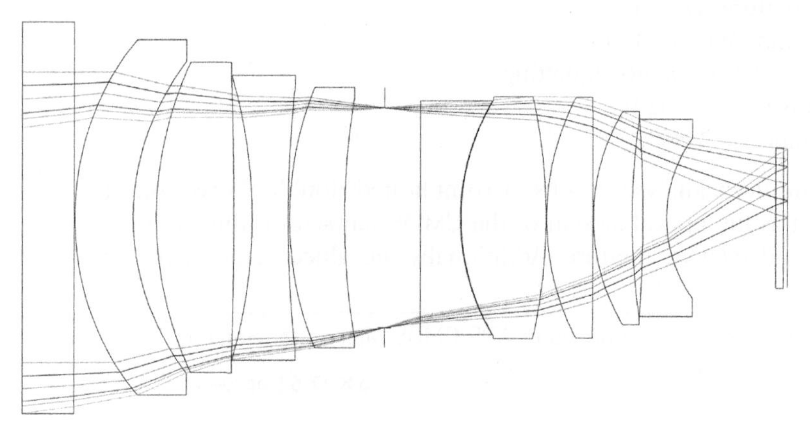

FIGURE 6.37
Lens structure of a star sensor.

radiative lanthanides glass are used in considering its application environment in space. The lens structure is shown in Fig. 6.37. This optical lens is composed of nine pieces of completely independent lens, covered by the quartz window in the front end to increase its space radiation resistance.

According to the on-orbit application features of a star sensor, to ensure the reliable on-orbit operation of the star sensor, the optical system shall pay attention to not only technical parameters (diapoint, aberration, and distortion required by traditional optical lens), but also temperature and on-orbit adaptation, especially their effects on system operation.

6.3.3.6 Effect of Temperature on Optical System

1. Uneven distribution of temperature

   The integrated star sensor is installed outside the satellite. Its bottom and the engine body of the satellite are for heat conduction, thus having relatively good temperature conditions. However, the lens and lens hood in its optical system extend longer, suffering more obvious temperature variation. Additionally, the heat from the power consumption of the system also can affect the temperature to a certain extent. Since the APS star sensor consumes a small amount of power (<2 W), its operating temperature only exerts slightly effect on the lens. According to the outer space law, the lens temperature of a star

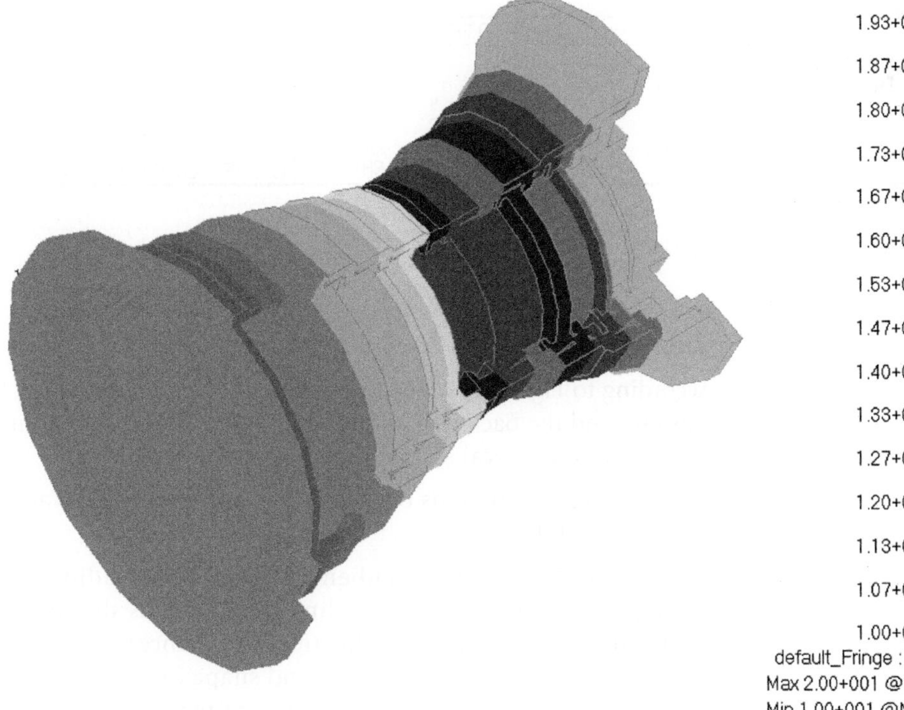

FIGURE 6.38
Temperature distribution with temperature gradient.

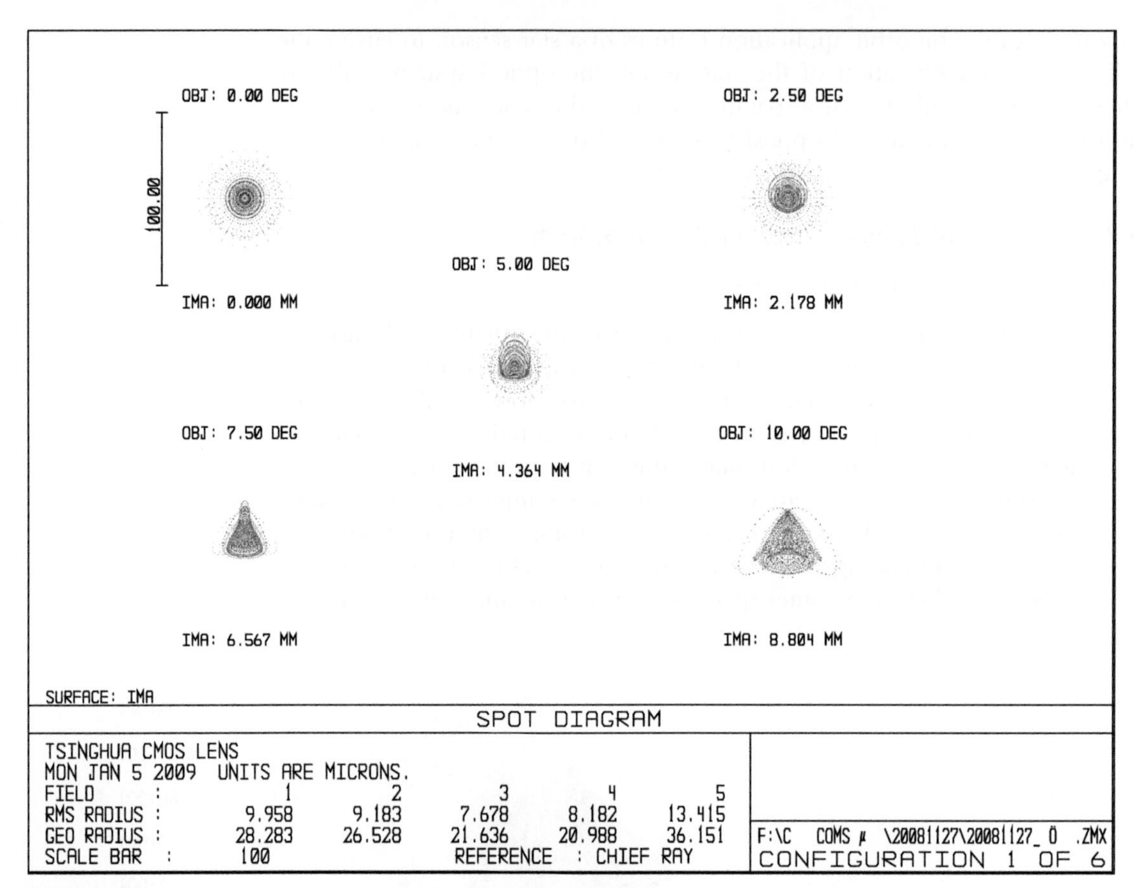

FIGURE 6.39
Diapoint under a 10°C axial temperature gradient.

sensor increases gradually from the inside to out, forming a temperature gradient of 10°C (Fig. 6.38)

According to Fig. 6.38, the temperature difference between the front lens end and the back lens end reaches 10°C. The corresponding diapoint of the optical system is shown in Fig. 6.39.

The RMS radius variations of the disc of confusion of FOV are presented in Table 6.6.

Conclusion: Temperature gradient influences lens radius, lens spacing, glass thickness, lens inclination, and bias slightly (smaller than system processing and adjustment tolerance), which can be viewed as insignificant to the size and shape of the disc of confusion. In other words, the system can maintain normal operations under this 10°C temperature gradient.

Table 6.6 Effect of 10°C Temperature Gradient (μm)

FOV (Degrees)	0	2.5	5	7.5	10
Initial RMS radius of the disc of confusion	9.943	9.316	8.094	8.525	14.106
RMS radius of the disc of confusion affected by temperature gradient	9.958	9.183	7.678	8.182	13.415
RMS radius variation of the disc of confusion	0.015	−0.133	−0.416	−0.343	−0.691

Table 6.7 Effects of Temperature Variation

Temperature Variation	FOV (Degrees)	RMS Radius of the Disc of Confusion (μm)	Energy Centroid Height (mm)	Focal Distance Variation (μm)
−40°C	0	9.893	0	−9
	2.5	9.152	2.1780	
	5	7.623	4.3632	
	7.5	7.888	6.5648	
	10	13.063	8.7972	
−20°C	0	9.892	0	−4.5
	2.5	9.202	2.1783	
	5	7.823	4.3639	
	7.5	8.167	6.5658	
	10	13.173	8.7984	
0°C	0	9.943	0	0
	2.5	9.307	2.1787	
	5	8.084	4.3645	
	7.5	8.504	6.5668	
	10	13.327	8.7998	
+20°C	0	10.044	0	+4.5
	2.5	9.466	2.1790	
	5	8.400	4.3652	
	7.5	8.892	6.5678	
	10	13.477	8.8011	
+40°C	0	10.193	0	+9
	2.5	9.674	2.1793	
	5	8.765	4.3658	
	7.5	9.324	6.5688	
	10	13.696	8.8024	

2. Temperature variation

The effects of temperature variation (±40 and ±20°C) on the disc of confusion, energy centroid, and focal distance are analyzed in Table 6.7.

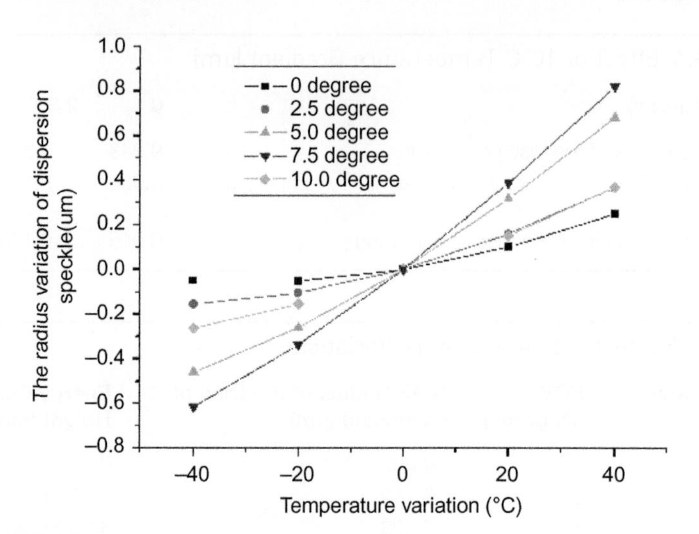

FIGURE 6.40
Relationship curves of temperature–RMS radius of the disc of confusion.

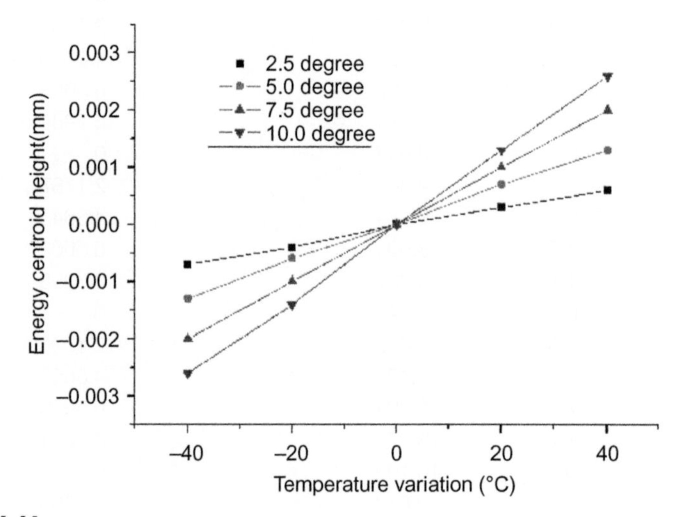

FIGURE 6.41
Relationship curves of temperature–energy centroid height.

The relation curves of the temperature–RMS radius of the disc of confusion and temperature–energy centroid height are shown in Figs. 6.40 and 6.41, respectively. Within a large temperature range, the energy centroid height changes within 3 μm (1/5 pixel), which will not influence the accuracy of the star sensor.

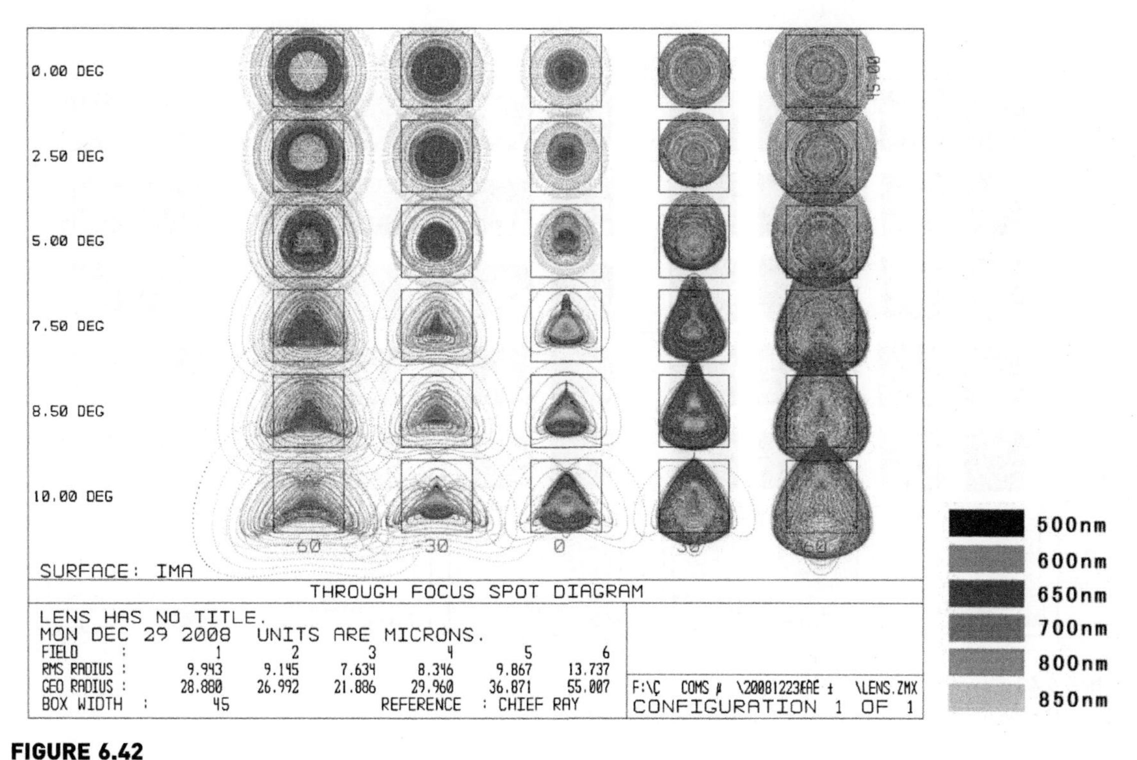

FIGURE 6.42
Focal distance diapoint.

3. Focal distance analysis

   The focal distance diapoint of the optical system is shown in Fig. 6.42. During the movement within $\pm$ 60 μm target surface, the circle of confusion largens and energy diffuses.

The variation of the focal distance diapoint is analyzed through a probe sampling simulation (Fig. 6.43). The sampling pixel of the disc of confusion is 15 μm $\times$ 15 μm, equal to a pixel of CMOS component. The practical image simulation takes the crosstalk effect of CMOS components (16% crosstalk in component manual) into account. Furthermore, the energy concentration degrees within the enclosing circle at different positions are exhibited in Fig. 6.43.

The simulation analysis results demonstrate that the star sensor often applies energy centroiding algorithm based by gray value. Under normal conditions, a gray value of 3×3 pixels is used as the basic algorithm. According to the above-mentioned imaging point analysis, a 3×3 pixel algorithm is easy to be accomplished with higher energy concentration degree within the whole temperature range when the image plane is installed at -30 μm.

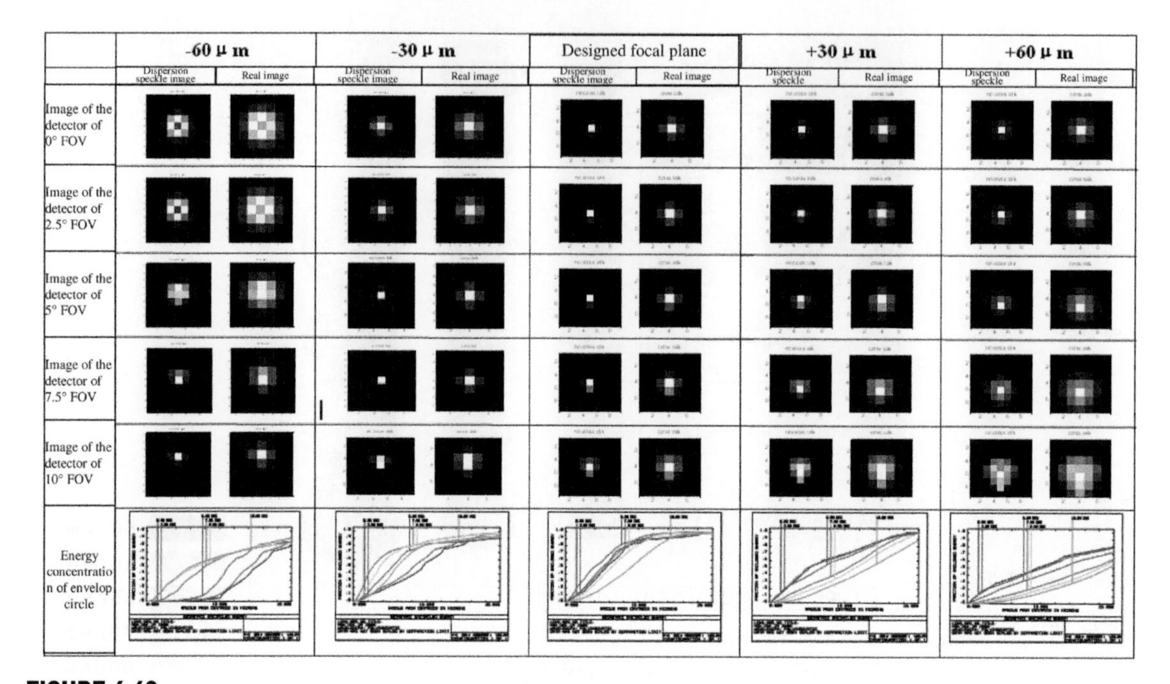

FIGURE 6.43
Relationship curve of optical surface position variation and star point imaging.

6.3.3.7 Electronics

With respect to the functional requirements, the star sensor needs peripheral configuration circuit module of Star1000 and FPGA module for Star1000 sensor exposure and read, DSP module for the purpose of fast star pattern identification, star point extraction and attitude algorithm, MCU module for flexible external interface (RS-232, RS-422 or CAN interface) and flexible multipattern switch, RAM/EPROM module for star pattern, intermediate variable and star catalogue storage as well as storage program and operation program, and power supply module for the whole system.

The electronics of the designed APS CMOS micro star sensor takes full advantages of the above hardware conditions and optimizes the whole procedure. It not only increases the high-performance processing functions of the star sensor, but also reduces the functional components and simplifies the electronics system design. The designed APS CMOS micro star sensor has two characteristics: (1) the whole hardware circuit design focuses on the FPGA and DSP system that are connected by the dual-port RAM (DPRAM), reducing the application of the large-capacity single-port RAM and bus switch in the FPGA in the original design; and (2) flow operation: images of the star sensor contain less useful information but more meaningless background

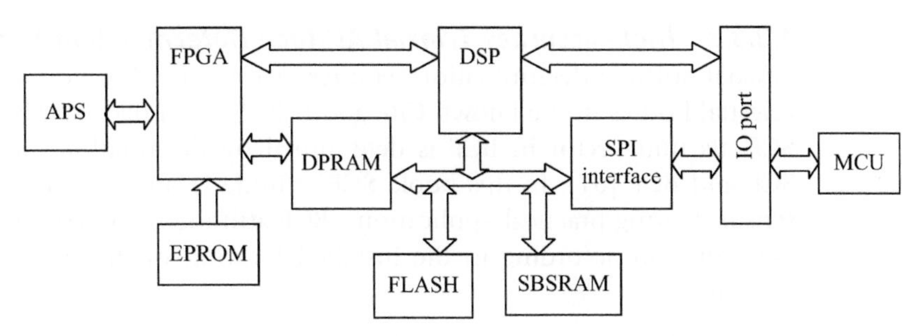

FIGURE 6.44
Data processing and circuit design.

noises. Therefore, data are processed directly in the FPGA. Effective pixels are extracted and stored in small-capacity DPRAM. DSP can access the DPRAM directly, making it possible for simultaneous reading and input. Such a flow operation can increase the overall updating rate, save large-capacity single-port RAM and bus switch, and lower power consumption. The corresponding circuit hardware structure is shown in Fig. 6.44.

6.3.3.8 Star Identification Software

Traditional star identification algorithms (e.g., triangle method [50–52] or grid method [53,54]) are disadvantageous due to large star pattern template data storage and slow searching. Recently, neural network algorithms [55–57] were developed, which are superior to traditional algorithms because of their smaller database capacity and better real-time and robustness. However, they require massive calculation and large training set [58], thus failing to achieve ideal practical application effects.

Mortari et al. put forward the fast k-vector range searching techniques [59,60], which determine possible star pairs based on known star diagonal distances. Xing Fei et al. (Tsinghua University) proposed the star pattern identification and matching method based on navstar domain and developed a fast 24-h-a-day autonomous star pattern identification method by combining with a k-vector searching technique. Based on the direct positioning of two navstars forming the star diagonal distance by using the k-vector, it uses the navstar domain algorithm to transform domains of five pairs of navstars in six pairs in the pyramid structured by four stars. Next, it combines domains of the sixth pair and the previous five pairs of navstars, and identify all four navstars in the pyramid structure simultaneously during the primary circulation. The direct positioning based on k-vector saves search time and simultaneous identification of four stars based on the navstar domain saves time for manifold cycles and comparison. Therefore, this algorithm is characteristic of fast processing, simple programming logic, and high reliability.

6.3.3.9 High-Accuracy Triaxial Attitude Determination Method

Triaxial attitude determination requires at least the direction vectors of two celestial bodies in the known CRF and BCF. Since a fixed star is measured in SCF, its unit vector in BCF is determined by the rotation matrix between SCF and BCF [61]. In this book, SCF attitude is output directly by the star sensor. During practical applications, BCF attitude is gained through coordinate rotation according to the installed location of the star sensor on the satellite.

To take advantage of multistar observation and quaternion attitude representation of a large FOV star sensor, we determine triaxial attitude by using the QUEST algorithm to solve the Wahba problem [62–65].

Let $\hat{W}_i$ represent the direction vector of fixed star in SCF; $\hat{V}_i$ represents the direction vector in CRF; a is positive weight, and n is the total number of observed fixed stars. The following coefficients are defined:

$$B = \sum_{i=1}^{n} a_i \hat{W}_i \hat{V}_i^T$$

$$S = B^T + B$$

$$\vec{z} = \sum_{i=1}^{n} a_i \hat{W}_i \times \hat{V}_i$$

$$\sigma = tr[B]$$

$$K = \begin{bmatrix} S - \sigma I & \vec{z} \\ \vec{z}^T & \sigma \end{bmatrix}$$

Then, the optimal attitude determined by the QUEST algorithm is:

$$q_{opt} = \frac{1}{\sqrt{1 + |y^*|^2}} \begin{bmatrix} y^* \\ 1 \end{bmatrix}$$

where

$$y^* = [(\lambda_{\max} + \sigma)I - S]^{-1}\,\vec{z}$$

Calculating $\lambda_{\max}$ from the characteristic equation of K by using iteration method:

$$\lambda^4 + a\lambda^3 + b\lambda^2 + c\lambda + d = 0$$

where

$$a = tr[K] = 0$$
$$b = -2(tr[B])^2 + tr\left[adj\left(B + B^T\right)\right] - \vec{z}\,z$$
$$c = -\,tr\left[adj(K)\right]$$
$$d = \det(K)$$

It can be confirmed that λ_{max} values are 1 when there is no measurement error. Therefore, the iteration operation is conducted from $\lambda_{max} = 1$.

After the star pattern identification, software of the star sensor prototype will calculate and output the attitude quaternion of the star sensor by using QUEST algorithm.

6.3.4 APS CMOS Micro Star Sensor System Software

Electronics system design is the basic implementation platform of the whole star sensor. Major functions stated in Chapter 5, Ground Tests of Micro/Nano Satellites, are achieved on this platform. The designed APS CMOS micro star sensor not only achieves low power consumption (<1.5 W), but also gives consideration to environmental requirements (e.g., space radiation and temperature) in its electronics design.

The implementation procedure of major functions of the designed APS CMOS micro star sensor on the electronics system is presented in Fig. 6.45.

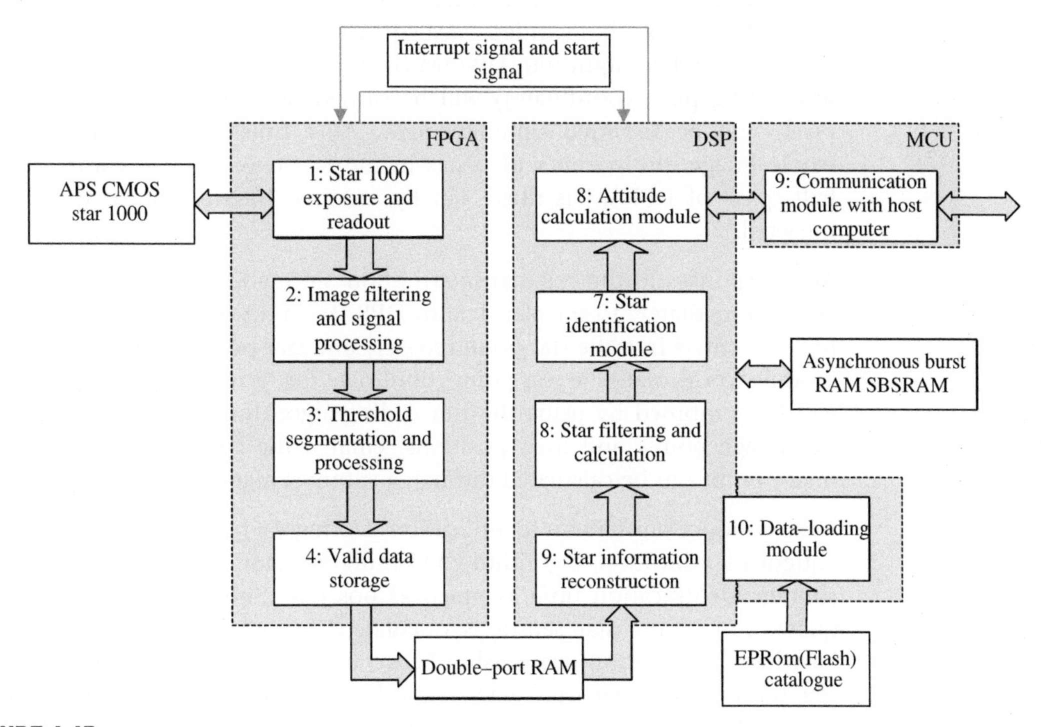

FIGURE 6.45
Workflow of the APS CMOS micro star sensor.

Modules 1, 2, 3, and 4 are implemented through FPGA; modules 5, 6, 7, 8, and 10 are implemented by the DSP; and module 9 is implemented mainly through the MCU. Data transmission between the FPGA and DSP is mainly achieved through 64K DPRAM. Meanwhile, external interruption of the DSP is connected with I/O of FPGA for communication and data exchange. Two-way communication between the DSP and MCU is achieved through an SPI (synchronous serial port). 512K EPROM (FLASH in experiment) and 2M SBSRAM are used to store star catalogue and some associated correction data.

6.3.4.1 General Workflow of a Circuit

The MCU communicates with the host computer through a serial port or CAN and will start the DSP upon receiving operation command from the host computer. At the beginning of operation, the DSP has to initialize the system and transfer star catalogue data stored in EPROM into SBSRAM of the DSP and RAM inside the DSP through direct memory access (DMA) of the DSP.

After the data initialization, the star sensor begins normal operation. Firstly, the FPGA begins to capture the image, starts the Star1000 probe, and collects gray values of pixels in order. Next, data filtering, threshold segmentation, and effective data input into DPRAM (including gray value of pixels and corresponding pixel coordinates) will be carried out. The number of effective pixels will be recorded simultaneously. After finishing data input of the whole image, the number of pixels in the last two bytes of corresponding half space of DPRAM is filled. Finally, external interruption of the DSP is triggered.

After responses to the external interruption are made, the DSP will start the DMA immediately to transfer data in DPRAM to RAM inside the DSP. Next, the DSP enters into the star point recovery and star point centroid calculation module to calculate the star point coordinate for recovery. All effective pixels will be combined by using the run coding algorithm to get the number of star image points and their positions. Finally, the coordinates of each star image point can be calculated through a subpixel algorithm.

Next, conduct star pattern identification. During 24-h-a-day autonomous star pattern identification, algorithm combining k-vector and navstar domain to shorten identification time as much as possible. Subsequently, the attitude quaternion of the star sensor is calculated through the QUEST algorithm based on the position relationship between the measured star point and the star point in the star catalogue. The calculated attitude quaternion will be delivered to the MCU which will further deliver it to the host computer through a serial port. Connections of major functions are shown in Fig. 6.46.

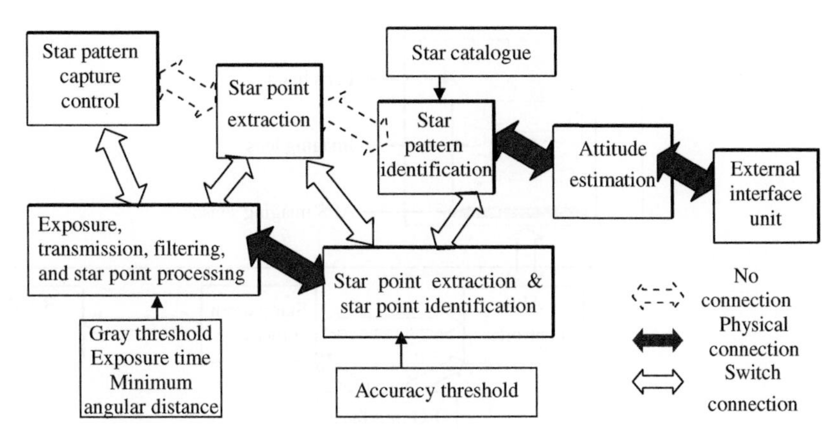

FIGURE 6.46
Connections of major functions of the APS CMOS micro star sensor.

6.3.5 APS CMOS Micro Star Sensor Prototype

The APS CMOS micro star sensor in this book is designed for microsatellites. Considering the resources on satellites and the integrated electronics system design (maybe using an onboard electronical integrated management), this APS CMOS micro star sensor gives adequate consideration to its two application modes: independent attitude component and cooperation with the electronic system on a microsatellite. The header and processing system of the star sensor are designed to be separable. Moreover, a Star1000 probe used in the optical system and header of the star sensor received radiation resistance reinforcement and is designed as an independent unit, thus making it more applicable in space. Since the electronics system generally has weak radiation resistance, the separable design is convenient for independent reinforcement.

The overall structure and physical pictures of the designed APS CMOS micro star sensor prototype in this book are shown in Figs. 6.47 and 6.48, respectively.

Based on this star sensor platform, all key technologies of the APS CMOS micro star sensor are realized, such as energy flow filtering, 24-h-a-day autonomous star pattern identification algorithm based on navstar domain and k-vector, etc.

The main performances of this APS CMOS micro star sensor are tested (Table 6.8).

6.3.6 Real Sky Test

The real sky test is a ground test that can test the most realistic sky performances of a star sensor [66,67]. It often installs a star sensor on the

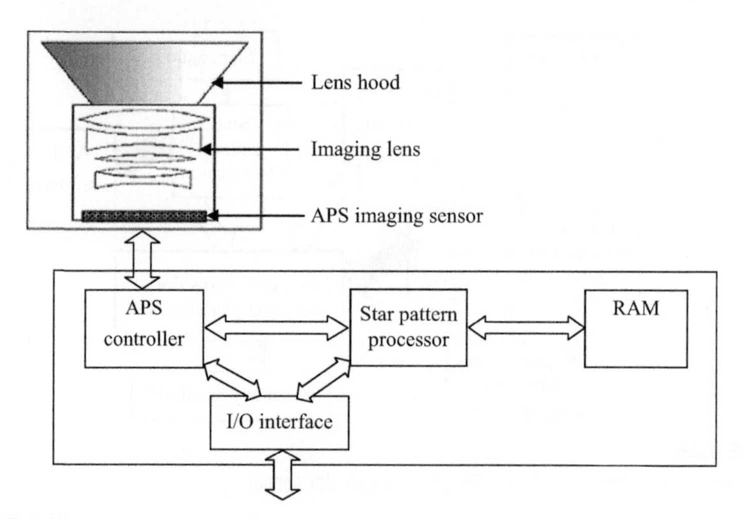

FIGURE 6.47
Overall structure of the APS CMOS micro star sensor.

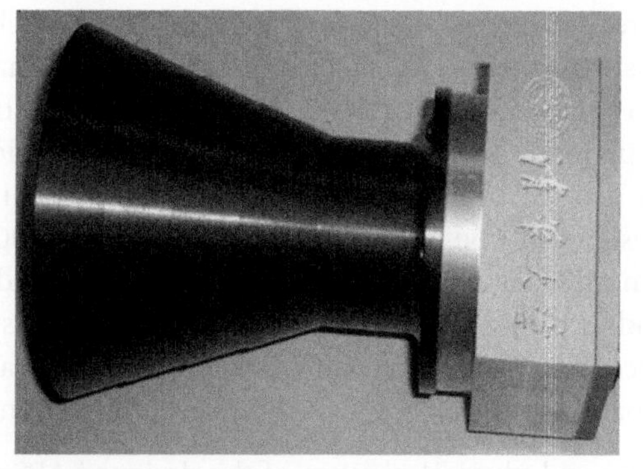

FIGURE 6.48
Physical picture of the APS CMOS micro star sensor.

Table 6.8 Performance Test Results of the APS CMOS Micro Star Sensor

Accuracy (3σ)	<7″ (pitch/yaw), 35″ (roll)
Sensitive magnitude (ground test)	$\geq 5.5\ M_V$
Data updating rate	5.2 Hz
24-h-a-day autonomous capture time	0.5 s
Weight (lens hood included)	1 kg
Power consumption	1.3 W

astronomical telescope and changes the pointing directions of the star sensor by rotating the astronomical telescope. This tests its identification performance in different sky regions. With high rotation accuracy, the astronomical telescope can measure the accuracy of the star sensor. However, astronomical telescopes are expensive and inconvenient, limiting their wide application, especially during R&D. Therefore, this book suggests testing the star pattern identification and accuracy of the star sensor by measuring the Earth's rotation. This can test the feasibility of the star sensor to a certain extent, especially in testing its 24-h-a-day star pattern identification [68].

The high accuracy and availability of the Earth's rotation provide convenient experimental conditions for a star sensor test. However, it should pay attention to the test climate because the moon and clouds will influence the test results significantly. The star sensor is fixed on the Earth, rotating with the Earth. Meanwhile, a turntable or adjustable tripod equipped to adjust its pointing direction and involve extensive observation region is fitted. The real sky test apparatus is shown in Fig. 6.49.

Appropriate weather, time, and place are the prerequisites of a real sky test. The test process and results are analyzed in the following text. This real sky test was conducted at Xinglong Observation Station of National Astronomical Observatories of China for 200 ms from 11:00 p.m. under 24-h-a-day autonomous identification mode. Fig. 6.50 represents some typical pictures.

According to the real sky test results, the star sensor probe can detect navstar ($M_V = 5.5$) images, indicating that it can detect navstars ($M_V > 5.5$) on satellite orbit without an atmospheric effect.

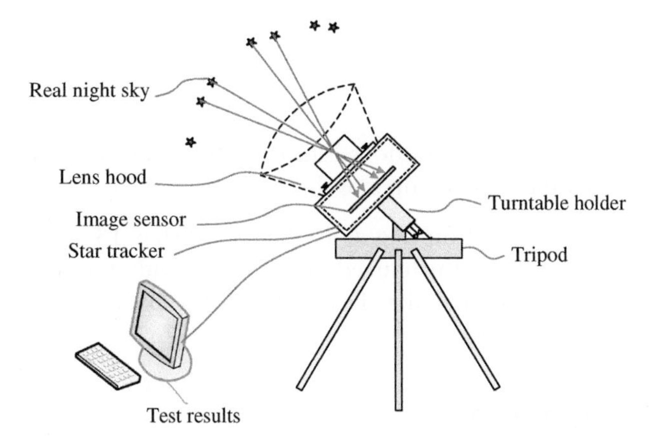

FIGURE 6.49
Real sky test apparatus.

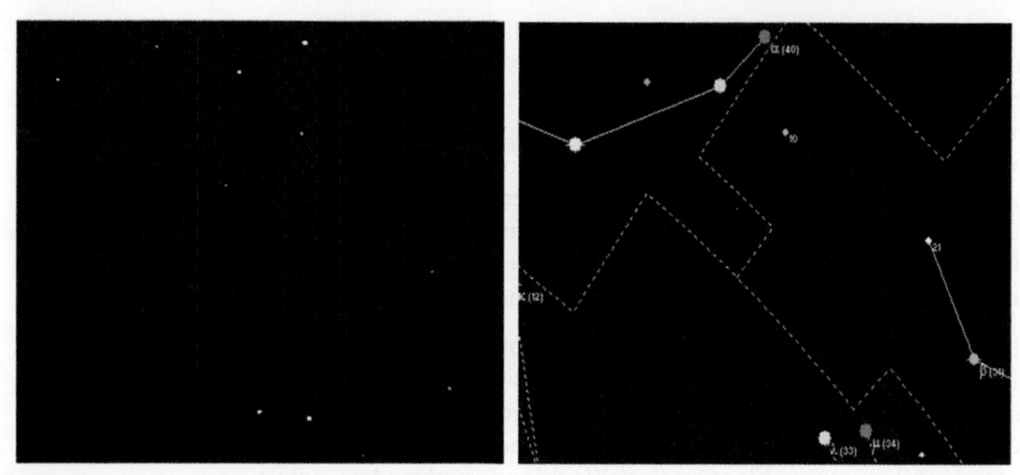

FIGURE 6.50
Captured star image (left) and corresponding Skymap (right).

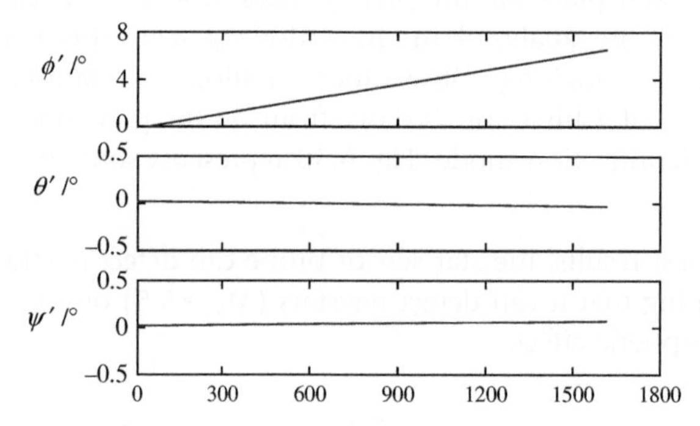

FIGURE 6.51
Earth rotation curves detected by the star sensor.

The star sensor attitude is estimated from the star pattern identification. The triaxial movement curve of the star sensor is shown in Fig. 6.51.

It can be seen from Fig. 6.51 that ϕ' satisfies 15 degrees/h basically, while θ' and ψ' remain unchanged. This agrees with the above theoretical analysis and confirms that the designed APS CMOS micro sensor can operate normally in real sky.

References

[1] D.D. Mazanek, R.R. Kumar, H. Seywald, M. Qu, GRACE mission design: impact of uncertainties in disturbance environment and satellite force models, Adv. Astronaut. Sci. 105 (II) (2000) 967–986.

[2] E.M.C. Kong, D.W. Miller, Optimal spacecraft reorientation for earth orbiting clusters: applications to Techsat 21, Acta Astronaut. 53 (11) (2003) 863–877.

[3] S. Richard, R. Patrice, Design of a micro-satellite for precise formation flying demonstration, in: Proceedings of the 5th IAA International Conference on Low-Cost Planetary Missions, The Netherlands, 2003.

[4] R.L. Ticker, D. McLennan, NASA's New Millennium space technology 5 (ST5) project, IEEE Aerosp. Conf. Proc. 7 (2000) 609–617.

[5] Y. Dong, R. Zhang, Autonomous satellite navigation by stellar sensor [J], Journal of Astronautics 16 (4) (1995) 36–41.

[6] H.-Y. Kim, N. Methods, For spacecraft attitude estimation. Ph.D. Thesis, Texas A&M University, 2002.

[7] D. Mortari, T.C. Pollock, J.L. Junkins, Towards the most accurate attitude determination system using star trackers, Adv. Astronaut. Sci. 99 (II) (1998) 839–850.

[8] S. Caihong, Development method and technology of light star sensor. Doctoral Dissertation on Chinese Beijing Astronomical Observatory, 2002.

[9] L. Baohua, Research on the fast star pattern identification algorithm for spacecraft attitude control. Doctoral Dissertation of Harbin Institute of Technology, 2006.

[10] J.R. Wertz, Spacecraft Attitude Determination and Control, Reidel, Dordecht, The Netherlands, 1984.

[11] A.S. Malak, Toward faster and more accurate star sensor using recursive centroiding and star identification. Ph.D. Thesis, Texas A&M University, 2003.

[12] Y. Wu, HXMT satellite attitude determination and control system design and experimental study. Doctoral Dissertation of Tsinghua University, 2004.

[13] M.C. Phenneger, J.L. Knack, Using the sun analog sensor (SAS) data to investigate solar array yoke motion on the GOES-8 and -9 spacecraft, SPIE 2812 (1996) 753–763.

[14] H. Zhengui, Spacecraft Attitude Dynamics, National University of Defense Technology Publishing, Changsha, 1997.

[15] F.F. Chen, J. Feng, Z. Hong, Digital sun sensor based on the optical vernier measuring principle, Meas. Sci. Technol. 17 (2006) 2494–2498.

[16] http://www.cast.ac.cn/cpyyy/jp_3.htm.

[17] http://www.cast.ac.cn/en/ShowArticle.asp?ArticleID=135.

[18] http://www.tpd.tno.nl/smartsite469.html.

[19] http://www.ufa.cas.cz/html/magion/DSSS-popisweb.html.

[20] H. Li, H. Yihua, Principle and development trendency of sun sensor, Electron. Compon. Mater. 25 (9) (2006) 5–7.

[21] Z. You, T. Li, Application of CMOS imaging sensor in space technology, Opt. Technol. 28 (1) (2002) 31–35.

[22] J.H. Hales, M. Pedersen, Two-axis MOEMS sun sensor for pico satellites, in: 16th Annual AIAA/USU Conference on Small Satellites, 2004, pp. 1–12.

[23] C.C. Liebe, S. Mobasser, MEMS based sun sensor, 2001, IEEE Aerosp. Conf. 3 (2001) 1565–1572.W.-C. Fang, A smart vision system-on-a-chip design based on programmable neural processor integrated with active pixel sensor, 2000 IEEE Int. Symp. Circuits Syst. 2 (2000) 128–131.

[24] M.L. Psiaki, Autonomous low-earth-orbit determination from magnetometer and sun sensor data, J. Guid. Control Dyn. 22 (2) (1999) 296–304.

[25] D. Tianhuai, Z. Zhiming, Principle experiments of micro-digital sun sensor, J. Tsinghua Univ. (Nat. Sci.) 45 (2) (2005) 186–189.

[26] http://www.selex-sas.com.

[27] C.C. Liebe, S. Mobasser, Y. Bae, et al., Micro sun sensor, IEEE Aerosp. Conf. 5 (2002) 2263–2273.

[28] O. Yadid-Pecht, B. Pain, C. Staller, C. Clark, E. Fossum, CMOS active pixel sensor star tracker with regional electronic shutter, IEEE J. Solid-State Circuits 32 (2) (1997) 285–288.

[29] Star1000 datasheet v6: http://www.fillfactory.com.

[30] G. Wahba, A least squares estimate of satellite attitude, SIAM Rev. 8 (3) (1966) 384–386.

[31] http://www.ballaerospace.com/aerospace/ct633.html.

[32] http://www.sodern.fr/.

[33] http://www.ems-t.com/.

[34] http://www.jena-optronik.de, 2004.

[35] B. Maurizio, J.L. Jørgensen, P.S. Jørgensen, T. Denver, Advanced stellar compass onboard autonomous orbit determination, preliminary performance, Ann. N.Y. Acad. Sci. 1017 (2004) 393–407. Available from: http://dx.doi.org/10.1196/annals.1311.022.

[36] C.C. Liebe, E.W. Dennison, B. Hancock, et al., Active pixel sensor based star tracker, IEEE Aerosp. Conf. 1 (1998) 119–127.

[37] http://telecom.esa.int/telecom/object/index.cfm?fobjectid = 26326.

[38] F. Xing, Z. You, G.F. Zhang, J. Sun, A novel active pixels sensor (APS) based sun sensor based on a feature extraction and image correlation (FEIC) technique, Meas. Sci. Technol. 19 (12) (2008) 125203 (9 pp.).

[39] F. Xing, Z. You, G.F. Zhang, J. Sun, APS sun sensor based future extraction and image correlation algorithm, 2010. Application No: 200910079564.8.

[40] C.G. Harris, M. Stephens, A combined corner and edge detector, in: Proceedings of the 4th Alvey Vision Conference, Manchester, UK, 31 August–2 September 1998.

[41] X. Fei, D. Ying, Y. Zheng, Laboratory calibration of star tracker with brightness independent star identification strategy, Opt. Eng. 45 (6) (2006) 063604.

[42] J. Enright, Godard, Advanced sun-sensor processing and designed for super-resolution performance. 2006 IEEE Aerospace Conference, 4–11 March, 2006.

[43] C.W. de Boom, N. van der Heiden, A novel digital sun sensor: development and qualification for flight, in: 54th International Astronautic Congress, IAC-03-A.P.20, September 29–October 3, Bremen, Germany, 2003.

[44] Z. Liu, Study on application technique of CMOS imager sensor in star tracker. Doctoral Dissertation, Changchun Institute of Optics, Fine Mechanics and Physics, Chinese Academy of Sciences, 2004.

[45] B.R. Hancock, R.C. Stirbl, T.J. Cunningham, et al., CMOS active pixel sensor specific performance effects on star tracker/imager position accuracy, SPIE 4284 (2001) 43–53.

[46] R. Hornsey, Noise in image sensors. Course notes presented at the Waterloo Institute for Computer Research, Waterloo, ON, May 1999.

[47] S. Cos, D. Uwaerts, J. Bogaerts, W. Ogiers, Active pixels sensor for star tracker: final resport. Doc. Nr: APS-FF-SC-05-023, Cypress Semiconductor Corp., 2006.

[48] R.V.F. Lopes, G.B. Carvalho, A.R. Silva, Star identification for three-axis attitude estimation of French-Brazilian scientific micro-satellite, in: Proceedings of the AAS/GSFC International Symposium on Space Flight Dynamics, AAS Paper 98-366, Greenbelt, MD, May 11–15, 1998, pp. 805–819.

[49] J.D. Vedder, Star trackers, star catalogs, and attitude determination—probabilistic aspects of system design, J. Guid. Control Dyn. 16 (3) (1993) 498–504.

[50] J. Kosik, Star pattern identification aboard an inertially stabilized spacecraft, J. Guid. Control Dyn. 14 (1) (1991) 230–235.

[51] J. Junkins, C. White, D. Turner, Star pattern recognition for real-time attitude determination, J. Astronaut. Sci. 25 (1997) 251–270.

[52] R. Van Bezooijen, A star pattern recognition algorithm for autonomous attitude determination. Automatic Control in Aerospace: IFAC Symposium, Tsukuba, Japan, 1989, pp. 51–58.

[53] C. Padgett, K. Kreutz-Delgado, A grid algorithm for star identification, IEEE Trans. Aerosp. Electron. Syst. 33 (1) (1997) 202–213.

[54] C.C. Liebe, Pattern recognition of star constellations for spacecraft applications, IEEE Aeronaut. Electron. Syst. Mag. 10 (1992) 2–12.

[55] C.S. Lindsey, T. Lindblad, A method for star identification using neural networks, SPIE 3077 (1997) 471–478.

[56] J.T. Aaron, Autonomous artificial neural network star tracker for spacecraft attitude determination. Ph.D. Thesis, University of Illinois at Urbana-Champaign, 2002.

[57] C. Li, K. Li, Y. Zhang, S. Jin, J. Zu, Star pattern identification based on nerual network technique, Chin. Sci. Bull. 48 (9) (2003) 892–895.

[58] J. Hong, A. Julie, Dickerson neural-network-based autonomous star identification algorithm, J. Guid. Control Dyn. 23 (4) (2000) 728–735.

[59] D. Mortari, B. Neta, k-Vector range searching techniques. Paper AAS 00-128 of the 10th Annual AIAA/AAS Space Flights Mechanics Meeting, Clearwaters, FL, January 23–26, 2000.

[60] D. Mortari, Search-less algorithm for star pattern recognition, J. Astronaut. Sci. 45 (2) (1997) 179–194.

[61] M.D. Shuster, S.D. Oh, Three-axis attitude determinatin form vector observations, J. Guid. Control Dyn. l4 (1) (1981) 70–77.

[62] F.L. Markley, Attitude determination using vector observations: a fast optimal matrix algorithm, J. Astronaut. Sci. 41 (2) (1993) 261–280.

[63] Y.B. Itzhack, REQUST: a recursive QUEST algorithm for sequential attitude determination, J. Guid. Control Dyn. 19 (5) (1996) 1034–1038.

[64] M.D. Shuster, Kalman filtering of spacecraft attitude and the QUEST model, J. Astronaut. Sci. 38 (3) (1990) 377–393.

[65] M.D. Shuster, A simple Kalman filter and smoother for spacecraft attitude, J. Astronaut. Sci. 37 (1) (1989) 89–106.

[66] A.R. Eisenman, J.L. Jørgensen, C.C. Liebe, Real sky performance of the prototype Ørsted Advances Stellar Compass, IEEE Aerosp. Appl. 2 (1996) 103–113.

[67] A.R. Eisenman, J.L. Jørgensen, C.C. Liebe, Astronomical performance of the engineering model Ørsted Advanced Stellar Compass, SPIE 2810 (1996) 252–264.

[68] B. Thomas, Characterizing a star tracker with built in attitude estimation algorithms under the night sky, SPIE 3086 (1997) 264–274.

CHAPTER 7

Miniature Inertial Measurement Unit

7.1 HISTORY AND DEVELOPMENT OF IMU

7.1.1 Traditional Inertial Devices and Their Development

Development of inertial technology can be divided into four stages according to their accuracy. The first generation was from the early 20th century to the time before World War II, and preliminary applications of inertial technology appeared. The second generation was from World War II to the 1960s, when the positioning resolution was about 1 km per hour. With the development of the electrostatic gyro technology, inertial technology entered the third generation, which was from the 1960s to the late 1970s, and the accuracy was increased by two to three orders of magnitude. The fourth generation was from the late 1970s to the present, during which the expected target accuracy of gyros can reach $1.5 \times 10^{-7\circ}$/h and the accuracy of accelerometers can reach 10^{-8} g. According to the technical methods, the development of inertial technology can be divided into three stages. The first stage is the development of mechanical gyroscopes (from the 1940s to the 1960s). The second stage is the development of strapdown gyroscope (from the 1960s to the 1980s). The third stage is the development of the micromechanical gyroscopes (since mid-1980s).

The combination of micromechanical technology and traditional inertial technology brought a great change to the inertial technology. Since the 1980s, Draper Laboratory (US), JPL (United States), Litton company (US), and SAGEM company (France) have carried out many researches into microgyroscopes, microaccelerometers, and other miniature inertial devices. Compared with traditional rigid rotor gyroscopes, the size, weight and cost of the miniature gyroscope are greatly reduced because there is no high-speed rotor and the corresponding support system. Meanwhile, the reliability, stability, and other aspects were also enhanced. In 1989, Draper laboratory first introduced a micro double-frame resonant gyroscope based on the Coriolis Effect. In 1993, they cooperated with Rockwell company and developed the micro comb-like resonant tuning fork gyroscope whose

Space Microsystems and Micro/Nano Satellites. DOI: http://dx.doi.org/10.1016/B978-0-12-812672-1.00007-2

effective size was only 1 mm and the expected performance was 10−100°/h. In 1986, BEI company developed the quartz tuning fork micromechanical gyroscope (QRS) and started production in early 1991. At present, WSC-6 satellite antenna stabilization systems with QRS have been equipped in the United States Navy and ships of Holland, Denmark, and other countries. Using Rockwell's digital processing technology, the QRS precision can reach 1°/h, which can be compared with mini fiber optical gyro, whereas the cost of QRS is much cheaper. Currently, more miniature inertial devices are being developed. For example, BEI's QRS11 micromechanical gyroscope are only $\phi 41.5 \times 16.5$ mm, whose resolution is higher than 0.004°/s and output noise is lower than $0.01°/s/\sqrt{Hz}$, the short-term stability is higher than 0.002°/s, which has been applied as an alternative to traditional inertial devices in spacecraft attitude measurement and control, missile midcourse guidance, as well as other military fields. In recent years, people have been carrying out research on micro fiberoptic gyroscope to effectively improve the accuracy of micro inertial devices. In 1976, Professor Vali of the University of Utah first conducted the first fiberoptic gyro display experiment [1]. In 1988, Draper laboratory made significant progress in the field of resonant fiber optic gyroscope. The zero stability of Northrop's resonant fiberoptic gyroscope was better than 1°/h [2]. At the same time, the development of a micro accelerometer was also carried out. In 1979, Stanford University first developed an open-loop micro accelerometer based on micromechanical processing technology. After that, the use of the silicon material increased the bearing capacity of the micromechanical accelerometer. Some micro inertial products are shown in Figs. 7.1−7.3.

The research field of the micro inertial sensor in China started late but developed rapidly. For example, AR series and HXA series gyro accelerometers of Chongqing Xin Chen Electronics Company are integrated at a high level. Tsinghua University developed micro gyroscopes based on vibratory wheels and vibratory rods, as well as torsion-type and comb-type micro silicon accelerometers, whose accuracies reached international levels.

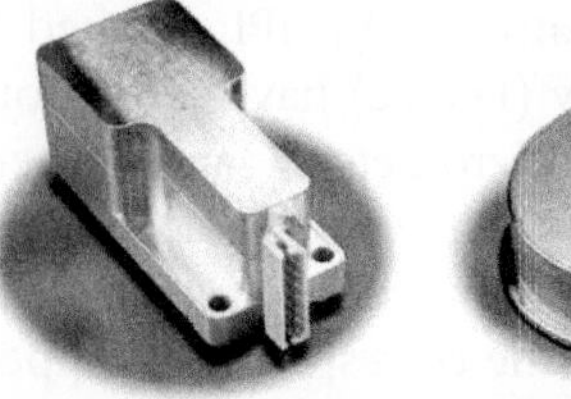

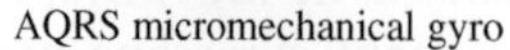

AQRS micromechanical gyro QRS14 micromechanical gyro

FIGURE 7.1
Micro gyros.

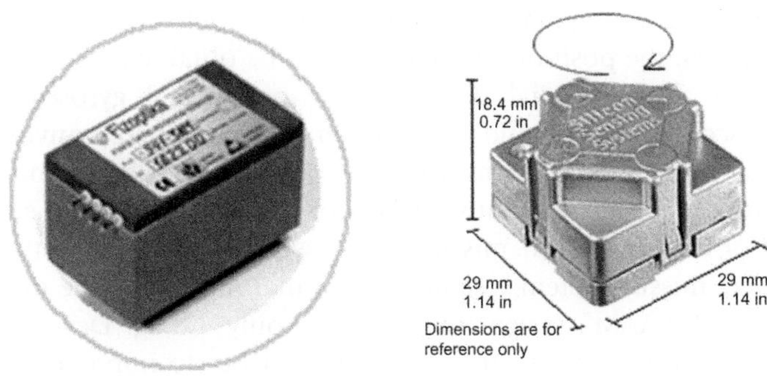

FIGURE 7.2
Micro gyros.

FIGURE 7.3
Micro-accelerometers in Columbia Laboratory.

7.1.2 Development of MIMU

Inertial navigation system (INS) is an autonomous system with good concealment, which is not dependent on any external information, nor radiates energy to external space, making it applicable in airspace, sea, or underground. Since the INS updates the data rapidly and possesses the advantages of high short-term accuracy and stability with small size and light weight, it can provide comprehensive navigation data, such as the location, speed, or the attitude of the carrier. Therefore INS plays a very important role in military field and civilian navigation [3]. In 1942, the German V-2 rocket installed the guidance system composed of inertial devices for the first time. With decades of progress of the inertial devices, they have played a more significant role in the navigation fields. Since the 1960s, inertial technology has greatly expanded applications in guidance systems, with "Hercules II" LGM-25C, "militia" series, "Flying Fish" MM-38, MX missiles, "AGM86C" cruise missiles and "Trident" strategic missiles [4].

The concept of the Inertial Measurement Unit (IMU) appeared in the 1970s, namely IMU measurements changed from the independent inertial measurement device. As independent inertial devices, gyroscopes and accelerometers can be employed to separately measure angular velocity, acceleration, or other parameters, the position of the object can be obtained through integral. Based on the above inertial devices, three orthogonal gyroscopes, three orthogonal accelerometers, and a computer for coordinate conversion can constitute a kind of IMU to measure the attitude, position, and velocity information of a carrier. Currently, IMU is becoming smaller and smaller. In 1974, IMU of the "Harpoon" missile system weighed 3.2 kg, and the IMU of the "Phoenix" missile system weighed 1.8 kg in 1980, while the mass of the IMU in the "AMRAAM" system in 1985 was only 1.4 kg. Development of micro and nanotechnologies offered new approaches to improve the inertial devices. Micro inertial measurement unit (MIMU) came into being, which integrates several micro inertial sensors, micro monitoring and control circuits, and a coordinate conversion circuit to obtain comprehensive inertial measurement parameters of moving objects (Fig. 7.4).

MIMU has been applied in many fields, such as geodetic survey, mineral exploration, vehicle location and earthquake prediction. BEI's MIMU product, MotionPak [5], consists of three micromechanical quartz rate sensors and three micromechanical silicon acceleration meters. Among them, the gyro owns a high resolution of 0.004°/s and weighs 60 g, while the accelerometer weighs 55 g with a resolution of 10 μg. The overall weight of the MotionPak is only 900 g. Besides, Draper laboratory developed a kind of low precision micro silicon MIMU with the size of $2 \times 2 \times 0.5$ cm^3, the weight of about 5 grams, and the gyro drift ratio about 10°/h. This MIMU is suitable for navigation in a short period.

At present, the major demand of MIMU in China is intelligent guidance, robots, geodetic survey, mineral exploration and other fields. Meanwhile, aircraft, microsatellite ADCs, and other space technologies will be important markets [6].

YH5000IMU

MotionPak

FIGURE 7.4
Foreign MIMU products.

7.1.3 Development of Optimal Estimation Theory and Its Application in MIMU

Applying the optimal estimation filter theory to MIMU technology can improve the navigation accuracy. In 1975, Gauss Karl proposed the least square method as the parameter estimation algorithm to determine the planetary orbits. Because the optimal index is used to optimize the accuracy of the measurement and it is not necessary to use the dynamic and statistics information related to the estimated parameter, the estimation algorithm is simple but the accuracy is not very high. Wiener filtering is a kind of linear minimum variance estimation algorithm, which is suitable for the processing of the stationary random processes in which useful signals and interference signals both have zero mean value. However, the design of the Wiener filter needs the information of the power spectrum and cross power spectrum of the useful and interference signal, making it difficult to guarantee the real-time processing. Besides, it is critically challenging to apply Wiener filtering to nonstationary processes.

In 1960, Calman proposed a recursive linear minimum variance estimation algorithm for the discrete-time system model [7]. Calman succeeded in introducing the concept of state variables into the filtering theory, replacing the usual covariance function by the state space model, connecting the description of the state space with the time update. Compared with the Wiener filtering, Kalman used the state equation in the time domain and differential equations characterizing the system state estimation and variance to obtain a recursive algorithm that was suitable for computer calculation under the minimum variance criterion. Because many practical engineering problems are nonlinear, nonlinear filtering algorithm are always required. Bucy and Snahara devoted to the study of Kalman filter theory in nonlinear systems and nonlinear measurement, and broadened the applications of Kalman filter theory [8,9]. The extended Kalman filtering algorithm includes linearization of the nonlinear equation and consequent estimation using the linear Kalman filtering method. The extended Kalman filtering algorithm is based on the linearization of the system and the observation equation.

However, the linearization of the system and the observation equation will inevitably cause a certain degree of linearization error, leading to a decrease in the final estimation precision. How to reduce the influence of linearization errors is the focus of the current study of filtering algorithm. L-Ddecomposition algorithm proposed by Fisher [10], Song's variable gain extended Kalman filter [11], and Ruokonen's parallel Kalman filtering technology extended and improved the Kalman filtering algorithm to a certain extent [12]. On the other hand, people tried to look for new nonlinear algorithms. Algrain proposed a mutual coupling linear Kalman filter (interlaced Kalman filtering) [13], which decomposes the complex nonlinear model into two pseudo-linear models that describe the linear and nonlinear characteristics of system state variables, respectively, and obtains a suboptimal estimation through parallel filtering and data fusion. On the basis of this method, Ruth proposed a suboptimal extended coupling Kalman filter [14], which successfully solved the problem of estimating the angular velocity of the satellite by using the differential equation of the observation vector. Julier and Uhlman proposed the unscented Kalman filter (UKF) [15]. They used a number of sampling points that could characterize the system state mean and variance and carried out nonlinear transformation through the UT transformation, which reached third-order accuracy approximation to the true mean and variance. UKF solved the problem of nonlinear approximation of EKF. E.A. Wan and R. van der Merwe applied the square root method to UKF [16], and further studied the algorithm.

For the multisensor combination system, Speyer, Bierman, and Kerr successively proposed decentralized filtering thinking from 1979 to 1985 [17]. The development of computer technology has created favorable conditions for the development of decentralized filter. Carlson's federal filter theory [18] provided the design theory for INS. Carlson constructed an augmented matrix and used the variance upper bound technique to eliminate the correlation between local filters. Finally, Carlson obtained the globally optimal estimation through data fusion algorithm based on the irrelevant theory. Because the algorithm structure is very suitable for parallel calculating, this algorithm has a wide range of engineering applications with the development of parallel computing technology.

7.2 SYSTEM INTEGRATION OF MIMU AND ATTITUDE DETERMINATION ALGORITHMS

7.2.1 MIMU Integration

Miniature inertial measurement units (MIMU) consists of micro-accelerometers and micro-gyroscope. The basic structure of an MIMU is shown in Fig. 7.5. It consists of three micro-gyroscopes and three micro-accelerometers, which were installed in three orthogonal surfaces of the cube. The sensitive axes x, y, and z are perpendicular to each other, and acceleration and angular velocity along the three directions are measured respectively. Here ω_x, ω_y, ω_z are outputs of micro-gyroscopes and a_x, a_y, a_z are outputs of micro-accelerometers. Fig. 7.6 is the prototype of an integrated MIMU.

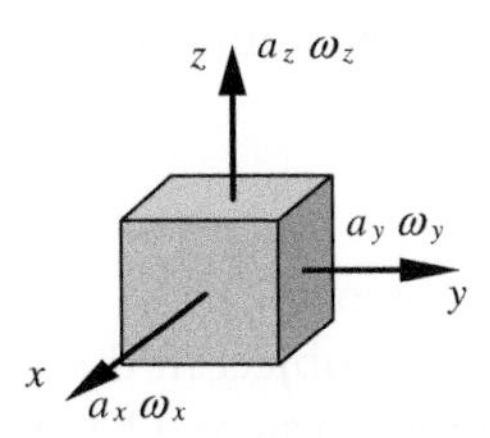

FIGURE 7.5
Basic structure diagram of an MIMU.

FIGURE 7.6
Prototype of an integrated MIMU.

7.2.2 Measurement Principles of an MIMU

7.2.2.1 *Establishment of the Coordinate System*

In order to clearly describe the position and the velocity of the carrier, we must select the correct reference coordinate system [19].

1. Geocentric equatorial inertial coordinate system (S_i)
   Origin O is in the center of the Earth, and plane $x_i y_i$ coincides with the Earth's plane. Axis x_i points to the vernal equinox, while axis z_i points to the Arctic Pole along with the Earth's rotation axis.
2. Earth coordinate system (S_E)
   Origin O is in the center of the Earth, and plane $x_E y_E$ coincides with the Earth's plane. Axis $O_E x_E$ points to the zero meridian, axis $O_E z_E$ points to the Arctic Pole along with the Earth's rotation axis. Earth

coordinate system S_E rotates at the Earth's rotation angular velocity ω_E.

$\omega_E = 15.041°/h$

3. Geographic coordinate system (S_e)
The position of the moving object in the Earth's surface O_e is set as the origin of the coordinate system, O_eX_e pointing to the east and O_ey_e pointing to the north, while O_ez_e, O_ex_e, and O_ey_e constitute the right-handed coordinate system, pointing to the sky along the direction of local Earth vertical, that is the "north-east-sky" coordinate system.
4. Orbit coordinate system (S_o)
Origin O_o is the satellite centroid, while axis O_ox_o is in the orbit plane and points to the direction of satellite motion, axis O_oy_o is perpendicular with the orbit plane and points to the "Right" which is opposite to the system moment of the momentum vector, axis O_oz_o plumb down towards the center of the Earth. This coordinate system is also the conventional second orbit coordinates.
5. Star coordinate system (S_b, carrier coordinate system)
In normal flight, axis O_bx_b coincides with axis O_bx_o, axis O_by_b coincides with axis O_by_o, and axis O_bz_b is along the extension direction of the gravity gradient rod which coincides with axis O_bz_o. In normal work, the satellite coordinate system S_b coincides with the orbit coordinate system S_o.

7.2.2.2 Measurement Principles

Miniature inertial measurement unit (MIMU) consists of three micro-gyroscopes and three micro-accelerometers, the output shafts of which are perpendicular to each other, to respectively sensing angular velocity and acceleration along three orthogonal directions. Data collection system can convert the output of gyroscope and accelerometer to digital signal in real time and can conduct the work of storage and real-time processing.

According to the attitude transfer principle, the attitude matrix between carrier coordinate system and the reference coordinate system can be expressed as follows:

$$C_e^b = \begin{bmatrix} \cos\psi\cos\theta & \sin\psi\cos\theta & -\sin\theta \\ \cos\psi\sin\theta\sin\gamma - \sin\psi\cos\gamma & \sin\psi\sin\theta\sin\gamma + \cos\psi\cos\gamma & \cos\theta\sin\gamma \\ \cos\psi\sin\theta\cos\gamma + \sin\psi\sin\gamma & \sin\psi\sin\theta\cos\gamma - \cos\psi\sin\gamma & \cos\theta\cos\gamma \end{bmatrix} \tag{7.1}$$

Here, θ, γ, and ψ respectively represents the pitch angle, roll angle, and heading angle of the carrier relative to the reference system. MIMU calculation

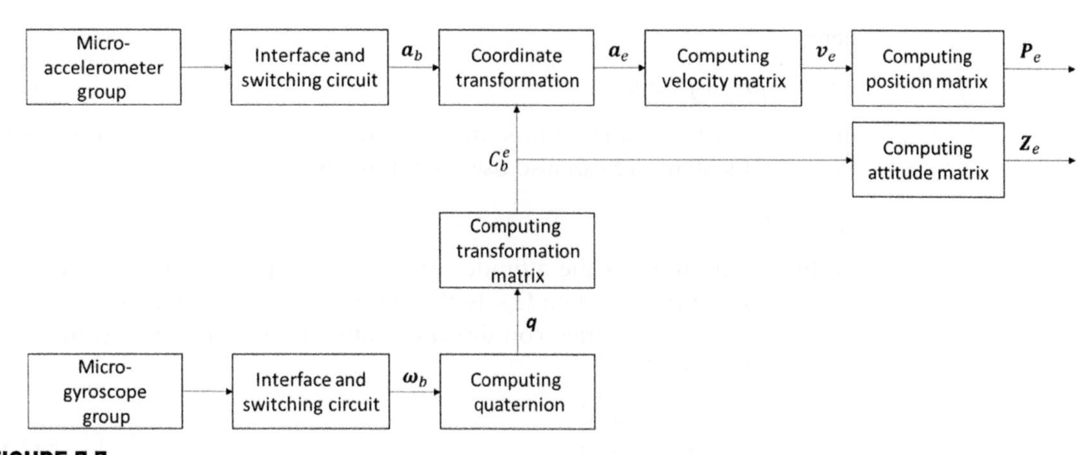

FIGURE 7.7
MIMU measurement process.

principle is shown in Fig. 7.7, subscripts b and e respectively represents the carrier coordinate system and reference coordinate system; $\boldsymbol{a} = \begin{bmatrix} a_x & a_y & a_z \end{bmatrix}^T$ is the acceleration vector, $\boldsymbol{V} = \begin{bmatrix} V_x & V_y & V_z \end{bmatrix}^T$ is the velocity vector, and $\boldsymbol{P} = \begin{bmatrix} P_x & P_y & P_z \end{bmatrix}^T$ is the position vector, among which x, y, and z respectively represents the horizontal axis, vertical axis, and vertical axis of carrier coordinate system; $\boldsymbol{\omega} = \begin{bmatrix} \omega_x & \omega_y & \omega_z \end{bmatrix}^T$ is the angular velocity vector, $\boldsymbol{Z} = \begin{bmatrix} \theta & \gamma & \psi \end{bmatrix}^T$ is the attitude angle vector; $\boldsymbol{q} = q_0 + q_1\boldsymbol{i} + q_2\boldsymbol{j} + q_3\boldsymbol{k}$ is the carrier rotation quaternion; $\boldsymbol{C}_b^e$ is the attitude matrix from the carrier coordinate system to reference coordinate system.

In practice, MIMU keenly senses the rotation of the carrier and converts it to the reference coordinate system by the software. The accelerometer group outputs the inertial acceleration $\boldsymbol{a}_b$ along the axis of the carrier coordinate system, which can be transformed into the acceleration $\boldsymbol{a}_e$ under the reference coordinate system, subsequently the speed and position under the reference coordinate system can be calculated through the second integral calculation. Gyroscope outputs the angular rate $\boldsymbol{\omega}_b$ under the reference coordinate system, then it is substituted into the transformation matrix formula to calculate $\boldsymbol{C}_b^e$ after quaternion calculation. Meanwhile, attitude angle vector $\boldsymbol{Z}$ can also be calculated. Quaternion differential equation characterizing rigid object rotation expresses as follows:

$$\dot{\mathbf{q}} = \frac{1}{2}\begin{bmatrix} 0 & -\omega_x & -\omega_y & -\omega_z \\ \omega_x & 0 & \omega_z & -\omega_y \\ \omega_y & -\omega_z & 0 & \omega_x \\ \omega_z & \omega_y & -\omega_x & 0 \end{bmatrix}\boldsymbol{q} \tag{7.2}$$

here:

$$q = q_0 + q_1\boldsymbol{i} + q_2\boldsymbol{j} + q_3\boldsymbol{k} \tag{7.3}$$

Additional, when using attitude matrix to describe the motion of carriers in the MIMU system, we can also use the following equation:

$$\dot{C}_b^e = C_b^e \boldsymbol{\Omega}_{eb} \tag{7.4}$$

In this equation, C_b^e is the attitude matrix from carrier coordinate system to reference coordinate system.$\boldsymbol{\Omega}_{eb}$ is the skew-symmetric matrix of rotational angular velocity of carrier coordinates relative to the reference coordinate system, which is:

$$\Omega_{eb} = \begin{bmatrix} 0 & -\omega_z & \omega_y \\ \omega_z & 0 & -\omega_x \\ -\omega_y & \omega_x & 0 \end{bmatrix} \tag{7.5}$$

Here, we can use angle increment algorithm to directly solve the attitude matrix directly, and the expression is:

$$C_b^e(t+\Delta t) = C_b^e(t)\left[\boldsymbol{I} + \frac{\sin\Delta\theta_0}{\Delta\theta_0}\boldsymbol{\Delta\theta}_{eb}^b + \frac{1-\cos\Delta\theta_0}{\Delta{\theta_0}^2}\left(\boldsymbol{\Delta\theta}_{eb}^b\right)^2\right] \tag{7.6}$$

Here, $\boldsymbol{\Delta\theta}_{eb}^b$ is the angular increment, which can be calculated by the following equation:

$$\boldsymbol{\Delta\theta}_{eb}^b = \int_t^{t+\Delta t} \boldsymbol{\omega}_{eb} dt \tag{7.7}$$

In specific calculations, we often use the fourth-order incremental algorithm:

$$C_b^e(t+\Delta t) = C_b^e(t)\left[\boldsymbol{I} + \left(1 - \frac{\Delta{\theta_0}^2}{6}\right)\boldsymbol{\Delta\theta}_{eb}^b + \left(\frac{1}{2} - \frac{\Delta{\theta_0}^2}{24}\right)\left(\boldsymbol{\Delta\theta}_{eb}^b\right)^2\right] \tag{7.8}$$

Using the above method to update the attitude transformation matrix in real time, and to do coordinate transformation of acceleration vector, the direction of the acceleration vector can be transformed from the carrier coordinate system to the reference coordinate system:

$$\boldsymbol{a}_e = C_b^e \boldsymbol{a}_b \tag{7.9}$$

Here $\boldsymbol{a}_b$ is the acceleration vector outputed by the accelerometer, $\boldsymbol{a}_e$ is the acceleration vector under the reference system, which can be expressed as follows:

$$\boldsymbol{a}_b = a_{xb}\boldsymbol{i} + a_{yb}\boldsymbol{j} + a_{zb}\boldsymbol{k} \tag{7.10}$$

$$\boldsymbol{a}_e = a_{xe}\boldsymbol{i} + a_{ye}\boldsymbol{j} + a_{ze}\boldsymbol{k} \tag{7.11}$$

The velocity components and the three-dimensional position of the carrier are obtained by two numerical integrations of the three components of a_e.

Meanwhile, we can calculate attitude angles of the carrier as follows:

$$
\begin{aligned}
\theta &= \sin^{-1}(T_{31}) \\
\gamma &= tg^{-1}\left(\frac{T_{32}}{T_{33}}\right) \\
\psi &= tg^{-1}\left(\frac{T_{21}}{T_{11}}\right)
\end{aligned}
\tag{7.12}
$$

Here, $T_{ij}(i, j = 1, 2, 3)$ is an element of the transformation matrix C_b^e.

Fig. 7.8 is our MIMU navigation algorithm process. In the above measurement process, accelerometers group and gyroscopes group constantly output acceleration and angular velocity of the carrier. On this basis, we can obtain the relative positional relationship between the measured points according to the principle of the strapdown inertial navigation system. Here, we use the computer to collect angular rate output signal of the gyroscopes, constantly updating the system's attitude transformation matrix and using the transformation matrix to transform the sampled acceleration signal into the reference

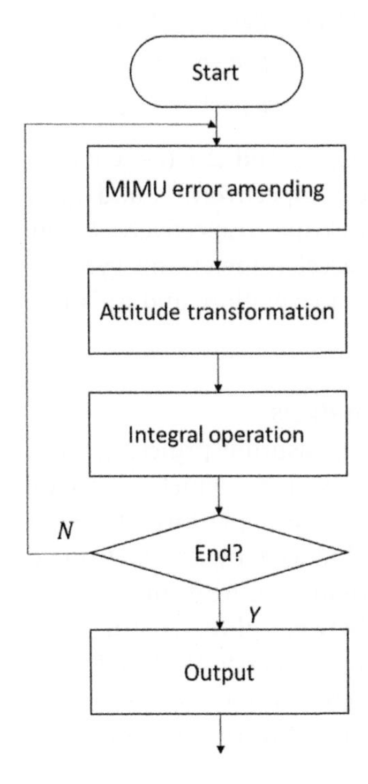

FIGURE 7.8

MIMU navigation algorithm process.

coordinate. Finally, the computer calculates the speed and position of the MIMU in the reference coordinate system through integral calculation.

7.2.3 Error Analyses for the MIMU Model

7.2.3.1 Inertial Sensor Error

When using MIMU to conduct inertial measurement, the main sources of measurement error include inertial element errors and measuring system errors. Among them, inertial element errors include: sensor's zero error, dynamic error, installation error, standard error, first power error, and quadratic error of specific force and other high-order error. In theory, we should consider all of these errors, and this is the content we are going to study in the next chapter of calibration technique. Here, when only taking the large error term into account, we get the following error model equation of the accelerometers and gyroscopes:

$$\begin{aligned} \delta a_x &= \theta^a_{xz} a_y - \theta^a_{xy} a_z + k_x a_x^2 + \Delta a_x + \Delta a_{rx} \\ \delta a_y &= \theta^a_{yx} a_z - \theta^a_{yz} a_x + k_y a_y^2 + \Delta a_y + \Delta a_{ry} \\ \delta a_z &= \theta^a_{zy} a_x - \theta^a_{zx} a_y + k_z a_z^2 + \Delta a_z + \Delta a_{rz} \end{aligned} \tag{7.13}$$

$$\begin{aligned} \delta \omega_x &= \theta^g_{xz} \omega_y - \theta^g_{xy} \omega_z + p_{xx} a_x + p_{xy} a_y + p_{xz} a_z + \varepsilon_x + \varepsilon_{rx} \\ \delta \omega_y &= \theta^g_{yx} \omega_z - \theta^g_{yz} \omega_x + p_{yx} a_x + p_{yy} a_y + p_{yz} a_z + \varepsilon_y + \varepsilon_{ry} \\ \delta \omega_z &= \theta^g_{zy} \omega_x - \theta^g_{zx} \omega_y + p_{zx} a_x + p_{zy} a_y + p_{zz} a_z + \varepsilon_z + \varepsilon_{rz} \end{aligned} \tag{7.14}$$

In the equation, $\Delta a_i (i = x, y, z)$ and $\Delta a_{ri} (i = x, y, z)$ are zero bias and random error of the accelerometer, respectively; ε_i and ε_{ri} are constant drift rate and random error of the gyroscope, respectively; θ^a_{ij} and $\theta^g_{ij} (i, j = x, y, z)$ are misalignment angles of the accelerometer and gyroscope, respectively. k_i is the squared coefficient of the acceleration; and p_{ij} is the coefficient related to the acceleration.

7.2.3.2 MIMU Error Analysis

According to the MIMU measuring principle, MIMU belongs to the strapdown inertial navigation system, which is capable of implementing the functions of inertial navigation platform through the attitude algorithm using the computer. Influenced by errors from gyro drift, accelerometer bias, and other errors originating from inertial devices, also limited by the mutual coupling effect between three channels in the calculation process, the long-term stability and accuracy of MIMU are not high. Here, based on semianalytic inertial system error theory, we give the MIMU system error equation, laying the foundation for constructing the integrated navigation system state equation.

1. Inertial system equations [20]

Select the geographic coordinate system as the reference system. The attitude matrix of the carrier system (b) relative to geographic coordinate system (e) is:

$$\mathbf{C}_e^b = \begin{bmatrix} \cos\psi\cos\theta & \sin\psi\cos\theta & -\sin\theta \\ \cos\psi\sin\theta\sin\gamma - \sin\psi\cos\gamma & \sin\psi\sin\theta\sin\gamma + \cos\psi\cos\gamma & \cos\theta\sin\gamma \\ \cos\psi\sin\theta\cos\gamma + \sin\psi\sin\gamma & \sin\psi\sin\theta\cos\gamma - \cos\psi\sin\gamma & \cos\theta\cos\gamma \end{bmatrix} \tag{7.15}$$

Here ψ, θ, γ, respectively represent the carrier's heading angle, pitch angle, and roll angle.

Using the direction cosine matrix to describe the movement of the carrier

$$\dot{\mathbf{C}}_b^e = \mathbf{C}_b^e \mathbf{\Omega}_{be} \tag{7.16}$$

Here C_b^e is the direction cosine matrix of the carrier from the carrier coordinate system to the geographic coordinate system, Ω_{be}^b is the skew symmetric matrix of the rotational angular velocity ω_{be}^b of the carrier coordinate relative to geographic coordinate system.

We use ω_{be}^e to represent rotational angular velocity of geographic coordinate system relative to inertial space. The component expression is:

$$\begin{aligned} \omega_x &= -\frac{V_y}{R+h} \\ \omega_y &= \frac{V_x}{R+h} + \omega_e \cos\varphi \\ \omega_z &= \frac{V_y}{R+h}\tan\varphi + \omega_e \sin\varphi \end{aligned} \tag{7.17}$$

Here ω_e, R, φ, and h represent the angular velocity of the Earth, the Earth's radius, latitude, and height of the carrier, respectively. V_x, V_y represent components of the carrier's velocity relative to the Earth along the east and the north in the geographic coordinate system.

ω_{bi}^b represents the absolute angular rate directly measured by rate-of-turn gyroscope, therefore:

$$\omega_{be}^b = \omega_{bi}^b - C_e^b \begin{bmatrix} -\dfrac{V_y}{R+h} \\ \dfrac{V_x}{R+h} + \omega_e \cos\varphi \\ \dfrac{V_y}{R+h}\tan\varphi + \omega_e \sin\varphi \end{bmatrix} \tag{7.18}$$

Apply Eq. (7.18) into Eq. (2.16), we can obtain the attitude matrix from the time integration.

Meanwhile, accelerometers measure the specific force along the carrier's three directions. Using the attitude matrix to conduct the coordinate transformation and eliminating the interference of Coriolis acceleration and centrifugal acceleration, we get the carrier's acceleration relative to the geographic coordinate system, and the mechanical choreography equations are:

$$\dot{V}_x = f_x + \left(\frac{V_y}{(R+h)\cos\varphi} + 2\omega_e\right)\sin\varphi\, V_y - \left(\frac{V_x}{(R+h)\cos\varphi} + 2\omega_e\right)\cos\varphi V_z$$
$$\dot{V}_y = f_y - \left(\frac{V_x}{(R+h)\cos\varphi} + 2\omega_e\right)\sin\varphi\, V_x - \frac{V_y V_z}{(R+h)}$$
$$\dot{V}_z = f_z + \left(\frac{V_x}{(R+h)\cos\varphi} + 2\omega_e\right)\cos\varphi\, V_x + \frac{V_y^2}{(R+h)} - g_e \tag{7.19}$$

Here $\dot{V}_x$, $\dot{V}_y$, $\dot{V}_z$ is respectively carrier's acceleration component relative to geographic coordinate system, f_x, f_y, and f_z respectively is the specific force component measured by the accelerometer, g_e is the acceleration of gravity.

According to the definition of latitude and longitude, we can get the location equation after integrating the speed terms:

$$\dot{\varphi} = = -\frac{V_y}{(R+h)}$$
$$\dot{\lambda} = \frac{V_x}{(R+h)\cos\varphi} \tag{7.20}$$
$$\dot{h} = V_z$$

2. MIMU system error equation.
   Due to the presence of various errors from the inertial device, according to the above mechanical arrangement equation, we can get the linear error model of the inertial navigation system MIMU using the perturbation equation.
   Error Angle equation:

$$\dot{\alpha} = \left(\frac{V_x}{R+h}\tan\varphi + \omega_e \sin\varphi\right)\beta - \left(\frac{V_x}{R+h} + \omega_e \cos\varphi\right)\gamma - \frac{1}{R+h}\delta V_y + \frac{V_y}{(R+h)^2}\delta h + \varepsilon_x$$
$$\dot{\beta} = -\left(\frac{V_x}{R+h}\tan\varphi + \omega_e \sin\varphi\right)\alpha - \frac{V_y}{R+h}\gamma + \frac{1}{R+h}\delta V_x - \omega_e \sin\varphi\delta\varphi - \frac{V_x}{(R+h)^2}\delta h + \varepsilon_Y$$
$$\dot{\gamma} = \frac{V_x}{R+h} + \omega_e \sin\varphi\alpha + \frac{V_y}{R+h}\beta + \frac{1}{R+h}\tan\varphi\delta V_x + \left(\frac{V_x}{R+h}\sec^2\varphi + \omega_e\cos\varphi\right)\delta\varphi$$
$$-\frac{V_x}{(R+h)^2}\tan\varphi\delta h + \varepsilon_Z \tag{7.21}$$

Speed error equation:

$$\dot{\delta V_x} = -f_z\beta + f_Y\gamma - \left(\frac{V_z}{R+h} - \frac{V_y}{R+h}\tan\varphi\right)\delta V_x + \left(\frac{V_x}{R+h}\tan\varphi + 2\omega_e\sin\varphi\right)\delta V_y$$
$$-\left(\frac{V_x}{R+h}\tan\varphi + \omega_e\cos\varphi\right)\delta V_z + \left(2\omega_e V_z\sin\varphi + \frac{V_y V_x}{R+h}\sec^2\varphi + 2\omega_e V_x\cos\varphi\right)\delta\varphi$$
$$+\left(\frac{V_x V_z}{(R+h)^2} - \frac{V_x V_y}{(R+h)^2}\tan(\varphi)\right)\delta h + \Delta a_x$$
$$\dot{\delta V_y} = f_z\alpha - f_x\gamma - 2\left(\frac{V_x}{R+h}\tan\varphi + \omega_e\sin\varphi\right)\delta V_x - \frac{V_z}{R+h}\delta V_y - \frac{V_y}{R+h}\delta V_z$$
$$-\left(\frac{V_x{}^2}{R+h}\sec^2\varphi + 2\omega_e V\cos\varphi\right)\delta\varphi + \left(\frac{V_x{}^2}{(R+h)^2}\tan\varphi + \frac{V_y V_z}{(R+h)^2}\right)\delta h + \Delta a_Y$$
$$\dot{\delta V_z} = -f_y\alpha + f_x\beta + 2\left(\frac{V_x}{R+h} + \omega_e\cos\varphi\right)\delta V_x + 2\frac{V_y}{R+h}\delta V_y - 2\omega_e V_x\sin\varphi\delta\varphi$$
$$+\frac{V_x{}^2 + V_y{}^2}{(R+h)^2}\delta h + \triangle a_Z \tag{7.22}$$

Position error equation:

$$\dot{\delta\varphi} = -\frac{1}{R+h}\delta V_y + \frac{V_y}{(R+h)^2}\delta h$$
$$\dot{\delta\lambda} = \frac{1}{R+h}\sec\varphi\,\delta V_x + \frac{V_x}{R+h}\sec\varphi\tan\varphi\,\delta\varphi - \frac{V_x}{(R+h)^2}\sec\varphi\,\delta h \tag{7.23}$$
$$\dot{\delta h} = \delta V_Z$$

Here, α, β, γ are error angles of the geographic coordinate system relative to the geographic coordinate system, respectively, δV_x, δV_y, δV_z are the spped errors, $\delta\varphi$, $\delta\lambda$, δh are the errors of geographic latitude, longitude, and altitude, and $[\Delta a_x \quad \Delta a_y \quad \Delta a_z]^T$ and $[\varepsilon_x \quad \varepsilon_y \quad \varepsilon_z]^T$ are, respectively, the accelerometer zero error and gyroscope drift rate. R is the radius of the Earth, ω_e is the angular velocity of the Earth, f_X, f_Y, f_Z are the corresponding specific force components.

Here, supposing that the carrier stays static relative to the ground, that is:

$$V_x = V_y = V_z = a_x = a_y = 0, \quad a_z = g$$

Using the change of latitude $\delta\phi \approx 0$, we can get the MIMU system simplified error equation under static base as follows:

$$\begin{bmatrix} \dot{\alpha} \\ \dot{\beta} \\ \dot{\gamma} \\ \delta\dot{V}_x \\ \delta\dot{V}_y \\ \delta\dot{\varphi} \\ \delta\dot{\lambda} \end{bmatrix} = \begin{bmatrix} 0 & \omega_e \sin\varphi & -\omega_e \cos\varphi & 0 & -\frac{1}{R} & 0 & 0 \\ -\omega_e \sin\varphi & 0 & 0 & \frac{1}{R} & 0 & -\omega_e \sin\varphi & 0 \\ \omega_e \cos\varphi & 0 & 0 & \frac{1}{R}\tan\varphi & 0 & -\omega_e \cos\varphi & 0 \\ 0 & -g & 0 & 0 & 2\omega_e \sin\varphi & 0 & 0 \\ g & 0 & 0 & -2\omega_e \sin\varphi & 0 & 0 & 0 \\ 0 & 0 & 0 & 0 & -\frac{1}{R} & 0 & 0 \\ 0 & 0 & 0 & \frac{1}{R}\sec\varphi & 0 & 0 & 0 \end{bmatrix} \cdot \begin{bmatrix} \alpha \\ \beta \\ \gamma \\ \delta V_x \\ \delta V_y \\ \delta\varphi \\ \delta\lambda \end{bmatrix} + \cdot \begin{bmatrix} \varepsilon_x \\ \varepsilon_y \\ \varepsilon_z \\ \Delta a_x \\ \Delta a_y \\ 0 \\ 0 \end{bmatrix} \tag{7.24}$$

3. MIMU system static error simulation
   Based on the static MIMU system error equation, we get:

Attitude errors	$\alpha = \beta = 0.01$ degree, $\gamma = 0.03$ degree
Speed error	$\delta V_x = \delta V_y = 0$
Latitude and longitude error	$\delta\varphi = \delta\lambda = 0.01$ degree
Accelerometer zero error	$\Delta a_x = \Delta a_y = \Delta a_z = 10^{-5}$ g (1σ)
Gyroscope drift rate	$\varepsilon_x = \varepsilon_y = \varepsilon_z = 0.01$ degree/h (1σ)

In addition, we set the radius of the Earth $R = 6367.65$ km, angular velocity of the Earth $\omega_e = 15.04107°/\text{h}$, local (Beijing) latitude $\phi = 39.9°$, local acceleration: $g = 9.78049$ m/s^2.

Using fourth-order Runge-Kutta method to solve Eq. (7.24), we can calculate the system static measurement error. The speed, position, and attitude angle error curve are shown in Figs. 7.9–7.15, respectively.

As shown in the figures, because of the mutual coupling between the three MIMU channels, the system error not only shows Shura oscillation cycle characteristics, but also has oscillation characteristics of the Foucault cycle and Earth cycle. Besides, the longitude error curve is in line with the longitude error equation, which possesses the open-loop characteristic.

7.3 RESEARCH ON INTEGRATED CALIBRATION OF MIMU

Because of the presence of errors such as installation error, zero deviation, and scale factor, the MIMU must be calibrated before use. Inertial device

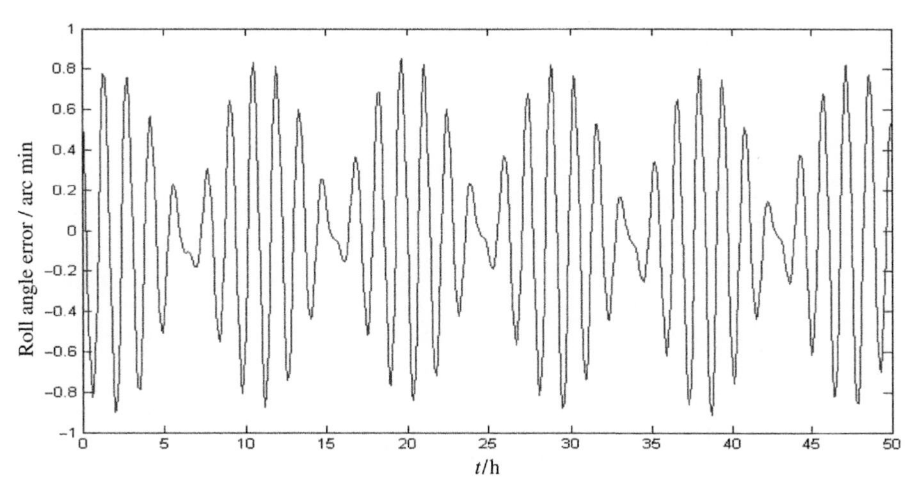

FIGURE 7.9
Roll angle error.

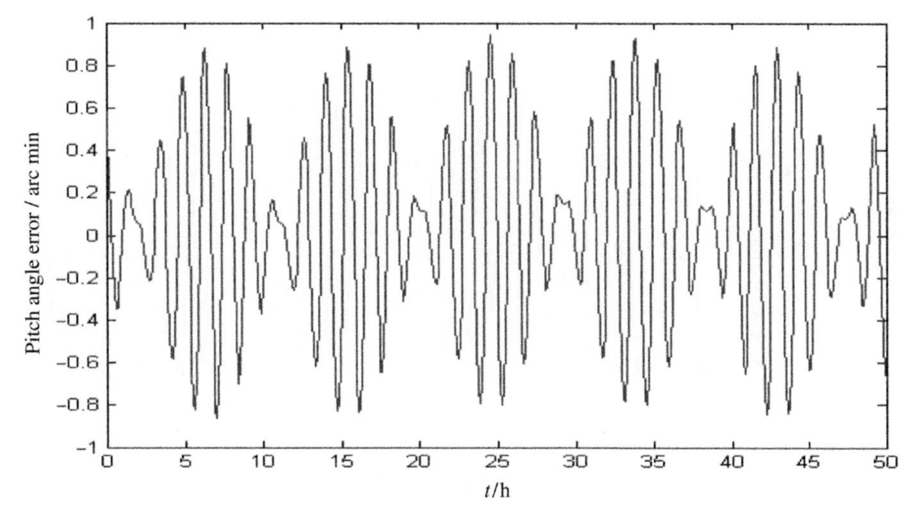

FIGURE 7.10
Pitch angle error.

errors consists of deterministic error and random error. Because of the nature of random error with its random properties, it can only be determined using a statistical method to obtain the rule of changing and then using the filter to compensate. In this chapter, we will only investigate the calibration and compensation of the deterministic errors.

MIMU calibration starts from the whole error model of the strap down system. By establishing the measurement model of the system angular velocity,

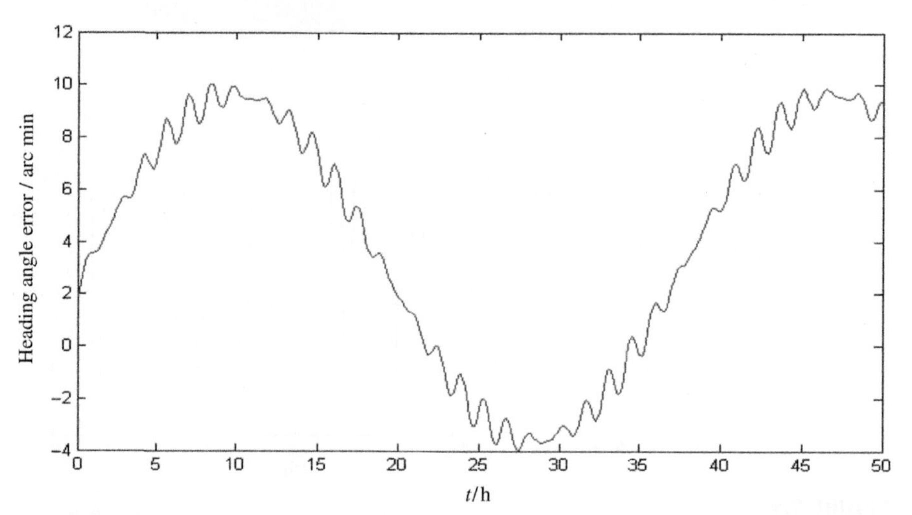

FIGURE 7.11
Heading angle error.

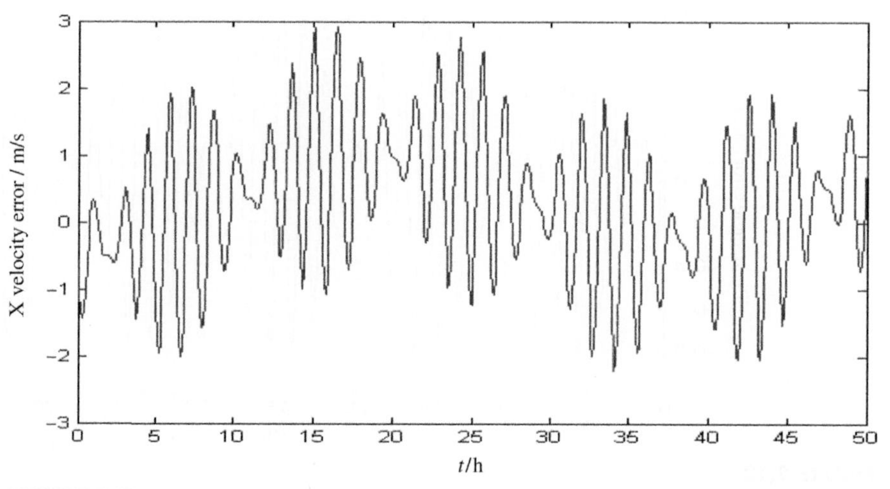

FIGURE 7.12
X velocity error.

linear acceleration, and doing the experiment using rate and multi-position, the overall calibration for the installation errors can be fulfilled, including scale factors, coefficients of drift and other error terms. The measurement accuracy can be greatly improved by making compensations for the raw outputs.

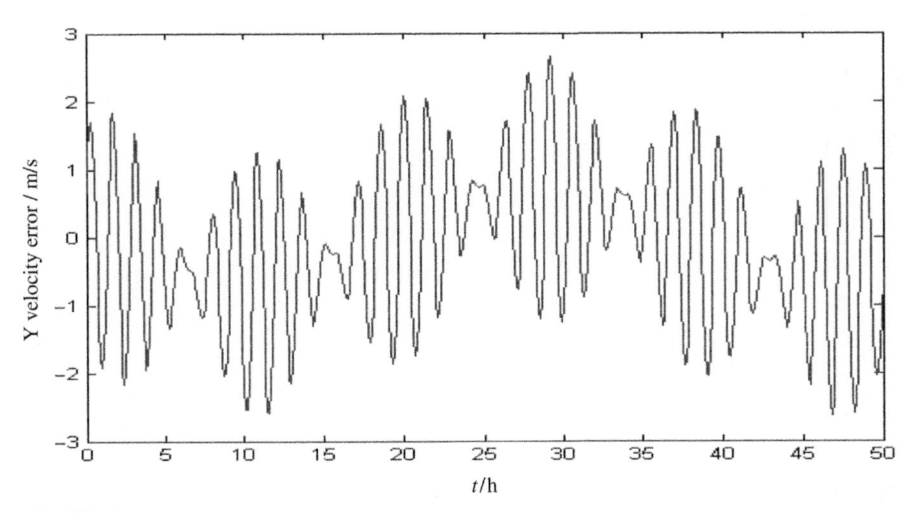

FIGURE 7.13
Y velocity error.

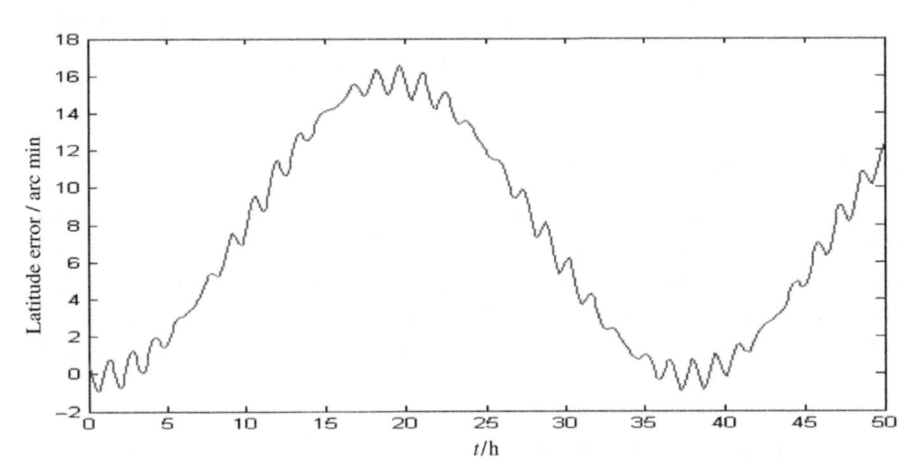

FIGURE 7.14
Latitude error.

7.3.1 Error Models of Inertial Devices

Due to the effect of environmental conditions, the inertial instrument (gyroscopes and accelerometers) of the strap down inertial measurement system will produce errors. Theoretically, without considering the error, the output of a gyroscope and accelerometer can be written as

$$G = K_g \omega_b \quad A = K_a a_b \tag{7.25}$$

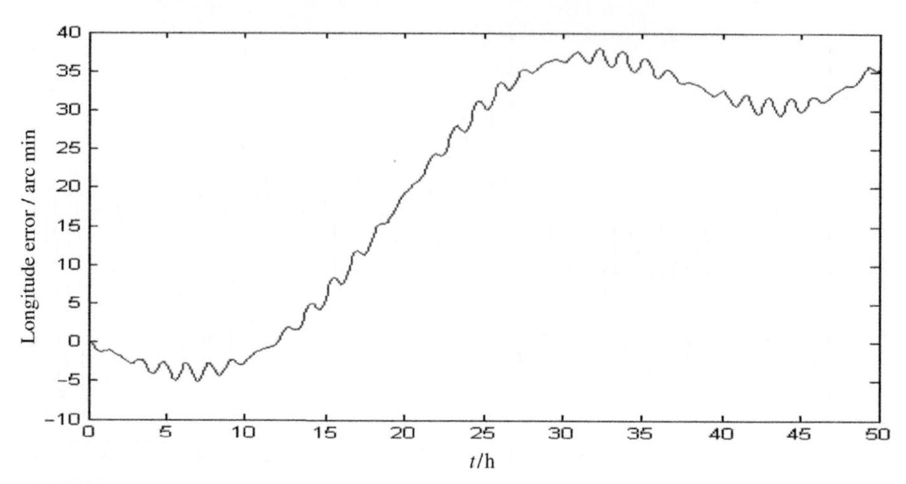

FIGURE 7.15
Longitude error.

where G and A—the respective outputs of the gyroscope and accelerometer;

K_g and K_a—the respective scale factors of the gyroscope and accelerometer;

ω_b—the movement velocity of the carrier

a_b—the acceleration of the carrier.

However, due to various interference factors existing in the instrument, actually the angular velocity and acceleration determined by the instrument are no longer simply the angular velocity and acceleration of the carrier. When the instrument converts the angular velocity and acceleration to output, it can also account for the error due to various kinds of interference in the process of transformation.

7.3.1.1 Research Error Model

1. The model equations of strapdown system

   First, to highlight the influence of the linear acceleration and angular velocity signals from the sensors, we suppose there is no scale factor error. Considering the line movement error, the angle motion error, ignoring the error due to the disturbance angle and the dynamic error, we can get the angular velocity output model:

$$\left.\begin{aligned} D_X &= K_g(X)[D_F(X) + \omega_{bX} + E_{XY}\omega_{bY} + E_{XZ}\omega_{bZ} + D_{1X}\omega_{bX} + D_{2X}\omega_{bY} + D_{3X}\omega_{bz} + \varepsilon_{gX}] \\ D_Y &= K_g(Y)[D_F(Y) + \omega_{bY} + E_{YZ}\omega_{bZ} + E_{YX}\omega_{bX} + D_{1Y}\omega_{bX} + D_{2Y}\omega_{bY} + D_{3Y}\omega_{bz} + \varepsilon_{gY}] \\ D_Z &= K_g(Z)[D_F(Z) + \omega_{bZ} + E_{ZX}\omega_{bX} + E_{ZY}\omega_{bY} + D_{1Z}\omega_{bX} + D_{2Z}\omega_{bY} + D_{3Z}\omega_{bz} + \varepsilon_{gZ}] \end{aligned}\right\} \tag{7.26}$$

where D_X, D_Y, D_Z—respectively, X_b, Y_b, Z_b—axis Gyro output voltage (V); $K_g(X), K_g(Y), K_g(Z)$—the gyro scale factor (V/degree/s); $D_F(X)$, $D_F(Y), D_F(Z)$—the gyro constant drift (degree/s); $\omega_{bX}, \omega_{bY}, \omega_{bZ}$—the respective carrier's rotational angular velocities around the X_b, Y_b, Z_b axis (degree/s); E_{ij}—the installation error coefficients around the carrier's j $(j = X_b, Y_b, Z_b)$ axis angular velocity relative to i $(i = X_b, Y_b, Z_b)$ axis output; D_{ij}—the orthogonal unbalanced drift coefficient of gyro $(i = 1, 2, 3, j = X, Y, Z)$; $\varepsilon_{gX}, \varepsilon_{gY}, \varepsilon_{gZ}$—the random drift (°/s).

The system acceleration measurement static output model is as follows:

$$\left.\begin{aligned} A_X &= K_a(X)[k_0(X) + a_{bX} + k_{1Y}(X)a_{bY} + k_{1Z}(X)a_{bZ} + \varepsilon_{aX}] \\ A_Y &= K_a(Y)[k_0(Y) + a_{bY} + k_{1Z}(Y)a_{bZ} + k_{1X}(Y)a_{bX} + \varepsilon_{aY}] \\ A_Z &= K_a(Z)[k_0(Z) + a_{bZ} + k_{1X}(Z)a_{bX} + k_{1Y}(Z)a_{bY} + \varepsilon_{aZ}] \end{aligned}\right\} \tag{7.27}$$

where A_X, A_Y, A_Z—respectively, the X_b, Y_b, Z_b—axis accelerometer output voltage (V); $K_a(X), K_a(Y), K_a(Z)$—the accelerometer scale factor (V/g_0); $k_0(X), k_0(Y), k_0(Z)$—the accelerometer bias (g_0); g_0—the gravitational acceleration; a_{bX}, a_{bY}, a_{bZ}—the X_b, Y_b, Z_b axial direction inertial acceleration (g_0);
$k_{1Y}(X), k_{1Z}(X), k_{1Z}(Y), k_{1X}(Y), k_{1X}(Z), k_{1Y}(Z)$—the installation error coefficient; $\varepsilon_{aX}, \varepsilon_{aY}, \varepsilon_{aZ}$—random drift (g_0).

2. Static scale factor error

   As for the gyroscope, in order to show the important role of the scale factor, we ignore the error of the angular velocity from gyroscope itself and some factors such as temperature. The output of the instrument not only reflects the scale factor K_g proportional to the angular velocity, but also reflects the scale factor $K_g^{(2)}$ that is in direct proportion to angular velocity squared, so the gyroscope output:

$$D = K_g(1 + \delta_g)\omega_b + K_g^{(2)}\omega_b^2 \tag{7.28}$$

where K_g—the meter scale coefficient;
δ_g—the relative value of scale factor error;
$K_g^{(2)}$—the scale factor in direct proportion to angular velocity square.
For acceleration, the rest can be deduced by analogy:

$$A = K_a(1 + \delta_a)a_b + K_a^{(2)}a_b^2 \tag{7.29}$$

7.3.2 Calibration of MIMU Error Coefficient

Based on the strapdown system output model, we focus on the discussion on calibration and compensation method of error coefficients. According to the working principle of the strapdown system, the calibration experiment can use three-axes turntable.

7.3.2.1 Rate Experiment

1. Steps:
   a. Establish three orthogonal frame positions of the turntable, with its outside frame vertical and the other two frames horizontal.
   b. Install the MIMU, making its OX_b, OY_b, OZ_b values respectively coincide with the middle, outer, and inner framework, as shown in Fig. 7.16.
   c. First turn the turntable 90° around the middle framework (OX_b), while keeping the inner framework (OZ_b) axis in the vertical position, then rotate it with constant angular velocity (choosing angular velocity according to the actual situation) around the outer frame. Write down the outputs of the three gyroscope axes during the rotation process.
   d. Remain the OX_b, OY_b axes remain in a vertical position respectively, and repeat the above steps.
2. The output of the gyroscope during the rate test
   a. When the inertial measurement unit is performing the rate experiment around the OZ_b axis, the output angular velocity and input acceleration input are, respectively:

$$\begin{cases} \omega_{bZ}(i) = \omega(i) + \omega_e \sin\varphi \\ \omega_{bX}(i) = \omega_e \cos\varphi \cos\varphi_3(t) \\ \omega_{bY}(i) = \omega_e \cos\varphi \sin\varphi_3(t) \end{cases} \quad \begin{cases} a_Z(i) = g_0 \\ a_X(i) = 0 \\ a_Y(i) = 0 \end{cases} \tag{7.30}$$

where $\omega(i)$ is the turntable's rotation angular velocity around the outer gimbal axis ($i = 1, 2, \ldots$) (degree/s); ω_e is the Earth's rotational angular velocity (degree/s); φ is the local latitude; $\varphi_3(t)$ is the angle between the middle framework and the north direction at time t.

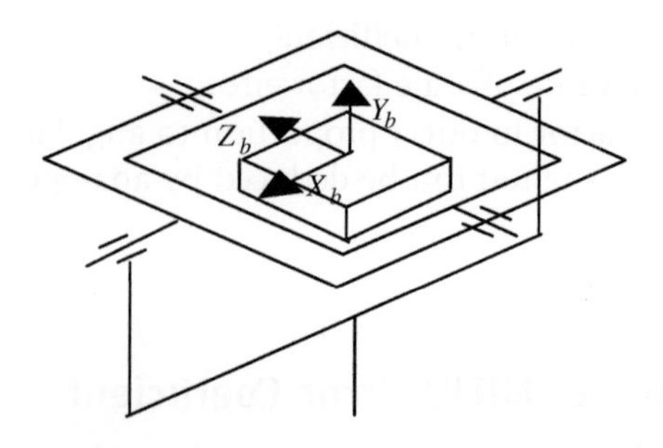

FIGURE 7.16
Schematic diagram of a three-axes turntable.

Substitute the above two equations into Eq. (3.2), Considering the nonlinearity of each axis scale factor in the equation, according to Eq. (3.4), and making $\delta_g = 0$:

$$\begin{aligned} D_X &= K_g(X)\omega_{bX} + K_g^{(2)}(X)\omega_{bX}^2 \\ D_Y &= K_g(Y)\omega_{bY} + K_g^{(2)}(Y)\omega_{bY}^2 \\ D_Z &= K_g(Z)\omega_{bZ} + K_g^{(2)}(Z)\omega_{bZ}^2 \end{aligned} \tag{7.31}$$

Using Eqs. (3.6) and (3.7), and considering that the installation error coefficient E_{ij} is small, the squared items can be ignored; when the turntable is rotated for 360°, $\varphi_3(t)$ varies from 0 to 2π. In this range, the integral of terms containing $\cos\varphi_3(t)$ or $\sin\varphi_3(t)$ are all 0. The strapdown system output for the three axes is:

$$\left.\begin{aligned} D_{XZ} &= \begin{bmatrix} [E_{XZ}(\omega(i)+\omega_e\sin\varphi) & [2E_{XZ}(\omega(i)+\omega_e\sin\varphi)D_F(X) \\ +D_F(X)] & +(D_F(X))^2] \end{bmatrix} \begin{bmatrix} K_g(X) \\ K_g^{(2)}(X) \end{bmatrix} \\ D_{YZ} &= \begin{bmatrix} [E_{YZ}(\omega(i)+\omega_e\sin\varphi) & [2E_{YZ}(\omega(i)+\omega_e\sin\varphi)D_F(Y) \\ +D_F(Y)] & +(D_F(Y))^2] \end{bmatrix} \begin{bmatrix} K_g(Y) \\ K_g^{(2)}(Y) \end{bmatrix} \\ D_{ZZ} &= \begin{bmatrix} \omega(i)+\omega_e\sin\varphi & [\omega(i)+\omega_e\sin\varphi \\ +D_F(Z)+D_Z(Z)g_0 & +D_F(Z)+D_Z(Z)g_0]^2 \end{bmatrix} \begin{bmatrix} K_g(Z) \\ K_g^{(2)}(Z) \end{bmatrix} \end{aligned}\right\} \tag{7.32}$$

where D_{XZ}, D_{YZ}, D_{ZZ} are the times of doing the rate experiment around the Z_b axis, gyro output voltage value (V) of the X_b, Y_b, Z_b channel, respectively.

b. When the IMU rate experiment is done around the OX_b axis, we obtain:

$$\begin{cases} \omega_{bZ}(i) = -\omega_e\cos\varphi\sin\varphi_3(t) \\ \omega_{bX}(i) = \omega(i) + \omega_e\sin\varphi \\ \omega_{bY}(i) = \omega_e\cos\varphi\cos\varphi_3(t) \end{cases} \quad \begin{cases} a_{bZ}(i) = 0 \\ a_{bX}(i) = g_0 \\ a_{bY}(i) = 0 \end{cases} \tag{7.33}$$

$$\left.\begin{aligned} D_{XX} &= \begin{bmatrix} [(\omega(i)+\omega_e\sin\varphi) & [(\omega(i)+\omega_e\sin\varphi) \\ +D_F(X)+D_X(X)g_0] & +D_F(X)+D_X(X)g_0]^2 \end{bmatrix} \begin{bmatrix} K_g(X) \\ K_g^{(2)}(X) \end{bmatrix} \\ D_{YX} &= \begin{bmatrix} [E_{YX}(\omega(i)+\omega_e\sin\varphi) & [2E_{YX}(\omega(i)+\omega_e\sin\varphi)(D_F(Y)+D_X(Y)g_0) \\ +D_F(Y)+D_X(Y)g_0] & +(D_F(Y)+D_X(Y)g_0)^2] \end{bmatrix} \begin{bmatrix} K_g(Y) \\ K_g^{(2)}(Y) \end{bmatrix} \\ D_{ZX} &= \begin{bmatrix} [E_{ZX}(\omega(i)+\omega_e\sin\varphi) & 2E_{ZX}(\omega(i)+\omega_e\sin\varphi)D_F(Z) \\ +D_F(Z)] & +(D_F(Z))^2 \end{bmatrix} \begin{bmatrix} K_g(Z) \\ K_g^{(2)}(Z) \end{bmatrix} \end{aligned}\right\} \tag{7.34}$$

c. When theinertial measurement unit rate experiment is done around the OY_b axis, we obtain:

$$\begin{cases} \omega_{bZ}(i) = \omega_e \cos\varphi \cos\varphi_3(t) \\ \omega_{bX}(i) = -\omega_e \cos\varphi \sin\varphi_3(t) \\ \omega_{bY}(i) = \omega(i) + \omega_e \sin\varphi \end{cases} \quad \begin{cases} a_{bZ}(i) = 0 \\ a_{bX}(i) = 0 \\ a_{bY}(i) = g_0 \end{cases} \tag{7.35}$$

$$\left.\begin{aligned} D_{XY} &= \begin{bmatrix} [E_{XY}(\omega(i)+\omega_e \sin\varphi) & [2E_{XY}(\omega(i)+\omega_e \sin\varphi)(D_F(X) \\ +D_F(X)+D_Y(X)g_0] & +D_Y(X)g_0)+(D_F(X)+D_Y(X)g_0)^2] \end{bmatrix} \begin{bmatrix} K_g(X) \\ K_g^{(2)}(X) \end{bmatrix} \\ D_{YY} &= \begin{bmatrix} [(\omega(i)+\omega_e \sin\varphi) & \\ +D_F(Y)+D_Y(Y)g_0] & [(\omega(i)+\omega_e \sin\varphi)+D_F(Y)+D_Y(Y)g_0]^2 \end{bmatrix} \begin{bmatrix} K_g(Y) \\ K_g^{(2)}(Y) \end{bmatrix} \\ D_{ZY} &= \begin{bmatrix} [E_{ZY}(\omega(i)+\omega_e \sin\varphi) & [2E_{ZY}(\omega(i)+\omega_e \sin\varphi)(D_F(Z) \\ +D_F(Z)+D_Y(Z)g_0] & +D_Y(Z)g_0)+(D_F(Z)+D_Y(Z)g_0)^2] \end{bmatrix} \begin{bmatrix} K_g(Z) \\ K_g^{(2)}(Z) \end{bmatrix} \end{aligned}\right\} \tag{7.36}$$

3. Integrated computation of rate test parameters
Respectively, we perform the rate experiment around the OX_b, OY_b, OZ_b axes, and under different n values of constant angular velocity. The gyro outputs are $D_{XX}(i), D_{YY}(i), D_{ZZ}(i)$ $(i = 1, 2, \ldots, n)$. If the gyroscope's three-channel drift coefficient is known, we can respectively calibrate the three-axes scale factors $K_g(X)$, $K_g^{(2)}(X)$, $K_g(Y)$, $K_g^{(2)}(Y)$, $K_g(Z)$, $K_g^{(2)}(Z)$ of the strapdown system using the third of Eq. (3.8), the first of Eq. (3.10), and the second equation of Eq. (7.36).
Taking a rate experiment around the OZ_b axis as an example, after n experimental runs, we can write the third Eq. (3.8) into the linear regression equation as follows:

$$\boldsymbol{U} = \boldsymbol{A}\boldsymbol{X} + \varepsilon \tag{7.37}$$

where $\boldsymbol{A}$ is the angular velocity vector; $\boldsymbol{U}$ is $n \times 1$ order measured value; $\boldsymbol{X}$ is the 2×1 order scale factor vector; ε is the $n \times 1$ order random vector.
Supposing the gyro output voltage $D_{ZZ}(1)$, $D_{ZZ}(2)$, ... and the gyroscope's drift coefficient $D_F(Z)$, $D_Z(Z)$ are already known, Eq. (3.13) can be employed to do a linear least squares regression calculation:

$$\hat{\mathbf{X}} = \left[\boldsymbol{A}^{\mathrm{T}} \cdot \boldsymbol{A}\right]^{-1} \boldsymbol{A}^{\mathrm{T}} \boldsymbol{U} \tag{7.38}$$

After calculating the three-axes scale factors $K_g(X)$, $K_g^{(2)}(X)$, $K_g(Y)$, $K_g^{(2)}(Y)$, $K_g(Z)$, $K_g^{(2)}(Z)$ of the strapdown system, we can calculate the installation error coefficients in all experiments from the first and second equations of Eq. (3.8), the second and third equations of Eq. (3.10) and the first and third equations of Eq. (3.12).

$$\left.\begin{aligned}
E_{XZ} &= \frac{D_{XZ} - D_F(X)K_g(X) - K_g^{(2)}(X)[D_F(X)]^2}{K_g(X)(\omega_3(i) + \omega_e \sin\varphi) + 2K_g^{(2)}(X)(\omega_3(i) + \omega_e \sin\varphi)D_F(X)} \\
E_{YZ} &= \frac{D_{YZ} - D_F(Y)K_g(Y) - K_g^{(2)}(Y)[D_F(Y)]^2}{K_g(Y)(\omega_3(i) + \omega_e \sin\varphi) + 2K_g^{(2)}(Y)(\omega_3(i) + \omega_e \sin\varphi)D_F(Y)} \\
E_{YX} &= \frac{D_{YX} - [D_F(Y) + D_X(Y)g_0]K_g(Y) - K_g^{(2)}(Y)[D_F(Y)+D_X(Y)g_0]^2}{K_g(Y)(\omega_3(i) + \omega_e \sin\varphi) + 2K_g^{(2)}(Y)(\omega_3(i) + \omega_e \sin\varphi)[D_F(Y) + D_X(Y)g_0]} \\
E_{ZX} &= \frac{D_{ZX} - D_F(Z)K_g(Z) - K_g^{(2)}(Z)[D_F(Z)]^2}{K_g(Z)(\omega_3(i) + \omega_e \sin\varphi) + 2K_g^{(2)}(Z)(\omega_3(i) + \omega_e \sin\varphi)D_F(Z)} \\
E_{XY} &= \frac{D_{XY} - [D_F(X) + D_Y(X)g_0]K_g(X) - K_g^{(2)}(X)[D_F(X)+D_Y(X)g_0]^2}{K_g(X)(\omega_3(i) + \omega_e \sin\varphi) + 2K_g^{(2)}(X)(\omega_3(i) + \omega_e \sin\varphi)[D_F(X) + D_Y(X)g_0]} \\
E_{ZY} &= \frac{D_{ZY} - [D_F(Z) + D_Y(Z)g_0]K_g(Z) - K_g^{(2)}(Z)[D_F(Z)+D_Y(Z)g_0]^2}{K_g(Z)(\omega_3(i) + \omega_e \sin\varphi) + 2K_g^{(2)}(Z)(\omega_3(i) + \omega_e \sin\varphi)[D_F(Z) + D_Y(Z)g_0]}
\end{aligned}\right\} \tag{7.39}$$

As each experimental formula contains an installation error coefficient to be calculated, we calculate the average of the installation error coefficients in n experimental runs.

7.3.2.2 Multiposition Experiment

This experiment uses the Earth's rotational speed and acceleration of gravity. We can obtain the error coefficient from inertial components' error model of the strapdown system through a static multiposition test. Make relevant arrangements around OX_b, OY_b, OZ_b axes and each interval represents a reading position. Making full use of the amplitude change in the Earth's rotation speed and acceleration of gravity at different positions, then reading gyro output voltage at each position, there are 24 positions including the zero position.

1. Steps:
   a. With outside frame vertical, middle frame horizontal and inner frame horizontal; among them the inner frame should point to the north, so at the beginning, the inner, middle, and outside frames' turning directions $\varphi_1, \varphi_2, \varphi_3$ are all 0. A strapdown system's three axes OX_b, OY_b, OZ_b point to the west, the sky, and north, respectively. Progressively rotate the inner framework and take the 45° interval as a reading position for a total of eight positions (1–8) including zero position. Read gyro output voltages at each position.
   b. With outside frame vertical, middle frame horizontal and pointing north, so at the beginning $\phi_1 = 0°$, $\phi_2 = -90°$, $\phi_3 = -90°$, with the strapdown system's three axes OX_b, OY_b, OZ_b respectively pointing to the north, west, and the sky. Progressively rotate the

middle framework and take the 45° interval as a reading position for a total of eight positions (9–16) including zero position. Read gyro output voltages at each position.

c. With the outside frame vertical, the middle frame is horizontal and pointing to the north, so at the beginning $\phi_1 = 90°\ ,\phi_2 = 0°\ ,\phi_3 = -90°$, with the strapdown system's three axes OX_b, OY_b, OZ_b pointing to the sky, south, and east, respectively. Progressively rotate the middle framework and take a 45° interval as a reading position for a total of eight positions (17–24) including zero position. Read gyro output voltages at each position.

2. Calculation of multiposition experiment parameters
When do calculation, we assume that the scale factor $K_g(X)$ and each installation error coefficient (E_{XY}, E_{XZ}) has been calibrated by experimental rate Omitting the scale factor nonlinearity, that is $K_g^{(2)}(X) = 0$, $\delta_g(X) = 0$ and omitting the dynamic error term (the unequal inertia item, angular acceleration) the first equation of Eq. (3.2) can be rewritten as:

$$U - WE = AX + \varepsilon \tag{7.40}$$

where $\boldsymbol{U}$ is the measuring angular velocity; $\boldsymbol{W}$ is the input angular velocity; $\boldsymbol{E}$ is the installation error coefficient; ε is the random drift; $\boldsymbol{A}$ is the input acceleration; $\boldsymbol{X} = [D_F(X)]$ is the drift coefficient.

Supposing all measurement of the 24 location experiments were independent of each other. Then Eq. (3.16) can perform a least squares linear regression, obtaining the optimal valuation of the drift coefficient:

$$\hat{\mathrm{X}} = \left[A^T \cdot A\right]^{-1} A^T (U - WE) \tag{7.41}$$

Therefore, the X_b channel gyro drift coefficient's optimal valuation $\hat{D}_F(X)$ can be calculated.

In the same way, respectively taking 24 positions of experimental data into the second and third equations of Eq. (3.2), and with above analyses and calculations, the optimal estimation of gyro drift coefficients in the Y_b, Z_b channels of the strapdown system can be obtained.

7.3.2.3 Calibration of the Accelerometer's Static Error Coefficient

Similar with gyro calibration, the calibration of each accelerometer's static error coefficient is conducted in a multi-position experiment at the same time, i.e., during the calibration of the accelerometer static error coefficients, we record the three channels of the accelerometer output voltage at the same time, and then use a linear regression method to calculate the accelerometer static error coefficients, including the scale factor, the second item, installation error, etc.

7.3.2.4 Rate Experiment and the Iterative Calculation of Multiposition Experimental Data

First, scale factors and installation errors of each channel are calculated from the rate experimental data. Using the known scale factor and the installation error, from the multiple location experiment data, calculate the drift coefficient for each channel gyroscope, and use the calculated gyroscope drift coefficient in the calibration calculation formula of the rate experiment, to calculate the scale factor and the installation error coefficients of each channel. Above process are conducted circularly, until each channel's relative scale factor error value is lower than the prescribed one. A block diagram of this process is shown in Fig. 7.17.

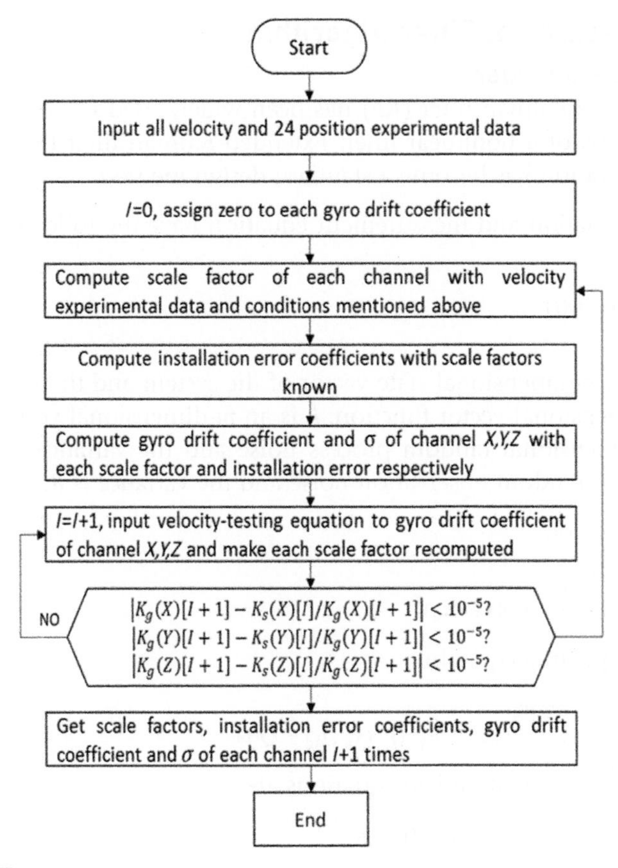

FIGURE 7.17
Calibration data experimental iterative process.

7.4 MIMU INTEGRATED NAVIGATION TECHNOLOGY

Due to measurement positioning errors accumulating over time, it is difficult for MIMU to complete long-term measurement. There are usually two solutions to this problem. First is increaing the properties of the miniature inertial device itself by using new materials, processes, and technologies.

However, the accuracy of the miniature inertial devices is limited by the working principle and the manufacturing conditions. Second is using integrated navigation technology, based on the filter design and other assisted measurement system in which measurement accuracy in long term is satisfied. This is an effective method and one of the main directions of navigation technology. In this chapter, we will carry out research on the MIMU and magnetometer, Global Positioning System (GPS)-integrated navigation and relative filtering technology.

7.4.1 Research of Filter Algorithm

7.4.1.1 Kalman Filter

Since 1960, the Kalman filter (KF) has been widely used in optimal filtering fields. In terms of a nonlinear filter, Extended Kalman filter (EKF) obtained by KF linearization has become a standard design method.

Filter state equation and measurement equation have the following standard form:

$$\begin{aligned} \dot{x}(t) &= f(x(t), t) + w(t) \\ z_k &= h(x_k) + v_k \end{aligned} \tag{7.42}$$

where $\boldsymbol{x}$ is an n-dimensional state vector of the system and the variance is $\boldsymbol{P}_k$; f is an n-dimensional vector function; h is an m-dimensional vector function; w_k is a p-dimensional random process noise and the variance is $\boldsymbol{Q}$; v_k is a q-dimensional random observation noise and the variance is $\boldsymbol{R}$.

1. Initialization

$$\begin{aligned} \hat{\mathbf{x}}_0 &= E[\boldsymbol{x}_0] \\ \boldsymbol{P}_0 &= E\left[(\boldsymbol{x}_0 - \hat{\mathbf{x}}_0)(\boldsymbol{x}_0 - \hat{\mathbf{x}}_0)^T\right] \end{aligned} \tag{7.43}$$

2. Time updating equation

$$\begin{aligned} \hat{\mathbf{x}}_k^- &= f(\hat{\mathbf{x}}_{k-1}) + w_k \\ \dot{\mathbf{P}}_k(-) &= F(\hat{\mathbf{x}}_k^-, t_k)\boldsymbol{P}_{k-1} + \boldsymbol{P}_{k-1}F^T(\hat{\mathbf{x}}_k^-, t_k) + \boldsymbol{Q}(t) \end{aligned} \tag{7.44}$$

3. The measurement update equations are

$$\begin{aligned} \boldsymbol{K}_k &= \boldsymbol{P}_k(-)H_k^T(H_k\boldsymbol{P}_k(-)H_k^T + \boldsymbol{R}_k)^{-1} \\ \hat{\mathbf{x}}_k &= \hat{\mathbf{x}}_k^- + \boldsymbol{K}_k(z_k - h(\hat{\mathbf{x}}_k^-, \overline{n})) \\ \boldsymbol{P}_k &= \boldsymbol{P}_k(-) - \boldsymbol{K}_k\boldsymbol{C}_k\boldsymbol{P}_k(-) \end{aligned} \tag{7.45}$$

where

$$F_k = \frac{\partial f}{\partial x}, \quad H_k = \frac{\partial h}{\partial x}$$

F_k, H_k are the linearized expressions of the state transfer equation and measurement transfer equation, respectively, known as the Jacobian matrix. The filtered value $\hat{\mathbf{x}}_k$ is the linear minimum variance estimation of state x_k, which is an unbiased estimation. Since the Kalman filter equation uses recursive forms, there is no need to store any measurement data, which greatly saves a lot of memory space and is very suitable for computer calculation.

7.4.1.2 Unscented Kalman Filter

From the introduction of the EKF algorithm above, we can see that the EKF only provides the first-order approximation of the optimal nonlinear estimation, and simply linearizes all the nonlinear models, and then applies the linear Kalman filter method. In reality, on one hand, if the time interval is not small enough, linearization might produce extremely unstable filtering. On the other hand, Jacobian matrix derivative is needed in the linearization process, which is not easy in most cases. In order to solve these problems, Julier and Uhlman presented a new generalized Kalman filter—the U-Kalman filter (UKF). Compared with EKF, the UKF selects a batch of sampling points that can express the mean and variance of the system state, and achieves a third-order approximation precision to real mean and variance through nonlinear transformation of sampling points.

1. Defects of the EKF
   In 1960, on the basis of the following two assumptions: (1) the state distribution is Gaussian distribution; (2) Gaussian distribution is still Gaussian distribution after linear operation, Kalman first proposed the Kalman filter. The state space method is used to describe the system. The algorithm adopts a recursive method to deal with multidimensional and nonstationary stochastic processes. As the original KF is only applicable to the linear system, EKF is proposed in order to make KF applicable to nonlinear conditions. EKF approximates nonlinear systems with a linear system through "linearization," replacing all nonlinear information with linear approximations.

   Supposing a nonlinear discrete system state equation:

$$X_k = f(X_{k-1}) + W_k \tag{7.46}$$

   where f is a state transition matrix and W_k is an r-dimensional system dynamic noise.

The observation equation:

$$y_k = h(X_K) + V_k \tag{7.47}$$

where y_k is an observation vector, H is the observation model, V_k is an r-dimensional measurement noise.

Assuming that noise W_k and V_k are irrelevant Gaussian white noises with a zero mean, also not relevant with the initial state X_0, for $k-1 \geq 1$, there are

$$\begin{aligned} E[W_{k-1}] = 0, E[W_{k-1}W_{j-1}^T] = Q_{k-1}\delta_{k-1,j-1} \\ E[V_{k-1}] = 0, E[V_k V_j^T] = R_k \delta_{k,j} \\ E[W_k V_j^T] = 0, E[X_0 W_{k-1}^T] = 0, E[X_0 V_k^T] = 0 \end{aligned} \tag{7.48}$$

Based on Gaussian random distribution, EKF recursive estimation is obtained as below:

$$\begin{aligned} \hat{X}_k = \hat{X}_k^- + K_k[y_k - \hat{y}_k^-] \\ P_{X_k} = P_{X_k}^- - K_k P \sim y_k K_k^T \end{aligned} \tag{7.49}$$

Although we do not assume that the model is linear, the formulas above are linear recursive, and the optimal iterative solution can be given by the formula below:

$$\hat{\mathbf{X}}_k^- = E[F(\hat{\mathbf{X}}_{k-1})] \tag{7.50}$$

$$K_k = P_{x_k y_k} P_{\tilde{y}_k \tilde{y}_k}^{-1}, \quad \tilde{y}_k = y_k - \hat{y}_k^- \tag{7.51}$$

$$\hat{y}_k^- = E[H(\hat{X}_k^-)] \tag{7.52}$$

$\hat{\mathbf{X}}_k^-$, $\hat{y}_k^-$ are the best predictions of X_k and y_k respectively.

In the case of linear conditions, KF can accurately obtain these values. But for nonlinear models, EKF obtains these values by approximation.

$$\hat{\mathbf{X}}_k^- \approx f(\hat{\mathbf{X}}_{k-1}) \tag{7.53}$$

$$K_k \approx P_{x_k y_k} P_{\tilde{y}_k \tilde{y}_k}^{-1} \tag{7.54}$$

$$\hat{\mathbf{y}}_k^- \approx H(\hat{\mathbf{X}}_k^-) \tag{7.55}$$

To determine the covariance, first the model should be linearized, $X_{k+1} \approx AX_k + BV_K$, $y_k \approx CX_k + DW_K$, and then the covariance matrix of the linear system can be determined, which is considered as the covariance of the nonlinear model. In other words, in EKF, Gaussian distribution is an approximation of the state distribution, and EKF gives the first-order approximation of the best estimate through the first-order linear transfer of the nonlinear system. EKF can also achieve the second-order accuracy. However, it is very difficult to achieve for its complexity. Therefore, usually in practice, the second order approximation is not considered. A First-order approximation probably introduces large errors into the real mean and covariance of a random variable after transformation, especially when the local linearization assumption is invalid, linearization produces very unstable filtering. In addition, in most cases the derivative of Jacobi matrix needs to be calculated in the EFF linearization process, which is not simple.

2. Unscented Transformation (UT)

UKF solves the approximation problem of EKF. The difference between EKF and UKF is that the UKF method uses UT to calculate the mean and variance of the estimated and measured predictions. The unscented transformation process is to generate a number of discrete sampling points from the estimated mean and variance, which, after the propagation of the state equation and the measurement equation, produce the mean and variance of the predicted value by weighting.

Sampling points are generated as follows: given the mean $\overline{\mathbf{X}}$ and variance $\boldsymbol{P}_{xx}$ of an n-dimentional random variable $\boldsymbol{X}$, through a nonlinear transformation $\boldsymbol{Y}=f(\boldsymbol{X})$, random variables $\boldsymbol{Y}$ can be obtained, and the estimations of its mean $\bar{y}$ and variance P_{yy} are:

a. Calculating the weighted sigma $2n+1$ sampling points

$$\begin{aligned}
\boldsymbol{X}_0 &= \overline{\mathbf{X}} \\
\boldsymbol{X}_0 &= \overline{\mathbf{X}} + (\sqrt{(n+\lambda)\boldsymbol{P}_{xx}})_i, \quad i=1,\ldots,n \\
\boldsymbol{X}_i &= \overline{\mathbf{X}} - (\sqrt{(n+\lambda)\boldsymbol{P}_{xx}})_{i-n}, \quad i=1,\ldots,2n \\
w_0^m &= \lambda/(n+\lambda), \\
w_0^c &= \lambda/(n+\lambda) + (1-\alpha^2+\beta), \\
w_i^m &= w_i^c = 1/[2(n+\lambda)], \quad i=1,\ldots,2n
\end{aligned} \tag{7.56}$$

where $\lambda=\alpha^2(n+k)-n$, α determines the degree of distribution of sampling points around the mean, k generally is 0, β contains distribution of prior information of X (for Gaussian distribution, $\beta=2$ is the optimal). $(\sqrt{(n+\lambda)\mathbf{P}_{xx}})_i$ is the i-th row of the matrix square root, and w_i^m, w_i^c are the weight of the mean and variance. When the state variable is single variable, $k=2$. Meanwhile,$0\le\alpha\le 1$, the adjustment of α can reduce the influence of the higher-order terms of the nonlinear equation.

b. Through the nonlinear transformation, the transformed samples are obtained

$$\boldsymbol{Y}_i = f(\boldsymbol{X}_i), i=0,\ldots,2n \tag{7.57}$$

c. Calculating the mean and covariance

$$
\begin{aligned}
\dot{Y} &= \sum_{i=0}^{2n} w_i^m Y_i \\
P_{yy} &= \sum_{i=0}^{2n} w_i^c (Y_i - - Y)(Y_i - - Y)^T
\end{aligned}
\tag{7.58}
$$

Fig. 7.18 is a schematic representation of the mean and variance estimation processes for a two-dimensional system: (1) the true mean value and variance are generated through Monte Carlo sampling; (2) the mean value and variance are generated through linearization, and linearization is the same with EKF; (3) for the use of UT. Obviously, UT is closer to true a nonlinear state than EKF linearization.

3. UKF design

Applying UT in the recursive process and remaining other parts the same, UKF filter can be obtained.

a. Initialization

Process noise and measurement noise can be augmented to state vectors, assuming that dimension is n, $\boldsymbol{x}^a$ is augmented state vector,

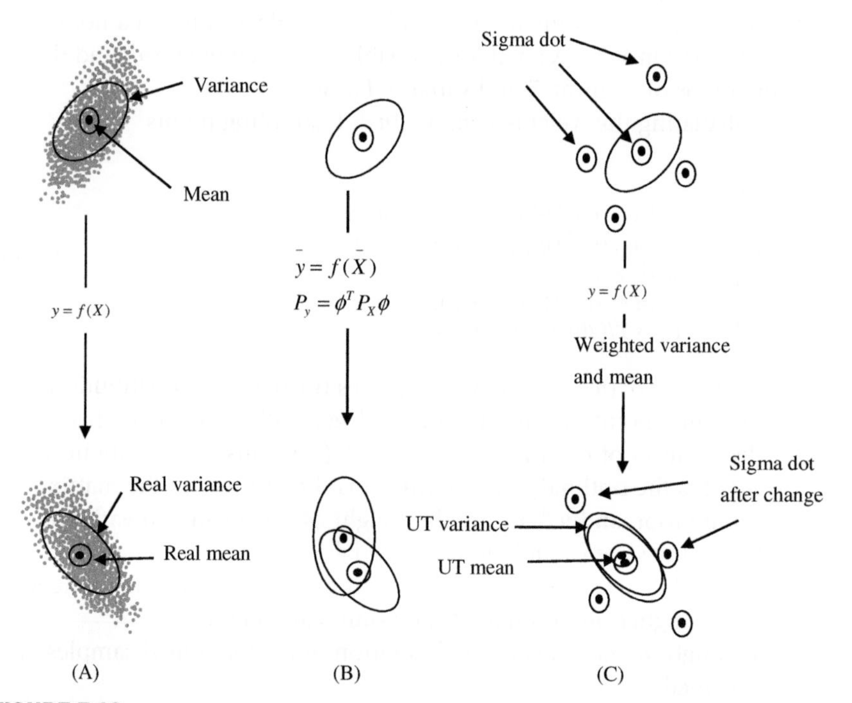

FIGURE 7.18
Mean and variance estimation of a two-dimentional system. (A) Monte Carlo sampling, (B) EKF, (C) UT.

w and v are the process noise and measurement noise. X^a is the corresponding sampling points vector. P_0 is the initial estimate of the original state vector covariance, P_v is the process noise variance, P_n is the measurement noise variance.

$$\begin{aligned}
x^a &= \left[x^T\ w^T\ v^T\right]^T \\
X^a &= \left[(X^x)^T\quad (X^w)^T\quad (X^v)^T\right]^T \\
\hat{x}_0 &= E[x_0],\quad P_0 = E\left[(x_0-\hat{x}_0)(x_0-\hat{x}_0)^T\right] \\
\hat{x}_0^a &= E\left[x_0^a\right] = \left[\hat{x}_0^T\quad 0\quad 0\right]^T \\
P_0^a &= E\left[\left(x_0^a-\hat{x}_0^a\right)\left(x_0^a-\hat{x}_0^a\right)^T\right] = \begin{bmatrix} P_0 & 0 & 0 \\ 0 & P_w & 0 \\ 0 & 0 & P_v \end{bmatrix}
\end{aligned}$$

b. Sampling points calculation

$$\begin{aligned}
X_{0,k-1}^a &= \hat{x}_{k-1}^a \\
X_{i,k-1}^a &= \hat{x}_{k-1}^a + \left(\sqrt{(n+\lambda)P_{k-1}^a}\right)_i, \quad i=1,\ldots,n \\
X_{i,k-1}^a &= \hat{x}_{k-1}^a - \left(\sqrt{(n+\lambda)P_{k-1}^a}\right)_i, \quad i=n+1,\ldots,2n \quad 2n+1 \text{ sampling points.}
\end{aligned}$$

c. Time updating equations

$$\begin{aligned}
X_{k|k-1}^x &= F\left(X_{k-1}^x, X_{k-1}^w\right) \\
\hat{x}_k^- &= \sum_{i=0}^{2n} w_i^m X_{i,k|k-1}^x \\
P_k^- &= \sum_{i=0}^{2n} w_i^c \left(X_{i,k|k-1}^x - \hat{x}_k^-\right)\left(X_{i,k|k-1}^x - \hat{x}_k^-\right)^T \\
Z_{k|k-1} &= H\left(X_{k|k-1}^x, X_{k-1}^v\right) \\
\hat{z}_k^- &= \sum_{i=0}^{2n} w_i^m Z_{i,k|k-1}
\end{aligned}$$

d. Measurement updating equations

$$\begin{aligned}
P_{\hat{Z}_k\hat{Z}_k} &= \sum_{i=0}^{2n} w_i^c \left(Z_{i,k|k-1} - \hat{z}_k^-\right)\left(Z_{i,k|k-1} - \hat{z}_k^-\right)^T \\
P_{X_k\hat{Z}_k} &= \sum_{i=0}^{2n} w_i^c \left(X_{i,k|k-1} - \hat{x}_k^-\right)\left(Z_{i,k|k-1} - \hat{x}_k^-\right)^T \\
K &= P_{X_k\hat{Z}_k} P_{\hat{Z}_k\hat{Z}_k}^{-1} \\
\hat{x}_k^+ &= \hat{x}_k^- + K\left(z_k - \hat{z}_k^-\right) \\
P_k^+ &= P_k^- - KP_{\hat{Z}_k\hat{Z}_k}K^T \\
P_{z_kz_k} &= \sum_{i=0}^{2L} w_i^c \left[Z_{i,k}(-) - \hat{z}_k^-\right]\left[Z_{i,k}(-) - \hat{z}_k^-\right]^T \\
P_{x_kz_k} &= \sum_{i=0}^{2L} w_i^c \left[X_{i,k}^x(-) - \hat{x}_k^-\right]\left[Z_{i,k}(-) - \hat{z}_k^-\right]^T
\end{aligned}$$

$$
\begin{aligned}
K_k &= P_{x_k z_k} P_{z_k z_k}^{-1} \\
\hat{\mathbf{x}}_k &= \hat{\mathbf{x}}_{\hat{k}}^{-} + K_k\left(z_k - \hat{\mathbf{z}}_k^{-}\right) \\
P_k &= P_k(-) - K_k P_{z_k z_k} K_k^T
\end{aligned}
$$

where z_k is the measured value of step k. It can be seen that the last three equations of the measurement updating equation are equivalent to the corresponding equations in the EKF, but the UKF improves the prediction accuracy. Therefore, it improves the filtering performance.

7.4.1.3 Federated Filter

With the increasing number of navigation systems available for carrier assembly, the increase in non-similar navigation subsystems increases the measurement information. The conventional centralized Kalman filter has some problems in data processing of an integrated multi-sensor system. First, when the combined information is redundant, the computational complexity will increase the third dimension of the filter dimension and cannot meet the real-time measurement requirements. Second, the increase in the navigation subsystems also increases the failure rate, and the entire navigation system is polluted when a subsystem fails and is not detected and isolated. Speyer, Bierman and Kerr, in order to solve this problem, proposed the idea of decentralized filtering. However, there are still some problems, for example, there is a large amount of information and correlation of each local state estimation solution needs to be considered in fusion. In 1988, Carlson [18] put forward the theory of Federated Filter, which provides integrated navigation system a design theory.

For linear time-invariant multi-sensor integrated system, it can be described in a discrete form below:

$$X(k+1) = f(k+1,k)X(k) + \Gamma(k)W(k) \tag{7.59}$$

where $X(k+1)$ is the system state variable; $f(k+1,k)$ is the first step transfer matrixes of the state; $\Gamma(k)$ is the system noise matrix; $W(k)$ is the white noise series with a zero mean, and $E\left[W(k)W^T(j)\right] = Q\delta(k,j)$.

Assuming that there are N sensors and each sensor makes an independent measurement, so there are corresponding N local filters and each filter can filter calculation independently. The models of i-th local filters are:

$$X_i(k+1) = f_i(k+1,k)X_i(k) + \Gamma_i(k)W_i(k) \tag{7.60}$$

$$Z_i(k+1) = H_i(k+1)X_i(k+1) + V_i(k+1) \tag{7.61}$$

where $Z_i(k+1)$ is the observation value of the i-th sensor; $H_i(k+1)$ is the observation matrix of the i-th sensor; $V_i(k+1)$ is the white noise series with a zero mean which is independent of $W(k)$, and $E\left[V_i(k)V_i^T(j)\right] = R_i\delta(k,j)$.

1. Principles
   Federal Filter is a kind of partitioned estimation and can be regarded as a special dispersion filtering method based on the information distribution. In Federal Filter, the subsystem includes external sensor 1, sensor 2, …, sensor N, and standard Kalman filters correspond to different sensors. Therefore the system has multiple local filters, and each local filter works in parallel. The information integration and sequential processing are disposed of through the global filter, and the global optimal state estimation is obtained from the information fusion of the filtering results of all local filters.
2. Information fusion algorithm based on the federal Kalman filter
   The purpose of information fusion is to analyze and combine the estimated information from multiple local filters in a certain way to get the global optimal estimation. Supposing the state estimation of the local filter is $\hat{\mathbf{X}}_i$, the covariance matrix of system noise is $\boldsymbol{Q}_i$ and the covariance matrix of state vector is P_i, $i = 1...N$; Let the parameters of global filter is $\hat{\mathbf{X}}_m$, $\boldsymbol{Q}_m$, $\boldsymbol{P}_m$. The calculation process of Federal Filter is as follows:
   a. Initialization
      Assuming that the initial value of initial global state is $\hat{\mathbf{X}}_0$, its covariance matrix is $\boldsymbol{P}_0$, and the system noise covariance matrix is $\boldsymbol{Q}_0$, which are distributed to each local filter and global filter through information distribution factor:

$$Q^{-1} = Q_1^{-1} + Q_2^{-1} + \cdots + Q_N^{-1} + Q_m^{-1}, \quad Q_i^{-1} = \beta_i Q^{-1}$$
$$P^{-1} = P_1^{-1} + P_2^{-1} + \cdots + P_N^{-1} + P_m^{-1}, \quad P_i^{-1} = \beta_i P^{-1}$$
$$P^{-1}\hat{X} = P_1^{-1}\hat{X}_{11} + P_2^{-1}\hat{X}_2 + \cdots + P_N^{-1}\hat{X}_N + P_m^{-1}\hat{X}_m, \qquad i = 1, 2, \ldots, N,$$

      where β_i conforms to the law of energy conservation:

$$\beta_1 + \beta_2 + \cdots + \beta_N + \beta_m = 1, \quad 0 \le \beta_i \le 1.$$

   b. Time updating of information

$$\hat{X}_i(k+1/k) = \phi(k+1/k)\hat{X}_i(k)$$
$$P_i(k+1/k) = \phi(k+1,k)P_i(k/k)\phi^T(k+1,k) + \Gamma(k+1,k)Q_i(k)\Gamma^T(k+1,k) \quad i = 1, 2, \ldots, N, m$$

   c. Measurement updating of information
      The measurement updating of the i-th local filter is

$$P_k^{-1}(k+1/k+1)\hat{X}_i(k+1/k+1) = P_k^{-1}(k+1/k+1)\hat{X}_i(k+1/k+1) + H_i^T(k+1)R_i^{-1}(k+1)Z_i(k+1)$$
$$P_k^{-1}(k+1/k+1) = P_k^{-1}(k+1/k) + H_i^T(k+1)R_i^{-1}(k+1)H_i(k+1)$$
$$i = 1, 2, \ldots, N$$

d. Information fusion

$$\hat{X}_g = P_g \sum_{i=1}^{N} P_i^{-1} \hat{X}i$$

$$P_g = \left(\sum_{i=1}^{N} P_i^{-1} \right)^{-1}$$

3. Structure analysis

In different methods of information distribution, the value of information distribution factor β_i ($i = 1, 2, \ldots, N, m$) is different. Federal Kalman Filters have different structures, and the general structure of Federal Kalman Filter is shown in Fig. 7.19.

a. No reset structure

Each local filter has all the information of its own and the output is determined by time updating and measurement updating, avoiding the influence of feedback reset. The output of the global filter is only determined by time updating, and it only makes a fusion of the input information, but it does not keep this information and has no feedback to local filters. Because there is no reset of the global optimal estimation of this method, local estimation accuracy is not high, but in exchange the system has good robustness of fault

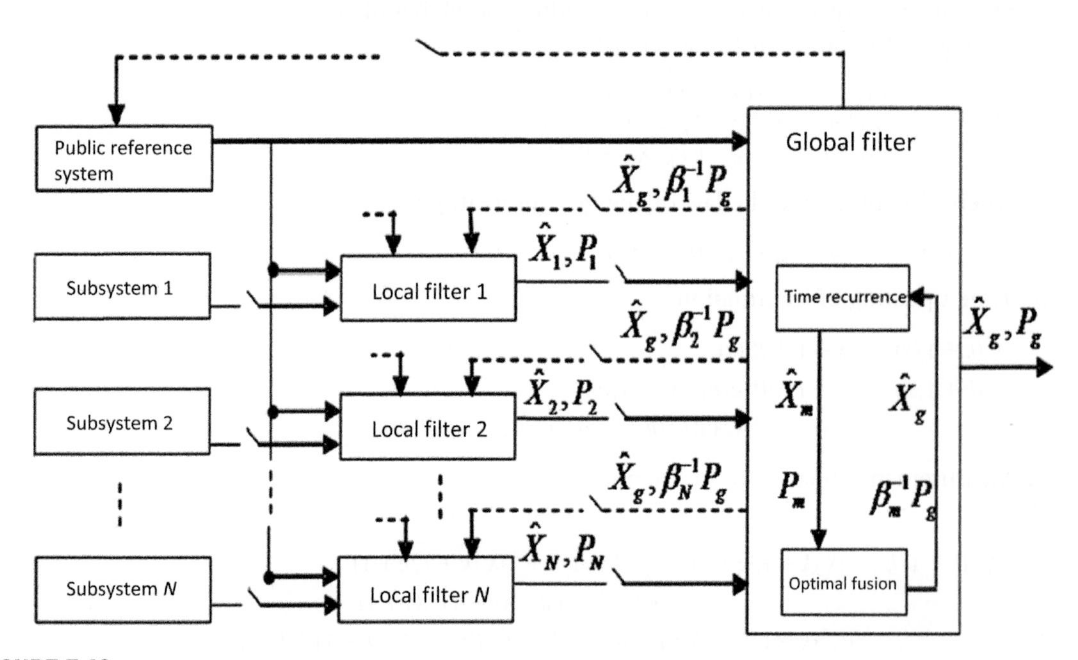

FIGURE 7.19
The structure of a general federal Kalman filter.

tolerance. Therefore, it is also known as a fault-tolerant Federal Filtering structure.

b. Zeros-reset structure
   Each local filter provides the global filter with new information obtained after the data were sent last time, which is processed under the condition that the estimation error variance of the public state is infinite. The main global filter has long-term memory function and keeps all the fusion information. This structure can process the data from each local filter at different times and has no feedback to the local filter. This structure is relatively simple to implement.
c. Fuse-reset structure
   Main global filter will return a feedback of distribution information after each fusion calculation. Each local filter needs to wait for the feedback information from the main filter before work. Because of the feedback effect, this structure can improve the accuracy. However, when the information transmits, the fault also transmits through the same route at the same time, so the fault-tolerant ability is a deficiency.
d. Rescale structure
   In the information fusion, each local filter only provides part of the information to the global filter, and keeps the rest of the information. When the fusion is completed, each local filter will expand the state estimation and variance to times of the reciprocal of the remaining information as the initial value for filtering, and then begin the next cycle of filtering.

7.4.2 Combination of MIMU and Magnetometer

Satellite autonomous navigation technology can realize independent operation without support from the system on the ground, which can not only effectively reduce the cost of ground support, but also improve the survivability of the satellite. Meanwhile, satellites tend to be miniaturized, which requires an autonomous navigation system that is miniaturized, cheap, and qualified for tasks.

A three-axes magnetometer is a reliable, low-cost attitude sensor with low volume and light weight. Because the geomagnetic field intensity vector is one of the vectors fixed with the Earth, different components of the geomagnetic data can be measured with the three-axes magnetometer, and in this way components and the sum of the magnetic field and all kinds of binding parameters can be obtained by computing the measured signals. With the improvement of the geomagnetic field model, the combination system constituted by the MIMU and magnetometer can effectively estimate gyro drift and improve the accuracy of integrated attitude determination. Especially for LEO small satellites, rich resources of the geomagnetic field can be taken advantage of and the requirement for attitude determination accuracy is not very high. Therefore, there is great practical significance to apply the measuring combination system of MIMU/magnetometer with low volume, light weight, reliable performance and a certain level of accuracy to autonomous attitude determination of small satellites.

Compared with a traditional single magnetometer attitude determination system, gyro output replaces the satellite dynamic model and there is no need to make recursion of dynamical equations, making filtering equations much simpler and decreasing the amount of calculation significantly. The original dynamic recursive accuracy is greatly influenced by the accuracy of the model itself, which influences the attitude-measuring accuracy. The introduction of gyro information improves the accuracy of measurement. On the other hand, gyro drift can be estimated through filtering algorithm, and the estimation information will feedback to the MIMU, so in this way more accurate angular velocity measurements can be obtained, which reduce the influence of MIMU system errors accumulating over time. What is more, as a kind of fully autonomous navigation system, MIMU also improves the anti-interference ability of this combination system.

7.4.2.1 Geomagnetic Field Model

The geomagnetic field is a geophysical field, composed of a basic magnetic field and a variable magnetic field. The basic magnetic field is derived from the interior of the Earth, and the variable magnetic field is related to the variation of the ionosphere and solar activities. The geomagnetic field is a vector field and geomagnetism has made great progress since Gauss introduced the spheric harmonic analysis method to geomagnetism and established a mathematical description of the geomagnetic field. In 1965, J. Cain and coworkers studied Gaussian analysis of global geomagnetic data and got the global geomagnetic field model. International Association of Geomagnetism and Aeronomy (IAGA) has explored a study of the International Geomagnetic Reference Field (IGRF) with a 5-year interval, and so far, there have been 21 pieces of IGRF (1900–2000) data for research. The research and production of IGRF make it possible to study long-term changes in the geomagnetic field.

Given that the potential function of the geomagnetic field is a function of spatial location [19]:

$$\nu(r,\theta,\varphi) = a\sum_{n=1}^{k}\sum_{m=0}^{n}\left(\frac{a}{r}\right)^{n+1}(g_n^m\cos m\varphi + h_n^m\sin m\varphi)P_n^m(\cos\theta) \tag{7.62}$$

where a is the Earth reference sphere equatorial radius, $a = 6378.2$ km, r is the distance from the center of the Earth to a space position, φ is the geographical longitude, θ is the geocentric colatitude and, P_n^m is the Schmidt function. g_n^m and h_n^m are Gaussian coefficients.

Because the intensity vector of the geomagnetic field B can be expressed as the negative gradient of the potential function of the geomagnetic field $B = -\nabla \nu$, the relationship between the geomagnetic field vector and space position can be obtained:

$$\begin{aligned} B_r &= -\frac{\partial \nu}{\partial r} = \sum_{n=1}^{k} \left(\frac{a}{r}\right)^{n+2} (n+1) \sum_{m=0}^{n} (g^{n,m} \cos m\varphi + h^{n,m} \sin m\varphi) P^{n,m}(\theta) \\ B_\theta &= -\frac{\partial \nu}{r \partial \theta} = -\sum_{n=1}^{n=k} \left(\frac{a}{r}\right)^{n+2} \sum_{m=0}^{n} (g^{n,m} \cos m\varphi + h^{n,m} \sin m\varphi) \frac{\partial P^{n,m}(\theta)}{\partial \theta} \\ B_\varphi &= -\frac{\partial \nu}{r \sin\theta \, \partial \varphi} = \frac{-1}{\sin\theta} \sum_{n=1}^{k} \left(\frac{a}{r}\right)^{n+2} \sum_{m=0}^{n} m(-g^{n,m} \sin m\varphi + h^{n,m} \cos m\varphi) P^{n,m}(\theta) \end{aligned} \tag{7.63}$$

According to the above three expressions, the geomagnetic field vector expression can be obtained under the Earth fixed coordinate system. Select a geomagnetic field model with $m = 10$, $n = 10$. Fig. 7.20 shows the geomagnetic field intensity curve with an orbital altitude of 400 km under the Earth coordinate system, and Fig. 7.21 shows the contour curve of geomagnetic field intensity with a magnetic field intensity unit of 1 nT in the graph.

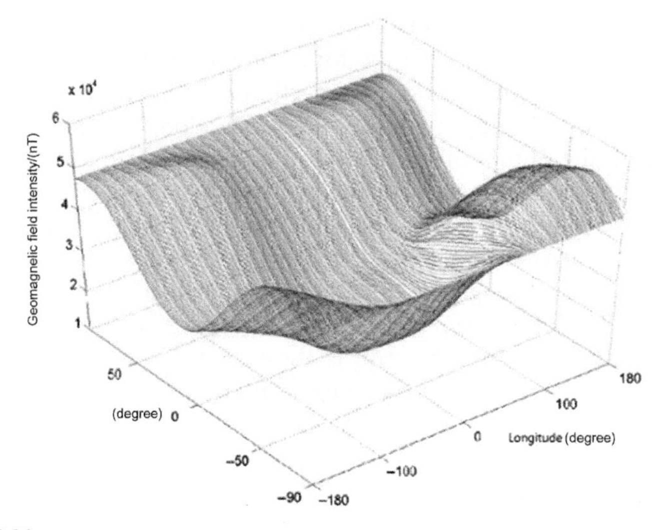

FIGURE 7.20
Geomagnetic field intensity curve.

7.4.2.2 Measurement Principles of the Magnetometer

The three-axes magnetometer measures the components of the Earth's magnetic field vector under the carrier coordinate system. The output of the magnetometer is the projection of the geomagnetic vector in the carrier coordinate system, and the attitude of the satellite can be obtained in comparison with known vector values in the reference coordinate system. The measurement at any single time contains only two-axes information and the three-axes attitude cannot be obtained by just a single measurement value, but using a series of sequential measurements within a certain period, we can obtain all the attitude angles through the measurement values and IGRF mathematical model. During the period of low solar activity cycle, the measurement accuracy of each axis of the satellite's attitude angle can reach <0.5 degree (1σ) by using the calibrated magnetometer.

Let S_b be the carrier coordinate system, S_o be the orbit coordinate system, B_X, B_Y, B_Z be the measured components of the geomagnetic field along three axes of carrier coordinate system S_b, B'_X, B'_Y, B'_Z be the components of the geomagnetic field along three axes of the orbit coordinate system, and heading angle, pitching angle, and roll angle be ψ, θ, γ, respectively. The relationship between the geomagnetic field in the orbit coordinate system and measured geomagnetic field is:

$$\begin{bmatrix} B_X \\ B_Y \\ B_Z \end{bmatrix} = \begin{bmatrix} 1 & 0 & 0 \\ 0 & \cos\gamma & \sin\gamma \\ 0 & \sin\gamma & \cos\gamma \end{bmatrix} \cdot \begin{bmatrix} \cos\theta & 0 & \sin\theta \\ 0 & 1 & 0 \\ -\sin\theta & 0 & \cos\theta \end{bmatrix} \cdot \begin{bmatrix} \cos\psi & \sin\psi & 0 \\ \sin\psi & \cos\psi & 0 \\ 0 & 0 & 1 \end{bmatrix} \cdot \begin{bmatrix} B'_X \\ B'_Y \\ B'_Z \end{bmatrix} \tag{7.64}$$

7.4.2.3 Integration Structures

There are generally two kinds of ways for combination: (1) the output correction combination, (2) the feedback correction combination. As we can see from Fig. 7.22, in the output correction combination, the filtering result is combined with the output of the inertial system, which can compensate output error without affecting the working state of the system. But this kind of combination without feedback is sensitive to the filter model error, so it requires a more accurate model. The feedback correction combination estimates the error first through the filter, and then feedback to correct the inertial system. Therefore, it can control the error of the inertial system by compensating the gyro, and the error is always within the assumed range of the linear model. However, the changing working condition of inertial components leads to a deviation from the best working area (Fig. 7.23).

7.4.2.4 Algorithm Model of Attitude Determination Combination

Due to the wide application and the simple design method, the Kalman filter has been successfully applied to the design of the navigation system of carriers with a high precision. Here, we use extended Kalman Filter (EKF) as the

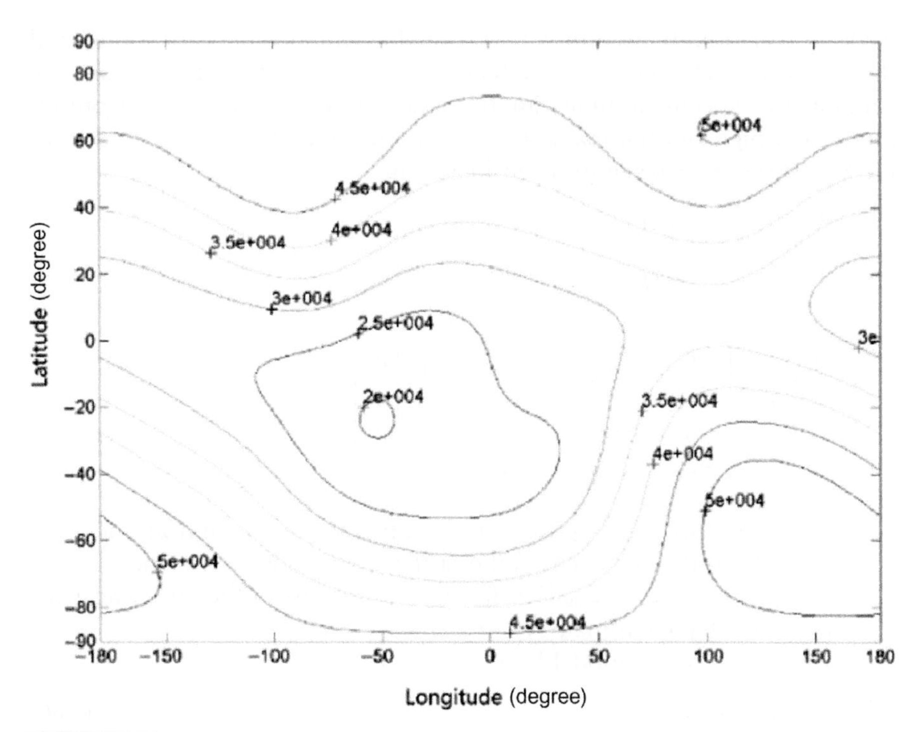

FIGURE 7.21
Contour curve of the geomagnetic field intensity.

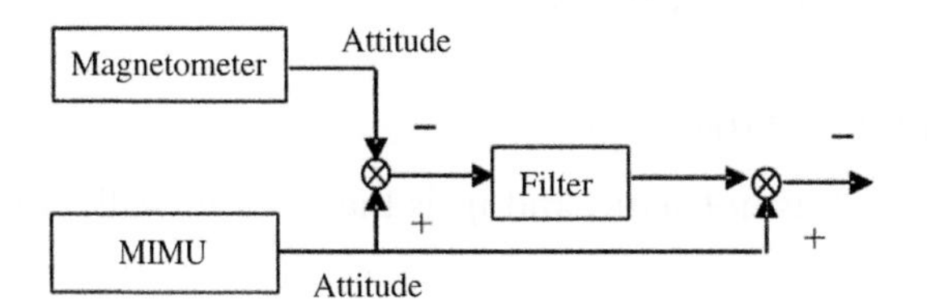

FIGURE 7.22
Output correction combination.

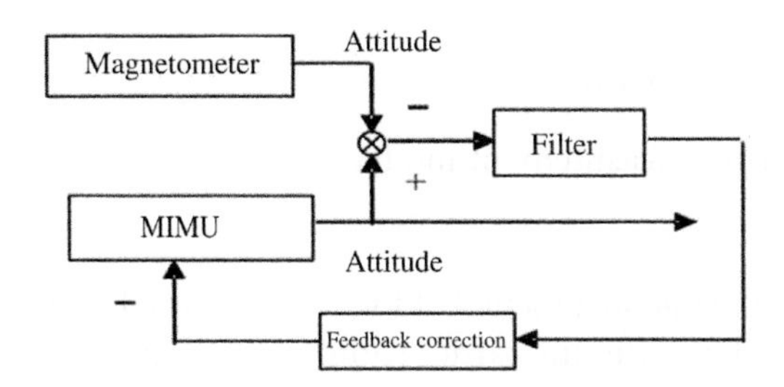

FIGURE 7.23
Feedback correction combination.

navigation algorithm. According to the satellite attitude kinematic model and three-axes magnetometer measurement model, the mathematical model of combination system attitude can be obtained. To simplify the algorithm model and highlight the main error of the MIMU system, only gyro drift of the MIMU system is taken into consideration.

1. Kinematic model

$$\dot{Q} = \frac{1}{2}\tilde{\omega}_{bi} \otimes Q$$

$$\begin{bmatrix} \dot{q}_1 \\ \dot{q}_1 \\ \dot{q}_1 \\ \dot{q}_1 \end{bmatrix} = \begin{bmatrix} 0 & \omega_{biz} & -\omega_{biy} & \omega_{bix} \\ -\omega_{biz} & 0 & \omega_{bix} & \omega_{biy} \\ \omega_{biy} & -\omega_{bix} & 0 & \omega_{bix} \\ -\omega_{bix} & -\omega_{biy} & -\omega_{bix} & 0 \end{bmatrix} \cdot \begin{bmatrix} q_1 \\ q_2 \\ q_3 \\ q_4 \end{bmatrix} \tag{7.65}$$

where $\mathbf{Q} = \begin{bmatrix} q_1 & q_2 & q_3 & q_4 \end{bmatrix}^T$ is the attitude quaternion from the inertial coordinate system to the carrier coordinate system, and $\sim \boldsymbol{\omega}_{bi} = \begin{bmatrix} \omega_{bix} & \omega_{biy} & \omega_{biz} \end{bmatrix}^T$ is the angular velocity of the carrier relative to the inertial coordinate system.

2. Gyroscope output model [21]

$$\omega_{bi} = \boldsymbol{u} - \boldsymbol{b} - \boldsymbol{\eta}_1 \tag{7.66}$$

where $\boldsymbol{\omega}_{bi}$ is the carrier's orbital angular rate in an ideal condition, $\boldsymbol{u}$ is the gyro's practical output, b is the gyro drift, and $\boldsymbol{\eta}_1$ is the Gaussian white noise error of gyro drift.

$$E\left[\eta_1(t)\right] = 0$$
$$E\left[\eta_1(t)\eta_1^T(t')\right] = Q_1(t)\delta(t - t')$$

The gyro drift is not static, and $\boldsymbol{\eta}_2$ is the random walk noise of gyro drift.

$$\frac{d}{dt}\boldsymbol{b} = \eta_2 \tag{7.67}$$

The characteristics of the stochastic process satisfy:

$$E\left[\eta_2(t)\right] = 0$$
$$E\left[\eta_2(t)\eta_2^T(t')\right] = Q_2(t)\delta(t - t')$$

3. Magnetometer measurement model:

$$B_b^b = C_i^b \cdot B_b^i \tag{7.68}$$

where B_b^b is the geomagnetic field vector measured by the magnetometer under the carrier coordinate system, B_b^i is the geomagnetic field vector calculated by IGRF under the inertial coordinate system, and C_i^b is the attitude matrix from the inertial coordinate system to the carrier coordinate system.

4. The design of the attitude determination combination filter
   The magnetometer and MIMU execute the attitude determination task, and the state variables of the system can be estimated by Kalman filter through the system state equation and magnetometer measurement equation. With the help of these results, we can compensate for the drift of MIMU system to improve the measurement precision of the system.
   Introduce the definition of error quaternion, and let it be the quaternion which is needed in the rotation from the attitude estimation to the real value.

$$\delta\bar{\mathbf{q}}^{-} = \bar{\mathbf{q}}^{-} \otimes \hat{\bar{\mathbf{q}}}^{-1} \approx \begin{bmatrix} \delta q \\ 1 \end{bmatrix} \tag{7.69}$$

$\bar{\mathbf{q}} = \begin{bmatrix} q & q_4 \end{bmatrix}^T$ is the real attitude quaternion, $\hat{\bar{\mathbf{q}}}$ is the attitude estimation quaternion, and $\delta\bar{\mathbf{q}}$ is the error quaternion.

According to the system model, assuming that the state vector and estimation error are

$$\text{State vector}: \ \boldsymbol{x} = \begin{bmatrix} \bar{\mathbf{q}} \\ b \end{bmatrix} \tag{7.70}$$

$$\text{Error vector}: \ \boldsymbol{\Delta x} = \begin{bmatrix} \delta q \\ b - \hat{b} \end{bmatrix} = \begin{bmatrix} \delta q \\ \boldsymbol{\Delta b} \end{bmatrix} \tag{7.71}$$

According to the motion equations of satellite:

$$\frac{d\bar{\mathbf{q}}}{dt} = \frac{1}{2}\bar{\boldsymbol{\omega}} \otimes \bar{q} \tag{7.72}$$

$$\frac{d}{dt}\hat{\bar{\mathbf{q}}} = \frac{1}{2}\hat{\bar{\boldsymbol{\omega}}} \otimes \hat{\bar{\mathbf{q}}} \tag{7.73}$$

so:

$$\frac{d}{dt}\delta\bar{q} = \frac{1}{2}\left[\hat{\bar{\omega}} \otimes \delta\bar{q} - \delta\bar{q} \otimes \hat{\bar{\omega}}\right] + \frac{1}{2}\delta\bar{\omega} \otimes \delta\bar{q} \tag{7.74}$$

where:

$$\delta\bar{\omega} = \begin{bmatrix} \omega - \hat{\omega} \\ 0 \end{bmatrix}$$

Ignore the second-order items, the state error equation is obtained as follows

$$\frac{d}{dt}\begin{bmatrix} \delta q \\ \Delta b \end{bmatrix} = \begin{bmatrix} [\hat{\omega}(t)\times] & -\frac{1}{2}I_{3\times 3} \\ 0_{3\times 3} & 0_{3\times 3} \end{bmatrix} \cdot \begin{bmatrix} \delta q \\ \boldsymbol{\Delta b} \end{bmatrix} + \begin{bmatrix} -\frac{1}{2}I_{3\times 3} & 0_{3\times 3} \\ 0_{3\times 3} & I_{3\times 3} \end{bmatrix} \cdot \begin{bmatrix} \eta_1 \\ \eta_2 \end{bmatrix} \tag{7.75}$$

where $[\hat{\omega}(t)\times]$ is the skew symmetric matrix of $\hat{\omega}$.

The sensitive axes of the magnetometer are parallel to the three axes of the satellite body coordinate system respectively, its measurement value B_b is the three-axes component of local magnetic field intensity under the carrier coordinate system. According to the magnetometer measurement equation, we can get:

$$\hat{\mathbf{B}}_m = C(\hat{\bar{q}})B_i \tag{7.76}$$

where $\hat{\mathbf{B}}_b$ is the estimation value of B_b, $C(\hat{\bar{\mathbf{q}}})$ is the attitude matrix derived from estimated quaternion, B_i is the magnetic field intensity calculated from IGRF.

Defining the measurement vector as

$$Z = B_b - \hat{\mathbf{B}}_b = (C(\delta - \mathbf{q}) - I)\hat{\mathbf{B}}_b \approx 2\left[\delta q \times\right]\hat{\mathbf{B}}_b + v \tag{7.77}$$

$\left[\delta q \times\right]$ is the skew symmetric matrix of δq, and v is measurement noise.

Through the simulation analysis, we can get that: the combination of MIMU and the magnetometer is feasible as a satellite attitude determination scheme, and it can satisfy the low attitude accuracy requirement. On one hand, this combination estimates the gyro drift, and it can effectively reduce the measurement error of the MIMU accumulated over time through correction. On the other hand, when there are magnetic anomalies, MIMU can work alone in a short time, which guarantees the requirement of satellite attitude determination. MIMU and the magnetometer rely on each other to achieve autonomous navigation and satisfy the requirement of the development of microsatellites in the future. Besides, the convergence rate of UKF is much faster than that of EKF, and UKF is not sensitive to initial errors, so UKF can be applied in the process of satellite rapid mobility in order to achieve high precision estimation in real time under the condition where the satellite attitude varies rapidly.

7.4.3 Integration of MIMU and GPS

7.4.3.1 The Meaning of MIMU/GPS Integration

Using inertial navigation system for navigation and positioning is completely autonomous, strongly confidential, all-weather, flexible. But the main disadvantage of the inertial navigation system is that the positioning error accumulates quickly over time, therefore it is difficult to conduct measurement for a long time. How to solve this problem has been a hotspot in the inertial navigation field. Adopting new materials, new fabrication processes, and new technologies to improve the accuracy of the existing devices, or developing new types of miniature inertial devices with high precision could solve the problem fundamentally, whereas it takes much manpower, material resources, financial resources, and time, and whether the accuracy of the inertial devices can be improved is limited by technology and many other factors. Combining the navigation system and other auxiliary measurement system to improve the measurement accuracy is an effective method.

Global positioning system (GPS) can provide 24-hour, global, all-weather location services with high-precision and low-cost measurement. Since its birth, its high precision and globalization have attracted people's attention. However, influenced by working conditions and other factors, GPS also has many shortcomings:

1. Poor autonomy. GPS is not an autonomous navigation system and relies on the satellite's radio signal.
2. Poor reliability of dynamic environment: GPS positioning requires at least four satellites' signals. During dynamic environment especially when flying with high mobility, it is possible that multiple satellites lose their lock at the same time. In addition, precision positioning using observation quantity of GPS carrier phase requires that no cycle slips occur. However, cycle slips often generate in dynamic environments due to the reduced signal-to-noise ratio and other reasons;
3. Susceptible to interference. Navstar's radio signal is vulnerable to be affected by the ionosphere, terrain shade, and other factors;
4. Update frequency of receiver's data is low, therefore it is difficult to meet the requirements of real-time measurement.

7.4.3.2 The Advantage of MIMU/GPS Integration

The MIMU and GPS combination system can complement each other and overcome the shortcomings of the individual system.

1. Improve the system accuracy. On one hand, GPS information of high precision can be used to correct the inertial navigation system and control the accumulated error. On the other hand, MIMU's characteristics of high positioning precision and high data sampling rate in short time provide auxiliary information for GPS. Using the auxiliary information, GPS receiver can keep track with low bandwidth. And during the period when the satellite coverage is poor, the inertial navigation system improves the ability of GPS to capture satellite signals again.
2. Strengthen the antijamming ability of the system. When the GPS receiver malfunctions, the inertial navigation system can position independently. When the GPS signal condition is improved significantly to allow tracking, the inertial navigation system provides information, such as the initial position and velocity for the GPS receiver, then the GPS can quickly get the GPS code and carrier wave.

The inertial navigation system signal can also be used to assist the GPS receiver antenna's direction to aim at the GPS satellite, which reduced the interference effect of the work system.

3. Solve the problem of real-time performance. Because the inertial navigation system has a high data update rate and the position and speed calculation measurement has certain accuracy in the short term, creating an integration system and regularly using GPS to update measurement of the inertial navigation system can ensure accuracy and improve the data update rate of the integration system at the same time.

7.4.3.3 Basic Principle of GPS

Twenty-one working satellites and three spare satellites form the GPS space system. The satellites are uniformly distributed in six uniformly spaced approximate circular orbit planes; the orbit altitude is 20,183 km, the operation cycle is 11 hours and 58 minutes. Therefore, on the Earth or any location at near-Earth space at any time, at least four satellites can be observed, which provide continuous three-dimensional position, three-dimensional velocity, and precision time information for all types of users, implementing global, all-weather continuous navigation and positioning. Users receive GPS satellites broadcast signals and obtain positioning observation through GPS receivers.

Here we describe the working principle of the pseudo range measuring method in absolute positioning field. Starting the GPS receiver at any time, users can contact at least four GPS satellites at the same time. We can calculate the accurate orientation of the current GPS receiver and the relationship between the receiver and the reference coordinate system through the accurate measurement of the actual distance between the receiver and the four satellites.

Because the GPS receiver does not carry atomic clocks, the measured distances between the receiver and GPS satellites introduce errors originating from the clock error, and the distance is called the pseudo range. As shown in Fig. 7.24, the pseudo PR_i range between receiver P and the i-th satellite Si can be determined by the following formula:

$$PR_i = R_i + C\cdot\Delta t_{Ai} + C\cdot(\Delta t_u - \Delta t_{si}) \tag{7.78}$$

Among them: $i = 1, 2, 3, 4$;

R_i—the real distance between the i-th satellite to the receiver P;
C—the speed of light;
Δt_{Ai}—the i-th satellite's transmission delay and other errors;
Δt_u—the receiver clock's errors relative to GPS system time;
Δt_{si}—the i-th satellite's errors relative to GPS system time.

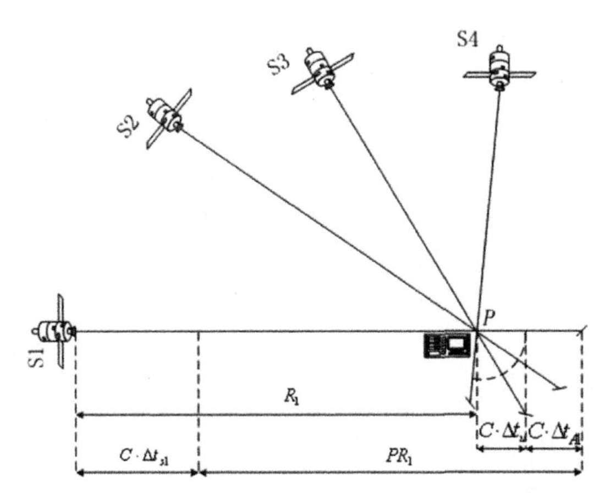

FIGURE 7.24
Principle diagram of the GPS pseudo range measurement [22].

Assuming the position of the satellite Si and the receiver P under the geocentric rectangular coordinate system is (X_{Si}, Y_{Si}, Z_{Si}) and (X, Y, Z), respectively, then

$$R_i = \sqrt{(X_{Si}-X)^2 + (Y_{Si}-Y)^2 + (Z_{Si}-Z)^2} \tag{7.79}$$

Apply Eq. (7.79) into Eq. (7.78), then

$$PR_i = \sqrt{(X_{Si}-X)^2 + (Y_{Si}-Y)^2 + (Z_{Si}-Z)^2} + C\cdot\Delta t_{Ai} + C\cdot(\Delta t_u - \Delta t_{si}) \tag{7.80}$$

In above formula, the satellite position (X_{Si}, Y_{Si}, Z_{Si}) and the satellite clock difference Δt_{si} are obtained through demodulating satellite cable and calculation. Radio wave propagation delay error Δt_{Ai} is corrected with a double-frequency measurement method, or can be estimated according to the wave propagation model by using the correction parameter provided by the satellite cable. The pseudo range PR_i is measured by the receiver. Linearizing Eq. (7.79) and using the iteration method to solve the linear equations, we can get the receiver's position (X, Y, Z).

7.4.3.4 The Composite Mode of MIMU and GPS

The combination of GPS receiver and MIMU usually has two forms. One is called location speed combination; the other is called the pseudo range and pseudo range rate combination.

Working principle of the location speed combination mode is shown in Fig. 7.25. In this mode, MIMU is reset using GPS information. Comparing the position and velocity information output by GPS with those calculated by MIMU and regard the difference as the measurement variable, the optimal

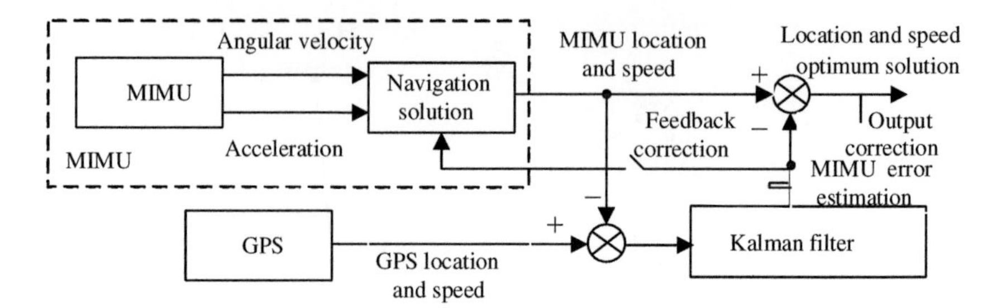

FIGURE 7.25
Location/speed integration model.

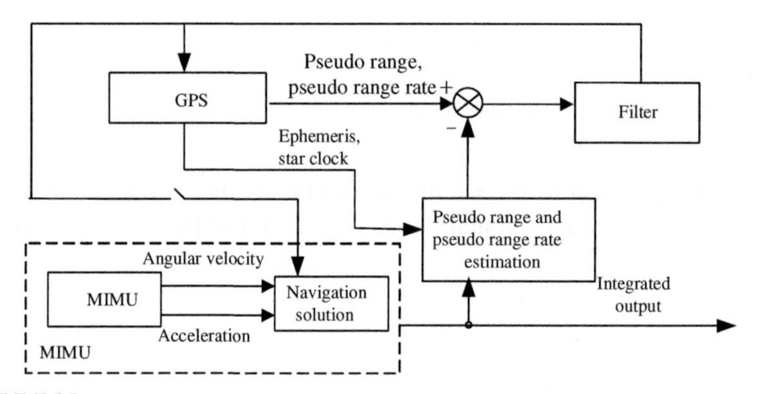

FIGURE 7.26
Pseudo range/pseudo range rate integration model.

estimation of the inertia system error obtained by filtering processing is used to correct the system, alleviating the disadvantage that the inertia system measurement error accumulates over time. According to the structure, location/speed integration model is divided into two modes, that is, with feedback and without feedback. Due to the independence of the GPS receiver, the combination system does not improve the GPS receiver's antijamming and dynamic characteristics.

In location/speed integration model, MIMU and GPS keep their independence. Meanwhile, he system structure is simple, where GPS receiver and the MIMU have strong independence. However, in this mode, measuring accuracy is lower than the pseudo range/pseudo range mode, also GPS antijamming performance is not been improved.

Working principle of the pseudo range/pseudo range rate integration model is shown in Fig. 7.26. In this mode, the position and speed of the MIMU output are used to estimate GPS's pseudo range and pseudo range rate, then it is compared with the pseudo range and pseudo range rate of the GPS's output, and the difference is taken as the input of the synthesis filter. Through the optimal estimation of the filter, MIMU and GPS can be corrected respectively using the output of the filter.

In pseudo range/pseudo range rate integration model, the integration system's measurement accuracy is high, also it can monitor the information integrity of the GPS receiver. However, the structure of the integrated navigation system is more complex.

7.4.4 Integration of MIMU, GPS, and Magnetometer

On the basis of the above research on the MIMU/magnetometer integration attitude determination and MIMU/GPS integrated positioning, we can also design the all integration of the navigation system MIMU/GPS/magnetometer. The system can effectively estimate the attitude, speed, position, gyro drift rate, and accelerometer zero offset and other error parameters, providing high-precision navigation parameters through compensation correction (Fig. 7.27).

7.4.5 Simulation of Integrated Navigation Based on MIMU

We take the MIMU/GPS integration navigation system as an example to understand the effect of the integration navigation (due to the limitations of space, the specific process is abbreviated).

7.4.5.1 *Centralized MIMU/GPS Integration*

Since the structure of pseudo range/pseudo range rate integration system is complex and the system is highly influenced by the hardware conditions, we adopt the conventional location speed feedback integration model to build the integration system. We regard the difference of position and velocity

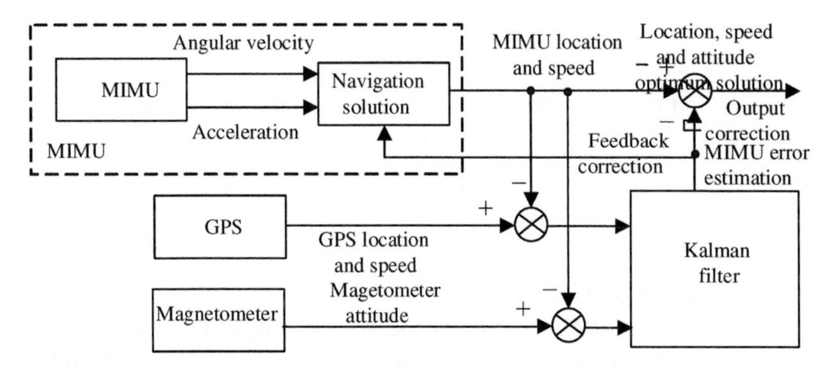

FIGURE 7.27

Schematic diagram of an MIMU/GPS/magnetometer integration mechanism.

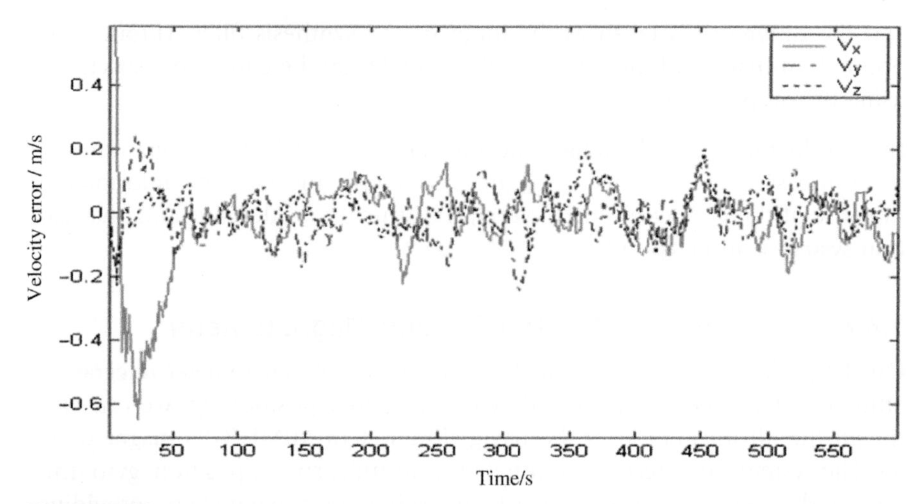

FIGURE 7.28
Position error curve.

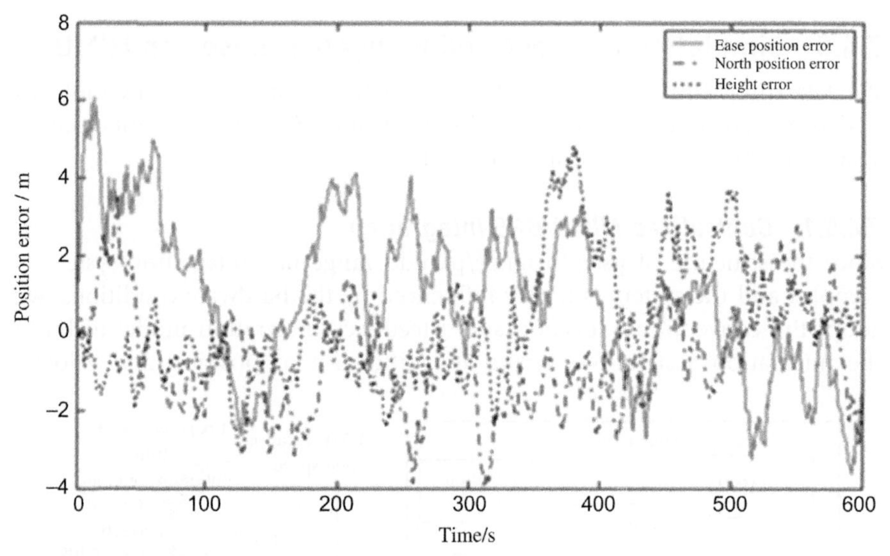

FIGURE 7.29
The velocity error curve.

information output by GPS and those calculated by MIMU as the input signal of the composite filter, and adjust the system output using the optimum estimate of MIMU navigation error obtained by the filter. Figs. 7.28 and 7.29 are the velocity error and position error curve of the MIMU/GPS integration, respectively. The location error of the combined system is less than

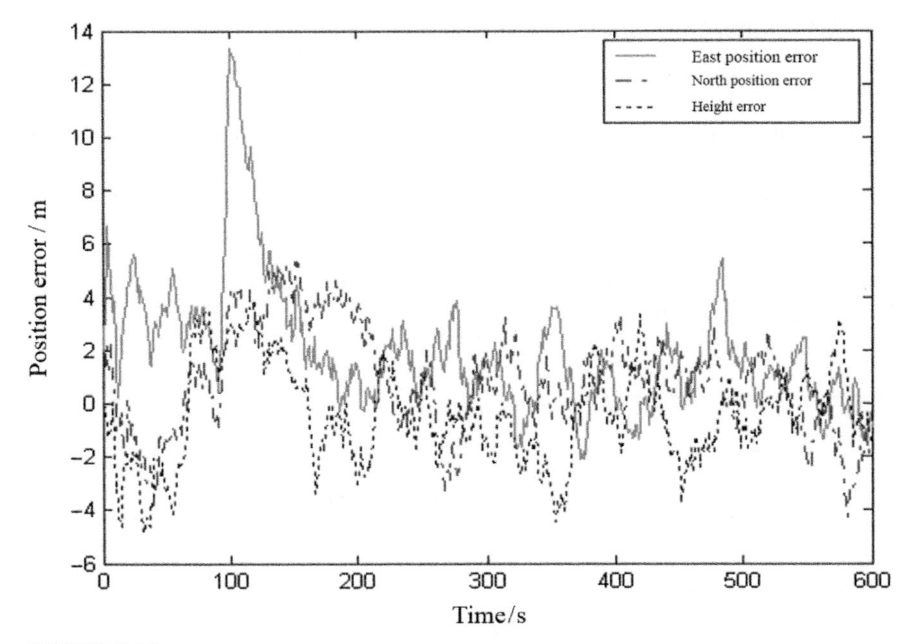

FIGURE 7.30
Eastward position error curve.

10 m (1σ) and the velocity error is less than 0.3 m/s (1σ), which are comparable to GPS accuracy.

Fig. 7.30 shows the MIMU/GPS integration GPS position error curve without data updating as GPS lock loss occurred during 90–100 seconds. At 100 seconds, the GPS data returned to normal. At this time the eastward position error of the integrated system was superior to 14 m, but for individual GPS navigation, position error produced by lock loss was 110 m. Obviously, the integrated system improves the reliability of the system. With the continuous renewal of valid GPS data, the system position error gradually returned to normal, further illustrating that GPS updating data can effectively reduce the MIMU positioning error accumulated over time.

In the experiments, it is also found that when the GPS velocity measurement signal jump occurred for a short time (measurement errors surged to 5 m/s), the severe changes led to filter divergence. In practical conditions, GPS receivers are prone to generate speed jumps. This unpredictable sudden situation often makes the composite filter unable to adapt to the change in time, leading to divergence.

In order to overcome this defect, the MIMU/GPS federated filtering structure is designed using the advantages of federal filtering with strong fault detection capability.

7.4.5.2 Federated MIMU/GPS Integrated System

The new composite structure is shown in Fig. 7.31. Two parallel processing subfilters are adopted. The two subfilters take the position error and velocity error value of the GPS and MIMU as measuring information to process. Because of the federated filtering structure with no feedback, the main filter directly integrates the output results of the two subfilters, achieving the optimal estimation of the system.

Figs. 7.32 and 7.33 show the location and velocity error curve obtained by the positioning subfilter alone.

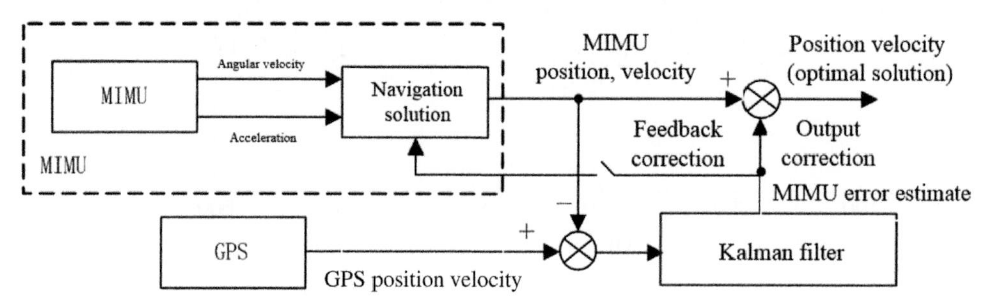

FIGURE 7.31
Federated MIMU/GPS integrated system.

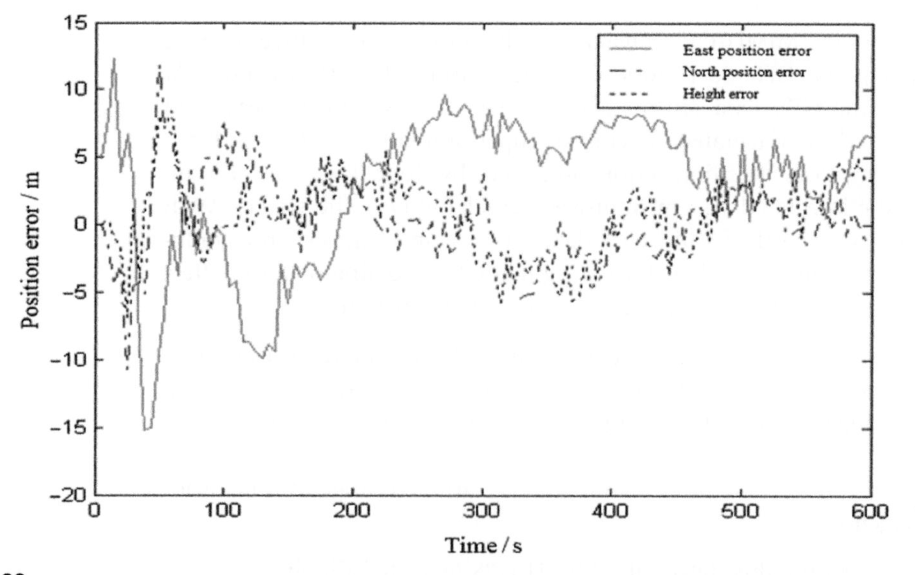

FIGURE 7.32
Position error curve.

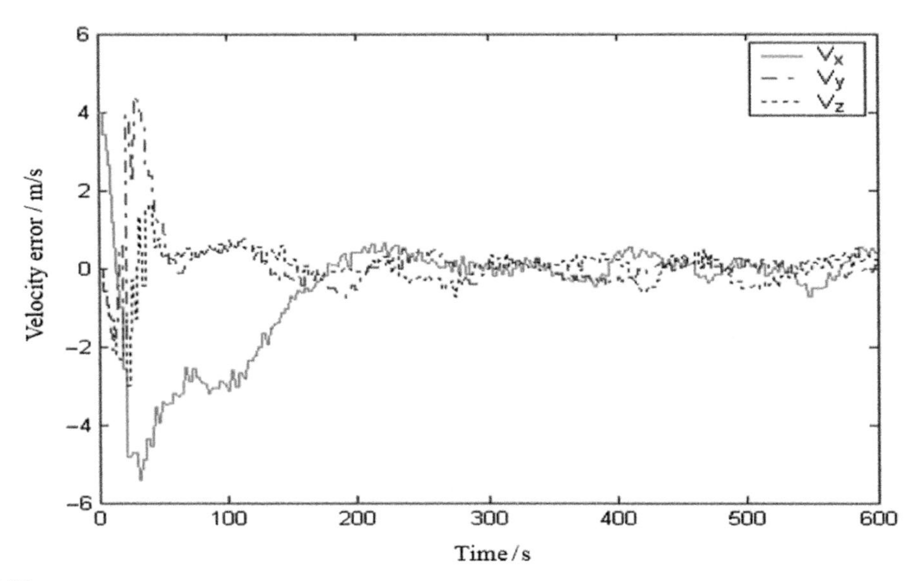

FIGURE 7.33
Velocity error curve.

Figs. 7.34 and 7.35 show the location and velocity error curve obtained by the velocity subfilter alone.

We can see that when the position subfilter worked alone, the location positioning error (1σ) was better than 10 m. As the location integration system adopted location information to update the measurement, so it only has the indirect observation ability when dealing with the velocity error and the speed estimation error is large. When the velocity subfilter worked alone, speed estimation had high precision, which was better than 0.3 m/s, but was almost unable to estimate the position.

Figs. 7.36 and 7.37 show the location and velocity error curve obtained by the federated filter.

The position/speed integration system has the advantages of the former subfilters. Compared with the previous centralized filtering, the overall filtering accuracy is equivalent to that of the centralized filter due to the change of updating time despite of the federated filtering structure without feedback.

The simulation analysis shows that the positioning error of the traditional centralized MIMU/GPS integrated is less than 10 m (1σ) and velocity error is less than 0.3 m/s (1σ), which is equivalent to GPS precision. Moreover, positioning error analysis of the integration system during the short period when GPS lock-lose occurred illustrated that the integration federated system can

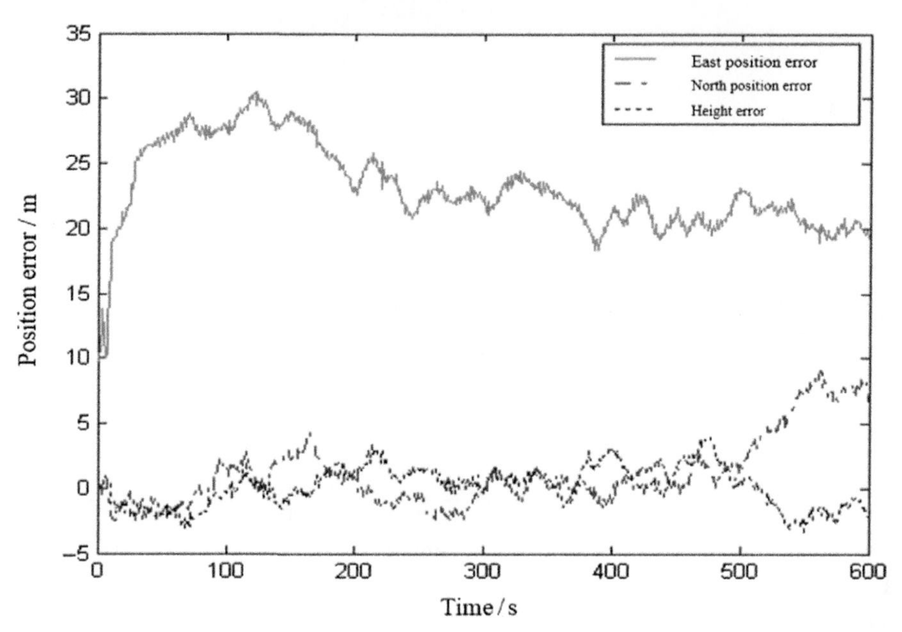

FIGURE 7.34
The location error curve.

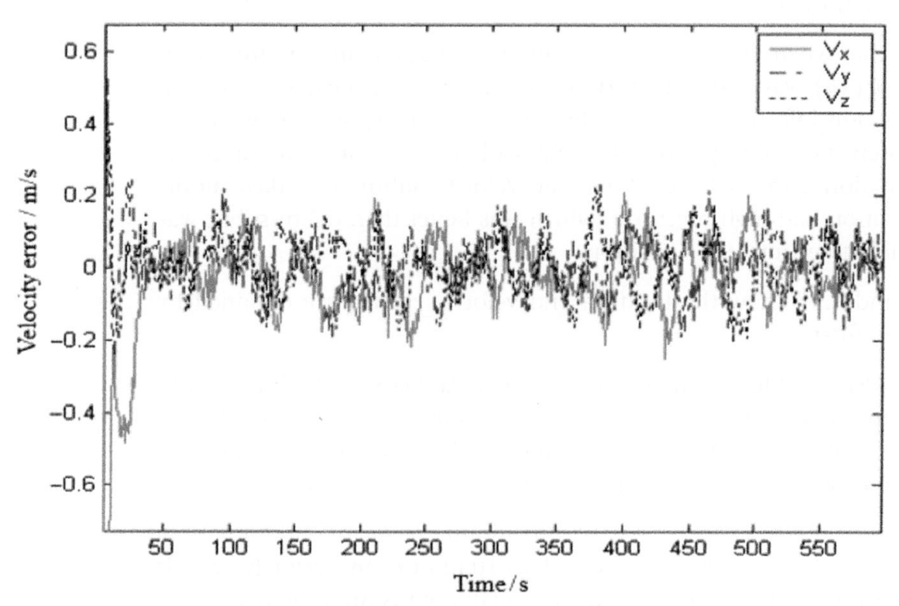

FIGURE 7.35
The velocity error curve.

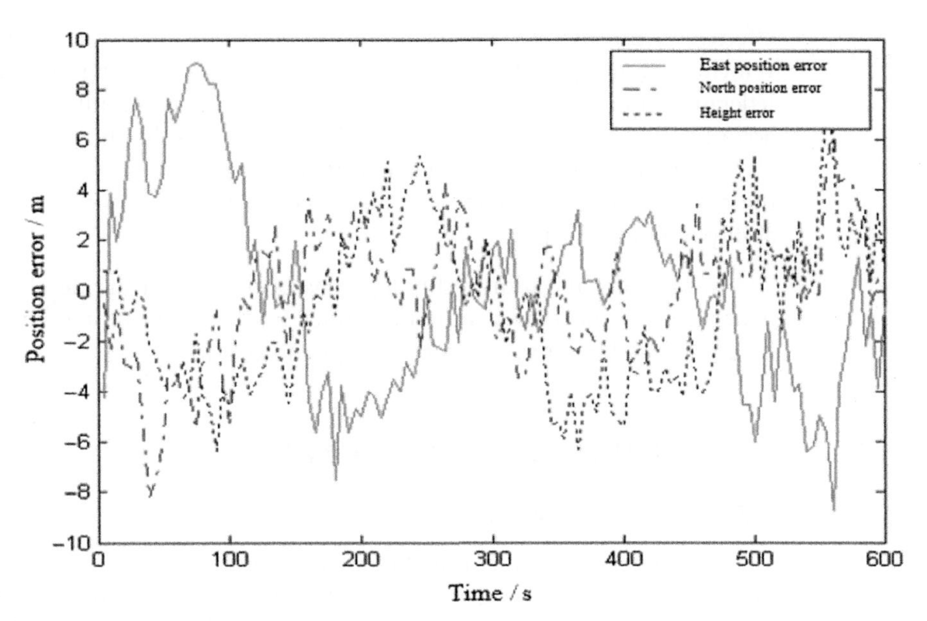

FIGURE 7.36
The location error curve.

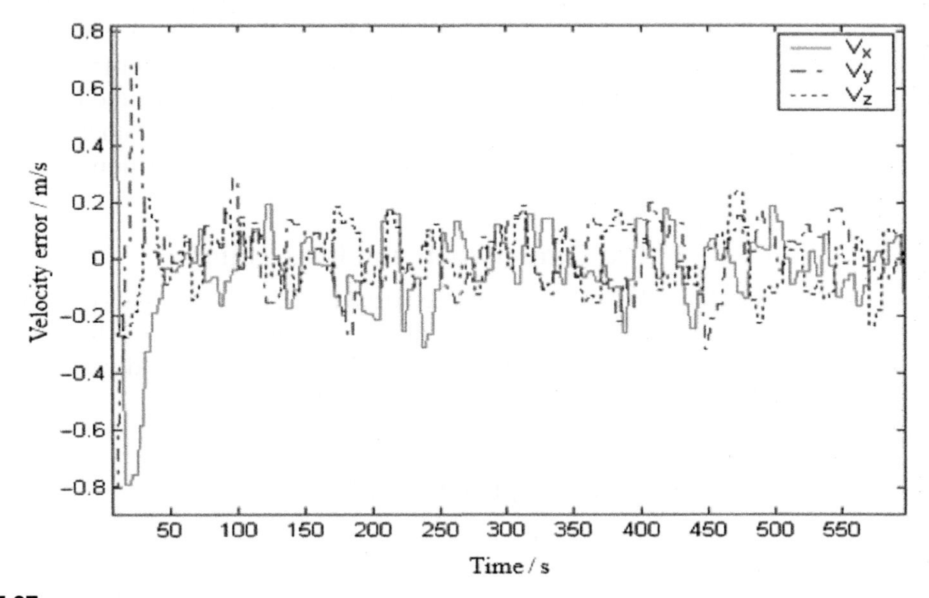

FIGURE 7.37
The velocity error curve.

effectively improve system reliability. In order to solve the problem that velocity data jumping from the GPS receiver leads to filtering divergence, the federated filter combination of position and velocity subfilters is designed according to the theory of federated filtering. The results show that the overall filtering accuracy of the federated filtering system is equivalent to that of the centralized filter system, and the federated filtering system has a certain ability of fault detection.

The integrated navigation technology has been used in the development of nanosatellites and played important roles in ensuring the stability and reliability of satellite attitude determination and control.

7.5 MIMU MODULE FLIGHT TEST

7.5.1 Test Objective

MIMU integrates multiple miniature inertial sensors to obtain comprehensive inertial parameters of the moving object, including the attitude, position, and speed information. MIMU is highly independent and less susceptible to outside interference with fast data updates and strong stability. MIMU module satellite-borne experiments aim to test and analyze embedded micro-accelerometers and micro-gyroscopes, and to establish the micro inertial measurement model, providing experiences for application of micro-accelerometers and micro-gyroscopes in space technology as well as for comprehensive integration of MIMU. During the MIMU flight test, we measured the micro-satellite attitude and orbit maneuver trajectory and then assembled the MIMU combining with the magnetometer, GPS and other devices, which promoted the study of microsatellite autonomous navigation.

7.5.2 MIMU Combination and Installation

In order to fully reduce the size, MIMU module was installed in the payload cabin that was at the bottom of the satellite, surrounded by microjet parts around MIMU. Figs. 7.38 and 7.39 show the virtual and practical assembly drawing of the MIMU in the module box, respectively.

7.5.3 Engineering Implementation

As MIMU is the carrying load of the "NS-1" nanosatellite, the strict and even harsh aerospace device standards raise new requirements for the reliability and stability of MIMU. According to "Aerospace vehicle standards" and "Aerospace Tsinghua NS-1 nanosatellite standards," the MIMU module has been tested in a series of system tests and environmental experiments.

FIGURE 7.38
Virtual assembly.

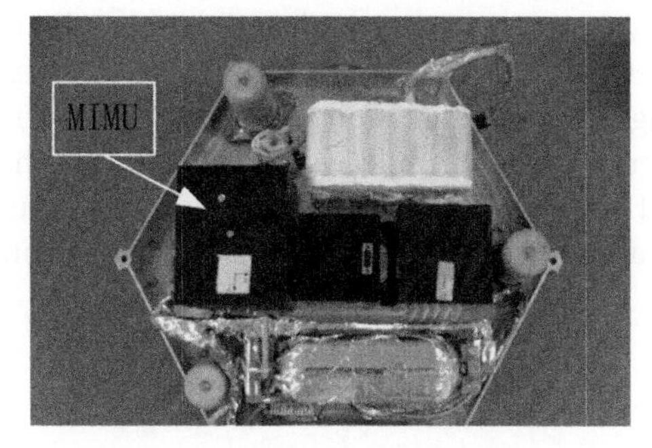

FIGURE 7.39
MIMU assembly in the entire NanoSat payload cabin.

7.5.3.1 Signal Processing Circuit Calibration

Since the interference in the circuit section will affect the measurement characteristics, there is a great need to calibrate the circuit in advance. Introduce known standard voltage and current signals as output signals of the gyroscope and the accelerometer respectively to processing circuits, then directly measure the output end of the circuits using a precision voltmeter. According to the relationship between the input and output of the gyroscope and accelerometer signal processing channels, we can carry out the calibration. The calibration results can be used to modify the MIMU measurements, which can minimize the impact of the circuit on the measurement results of the system.

7.5.3.2 Overall Calibration of MIMU System

According to the section "Research on MIMU entire calibration technology," we established the mathematical model of the combined sensor and conducted the whole calibration. As shown in Fig. 7.40, we carried out speed and position tests through the three-axes turntable to calibrate the MIMU. We obtained 33 error coefficients in total, including 3 gyro scale factors, 6 gyro installation error coefficients, 3 gyro constant drift coefficients, 9 gyro orthogonal imbalance drift coefficients, 6 accelerometer installation error coefficients, 3 accelerometer zero deviation coefficients, and 3 accelerometer scale factors.

7.5.3.3 Calibration Verification Measurement

According to above calibration results, we carried out the validation experiment. The performance test of the MIMU was carried out using a single-axis turntable (accuracy: 0.0001°/s). The calibration results were verified by comparing the MIMU measurement signals before and after calibration under the standard input rotational speed. Experimental results showed that after calibration, the measurement performance and 3D positioning accuracy of the MIMU were greatly improved compared with the results before calibration. There was an order of magnitude improvement in the measurement performance, positioning accuracy was kept in 15 cm during the 3.5-m travel in 5 seconds. However, the experimental results also showed that the positioning error of the MIMU was still increasing rapidly with the accumulation of time despite the previous calibration. Without the support of other measurement

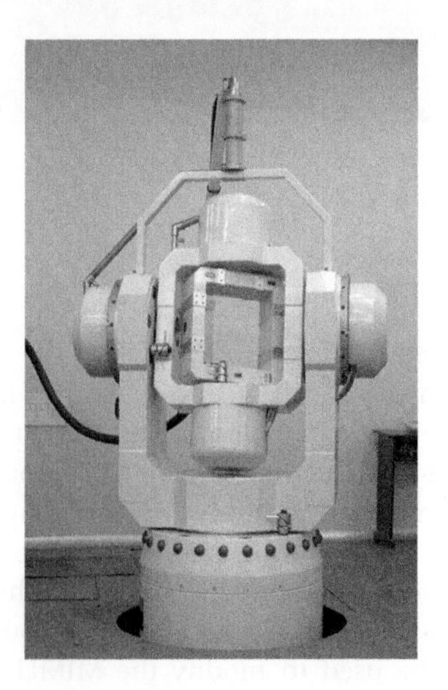

FIGURE 7.40
Three-axes turntable.

systems with long-term reliable measurement accuracy, MIMU can only be employed to position targets in a short time.

7.5.3.4 Space Environment Test

According to the requirements of the space environment and "requirements of the satellite environment test," MIMU has been tested in a series of environmental experiments embedded in the nanosat, including the thermal cycle, thermal vacuum, vibration, and so on (Figs. 7.41–7.43).

Circulating temperature range was −25°C to +55°C. Heat insulation in the high and low extreme temperatures lasted for 2 hours respectively in each cycle, and there were four cycles, namely a total of 20 hours. The vibration test was divided into three parts: sinusoidal vibration, random vibration, and half sine shock. Thermal vacuum test conditions were: vacuum degree was 1.5×10^{-3} Pa; there were four cycles with the temperature range of −65°C to +55°C and heat insulation in the high and low extreme temperature lasted for 6 hours respectively in each cycle. The successful completion of environmental experiments proves the reliability of the MIMU system, which met the requirements of space flight.

The experimental satellite was successfully launched from the Xichang satellite launch site in April 2004, and tests of the main subsystems including the MIMU were completed. Telemetry data showed that the MIMU system was working properly and the space experiment was a complete success.

FIGURE 7.41

Thermal cycle test of the entire satellite.

FIGURE 7.42
Vibration test.

FIGURE 7.43
Thermal vacuum test.

References

[1] G.A. Pavlath, Fiber-optic gyroscope, in: Proceedings of the IEEE LEOS Annual Meeting, 1994, pp. 237–238.

[2] M.S. Perlmutter, A tactical fiber optic gyro with all-digital signal proceeding, in: Proceedings of SPIE, 1994, pp. 192–205.

[3] G. Mao, Q. Gu, et al., Review of micro inertial measurement unit development, Navigation 35 (2) (1999) 8–14.

[4] M. Kraft, Micro-machined inertial sensors: the state-of-the-art and a look into the future, Meas. Control 33 (6) (2000) 164–168.

[5] A.M. Madni, D. Bapna, et al., Solid-state six degree of freedom motion sensor for field robotic applications, in: IEEE International Conference on Intelligent Robots and Systems, Victoria, BC, Canada, October 1998, pp. 1389–1398.

[6] A.I. Tkachenko, Satellite correction equations for a strapdown inertial system, J. Autom. Inf. Sci. 29 (1) (1997) 91–99.

[7] R.E. Kalman, A new approach to linear filtering and prediction problem, J. Basic Eng. (ASME) 82D (1960) 35–46.

[8] R.E. Kalman, C.S. Bucy, New results in linear filtering and prediction theory, J. Basic Eng. (ASME) 83D (1961) 95–108.

[9] Y. Sunahara, An approximate method of state estimation for nonlinear dynamical systems, in: Proceedings of Joint Automatic Control Conference, University of Colorado, 1969, pp. 161–172.

[10] J.L. Fisher, A factorized extended Kalman filter, in: Proceedings of SPIE, 1985, pp. 119–129.

[11] T.L. Song, J.L. Speyer, The modified gain extended Kalman filter and parameter identification in linear system, Automation 22 (1) (1986) 59–75.

[12] T. Ruokonen, Failure detection performance analysis of the optimal nonlinear filter for identification problem, in: Proceedings of the American Control Conference, 1989, pp. 876–878.

[13] M.C. Algrain, J. Saniie, Interlaced Kalman filtering of 3-D angular motion based on Euler's nonlinear equations, IEEE Trans. AES 30 (1) (1994) 175–185.

[14] A. Ruth, Satellite angular rate estimation from vector measurements, J. Guid. Control Dyn. 21 (3) (1998) 450–457.

[15] S. Julier, J.K. Uhlmann, Reduced sigma point filtering for the propagation of means and covariances through nonlinear transformations, in: Proceedings of the American Control Conference, May 2002, pp. 887–892.

[16] E. Wan, R. van der Merwe, The unscented Kalman filter for nonlinear estimation, in: Proceedings of IEEE Symposium 2000 (AS-SPCC), Alberta, Canada, October 2000, pp. 153–158.

[17] T.H. Kerr, Decentralized filtering and redundancy management for multi-sensor navigation, IEEE Trans. Aerosp. Electron. Syst. AES-23 (1) (1987) 83–119.

[18] N.A. Carlson, Federated filter for fault tolerant integrated navigation systems, in: Proceedings of IEEE PLANS'88, 1988, pp. 110–119.

[19] D.H. Titterton, J.L. Weston, Strapdown Inertial Navigation Technology, Peter Peregrinus Ltd, London, 1997, pp. 19–318.

[20] Z. Sun, Strapdown Inertial Measurement System, Ministry of Aeronautics and Astronautics Beijing Institute of Control Instruments, Beijing, 1988, pp. 451–536.

[21] E.J. Lefferts, F.L. Markley, Kalman filtering for spacecraft attitude estimation, J. Guid. Control Dyn. l5 (3) (1982) 417–427.

[22] Z. Qiu, Design of Nanosatellite GPS Receiver and Its Application, Department of Precision Instrument Tsinghua University, Bejing, 2003, pp. 16–20.

CHAPTER 8

Micropropulsion

8.1 SUMMARY

8.1.1 The Necessity for the Study of Micropropulsion

Building a constellation or formation of flying microsatellites should be cheaper, more robust, and more versatile than building a single huge satellite. Typical space missions well adapted for microsatellite applications are space science, asteroid missions, or multispacecraft observer clusters. Propulsion technologies become a key point in the miniaturization of satellites because micro- or nanosatellites would need very small and very accurate forces to realize their stabilization, pointing, and station keeping. The level of thrust and the impulse precision required for microsatellite maneuver can not be reached with conventional propulsion systems. As a result, the micropropulsion field has been a worldwide active field of research.

First of all, the volume and weight of subsystems need to be reduced as the volume and weight of satellites decrease. As a propulsion system equipped on a small satellite, it should first satisfy the requirements of the volume and weight of the whole satellite. And according to experience, the volume and weight of traditional propulsion systems are large, the weight and size is often the largest share of the satellite subsystems; therefore, reducing the weight and size of the system becomes the bottleneck in making more compact satellites. A new micropropulsion system to meet the demand of satellites must be developed to build microsatellites with orbit maneuver function.

Secondly, development and application of the distributed space satellite network represented by microsatellite constellation and satellite formation flying technology, require the development of a micropropulsion system. Small satellite formation, microsatellite constellation can accomplish a great deal of work that complex and expensive large satellites can not complete, such as the composition of the carrier distributed satellite-borne radar,

Space Microsystems and Micro/Nano Satellites. DOI: http://dx.doi.org/10.1016/B978-0-12-812672-1.00008-4

satellite three-dimensional imaging, and high-resolution synthetic aperture remote sensing. To accomplish these tasks, it is an essential requirement to have high-precision maintenance of the relative position and attitude control for microsatellites. Microsatellites during formation flying should regularly hold their position to maintain a desired formation, the goal is to maintain the relative position between satellites, rather than keeping the absolute position of each satellite, the impulse needed is very small, the difference in aerodynamic drag from the strongest and weakest between the satellites in a constellation is the only thing need to be overcome. During each mission, it is estimated that every 10–100 seconds, 1 μNs to 1 mNs impulse need to be provided to overcome the difference in aerodynamic drag mentioned above. Existing conventional propulsion systems find it difficult to provide such small and precise impulses [1,2].

In addition, micropropulsion systems can be used as attitude control actuators for microsatellites, especially nano/pico satellites which are lighter than 20 kg. If they are used reasonably, they will decrease the number of components of the control system to improve integration of nano/pico satellites. Nano/pico satellites, because of their small size and weight and small moment of inertia, suffer from low torque disturbance in space. Depending on the accuracy and stability requirements for its attitude control, the thrust required to overcome interference torque is generally at the mN level.

The current thrust propulsion systems tend to be large, hardly able to meet these requirements. Therefore, we need to study high-precision small-thrust propulsion systems.

At the same time, the development of micro/nanotechnology, microelectronics, and MEMS technology, provides technical support for research into the micropropulsion system. With the development of these technologies, more and more microsystem applications are used in the aerospace equipment, greatly improving the integration and functional density of satellites. The application of these new systems also led to the further development of micro/nanotechnology.

All in all, the development of micro/nanotechnology and the requirement of microsatellite technology, provide technical support and an application market for research and development of micropropulsion systems.

8.1.2 Overview of Micropropulsion

Compared with conventional propulsion systems, micropropulsion systems have the following two characteristics in general.

1. They are capable of generating a relatively small thrust (100 mN magnitude or less) and impulse (μNs–mNs).
2. Volume and weight are relatively small. Generally, weight is in kilograms or lower orders of magnitude.

A propulsion system only with characteristic (1) above is usually called a micropropulsion system.

The propulsion system is actually a mechanical system which turns different forms of energy into kinetic energy. According to the kind of energy utilized, they are roughly divided into four categories [3]:

1. Gas thruster system: using the potential energy of high-pressure gas stored in a high-pressure vessel (nitrogen, helium, hydrogen, etc.), accelerating the propellant gas, so that it is ejected at high speed from the nozzle, to produce thrust.
2. Chemical propulsion system: catalytic reaction and combustion utilizing the propellant that release inherent chemical energy, are used to accelerate the reaction product (usually gas). The reaction product is sprayed through the nozzle at high speed, to produce thrust.
3. Electric propulsion system: using the means of electric heating, electromagnetic, or electrostatic to accelerate the propellant, and emerge with a high-velocity jet-flow.
4. Nuclear propulsion system: energy produced by nuclear fission (or fusion) making the propellant working fluid (typically an inert reaction substance such as hydrogen or helium) heated to a high temperature, then the reaction is ejected from the high-speed spray nozzle, producing thrust.

Traditional gas thruster systems, chemical propulsion systems, and electric propulsion systems have been studied extensively around the world and there are many practical flying experiences. Nuclear propulsion systems are currently being studied in a new type of propulsion system, such as the Pennsylvania State University is studying antiproton catalyzed microfission/ fusion propulsion. In addition, the United States, Europe, and Japan and many other countries have also proposed new concepts like laser propulsion and solar propulsion [3,4].

The micropropulsion system is a special propulsion system. At present domestic and international research mainly concentrates on miniaturization of the traditional gas thruster propulsion, chemical propulsion, and electric propulsion systems, and miniaturization based on advanced MEMS technology. Descriptions of each micropropulsion system and the principles and research into them are introduced here.

8.1.2.1 Gas Thruster

The gas thruster is currently the widely used and the most mature one for micro/nano satellite. It mainly consists of a propellant tank, solenoid valve, pipeline, and a nozzle. There are many propellants available for gas thruster, such as liquid nitrogen, liquid ammonia, freon, helium, or butane, which produce thrust by the pressure of the medium itself in the propellant tank. It works simply, to produce thrust of 10–100 mN magnitude, and it is successfully used in many small satellites. Fig. 8.1 shows the University of Surrey SNAP-1 gas thrusters which were used successfully on a satellite launched in 2000; its specific performance indicators are described in Table 8.1 [5,6].

8.1.2.2 Pulsed Plasma Thruster (PPT)

The PPT is one of the earliest researched electric thrusters. In the 1950s pulsed plasma thruster technology began to be developed and applied. In 1964 in the Soviet Zond-2 satellite, it was used in flight tests for the first time [7]. Because it relies on an electromagnetic field to accelerate the plasma to produce thrust, it is an electromagnetic electric thruster. Fig. 8.2 illustrates the PPT in a schematic diagram.

The PPT system comprises: a solid propellant rod, propellant feed supply spring, propellant thrust ring, spark generator, and the yin and yang electrodes. Customary propellants contain polypropylene, Delrin, Teflon, and the like. Experimental results show that the best of these is Teflon, which is

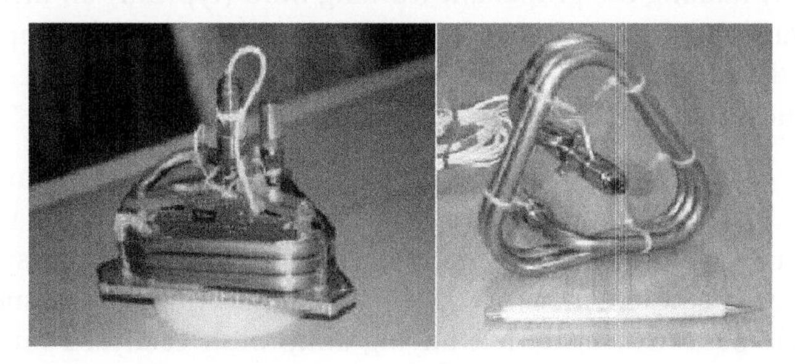

FIGURE 8.1
Snap-1 gas-thruster [5].

Table 8.1 Typical Gas Thruster System Performance [6]

Propulsion System Type	Snap-1 Gas-Thruster (Butane)
Thrust form	Continuous
Thrust (mN)	45 at 0°C, 120 at 40°C
Specific impulse (s)	>60
Mass (kg)	<0.5

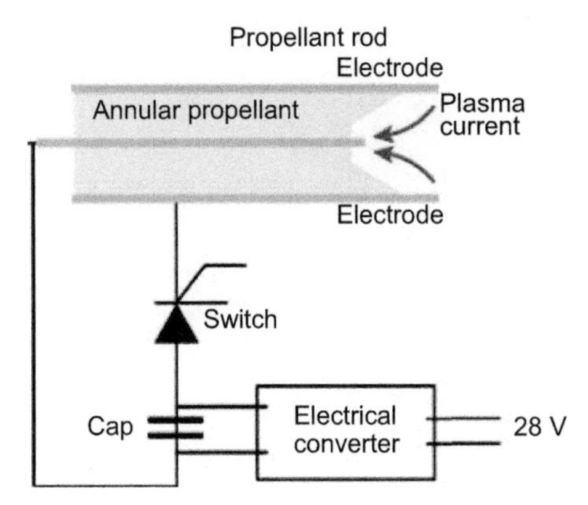

FIGURE 8.2
Pulsed plasma thruster principle [8].

suitable for long-life use of PPT. The work principle of PPT is that a circuit charges the capacitor to a sufficiently high voltage ($\sim$2 kV), closes the thyristor switch, high voltage discharges between the cathode and the anode rod sleeve, high temperature heats the exposed front end portion of the propellant to make it decompose, the decomposed gas is ionized, the joint action of its own magnetic pressure and the gas dynamic pressure in the thin layer of ionized gas, and gas is accelerated to produce thrust.

From the first flight test, there had been no attention to PPT because the thrust generated is small and it has no application background. However, from the beginning of the 1990s, due to the development of small satellites and satellite formation flying technology, PPT has drawn attention from researchers for its simple design, low cost, robustness, and capability of producing low-level thrust, suitable for formation flying satellites for microapplications. AFRL is developing a micro-PPT (μPPT, Fig. 8.3) whose mass is less than 100 g. It is desired to be applied in micro/nanosatellite formation flying and a special attitude control system [8]. Table 8.2 gives the performance parameters of two typical PPT systems.

8.1.2.3 Field Emission Electric Propulsion (FEEP)

FEEP is an electrostatic electric propulsion, which relys on a high-voltage electrostatic field accelerating charged ions to produce thrust. At present, no information is available about FEEP for flight tests. Typical FEEP principles are shown in Fig. 8.4.

FEEP mainly includes a transmitter (containing a propellant reservoir chamber), an accelerating electrode, and a neutralizer. The propellant used includes low melting point metals, such as cesium, indium, etc. Solid propellant is stored in the transmitter storage chamber until needed, when the

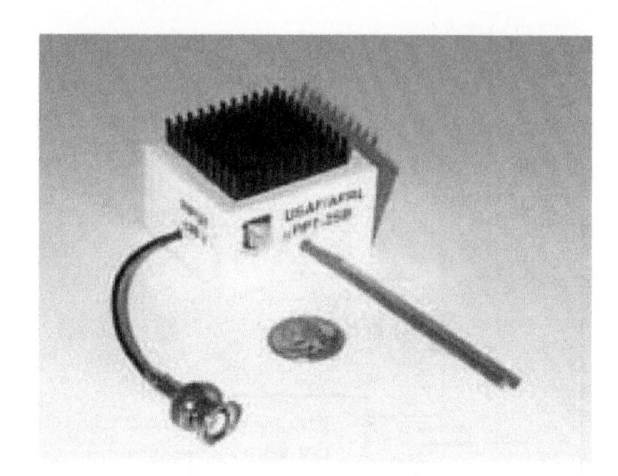

FIGURE 8.3
UPPT prototype developed by AFRL [8].

Table 8.2 Typical PPT Propulsion System Performance

Propulsion System Type	DawgStar PPT [9]	μPPT at AFRL [8]
Thrust form	Pulse	Pulse
Thrust (μN)	<112	~10
Specific impulse (s)	500	
Minimum impulse bit (N s)	5.6e-5	
Thrust to power ratio (μN/W)	8.3	
Mass (kg)	3.95	<0.1

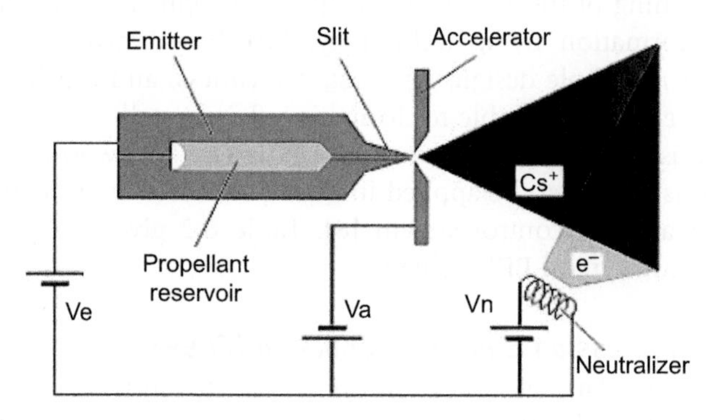

FIGURE 8.4
FEEP ion thruster principles [10].

reservoir chamber is heated to liquefy the propellant. Due to capillary action, the liquid metal flows to the emitter outlet slit. High voltage is applied (8–15 kV) between the emitter electrode exports and the accelerating electric field, causing metal ions to overcome the surface tension and accelerate and

Table 8.3 Typical FEEP Propulsion System Performance [9]

Propulsion System Type	Slit Emitter FEEP (Cesium)	FEEP (Indium)
Thrust form	Continuous	Continuous
Thrust (μN)	0.1–1200	1–100
Specific impulse (s)	7000–11,000	10,000
Minimum impulse bit (N s)	1e-9	<1e-8
Thrust to power ratio (μN/W)	16–20	15
Mass (kg)	2.2	2.5

thus to produce thrust. An electric field is formed by applying a positive voltage to the transmitter and a negative voltage to the accelerator, and electron beams emit discharge metal ions, reducing the pollution plume.

Cesium has a low melting point (~29°C), low escape energy (2.14 eV), and a high atomic mass (~2.207 × 10−25 kg), and is an ideal working fluid for propellant FEEP. FEEP thrust is proportional to the magnitude of the operating voltage, it is a function of the length of the accelerating electrode slit. The slit width is generally 1 μm, with a length of 1–15 mm is generally capable of producing thrust ranging from 0.1 to 100 μN [11].

As for small satellite applications, the biggest problem with FEEP is the excessively high operating voltage. Table 8.3 gives two types of typical FEEP system performance parameters.

8.1.2.4 Colloid Thruster

The colloid thruster is another electric propulsion system that was studied earlier, and is categorized as electrostatic. In the 1960s and 1970s, a lot of research was undertaken in the international community into this system. However, because of its higher operating voltage and the huge volume of the system needed to obtain a large advance, its study was not extensive. In contrast to PPT, until now the international community had not widely heard of the colloid thruster flight test. Recently, with the development of technology, the study of colloid thrusters has begun to heat up again. Fig. 8.5 is a schematic for principles of colloid thrusters.

The colloid thruster comprises four parts: charged particle emitters, extraction electrodes, accelerating network, and neutralizer. It works similar to FEEP and also uses high voltage to accelerate charged particles to produce thrust. It uses more conductive liquid as working fluid, such as glycerol or sodium iodide, etc. In the strong electric field formed under 5–10 kV voltage, propellant is extracted to form the mist stream of charged particles, and accelerated by network acceleration. In order to prevent ionic contamination caused by

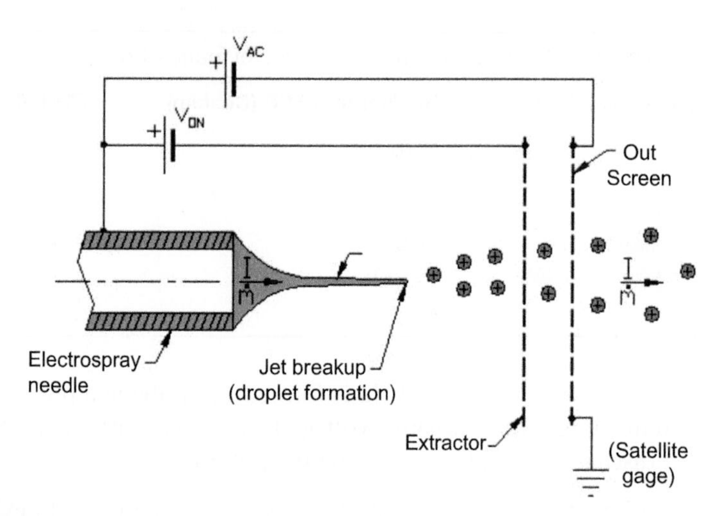

FIGURE 8.5
Colloid thruster principles [9].

Table 8.4 Typical Colloid Thruster System Performance [9]

Propulsion System Type	Colloidal Advance
Thrust form	Continous
Thrust (μN)	0.1–1200
Specific impulse (s)	7000–11,000
Minimum impulse bit (N s)	1e-9
Thrust to power ratio (μN/W)	16–20
Mass (kg)	2.2

the ejected particles on the satellites, a neutralizer is needed to discharge. Extraction electrodes and the acceleration electrodes may be either positive potential or negative potential, the electrodes may generate different polarity of particle stream.

Typical colloid thruster performance parameters are listed in Table 8.4.

8.1.2.5 Hall Thruster

Hall thrusters belong to electrostatic electric propulsion. They have been widely researched in Russia, Europe, Japan, and the United States. Their technologies and applications are mature among electric propulsions. In 1974 it was first used for orbit control in the Meteor-8 satellite, before going on to be used on many satellites. Its general thrust is in mN welterweight, which means Hall propulsion technology needs further development. Fig. 8.6 illustrates a Hall thruster.

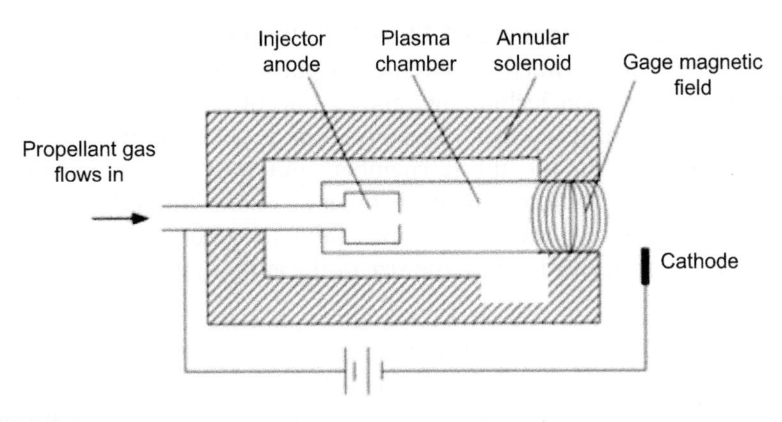

FIGURE 8.6
Hall thruster principles [9].

Table 8.5 Typical Hall Thruster System Performance [9]

Propulsion System Type	Hall Advance (Xenon)
Thrust form	Continuous
Thrust (μN)	>4e-3
Specific impulse (s)	>1200
Minimum impulse bit (N s)	1e-3
Thrust to power ratio (μN/W)	60
Mass (kg)	0.9

The basic structure of a Hall thruster is composed of an annular solenoid, injector anode, a plasma chamber, and a cathode. Electrons emitted from the cathode initially move to the anode in the electric field strength. They are captured during movement by a magnetic field in the plasma chamber. Propellant gas (usually xenon) is injected from the anode to plasma chamber, where it is ionized due to the effect of the electrons. Once the anode gas is ionized, it will be accelerated and fly out in the electric field, producing thrust. Typical Hall thruster performance parameters are listed in Table 8.5.

8.1.2.6 MEMS Thruster

The propulsion systems described above are mainly miniaturized large propulsion systems instead of microminiaturization. With the development of MEMS technology, there is now a microminiaturized propulsion system based on MEMS technology, known as the MEMS thruster in the 1990s. There are mainly two categories of MEMS thruster at home and abroad: MEMS electric propulsion and MEMS chemical propulsion.

8.1.2.6.1 Electrothermal MEMS Electrical Propulsion

The structure of electrothermal electric propulsion is simple, easy to be implemented by MEMS technology, and is one of the hotspots of MEMS electrical propulsion. It works in two ways, one is using electrical resistance heaters to heat gas in the thrust chamber, and then discharging the gas through a nozzle to produce thrust. In order to reduce the leakage rate, the pusher is generally still using a solenoid valve to control the air flow off, which makes it difficult to reduce the size of the entire system.

Fig. 8.7 shows the structure of an MEMS micro gas electric thruster developed by MIT's Robert L. Bayt. The thruster is a supersonic micropropulsion system. A de Laval nozzle that accelerates the fluid from a plenum is designed and fabricated by the Deep Reactive Ion Etching(DRIE) of silicon. The thruster's performance is greatly improved by heating the flow in the plenum. This is achieved using integrated microheaters, which are fabricated in a novel manner, using silicon both as the structural material and the electrical heater element. Experiments show that when the thruster works at 420°C, its specific impulse can reach 83 seconds [12].

Another type of MEMS electric thruster is a subliming solid micro-thruster or a vaporizing liquid micro-thruster. Its operating principle is that the propellant phase transition which is heated by the heating resistor generates gas, then the gas exits through the specially shaped nozzle with high speed, and thus producing thrust. Suitable propellants are ammonia salt, water, ammonia, hydrazine, etc. This greatly reduces the requirement for sealing of the valve, so an MEMS valve is generally used, reducing the quality and size of the system. At the same time, compared with the cold gas thruster, its specific impulse increases, but is still small, and consumes higher power.

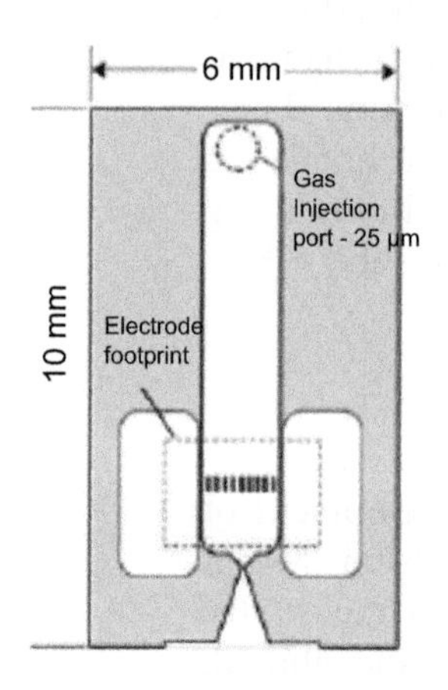

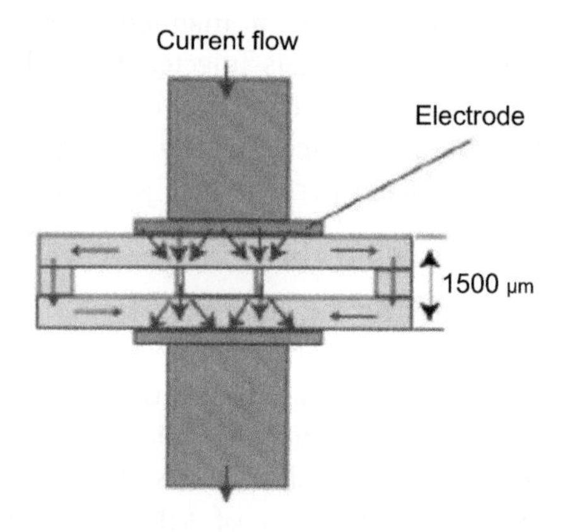

FIGURE 8.7

Resistance heating gas microthruster [12].

Figs. 8.8 and 8.9 are, respectively, a vaporizing liquid micro-thruster and a subliming solid micro-thruster developed by JPL. A vaporizing liquid micro-thruster is composed of a film heater, chamber, and micro-nozzle. Its performance figures include: specific impulse 75–125 seconds, thrust 0.5 mN, power <5 W, efficiency ≥50%, the mass of a few grams. A subliming solid microthruster consists of the propellant tank, the microvalve, microfilter, micro-nozzle, and other components. Its performance figures include: specific impulse 50–75 seconds, thrust 0.5 mN, power <2 W/mN, the mass of a few grams.

In 1999, the Micro-Nano Technology Research Center of Tsinghua University developed China's first MEMS-based vaporizing water thruster [14]. It consists of two silicon wafers as shown in Fig. 8.10. The nozzle and the holes for wire bonding are fabricated on the top wafer by bulk silicon etching. The vaporizing chamber and the microchannel are etched into bulk silicon on the front side of the bottom wafer. The heating resistor, internal leads, and

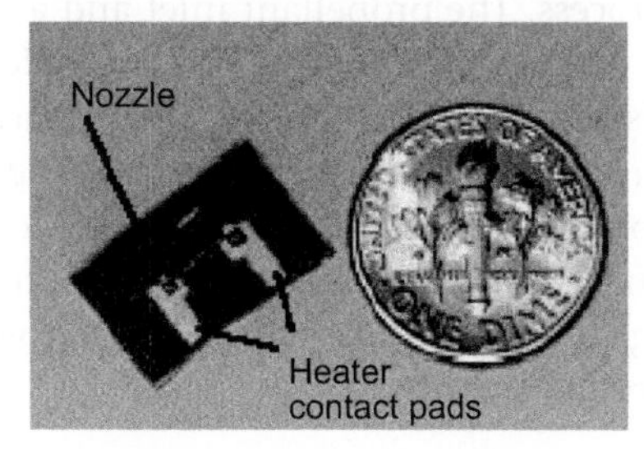

FIGURE 8.8
Liquid vaporization MEMS electric thruster [13].

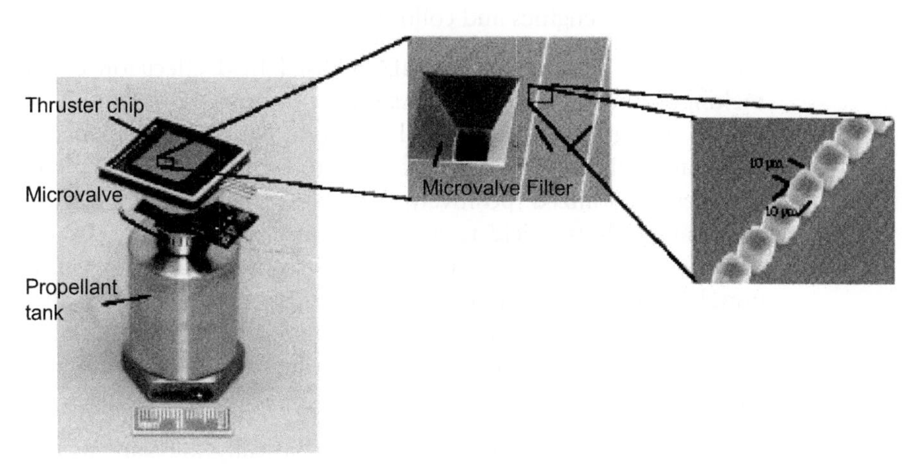

FIGURE 8.9
Solid sublimation MEMS electric thruster [13].

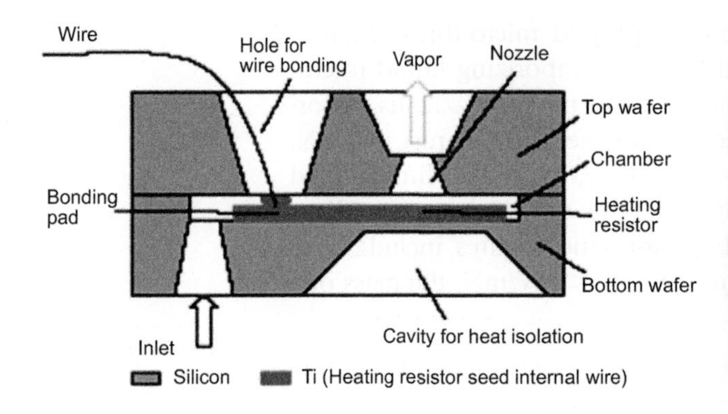

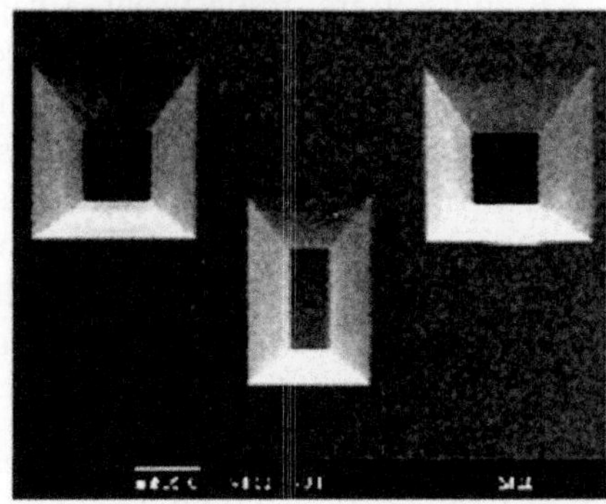

FIGURE 8.10
Evaporating water thruster [14].

bonding pads are formed by metal Ti at the bottom of the chamber to simplify the fabrication process. The propellant inlet and a cavity are etched into the bulk silicon from the back side of the bottom wafer. The cavity formed under the heating resistor is aimed to reduce heat capacitance of the bottom silicon layer, which in turn decreases the heat loss. The vaporizing chamber is connected with a propellant tank through a microchannel and an inlet. Wires are bonded with the pads through the holes in the top wafer, and the holes are filled with glue to seal water. At a working voltage of 30.2 V, the maximum thrust can be generated at 2.8 N.

8.1.2.6.2 Electrostatic MEMS Electrical Thruster

The electrostatic MEMS electric thruster has two types of working principles: field-effect ion engines and colloid thrusters.

The working medium of the MEMS-based field effect ion engine is a low-melting liquid metal, such as cesium or indium. Its structural principles are shown in Fig. 8.11 [15]; several electrode layers are machined on a silicon substrate, with different high pressures to form a strong electric field, the liquid metal is ionized, positively charged metal atoms are ejected out of the accelerating electric field to generate thrust, and electrons are discharged by the external power source. It has the advantages of low power, high specific impulse (even up to thousands of seconds), the thrust is small and precise, and it is easy to control. However, its working voltage is relatively high. In addition, the cesium atoms are bombarded on other structural surfaces and deposited cesium causes pollution.

The MEMS microcolloid thruster works basically the same as the field effect particle thruster, but the liquid propellant is nonmetallic, accelerating tiny charged droplets, its operating voltage is higher, and the thrust density is

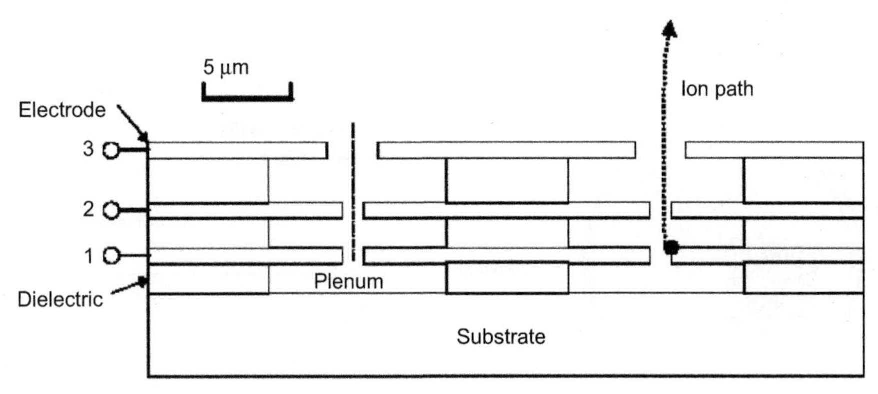

FIGURE 8.11
FET ion engine working principles [15].

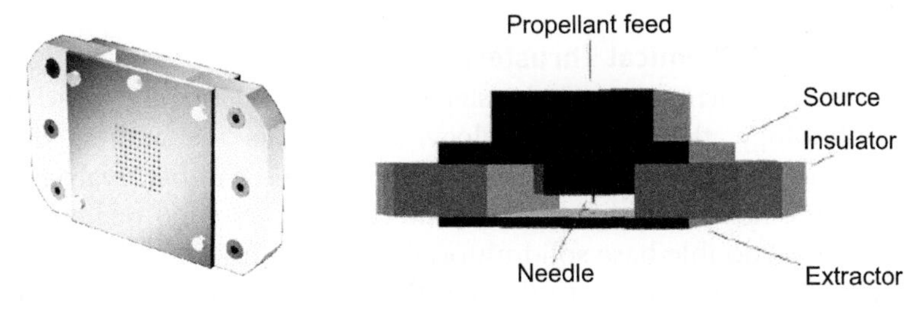

FIGURE 8.12
Advanced microcolloid thruster [16,17].

also increased. However, its specific impulse is smaller than the field effect ion engine and several times higher than the electrothermal thruster. Fig. 8.12 shows a microcolloid thruster developed by Stanford University, it is used for the satellite orbit maintaining in emerald constellation [18]. Thruster units are divided into three for the MEMS structure, these being the source, the insulating layer, and the extraction electrode. The material of the source and extraction electrode is copper, and the insulating layer is 0.5 cm thick boron nitride. A stainless steel micronozzle extends from the source electrode, the propellant is glycerol or isopropyl alcohol. Positive voltage is applied on the source during working, and a negative voltage is applied to the extract pole. Under the influence of an electric field, the propellant droplets are accelerated to produce thrust. The thruster can implement small-scale position control and can provide the order of one-thousandth of a Newton thrust vectoring, a specific impulse of about 1000 seconds [16,17].

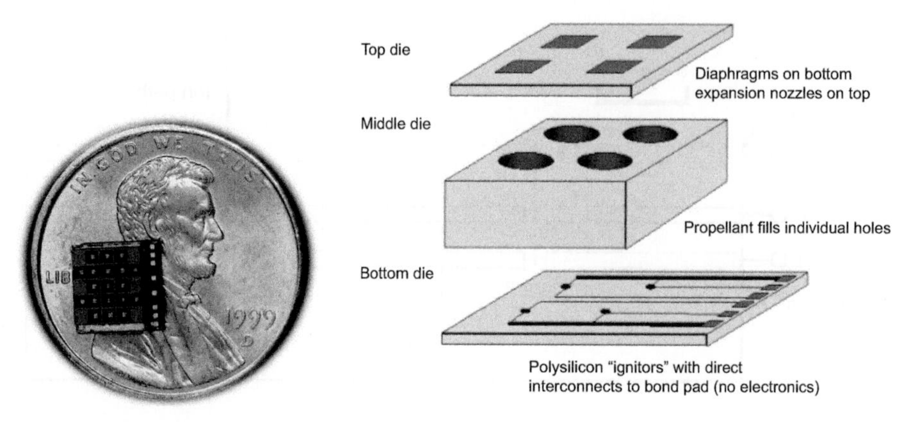

FIGURE 8.13
Prototype and unit structure of a digital microthruster [19].

8.1.2.6.3 MEMS Chemical Thrusters

A typical MEMS microchemical thruster is a digital microarray, which uses MEMS technology, the SOC technology, which puts the address driving circuit, ignition, working fluid tank, combustion chamber, valves, and microminiature nozzles integrated into one chip. It uses solid or liquid propellants, such as double-base solid nitrocellulose. The impulse of such thrusters is higher than that of electrothermal thrusters, but the advance is lower than the electrostatic and electromagnetic thrusters, with the typical specific impulse being 100–300 seconds.

Fig. 8.13 shows a digital micropropulsion system developed by Darpa and cooperated with by TRW. It consists of three layers of silicon and glass in a double-standard 24-pin ceramic electronic package, a total of 15 separate thrusters in the middle of the 3×5 arrangement, and a visible bond connected to the heating resistor on each thrust [20]. Its internal structure is divided into three layers: the top layer of the nozzle; an intermediate layer of propellant storage unit which may raise various types of propellant; and the bottom for the ignition, namely the heating resistor.

The US Honeywell Center and Princeton University cooperative research MEMS micropropulsion unit trillion cell array is shown in Fig. 8.14. The MEMS array consists of 512×512 independent propulsion unit arrays with spacing of $51\ \mu m \times 51\ \mu m$ integrated on 1.3 in. × 1.3 in. silicon wafers. Each unit has a separate heating wire, which is arranged coaxially in the cavity with fuel injected, and is integrated with an RICMOS circuit, so that each unit can be individually addressed and ignited. Two-stage ignition is used to detonate the fuel combustion to produce thrust. One nanogram of thermal

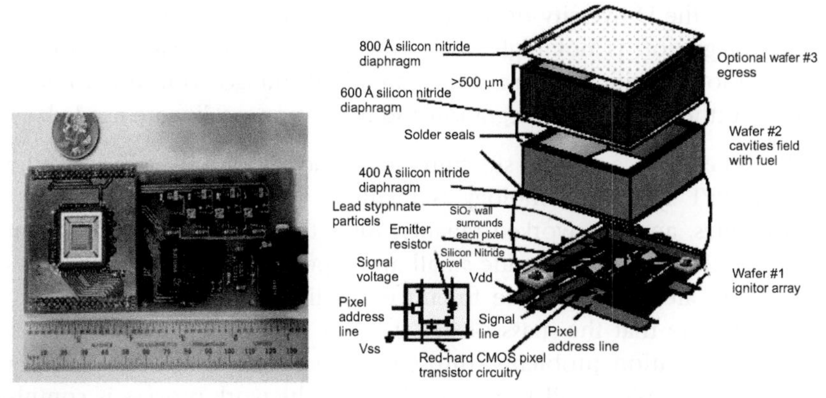

FIGURE 8.14
MEMS trillion cell array miniature thruster prototype and its unit structure [1].

Table 8.6 Typical MEMS Thruster Systems Performances [9].

Propulsion System Type	MEMS Chemical Propulsion	MEMS Electrostatic Electric Propulsion	MEMS Liquid Evaporates Electric Propulsion
Thrust form	Pulse	Pulse	Pulse
Thrust (μN)	1–100,000	1–100,000	~500
Specific impulse (s)	100–300	500–2000	50–75
Minimum impulse bit (N s)	1e-9	1e-9	
Thrust to power ratio (μN/W)	1000	>1000	500
Mass (kg)	2.4e-3		

explosion TNR is first heated, the deflagration of which releases a lot of heat. With this heat, a mixture of nitrocellulose in the cavity is ignited, and the nitrocellulose is quickly vaporized and ejected, generating thrust [1].

Performance parameters of three typical MEMS thrusters are shown in Table 8.6.

8.1.3 Comparison of Different Micropropulsion Systems

From the development of the above-mentioned micropropulsion system at home and abroad, the various micropropulsion systems are being studied and the gradual application of the principles of different micropropulsion systems have their own advantages and disadvantages.

Gas thruster systems are currently the most sophisticated propulsion systems for small satellites, and their structure is simple, reliable, and has already been successfully used in many launched tiny satellites, such as the Snap-1 satellite developed by the University of Surrey. However, the gas thruster has a large volume, integration is not high, with a need for bulky massive propellant tanks, with low specific impulse, which are disadvantageous factors for small satellites. Further enhanced integration and mitigation quality are needed.

The PPT system has a simple structure, and is low cost with good stability. Especially, PPT system technology is relatively mature, has been used on many satellites, and has worked reliably. Also, the PPT system has a higher specific impulse, and can provide small and repeatable pulse impulse. These characteristics are more suitable for microsatellite formation. However, the shortcomings are that the mass of PPT systems is commonly large, and the plume contamination problems, thrust to power ratio is low. In addition, the physical principles of PPT are complex, and the work process is complex with electromagnetic and electric effects requiring further study. A micro-pulsed plasma thruster developed by University of Washington, which was funded by a program of university satellites, will be used in the UW DawgStar in ION-F Constellation.

FEEP also has many advantages that make it suitable for small satellites for relative position holding and track keeping. It has a high specific impulse, small thrust, and it can provide a slight pulse impulse. Although there is no formal FEEP used in space flight tests, indium ion emission technology was on the MIR satellite launched in 1991, the GEOTAIL satellite launched in 1992, and the EQUATOR-S satellite launched in 1997, carried out the experiment, therefore this technology has achieved space flight requirements. FEEP's drawback is that its pushing power ratio is low, and a high operating voltage ($\sim$10 kV) is needed. It also has the problem of pollution plumes. In addition, due to the life limit of the emission cathode tube ($\sim$28,000 hours), its working life is limited. Of course, this is not a problem in the microsatellite in terms of the short working life.

The colloid thruster system is a kind of propulsion system which can be used in small satellites. It is a simple design, resulting in low cost and high stability. It uses an inert propellant, conducive to the preservation and long-term task. It is promoted further because it can provide a wide range of thrust and specific impulses. Its drawback is that it has a low ratio of pushing power, the operating voltage is higher than FEEP at the same time, and the presence of a plume of pollution and the cathode emitter has a lifetime limit.

The advantages of the Hall thruster system include its higher specific impulse and high thrust power ratio. Meanwhile, the Hall thruster is the most mature system of the electric thruster systems. It is used in a large number of satellites. There is a lower risk of a contamination plume. Its disadvantage is the poor quality and the large, complex structure, and it is also limited by the cathode emitter lifetime, which is unfavorable for use on these microsatellites.

The MEMS microthruster system is a new direction in development, and is combined with MEMS technology, creating a new method for propulsion systems research in microsatellites, particularly for nano/pico satellites. The main advantages of the MEMS are its quality and small size, high integration, low power consumption, volume manufacturing, and low cost. For different principles of MEMS thruster, they also have distinct advantages and disadvantages. The electrothermal MEMS electric thruster has no pollution, any propellant can be use in principle, however its drawbacks are its low specific impulse and there are often leaks. The electrostatic MEMS electric thruster, such as the colloidal thruster or ion thruster, has a high specific impulse, low power, and high push to power ratio. However, the presence of high-voltage, complex control systems and existing plumes of pollution are disadvantages. For the MEMS chemical thruster, in particular the solid chemical thruster array, since it has no moving parts, there is no leakage and it has high reliability. In addition, it is easy to adjust the thrust to provide small and precise impulses. Its low specific impulse is insufficient compared with electric propulsion. In short, with the MEMS technology, the development of microthrusters is a new technology, but it is not mature and needs further research and testing. The MEMS microchemical thruster array is ready for flight testing in the United States in the planned TechSat21 satellite formation flight. MEMS microcolloid thrusters developed by Stanford University will also be used in emerald constellation inclusive orbiting satellites maintenance [16–18].

Micropropulsion should mainly be chosen based on functional requirements and the actual satellite mission need. Fig. 8.15 shows the thrust, total impulse, and mobility provided by various microthrusters.

8.2 DESIGN AND SIMULATION OF MEMS-BASED SOLID PROPELLANT PROPULSION

8.2.1 Structure and Principles

The bottom layer is in Fig 8.16 is the ignition layer (hereinafter also referred to as the head), to produce the ignition resistor array on a glass substrate, the resistor is a Pt thin-film resistor.

The intermediate layer is a combustion chamber, where single-crystal Si is used as the structural material. As a result of the bottom ignition program, the gas-guide holes are designed with a symmetrical distribution around the combustion chamber. The presence of gas-guide holes enables the ignited gas to be discharged from the combustion chamber in a way that ensures the

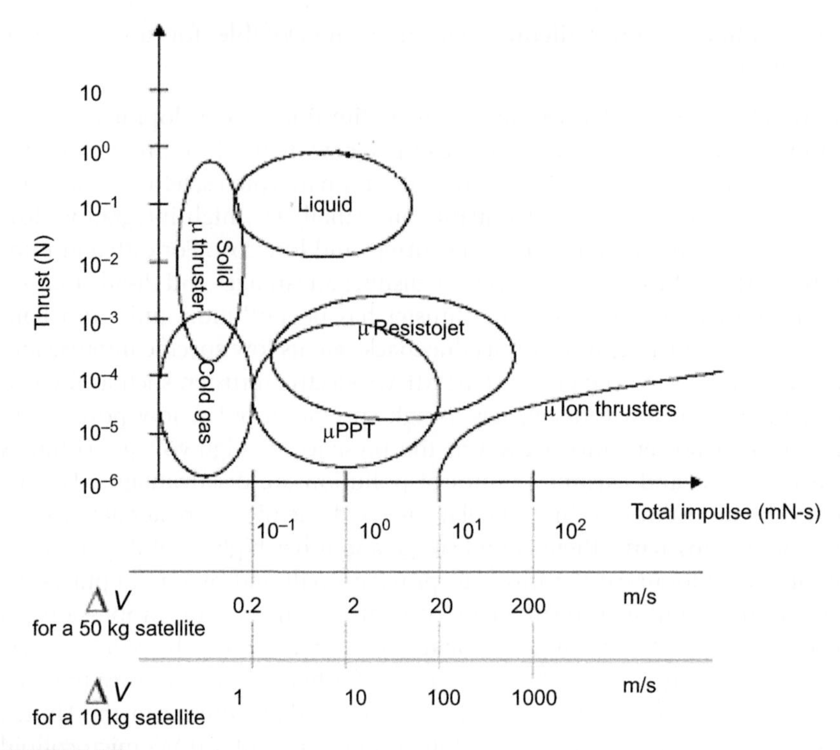

FIGURE 8.15
Part of a microthruster performance comparison [21].

smooth flow of gas. Meanwhile, the structure can increase the pressure in the combustion chamber, which is conducive to combustion of the propellant.

The top layer is the nozzle, using silicon as structural material. It is convergent/divergent exhaust nozzle.

The entire propeller chip is arranged in unit array form, and the size of the array is selected according to actual need, as well as to the level of technology limitations. Each propulsion unit can be independently ignited, or ignition can be combined. The array structure is shown in Fig. 8.17. Fig. 8.18 shows a thruster array chip package model. The minimum impulse pulse can be determined by resizing the combustion chamber. The entire pusher has no moving parts, which makes it highly reliable.

8.2.2 Structure Mechanics and Heat Transfer Simulation for Combustion Chambers

Propellant combusts in the combustion chamber to produce high-pressure gas, the gas affects the chamber wall leading to the combustion chamber deforming and producing mechanical and thermal stresses. To ensure the

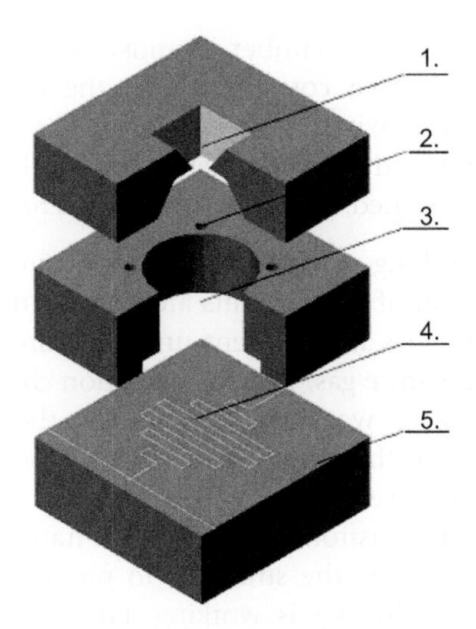

FIGURE 8.16
Schematic of a propulsion unit: 1, convergent divergent nozzle; 2, gas-guide hole; 3, combustor; 4, firing resistor; 5, glass substrate.

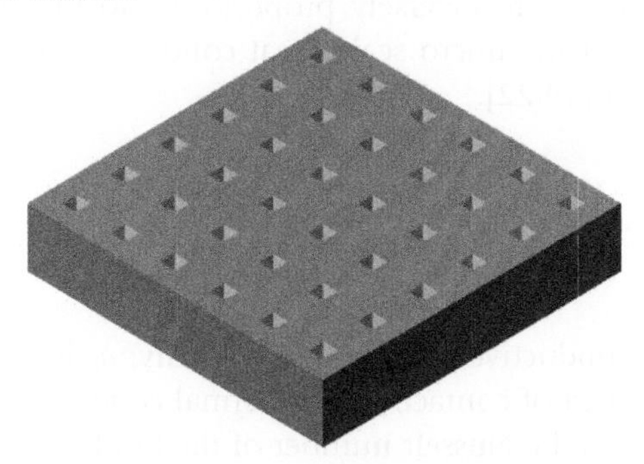

FIGURE 8.17
Propulsion array structure schematic diagram.

FIGURE 8.18
Thruster package model.

reliability of the combustion chamber, it must be simulated on structural mechanics and thermal force coupling when the combustor is working. Through calculating the maximum stress $\sigma_{\max}$ of the combustion chamber wall, it should be compared with the yield stress of the silicon material σ_s. When the structure is designed, $\sigma_{\max} < \sigma_s$ should be ensured.

When the thruster is working, not all units are simultaneously lit, often a single unit or a combination of several units are lit. A working unit is bound to generate heat which reaches the adjacent units that are not working. Under the effect of high-temperature gas, if the combustion chamber temperature of an adjacent unit that is not working rises to above the decomposition temperature of the propellant, the propellant will decompose even if mistakenly lit, which is called "hot crosstalk." In order to ensure that there is no crosstalk between adjacent combustion chambers, heat transfer simulation should be conducted when designing the structure, to provide a guarantee of thermal reliability while the thruster is working. Eqs. (8.1) and (8.2) give the heat conduction and convection transfer formulas in microsystems; it can be seen from these formulas that per unit area thermal convection and the thermal conductivity speed are inversely proportional to the characteristic scale of the systems. Under the micro scale, heat conduction and heat convection are greatly enhanced [19,22].

$$\frac{q_k}{A} = \frac{k}{L}(\Delta T) \tag{8.1}$$

$$\frac{q_c}{A} = N_u \frac{k}{L}(\Delta T) \tag{8.2}$$

wherein, q_k is the conductive heat transfer velocity; q_c is the convective heat transfer rate; is the area of contact; is the thermal conductivity of the material or the fluid; and N_u is the Nusselt number of the fluid.

A problem of integration of the array, namely the number of cells per unit area, also needs to be considered while designing the thruster array. The main factors affecting the degree of integration are stress, heat transfer, and process. A structural mechanics and thermal coupling force simulation study is used to investigate the influence of stress and heat transfer on the degree of integration of the array, generally using finite element method (FEA) simulation analysis.

8.2.3 Process Flow and Results

The solid chemical thruster array is a three-tiered structural, ignition, combustion chamber and nozzle that are processed respectively on the glass and silicon substrates, and then bonded or integrated with adhesive. Wafers are 300 μm thick double-sided polishing (100) single-crystal silicon, and glasses are 500 μm thick Pyrex7740 special bonding glass. The process of manufacture is as follows (Fig. 8.19):

1. The underlying layer igniter
   a. A glass substrate is selected, and a common lithography machine is used to photoetch with AZ1500 photoresist. The wire reticle, a pad, and a resistor pattern are transferred to the photoresist, and shallow grooves are etched on the glass with HF acid buffer.
   b. Cr, Pt, and Au are consecutively sputtered by magnetron sputtering. The thickness of sputtering is about 10 nm, 50 nm, and 50 nm, respectively.
   c. The metal layer, other than the wire, the pads, and the resistive patterns, is peeled off using an ultrasonic stripping process, forming the leads, pads, and resistive patterns.
   d. The etched Au pattern is transferred to the photoresist by the second photolithography process, forming etching windows.
   e. Au is etched on to the surface of resist by a metal etching process to form platinum resistance.
2. An intermediate layer of the combustion chamber
   a. Using lightly doped n-type (100) silicon wafer with a thickness of 300 μm, a silicon oxide layer of 900 nm is grown by a thermal growth process, serving as the etching mask layer.
   b. The diffusion window pattern is transferred to the photoresist by lithography, forming a silicon dioxide etching window, and a window is formed by etching silicon dioxide with an HF acid buffer layer.
   c. The P-type island is formed in the n-type silicon through the use of diffusion window B under the process of diffusion, and the silicon dioxide masking layer is etched.
   d. A 900 nm thick layer of silicon dioxide is grown by a thermal growth process, and is used as a masking layer.
   e. The window of the etching blind hole is transferred to a photoresist by lithography to form a corrosion window. The silicon dioxide of the window is etched with an HF acid buffer layer, exposing the underlying silicon.
   f. Blind holes of thruster are etched using a photoresist mask and silicon dioxide mask under the ICP etching process.
   g. The pattern of the through-hole is transferred onto the photoresist by lithography, forming etching windows. The oxide layer is etched with HF acid buffer to form an etching window.
   h. Combustion chamber through-holes of the thruster are etched using a photoresist mask and silicon dioxide mask under the ICP etching process. The silicon dioxide mask layer is etched away with an HF acid buffer.

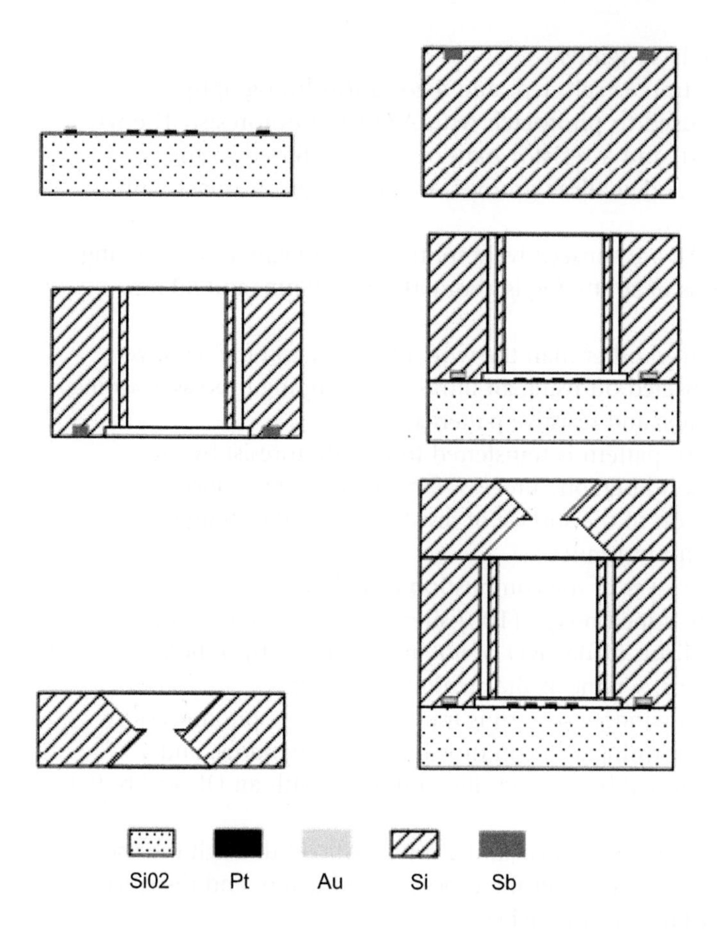

FIGURE 8.19

Fabrication process of the microthruster array (A) Deposit metal on glass to form igniter resistor, (B) diffuse Sb in P type silicon to form N type island, (C) ICP etching process to form combustion chamber, (D) bonding the combustion chamber layer and the igniter layer, (E) double-side etch to form nozzle, (F) Nozzle layer is bonded to the combustion chamber filled with solid propellants.

3. Top nozzle layer
   a. The silicon oxide layer (6000 Å) is grown on (100) P-type silicon substrate with thickness of 300 μm by double-sided thermal oxidation as an etch mask layer.
   b. The etching pattern of the inlet is transferred to the photoresist by lithography, forming a nozzle inlet window, and a silicon dioxide window is formed by etching with an HF acid buffer layer.
   c. The etching pattern of the outlet is transferred to the photoresist by lithography, forming a nozzle outlet window, and a silicon dioxide window is formed by etching with an HF acid buffer layer.
   d. The nozzle is etched with anisotropic etching solution KOH or EPW to form the nozzle shape.

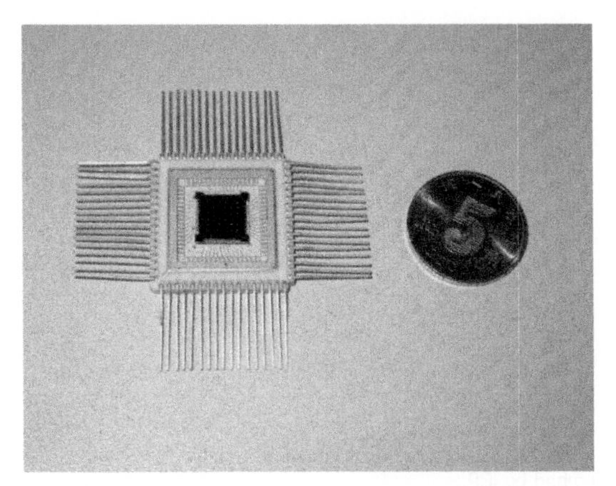

FIGURE 8.20
Packaged thruster chip.

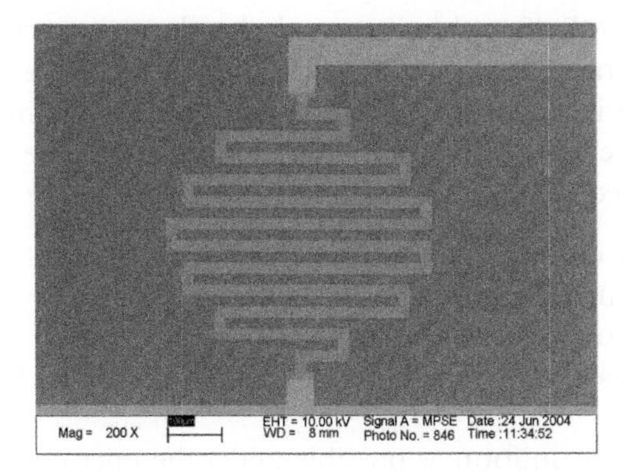

FIGURE 8.21
Ignition resistance.

e. The silicon-based combustion chamber layer and the glass-based ignition layer are bonded together by a silicon-glass anodic bonding process. The chip is divided into individual dies.
f. Two bonded layers of ignition and combustion are filled with solid composite propellant. The nozzle layer is bonded to the combustion chamber using epoxy glue.

Fig. 8.20 shows a prototype of solid chemical microthruster chip, and Figs. 8.21 and 8.22 are partial SEM photographs of the chip die.

8.2.4 Propellant

The selection of propellant is important in the design and manufacture of MEMS microchemical thrusters. The selection of different propellants affects

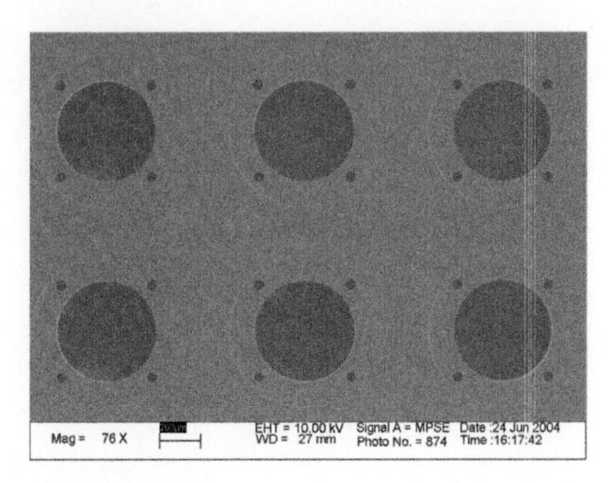

FIGURE 8.22
Combustion chamber etched by ICP.

the difficulty of MEMS technology and the thruster-specific impulse and thrust limit performance. Both the general principles of the propellant selection and the special requirements of an array of MEMS thruster should be taken into consideration when selecting the propellant. The principles below should be followed:

- High energy density;
- Good combustion performance;
- Conducive to the realization of the structure;
- Economic, easy injection, without too much difficulty for MEMS technology;
- Component as uniform as possible under the micro-scale;
- Stable performance, with good physical stability and chemical stability.

Chemical propellants are generally divided into two categories: liquid propellants and solid propellants. Relative to the solid propellant, the specific impulse of the liquid propellant is higher, but reducing and oxidizing agents need to be stored separately, and to be supplied by specialized equipment at work, when mixed in the combustion chamber to ignite, to achieve combustion. The system design includes micropumps and microvalves, etc., which increases the complexity of the system. The performance capabilities of subsystems should be considered, such as leakage of the microvalves. The liquid propellant is only used on occasions where specific impulse is strictly required for combustion to be controlled. The solid propellant can be conveniently stored, and the reducing agent can be stored by mixing with the oxidizer at room temperature. The high chemical stability, high specific impulse density, and small size, can greatly simplify system design, and this is not difficult for MEMS technology to control. According to the basic principles mentioned above, considering the performances of the two types of

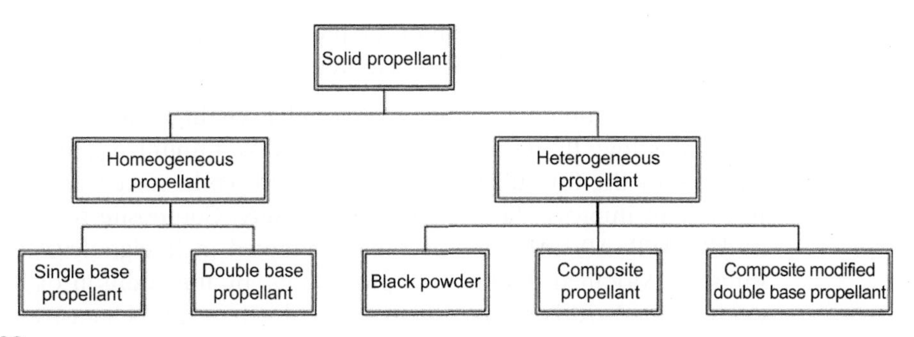

FIGURE 8.23
Classification of solid propellants.

propellant, solid propellant is most suitable for use in the MEMS propulsion system.

Solid propellant can be generally divided into two categories (Fig. 8.23): homogeneous propellant and heterogeneous propellant. The fuel and oxidizer are uniformly mixed in homogeneous propellant to form a colloidal solution. Its elements and performance over the entire substrate is uniform. Single-base propellant has a nitrocellulose (NC)-based gel structure. Double-base propellant (DB) has a gel structure based on the basic components of nitrocellulose (NC) and nitroglycerin (NG). Heterogeneous propellant is different from homogeneous propellant. Its composition and properties are not uniform. Black powder is a typical heterogeneous propellant that was used a long time ago but has now been abandoned; black powder is made by a mechanical mixture of sulfur, charcoal, and potassium nitrate. Modern composite propellant is made up under similar principles, generally consisting of the oxidant, adhesives, metal fuel, curing agents, plasticizers, etc., and has been widely adopted. Some heterogeneous components are added on the basis of double-base propellants to improve the performance of double-base propellant, and this kind of propellant is called a composite-modified double-base propellant (CMDB), and therefore belongs to category of heterogeneous propellants [23–26].

Considering propulsion unit thrust and impulse uniformity, double-base propellants are an ideal choice. However, the ignition of double-base propellant is more difficult, with more sticking due to the pressure in the combustion chamber. Meanwhile, the energy of double-base propellant is relatively low, due to the relatively large loss of silicon combustion heat, and likely to cause incomplete combustion of the propellant and large energy losses, making it difficult to meet the basic requirements of the impulse.

The energy of composite propellants is relatively high, and the ignition temperature is low. By means of a certain process, minimizing the size of solid particles in the formulation and being sufficiently stirred and mixed, it is possible to improve the propellant uniformity. Composite propellants, such as AP/HTPB, AP/GAP, etc., have been adopted in foreign-developed MEMS thrusters. Modified double-base propellants, by the addition of high-energy easily ignited oxidant ingredients in double-base propellants, can improve the performance of those double-base propellants, while ensuring uniformity of propellant components to some degree [27]. Thus, the composite propellants and modified double-base propellants are ideal propellants for MEMS solid chemical thrusters.

8.3 PERFORMANCE MODELING AND ANALYSIS

8.3.1 Modeling and Heat Transfer Analysis of the Pt Resistor Igniter

As to traditional solid chemical rocket engines, igniter design is aimed at analyzing the physical and chemical processes from ignition powder deflagration to propellant ignition. The influence of ignition powder gas temperature, ignition air density, and flow velocity on the ignition delay time are considered. Since the ignition process is very complex, the establishment and solution of the whole process simulation model are generally difficult. In the design and analysis of thruster ignition, it is often assumed that the ignition process is a one-dimensional unsteady heat transfer process. The ignition temperature criteria are used and all phase transitions and chemical reactions before ignition are ignored to establish the unit heat conduction model with convection boundary conditions and to find the analytical solution. According to the solution, the impact of various factors on the ignition delay is qualitatively analyzed and the approximate quantitative values are given [28].

For the igniter design of the MEMS solid chemical thruster, the ignition delay time is not only considered, but also the influence of the resistance bridge wire and the ignition voltage on the temperature of propellant before ignition. Therefore, it is necessary to establish a heat conduction model for the MEMS igniter. In order to analyze the influence factors on the temperature distribution in the ignition phase qualitatively and approximately quantitatively, the influence of temperature on the surface of propellant especially, and also finding the main influence factors on the ignition delay times, a one-dimensional heat conduction model has been established.

In the thruster unit, propellant, a film resistor bridge wire, and the glass substrate constitute a sandwich structure, while the resistance bridge wire is a heating element. We can consider the propellant and the glass substrate as two one-dimensional semiboundless heat conductors and the resistance bridge wire heat rate as the convective boundary condition of the common boundary of these two objects.

Electric power for resistant bridge wire:

$$P = \frac{U^2}{R} = \frac{U^2}{R_s}\frac{b}{L} \tag{8.3}$$

where U is the voltage on the resistance bridge wire, R_s is the resistivity, b is the width of the resistant bridge wire, and L is the equivalent length of the bridge wire. The electric power can be divided into two parts, one part transferred to propellant P_p and the other transferred to the glass substrate. P_{gl}:

$$P = P_p + P_{gl} \tag{8.4}$$

The heat flux density of propellant boundary can be described as:

$$q_p = \frac{P_p}{S} \tag{8.5}$$

One-dimensional heat-conduction equation on the semi-unbounded region is described as below, as to the propellant:

$$\frac{\partial T_p}{\partial t} = a_p \frac{\partial^2 T_p}{\partial t^2} \tag{8.6}$$

where $a_p = \lambda_p/\rho_p c_{pr}$ refers to the thermal diffusivity of propellant, and λ_p, ρ_p, c_{pr} are the thermal conductivity, density, and specific heat capacity of the propellant, respectively.

The initial condition is:

$$T_p(z,t) = T_0, \quad (0 \le z < \infty, t = 0) \tag{8.7}$$

The second boundary condition is:

$$-\lambda_p \frac{\partial T_p}{\partial z} = q(t) = \begin{cases} 0, & (t = 0) \\ q_p, & (t > 0) \end{cases} \tag{8.8}$$

where T_0 is the initial temperature and q_p is the heat flux of ignition resistance wire transferred to the propellant.

Both sides of Eq. (8.6) are multiplied by $-\lambda_p$, and differentiated with respect to z. This obtains:

$$\frac{\partial}{\partial t} = \left(-\lambda_p \frac{\partial T_p}{\partial z}\right) = a_p \frac{\partial^2}{\partial z^2}\left(-\lambda_p \frac{\partial T_p}{\partial z}\right) \tag{8.9}$$

While $q = -\lambda\frac{\partial T_p}{\partial z}$ is heat flux density, Eq. (8.9) can be converted into Eq. (8.10) with the excess heat flux density introduced.

$$\frac{\partial q_\theta}{\partial t} = a_p \frac{\partial^2 q_\theta}{\partial z^2} \tag{8.10}$$

The corresponding boundary conditions are converted into Eq. (8.11).

$$q_\theta(t) = \begin{cases} q_p, & (t=0) \\ 0, & (t>0) \end{cases} \tag{8.11}$$

Methods of solving equation involve separation of variables, Laplace transform, and so on. We choose the method of separation of variables. We assume that the solution of the equation has the following form:

$$q_\theta = X(t)Y(z) \tag{8.12}$$

where $X(t)$ has only one variable t, simplified as X. $Y(z)$ has the variable z, simplified as Y. We can obtain Eq. (8.13) by substituting Eq. (8.12) into Eq. (8.10).

$$\frac{1}{a_p X}\frac{\partial X}{\partial t} = \frac{1}{Y}\frac{\partial^2 Y}{\partial z^2} \tag{8.13}$$

In Eq. (8.13), the left side is a function of the time variable, and the right is a function of the spatial variable. To ensure this equation is tenable, both sides must equal a constant. From the physics point of view, q should have meanings when $t \to \infty$. Therefore, we let $\frac{1}{a_p X}\frac{\partial X}{\partial t} = \frac{1}{Y}\frac{\partial^2 Y}{\partial z^2} = -\beta^2$ and Eq. (8.13) can be divided into two independent ordinary differential equations.

$$\begin{cases} \frac{\partial X}{\partial t} + a_p X \beta^2 = 0 \\ \frac{\partial^2 Y}{\partial z^2} + Y\beta^2 = 0 \end{cases} \tag{8.14 and 8.15}$$

The solutions is as follows:

$$X = C_1 e^{\int -a_p B^2 dt} = C_1 e^{-a_p \beta^2 t} \tag{8.16}$$

$$Y = D_1 e^{i\beta z} + D_2 e^{-i\beta z} \tag{8.17}$$

Substituting Eqs. (8.16) and (8.17) into Eq. (8.12) gives Eq. (8.18):

$$q_\theta = C_1 e^{-a_p \beta^2 t}\left(D_1 e^{i\beta z} + D_2 e^{-i\beta z}\right) \tag{8.18}$$

This is the particular solution of the equation, because the solution is tenable regardless of the value of β. Since Eq. (8.10) is a linear equation, satisfying the superposition principle, the general solution is described below.

$$q_\theta = \sum_{(\beta)} C_{1\beta} e^{-a_p \beta^2 t}\left(D_{1\beta} e^{i\beta z} + D_{2\beta} e^{-i\beta z}\right) \tag{8.19}$$

Without any boundary conditions, $\beta \in (-\infty, +\infty)$, $\left(D_{1\beta}e^{i\beta z} + D_{2\beta}e^{-i\beta z}\right)$ can be simplified as $D_{\beta}e^{i\beta z}$. The summation of Eq. (8.19) can be substituted with integration of β because of the continuity; this gives Eq. (8.20):

$$q_\theta = \int_{-\infty}^{+\infty} C_{1\beta}e^{-a_p\beta^2 t}D_\beta e^{i\beta z}d\beta = \int_{-\infty}^{+\infty} C(\beta)e^{i\beta z - a_p\beta^2 t}d\beta \tag{8.20}$$

according Eq. (8.11), when $t=0$, $q_\theta = q_p$, then Eq. (8.20) is converted to Eq. (8.21):

$$q_p = \int_{-\infty}^{\infty} C(\beta)e^{i\beta z}d\beta \tag{8.21}$$

According to the inverse Fourier transform formula, the following result is obtained:

$$C(\beta) = \frac{1}{2\pi}\int_{-\infty}^{+\infty} q_p e^{-i\beta\zeta}d\zeta \tag{8.22}$$

Substituting Eq. (8.22) into Eq. (8.20), we obtain the following results:

$$\begin{aligned} q_\theta &= q_p\int_{-\infty}^{+\infty}\int_{-\infty}^{+\infty}\left(\frac{1}{2\pi}e^{i\beta(z-\zeta)-a_p\beta^2 t}\right)d\beta d\zeta \\ &= q_p\int_{-\infty}^{+\infty}\frac{1}{2\sqrt{a\pi t}}e^{-\frac{(z-\zeta)^2}{4a_p t}}d\zeta \\ &= \frac{q_p}{2\sqrt{a\pi t}}\left(\int_0^{+\infty} e^{-\frac{(z-\zeta)^2}{4a_p t}}d\zeta + \int_{-\infty}^{0} e^{-\frac{(z-\zeta)^2}{4a_p t}}d\zeta\right) \\ &= \frac{q_p}{2\sqrt{a\pi t}}\int_0^{+\infty}\left(e^{-\frac{(z-\zeta)^2}{4a_p t}} - e^{-\frac{(z-\zeta)^2}{4a_p t}}\right)d\zeta \end{aligned} \tag{8.23}$$

Because q_p is a constant, Eq. (8.24) follows with the first integration $\zeta = z + 2\beta\sqrt{a_s t}$ and the second integration $\zeta = -z + 2\beta\sqrt{a_s t}$ [29].

$$\begin{aligned} q_\theta &= \frac{q_p}{\sqrt{\pi}}\int_{-\frac{z}{2\sqrt{a_p t}}}^{+\infty} e^{-\beta^2}d\beta - \frac{q_0}{\sqrt{\pi}}\int_{\frac{z}{2\sqrt{a_p t}}}^{+\infty} e^{-\beta^2}d\beta \\ &= \frac{q_p}{\sqrt{\pi}}\int_{-\frac{z}{2\sqrt{a_p t}}}^{\frac{z}{2\sqrt{a_p t}}} e^{-\beta^2}d\beta \\ &= \frac{2q_p}{\sqrt{\pi}}\int_0^{\frac{z}{2\sqrt{a_p t}}} e^{-\beta^2}d\beta \\ &= q_p\cdot \mathrm{erf}\left(\frac{z}{2\sqrt{a_p t}}\right) \end{aligned} \tag{8.24}$$

where $\mathrm{erf}(x) = \frac{2}{\sqrt{\pi}}\int_0^x e^{-\beta^2}d\beta$ is an error function.

We obtain the following equation from Eq. (8.24):

$$q = q_p - q_\theta = q_p\left[1 - \operatorname{erf}\left(\frac{z}{2\sqrt{a_p t}}\right)\right] = q_p \cdot \operatorname{erfc}\left(\frac{z}{2\sqrt{a_p t}}\right) \tag{8.25}$$

The extra temperature is indicated in Eq. (8.26):

$$T_\theta = T_p - T_0 \tag{8.26}$$

Therefore the temperature distribution function is describe as follows:

$$\begin{aligned} T_p &= T_0 + T_\theta = T_0 + \left(-\int_z^{+\infty} \frac{q}{\lambda_p} dz\right) \\ &= T_0 + \left(-\frac{q_p}{\lambda_p}\int_z^{+\infty} \operatorname{erfc}\left(\frac{z}{2\sqrt{a_p t}}\right)\right) \\ &= T_0 + \frac{2q_p\sqrt{a_p t}}{\lambda_p}\operatorname{ierfc}\left(\frac{z}{2\sqrt{a_p t}}\right) \end{aligned} \tag{8.27}$$

where $\operatorname{ierfc}(x) = \frac{1}{\sqrt{\pi}}e^{-x^2} - x \cdot \operatorname{erfc}(x)$ is the integral extra error function [29].

Therefore the propellant temperature distribution function is indicated as:

$$\begin{aligned} T_p(z,t) &= T_0 + \frac{2q_p\sqrt{a_p t}}{\lambda_p}\left[\frac{1}{\sqrt{\pi}}e^{-\frac{z^2}{4a_p t}} - \frac{z}{2\sqrt{a_p t}} \cdot \operatorname{erfc}\left(\frac{z}{2\sqrt{a_p t}}\right)\right] \\ &= T_0 + \frac{2q_p}{\lambda_p}\sqrt{\frac{a_p t}{\pi}}\left[e^{-\frac{z^2}{4a_p t}} - \frac{z}{2}\sqrt{\frac{\pi}{a_p t}} \cdot \operatorname{erfc}\left(\frac{z}{2\sqrt{a_p t}}\right)\right] \end{aligned} \tag{8.28}$$

Eq. (8.28) is the solution to Eq. (8.6).

When $z = 0$ follows, $T_p(0,t)$ is the temperature of the contact surface boundary between the propellant and resistance:

$$T_p(0,t) = T_0 + \frac{2q_p}{\lambda_p}\sqrt{\frac{a_p t}{\pi}} \tag{8.29}$$

Resistance wire fusing is not under consideration when finding the solution. The real situation is that the resistance fuses when $T_{melt} > 773.15\text{K}$, and the heat flow density $q = 0$.

At the same time, the derivations above are under the condition of unburned propellant and no phase transitions and chemical reactions. The real ignition point of propellant is $T_{ig} = 543.15 - 653.15\text{K}$. According to the ignition temperature criterion, we believe that when $T(0,t) > T_{ig}$, the propellant starts to burn.

Based on the consideration above, Eq. (8.29) should be transformed into Eq. (8.30):

$$T_p(0,t) = T_0 + \frac{2q_p}{\lambda_p}\sqrt{\frac{a_p t}{\pi}}, \quad T_p(0,t) \le T_{ig} \tag{8.30}$$

As for the glass substrate, a one-dimensional temperature can be formed similar to the propellant:

$$\frac{\partial T_{gl}}{\partial t} = a_{gl}\frac{\partial^2 T_{gl}}{\partial t^2} \tag{8.31}$$

where $a_{gl} = \frac{\lambda_{gl}}{\rho_{gl} c_{gl}}$ is the propellant thermal diffusivity, and λ_{gl}, ρ_{gl}, c_{gl} are the thermal conductivity, density, and special heat of the glass substrate, respectively.

Initial condition:

$$T_{gl}(z,t) = T_0, \quad (0 \le z < \infty, t = 0) \tag{8.32}$$

Second boundary condition:

$$-\lambda_{gl}\frac{\partial T_{gl}}{\partial z} = q(t) = \begin{cases} 0, & (t=0) \\ q_{gl}, & (t>0) \end{cases} \tag{8.33}$$

where

$$q_{gl} = \frac{P_{gl}}{S} \tag{8.34}$$

This is the heat flow density from the resistance wire to the glass substrate. T_0 is the initial temperature.

Similarly, we can obtain Eq. (8.35):

$$T_{gl} = T_0 + \frac{2q_{gl}\sqrt{a_{gl}t}}{\lambda_{gl}}\operatorname{ierfc}\left(\frac{z}{2\sqrt{a_{gl}t}}\right) \tag{8.35}$$

When $z = 0$, the boundary temperature is:

$$T_{gl}(0,t) = T_0 + \frac{2q_{gl}}{\lambda_{gl}}\sqrt{\frac{a_{gl}t}{\pi}}, \quad T_{gl}(0,t) \le T_{melt} \tag{8.36}$$

The common boundary of propellant and the glass substrate makes Eq. (8.37) tenable.

$$T_{gl}(0,t) = T_p(0,t) \tag{8.37}$$

With substituting Eq. (8.30) into Eq. (8.35), the result in Eq. (8.38) is obtained:

$$\frac{q_p}{q_{gl}} = \frac{\lambda_p\sqrt{a_{gl}}}{\lambda_{gl}\sqrt{a_p}} = \xi \tag{8.38}$$

According to Eqs. (8.3), (8.4), and (8.34), the following is tenable:

$$q_{gl} + q_p = \frac{P}{S} = \frac{P}{bL} \tag{8.39}$$

Eq. (8.40) is the solution to Eqs. (8.38) and (8.39):

$$q_p = \frac{\xi}{1+\xi}\frac{P}{bL} \tag{8.40}$$

Substituting Eqs. (8.3) and (8.40) into Eq. (8.30) gives the propellant boundary temperature function of time:

$$T_p(0,t) = T_0 + \frac{2\xi}{1+\xi}\frac{U^2}{\lambda_p R_s L^2}\sqrt{\frac{a_p t}{\pi}}, \quad T(0,t) \le T_{ig} \tag{8.41}$$

Considering the reality of three-dimensional thermal conduction, the model correction coefficient k is added to obtain Eq. (8.42), where $0<k<1$:

$$T_p(0,t) = T_0 + k\cdot\frac{2\xi}{1+\xi}\frac{U^2}{\lambda_p R_s L^2}\sqrt{\frac{a_p t}{\pi}}, \quad T(0,t) \le T_{ig} \tag{8.42}$$

Therefore the thruster ignition delay can be described as:

$$\begin{aligned} t_{delay} &= \frac{\pi(1+\xi)^2\lambda_p^2 R_s^2 L^4}{4k^2\xi^2 U^4 a_p}(T_{ig}-T_0)^2 \\ &= \frac{\pi(1+\xi)^2\lambda_p^2 b^2 L^2}{4k^2\xi^2 P^2 a_p}(T_{ig}-T_0)^2 \end{aligned} \tag{8.43}$$

When $T_p(0,t) = T_{ig}$, the ignition power and the ignition voltage can be described as follows from Eqs. (8.3) and (8.42):

$$P = \frac{(1+\xi)\lambda_p bL(T_{ig}-T_0)}{2k\xi}\sqrt{\frac{\pi}{a_p t}} \tag{8.44}$$

$$U = \left[\frac{(1+\xi)\lambda_p R_s L^2(T_{ig}-T_0)}{2k\xi}\sqrt{\frac{\pi}{a_p t}}\right]^{\frac{1}{2}} \tag{8.45}$$

There are several ways to reduce the ignition delay time under the conditions of a particular initial temperature and unchanged propellant properties from Eqs. (8.41) to (8.42).

1. Increase the ignition voltage, because $t \propto \frac{1}{U^4} \propto \frac{1}{P^2}$. This is the most obvious method.
2. Reduce the resistance value of the resistance bridge wire because of $t \propto R_s^2$. In case of a certain voltage, this method increases the ignition power.

3. Reduce the width and length of the resistance wire. This reduces the area of the resistance and increases the heat flow density in a certain ignition power. If the contact area of the resistance wire and propellant is too small, it is not suitable for the propellant end face to be ignited at the same time.
4. Increase ξ. Because $\xi \propto \sqrt{\frac{1}{\rho_{gl} c_{gl} \lambda_{gl}}}$, the essence is to reduce the thermal conductivity, density and specific heat of the base material to reduce the heat loss from the base material.

From Eqs. (8.44) to (8.45), we can conclude that the ignition power and the ignition voltage have several influence factors with certain propellant properties and ignition delay time regardless of heat loss caused by the resistance divider.

The higher the initial temperature, the less the ignition power and ignition voltage.

The ignition power is reduced by reducing the width and length of the resistance wire. A shorter resistance wire and lower electrical resistivity creates a lower thruster ignition voltage. The width of the resistance wire makes no difference to the ignition voltage.

Increasing ξ reduces the thermal conductivity, density, and specific heat of the base material in order to reduce the heat loss of the base material and the ignition power and ignition voltage.

8.3.2 Interior Ballistic Lumped Parameter Model and Simulation

Simulated calculations are taken on a one-dimensional lumped parameter model established on the energy conservation equation, mass conservation equation, and the ideal gas state equation. The advantage of this model is in influencing shock waves on gas flow and heat transfer in the combustion chamber.

The object of the model is the gas in the combustion chamber from the burning surface to the nozzle throat. In order to ensure accuracy and convenience, the following conditions are assumed to be true:

- The gas temperature and pressure are uniform and approximately invariable.
- The gas meets the ideal gas law.
- The gas in the nozzle expansion is one-dimensional isentropic flow.
- The change of gas pressure in the combustion chamber with propellant burning rate is ignored.

8.3.2.1 Mass Conservation Equation

The mass of the gas in the combustion chamber is changing all the time. Mass m meets the following equation:

$$\frac{dm}{dt} = q_{in} - q_{out} \tag{8.46}$$

where q_{in} and q_{out} are the mass of gas going into the combustion chamber and the mass of gas going out from the nozzle throat per unit time.

8.3.2.1.1 Incoming gas flow

The mass of gas flowing into the combustion chamber per unit time, propellant burning rate $\partial x/\partial t$, burning cross-sectional area $A_{c(x)}$, and solid propellant density ρ are met in Eq. (8.47), where x is the coordinates of the burning cross-sectional area:

$$q_{in} = \frac{\partial x}{\partial t} A_{c(x)} \rho \tag{8.47}$$

The propellant burning rate can be described as in Eq. (8.48) according to experience:

$$\frac{\partial x}{\partial t} = aP^n \times 10^{-3} \tag{8.48}$$

where a is the coefficient of propellant burning rate (mm/s), n is the pressure index of the propellant burning rate, and P is the pressure of the combustion chamber.

8.3.2.1.2 Out going gas flow

The mass of the gas out of the nozzle throat per unit time is related to the flow rate of the gas in the throat. The calculations can be divided into hypersonic flow and subsonic flow.

For convergence–divergence nozzles, the critical pressure ratio of the gas in the nozzle b_c meets the following equation:

$$b_c = \left(1 + \frac{\gamma - 1}{2} M_e^2\right)^{-\gamma/(\gamma-1)} \tag{8.49}$$

where gas Mach number of the nozzle outlet M_e can be decided by Eq. (8.50).

$$\frac{A_e}{A_t} = \frac{1}{M_e}\left(\frac{2}{\gamma+1}\right)^{(\gamma+1)/2(\gamma-1)} \left(1 + \frac{\gamma-1}{2} M_e^2\right)^{(\gamma+1)/2(\gamma-1)} \tag{8.50}$$

When the environment pressure $P_a/P \leq b_c$, the gas at the throat can reach sonic levels. The mass flow of the gas can be described as follows:

$$q_{out} = \sqrt{\frac{\gamma}{r}}\left(\frac{\gamma+1}{2}\right)^{(\gamma+1)/2(1-\gamma)} \frac{P}{\sqrt{T}} A_t \tag{8.51}$$

When $P_a/P > b_c$, the gas at the throat can reach subsonic and the mass flow of the gas meet Eq. (8.52):

$$q_{out} = \sqrt{\frac{\gamma}{r}}\left(\frac{\gamma+1}{2}\right)^{(\gamma+1)/2(1-\gamma)} \frac{P}{\sqrt{T}} A_t \sqrt{1 - \left(\frac{(P_e/P)-b_c}{1-b_c}\right)^2} \tag{8.52}$$

where A_e is the cross-sectional area of the nozzle outlet, A_t is the cross-sectional area of the nozzle throat, γ is the specific heat ratio of the gas, r is the gas constant, T is the gas temperature in combustion chamber, and P_e is the gas pressure at the nozzle outlet.

8.3.2.2 Energy Conservation Equation

We assume that the gas in the combustion chamber has uniform temperature, pressure, and density. According to the energy conservation equation, the gas meets the following equation:

$$m\frac{dT}{dt} + T\frac{dm}{dt} = -\gamma T q_{out} + \frac{\phi}{C_v} \tag{8.53}$$

where, C_v is gas constant volume specific heat capacity, ϕ is the heat change of the gas in the combustion chamber, consisting of heat caused by propellant burning ϕ_{in}, heat loss of gas convection ϕ_{con}, and heat loss of gas thermal radiation ϕ_{rad}. These parameters meet Eq. (8.54).

$$\phi = \phi_{in} - \phi_{con} - \phi_{rad} \tag{8.54}$$

Heat caused by propellant burning meets the following equation:

$$\phi_{in} = q_{in} h_c \tag{8.55}$$

where h_c is propellant burning entropy which can be obtained by experience or thermodynamic calculation.

As to heat loss of gas convection, the flowing equation is tenable:

$$\phi_{con} = hS_{c(x)}(T - T_w) \tag{8.55}$$

where h is the convective heat transfer coefficient, $S_{(x)}$ is heat transfer surface area, and T_w is the wall temperature of the combustion chamber.

Heat loss due to radiation heat transfer is illustrated in Eq. (8.56):

$$\phi_{\mathrm{rad}} = \sigma \varepsilon S_{r(x)}(T^4 - T_{\mathrm{w}}^4) \tag{8.56}$$

where σ is Boltzmann constant of blackbody radiation, ε is the coefficient of thermal radiation, and $S_{(x)}$ is the heat transfer surface area.

8.3.2.3 The Ideal Gas State Equation

The differential Eqs. (8.46) and (8.47) have three unknowns: m, T, and x. There are three equations needed to obtain the numerical solution. If the gas is an ideal gas then it fulfills Eq. (8.57):

$$P = \frac{mrT}{V_{(x)}} \tag{8.57}$$

where,

$$r = \frac{R}{m_g} \tag{8.58}$$

and m_g is the average molecular weight of the gas.

Eqs. (8.46), (8.53), and (8.57) compose first-order ordinary differential equations, having a unique solution with initial conditions x_0, m_0, and T_0.

8.3.2.4 Calculation of Thrust and Impulse

By solving the first-order ordinary differential equations, we can get the relation of T, P with time. We assume that the gas at the nozzle expansion is one-dimensional isentropic flow and we have the thrust description in Eq. (8.60)

$$F = \frac{1}{2}\left(P_e A_e\left(1 + \gamma M_e^2\right) - A_e P_a\right)(1 + \cos\theta_e) \tag{8.59}$$

θ_e is divergence angle in the nozzle, equal to $\theta_e = 35.3$ degrees. From Eq. (8.59) P_e and M_e are needed to solve the thrust F.

The impulse of the thruster unit is decided by the following equation:

$$I = \int_0^t F dt \tag{8.60}$$

8.3.2.5 Determination of P_e and M_e

Three conditions are discussed according to the laws of isentropic flow of the interface tube.

1. $P_a/P > b_c$, the flow is subsonic in the divergent of nozzle.

$$P_e = P_a \tag{8.61}$$

$$M_e = \sqrt{\frac{2}{\gamma - 1}\left(\left(\frac{P_e}{P}\right)^{(1-\gamma)/\gamma} - 1\right)} \tag{8.62}$$

2. $P_a/P \le b_c$, the flow is supersonic in the divergent of nozzle.

$$P_e = P\left(1 + \frac{\gamma - 1}{2}M_e^2\right)^{-\gamma/(\gamma-1)} \tag{8.63}$$

M_e can be determined by Eq. (8.64).

$$\frac{A_e}{A_t} = \frac{1}{M_e}\left(\frac{2}{\gamma+1}\right)^{(\gamma+1)/2(\gamma-1)}\left(1 + \frac{\gamma - 1}{2}M_e^2\right)^{(\gamma+1)/2(\gamma-1)} \tag{8.64}$$

3. $P_a/P \le b_c$, the flow is supersonic flow with right shock in the divergent of nozzle. Therefore we should ensure the position of the shock waves. We assume the shock waves are normal shock and the cross-sectional area is A_s, the Mach number of the upstream gas of shock is M_{n0}.

$$\frac{A_s}{A_t} = \frac{1}{M_{n0}}\left(\frac{2}{\gamma+1}\right)^{(\gamma+1)/2(\gamma-1)}\left(1 + \frac{\gamma - 1}{2}M_{n0}^2\right)^{(\gamma+1)/2(\gamma-1)} \tag{8.65}$$

The Mach number of the downstream gas of shock can be described as:

$$M_{n1} = \sqrt{\frac{1 + ((\gamma - 1)/2)M_{n0}^2}{\gamma M_{n0}^2 - ((\gamma - 1)/2)}} \tag{8.66}$$

The stagnation pressure of downstream gas of shock is:

$$P_{i1} = \left(\frac{2\gamma}{\gamma+1}M_{n0}^2 - \frac{\gamma-1}{\gamma+1}\right)^{-1/(\gamma-1)}\left(\frac{2}{\gamma+1}\frac{1}{M_{n0}^2} + \frac{\gamma-1}{\gamma+1}\right)^{-\gamma/(\gamma-1)} P_{i0} \tag{8.67}$$

where P_{i0} is the stagnation pressure of upstream gas of shock and $P_{i0} = P$.

$$\frac{A_e}{A_s} = \frac{M_{n1}}{M_e}\left(\frac{1 + \frac{\gamma-1}{2}M_e^2}{1 + \frac{\gamma-1}{2}M_{n1}^2}\right)^{(\gamma+1)/2(\gamma-1)} \tag{8.68}$$

Using Eq. (8.68), we can solve M_e. Substituting M_e into Eq. (8.69) we obtain P_e:

$$P_e = \left(1 + \frac{\gamma - 1}{2}M_e^2\right)^{-\gamma/(\gamma-1)} P_{i1} \tag{8.69}$$

If

$$P_e = P_a \tag{8.70}$$

the assumptions are tenable. Otherwise, Eqs. (8.65)–(8.70) are executed to find the solution by adding a small incremental to A_s until $A_s = A_e$, and then turn to condition 2.

According to the theory model above, SIMULINK is used to simulate the performance with unique structures and parameters. The SIMULINK flow chart is illustrated in Fig. 8.24.

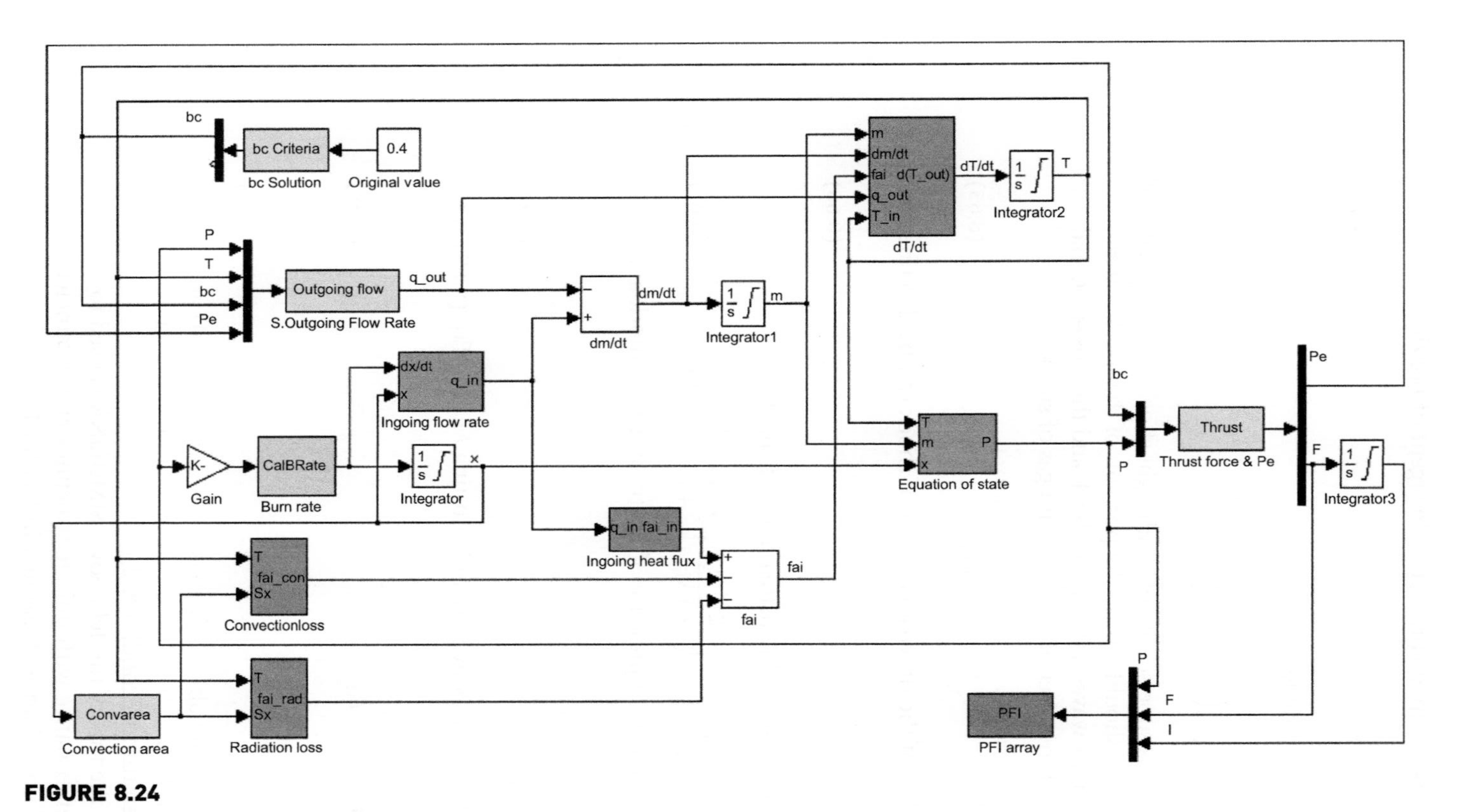

FIGURE 8.24

Model of one-dimensional gas flow in SIMULINK.

8.4 TEST OF MICROPROPULSION

8.4.1 Summary of Micropropulsion Measurement

The testing of micropropulsion is a very difficult task because of its characteristics. The thrust of micropropulsion is too small, from several μN to dozens of mN, therefore the challenges are mainly manifested in the following aspects.

1. Too small thrust. Relative to the weight of the equipment installed on the test platform, the thrust and impulse is much smaller. This means that the test platform is very sensitive to small thrusts. Therefore the design of the test system is of high impostance.
2. Too many elements which may influence the accuracy. Thrust between several μN and dozens of mN can be submerged by many elements which normally could be ignored. The friction and circuit connection could also influence the accuracy.

The resolution, error, and thrust measurement range should be fully analyzed when the test system is designed.

The test contains average thrust measurement and impulse measurement based on the thrust bench. The essence is to observe the movement of the thrust bench to measure the thrust and impulse indirectly. The measurement methods vary according to the measurement range. As for a thrust of dozens of mN, weighting benches and swing benches are used while a thrust of several μN is needed for a torsion pendulum, double pendulum, or four-arm counterweight structure.

This measurement system involves time scales. If the natural frequency of the platform is much higher than the frequency of the thruster impulse, the platform can be used to measure the average thrust and the impulse in the time integral of the thrust. If the frequency of the platform is much smaller than the frequency of the thruster impulse, the impulse can be measured and the damper or spring effect can be ignored. However, if the natural frequency is near the frequency of the thruster impulse, the influence of the natural frequency of the platform on the thrust measure should be eliminated. Since the vibration amplitude is related to the natural frequency, it should be guaranteed that the platform natural frequency is near the thruster impulse frequency when designing the platform.

Table 8.7 lists the basic types, principles, and realization forms of microimpulse systems.

8.4.2 Measurement System Based on Laser Interference and Rigid Pendulum Principle

The thrust of the MEMS solid chemical thrusters is impulse thrust. The thrust action time is very short, from several to dozens of microseconds. The unit

Table 8.7 Basic Measurement Schemes of Microimpulse Systems [30]

Basic Types	Basic Principle	Realization Form
Balance structure	Balance the weight of the thruster and its accessories and use the force caused by measuring elements to balance the thrust of the micropropulsion	Balance
Rigid pendulum	Twist moment balance theory	Positive pendulum, straight pendulum, inverted pendulum
Flexible pendulum	Twist balance theory or thrust balance theory	Twisted wire hanging pendulum, parallelogram, structure, single pendulum
Rotary table with no friction	The platform is supported frictionlessly, measuring the torque caused by microthruster or force of equilibrium torque to implement measurement	Floating platform, rotating shaft support platform

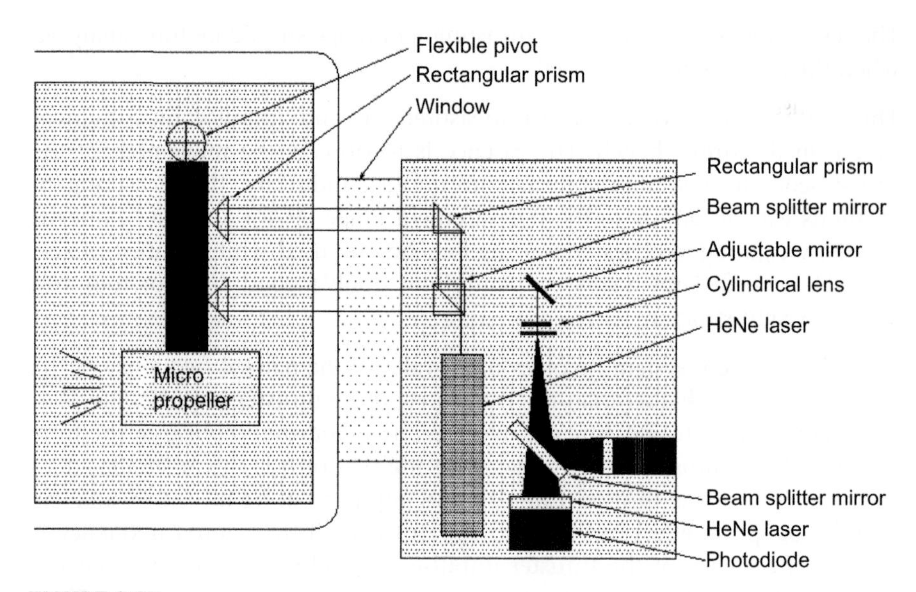

FIGURE 8.25

Laser interferometer rigid pendulum system.

impulse is hundreds of μNs in a vacuum and dozens of μNs at atmosphere pressure. Therefore a laser interferometer rigid pendulum is selected, as illustrated in Fig. 8.25. The thruster is fixed on the pendulum and the impulse thrust causes the pendulum to vibrate. The amplitude of the laser interferometer measurement reflects the impulse.

Department of Precise Instruments, Tsinghua University, has built a MEMS thruster impulse measurement system based on laser interferometer working

FIGURE 8.26
Test platform.

at atmosphere pressure. The pendulum amplitudes are recorded by computer at 10 Hz. The natural frequency of the vibration isolation platform is 3 Hz. The thruster is fixed on a precise platform with a step progress of 50 nm and automatically moved to the pendulum controlled by computer to guarantee the distance between the pendulum and nozzles. The measurement platform is shown in Fig. 8.26.

8.4.3 Microimpulse Test and Data Analysis of the MEMS-Based Solid Propellant Propulsion

Measurements are taken on a 36-array solid chemical microthruster using the test system above. The impulsion of a single unit is too small and easy to be submerged by free oscillation of a simple pendulum. Therefore multiple units are ignited at the same time. Ten test results are shown in Fig. 8.27A–J (Table 8.8).

Ten measurements are taken and 43 unit propulsions are obtained. The average unit impulse is 7.603 μNs. The standard deviation is 1.683 μNs. The minimum value is 4.561 μNs and the maximum value is 10.859 μNs.

The measurement results have a certain discreteness. The main causes of these results are:

1. Experiments are performed in the lab atmospheric environment and are influenced by atmospheric flow, causing random error.
2. Composite propellant is not totally uniform, causing uneven hole filler loading.
3. Incomplete combustion.

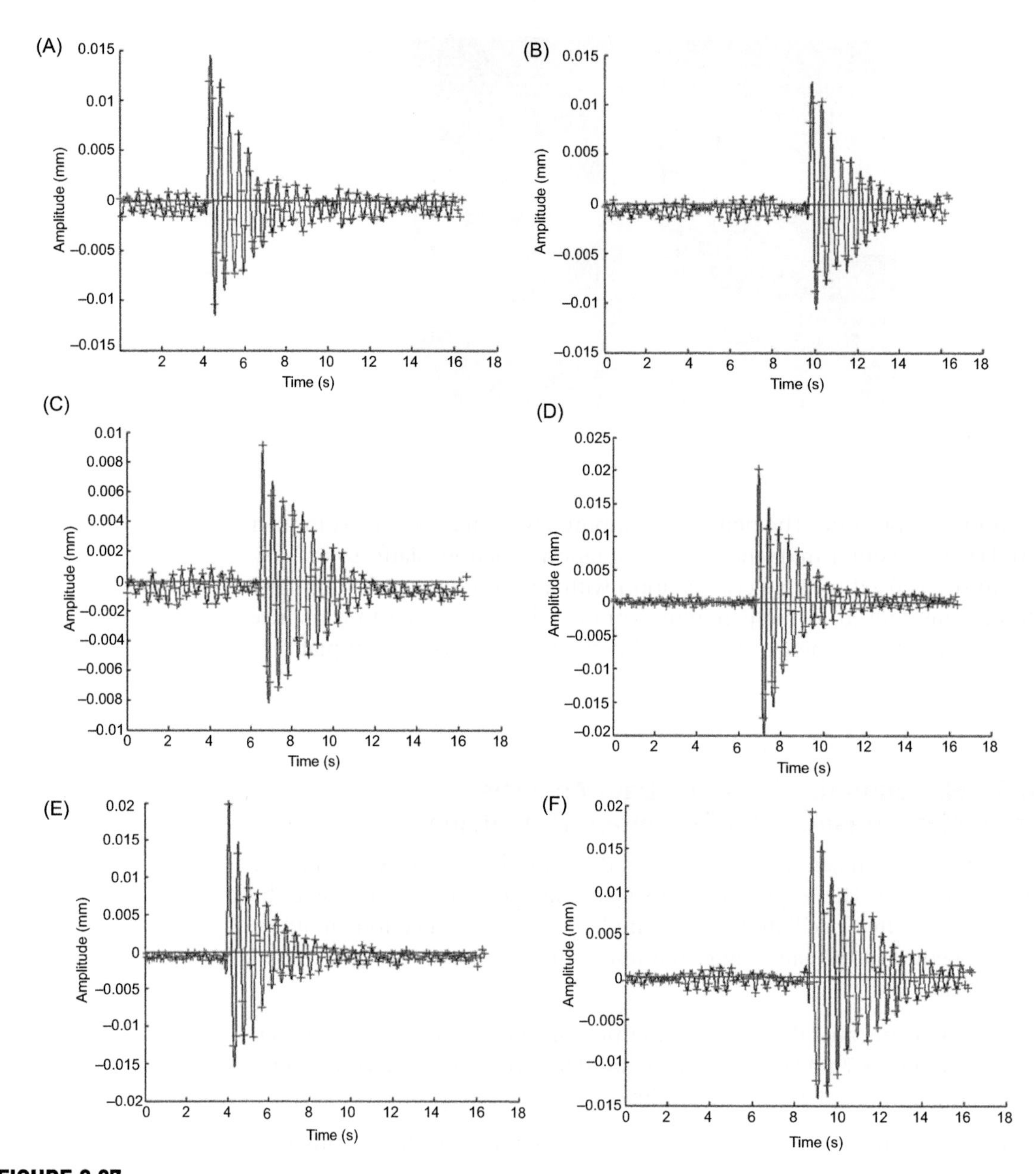

FIGURE 8.27

Impulse measurement curves. (A and C) 3 units, (B, D–F) 4 units, (G–I) 5 units, (J) 7 units.

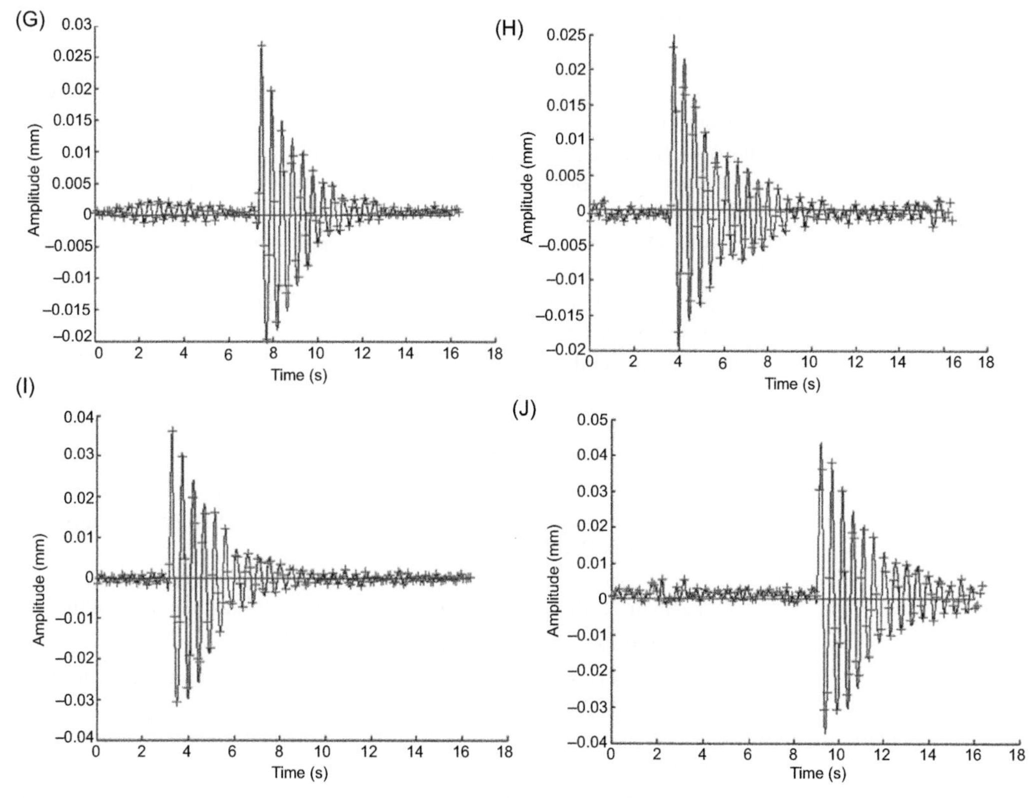

FIGURE 8.27
(Continued)

Table 8.8 Micropropulsion Test Results

Number of Units	Maximum Amplitude (μm)	Simple Pendulum Amplitude (μm)	Measurement Error	Average Single-Hole Impulse (μNs)
3	14.43	1.61	0.1116	7.2169
3	12.31	1.56	0.1267	6.15663
3	9.12	1.64	0.1798	4.5612
4	19.87	1.13	0.0569	7.4532
4	20.56	0.7	0.0340	7.7121
4	19.19	1.59	0.0829	7.1982
5	26.82	2.06	0.0768	8.0481
5	24.93	1.64	0.0658	7.4810
5	36.16	1.81	0.0501	10.8509
7	43.63	5.39	0.1235	9.3518

References

[1] D.W. Youngner, S.T. Lu, E. Choueiri, et al., MEMS mega-pixel micro-thruster arrays for small satellite stationkeeping, in: 14th Annual/USU Conference on Small Satellites, AIAA, SSC00-X-2, Logan, UT, 2000.

[2] C. Rossi, M.D. Rouhani, D. Esteve, Prediction of the performance of a Si-micromachined microthruster by computing the subsonic gas flow inside the thruster, Sens. Actuators A 87 (2000) 96–104.

[3] G.P. Sutton, O. Biblarz, Rocket Propulsion Elements, John Wiley & Sons, Inc., New York, 2001.

[4] F. Tang, Theoretical and experimental study of a silicon-based microthruster. Thesis, Tsinghua University, 2003.

[5] http://www.sstl.co.uk/documents/SNAP-1%20Propulsion.pdf.

[6] C. Underwood, G. Richardson, J Savignol, SNAP-1: a low cost modular COTS-based nano-satellite—design, construction, launch and early operations phase, in: 15th AIAA/USU Conference on Small Satellites, SSC01-VI-7, 2001.

[7] R.L. Burton, P.J. Turchi, Pulsed plasma thruster, J. Propul. Power 14 (5) (1998) 716–735.

[8] G. Spanjers, New satellite propulsion system has mass below 100 grams. <http://www.afrl-horizons.com/Briefs/Dec01/PR0109.html>, 2003.

[9] R. Leach, K.L. Neal, Discussion of micro-newton thruster requirements for a drag-free control system, in: 16th Annual/USU Conference on Small Satellites, SSC02-VIII-1, 2002.

[10] http://www.centrospazio.cpr.it/FEEPPrinciple.html

[11] S. Marcuccio, A. Genovese, M. Andrenucci, Experimental performance of field emission microthrusters, J. Propul. Power 14 (1998) 774–781.

[12] R.L. Bayt, Analysis, fabrication and testing of a MEMS-based micropropulsion system. Ph.D. Thesis of MIT, 1999.

[13] J. Anderson, J. Blandino, J. Brophy, et al., Phase-change micro-thrusters. <http://www.islandone.org/APC/Micropropulsion/01.html>, 2003.

[14] X.Y. Ye, F. Tang, H.Q. Ding, et al., Study of a vaporizing water micro-thruster, Sens. Actuators A 89 (1–2) (2001) 159–165.

[15] S.W. Jason, H. Helvajjian, W.W. Hansen, et al., Microthrusters for nanosatellites, in: The 2nd International Conference on Integrated Micro Nanotechnology for Space Applications, Pasadena, CA, April 11–15, 1999.

[16] F.M. Pranajaya, M. Cappelli, Progress on colloid micro-thruster research and flight testing. Reach Report, 1999, pp. 1–8.

[17] M. Martin, H. Schlossberg, J. Mitola, et al., University nanosatellite program. IAF Symposium, Redondo Beach, CA, April 19–21, 1999, pp. 1–8.

[18] K. Luu, M. Martin, M. Stallard, et al., University nanosatellite distributed satellite capabilities to support TechSat 21, in: 13th AIAA/USU Small Satellite Conference, Logan, UT, August 23–26, 1999.

[19] D. Teasdale, Solid propellant microrockets. Master Thesis of UCB, 2000.

[20] H. David, W. Siegfried, B. Ronald, et al., Digital micropropulsion, Sens. Actuators A 80 (2) (2000) 143–154.

[21] C. Rossi, Micropropulsion for space: a survey of MEMS-based micro thrusters and their solid propellant technology. Research Report, CNRS, France, 2003, p. 274.

[22] Z. Yu, Y. Lu, Heat Transfer Theory, Higher Education Press, Beijing, China, 1995.

[23] L. Hou, Composite Solid Propellant, Aerospace Press, Beijing, China, 1992.

[24] Z. Zheng, Solid Rocket Propellant, National University of Defense Technology Press, Beijing, China, 1981.

[25] Key Laboratory of Explosive Fire, Fire dynamite manual, 1980.

[26] A. Dalvina, in: D. Zhang (Ed.), Technology of Solid Rocket Propellant, Aerospace Press, Beijing, China, 1997.

[27] J. Zhang, Solid Propellant Chemistry and Technology, National University of Defense Technology Press, Beijing, China, 1987.

[28] C. Fan, F. Li, A study on igniter to shorten ignition delay time for micro-solid propellants rocket motor, J. Propul. Technol. 3 (1995) 42–45.

[29] E.R.G. Eckert, R.M. Drake, Analysis of Heat and Mass Transfer, McGrawHill Book Co, Tokyo, 1972.

[30] X. Liu, N. Fan, K. Li, The state-of-the-art and development tendencies of the thrust measurement, Obs. Control Technol. 23 (5) (2004) 18–20.

CHAPTER 9

Magnetometer Technology

9.1 SUMMARY

9.1.1 Concept, Function, and Application of Magnetometers

The magnetometer, also known as a magnetic sensor, is a sensor for measuring magnetic induction (magnetic field intensity), which is an important sensor component in all types of aircraft and spacecraft. It also has been widely used in other fields, such as industry, agriculture, national defense, as well as biology, medicine, aerospace, interplanetary research, etc., and currently almost no field of technology is immune from magnetic field measurement [1,2].

The magnetometer plays a more important role in the development, research, operation, management, and maintenance of defense equipment and weapons. For example, magnetic sweeping, ship degaussing, weapons search, magnetic wave communication, magnetic detection, magnetic guided missiles, as well as underwater mines, landmines, bomb detectors, and magnetic navigation, etc., are all inseparable from magnetic field measurement techniques. In addition, the magnetometer has the characteristic that other types of sensors do not have of being able to work normally under severe and limited conditions.

In the field of aeronautics, the magnetometer can be used to measure the geomagnetic field vector information of the position of the aircraft body, such as airplanes and satellites. And, according to the reference model for the Earth's magnetic field and local magnetic field, the angle information of a certain precision can be obtained through an algorithm, therefore, the magnetometer is widely used in aircraft attitude determination systems, especially in microsatellites, such as nanosatellites and picosatellites, etc. [3,4].

9.1.2 Principle and Classification of Magnetometers

Magnetometers can be classified according to different criteria.

According to its physical effects, magnetometers can be classified as follows: sensors made according to Faraday's electromagnetic induction law are called

Space Microsystems and Micro/Nano Satellites. DOI: http://dx.doi.org/10.1016/B978-0-12-812672-1.00009-6

induction magnetometers; magnetometers working by the principle that current in the magnetic field can generate a Lorentz force are called magnetic magnetometers; where the resistivity of the conductor changes in the magnetic field, this type of sensor is called a magnetoresistive magnetometer; magnetometers based on the Faraday magneto-optical effect are called magneto-optical magnetometers, such as the optical pump magnetometer; magnetic sensors based on the Josephson effect are called superconducting quantum interference devices (SQUID), etc. [2].

More often, from the point of view of the detection technology, classification can be carried out according to different measuring methods. There are similarities as well as differences between this classification and the classification above. Some magnetometers' physical effects are the same, but they are different in their detection method, which is because the magnetic field measuring methods are built on the basis of various physical effects and physical phenomena related to the magnetic field. There are dozens of methods to measure the magnetic field at present. Some of these are basic, widely used, and have broad development and can be divided into the following types: force and moment method, measuring by the Lorentz force of magnet or carrying fluid acting in the magnetic field; electromagnetic induction method, based on Faraday's law of induction, can measure DC, AC, and pulse magnetic fields, usually including a ballistic galvanometer, flux meter, electronic integrator, rotation coil magnetometer, vibrating coil magnetometer, etc.; Hall effect method; magnetoresistance effect method; magnetic resonance method, weakly connected superconducting effect method (SQUID); magnetic flux gate method; magneto-optical method; magnetostrictive method; etc. [5–7].

They can also be classified in regard to some other characteristics.

According to the ability to measure the vector information of the magnetic field, there are vector sensors which are available to measure the magnetic field along the magnetometer sensitivity axis, and scalar (total) sensors, which can only measure the magnitude of the magnetic-field vector.

There are also point field sensors that can measure a point of magnetic field information in space and magnetometers that can only measure the average magnetic field information in a certain region in space, which is related to the size of the sensitive element of the magnetometer. However, there are no uniform criteria for this classification. If it is small enough for the size of the sensitive direction of the sensitive element compared to the range measurement, we can consider that it is capable to measurement as a point magnetic field.

In addition, there is another important classification method, which is based on the sensitive range and the sensitive resolution of various specific magnetometers. The magnetic field to be measured can generally be divided into three

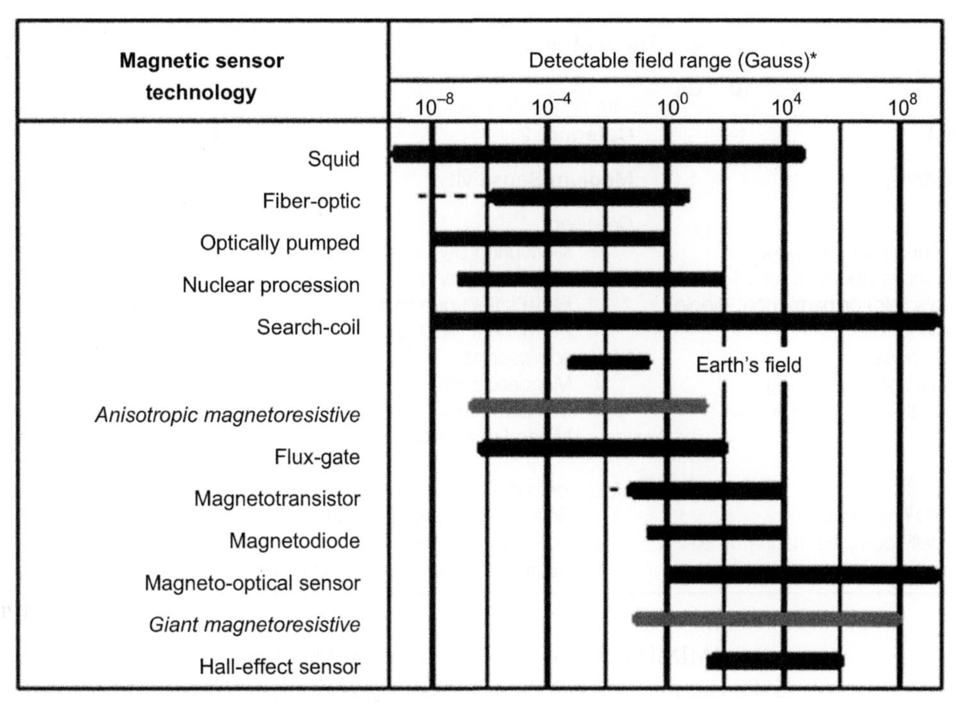

FIGURE 9.1
Sensitive range of different types of magnetometer.

categories as follows: weak magnetic field (<1 mGs [0.1 nT]); medium magnetic field (1 mGs–10 Gs); strong magnetic field (>10 Gs) [8]. This classification is based on the intensity of the geomagnetic field. Considering the needs of the task, magnetometers that measure the above magnetic fields are also divided into three types: high-sensitivity magnetometers; medium-sensitivity magnetometers; and low-sensitivity magnetometers. Fig. 9.1 and Table 9.1 compare the sensitivities and applications of some major magnetometers.

Table 9.2 compares various types of magnetic field measurement methods.

9.2 GEOMAGNETIC FIELD MODEL

The devices used to obtain satellite attitude are mainly inertial measurement units (IMUs), GSPS, magnetometers, sun sensors, infrared earth profilers, star trackers, etc. The devices which satisfy the requirements of microsatellites such as nano- or picosatellites are mainly miniature inertial measurement

Table 9.1 Classification and Main Applications of Magnetometers [8]

10^{-5} Gs	1 Gs	
Category 3	**Category 2**	**Category 1**
High Sensitivity	**Medium Sensitivity**	**Low Sensitivity**
Definition • Measuring field gradients or differences due to induced (in Earth's field) or permanent dipole moments	*Definition* • Measuring perturbations in the magnitudes and/or direction of Earth's field due to induced or permanent dipoles	*Definition* • Measuring fields stronger than the Earth's magnetic field
Major applications • Brain function mapping • Magnetic anomaly detection	*Major applications* • Magnetic compass • Munitions fuzing • Mineral prospecting	*Major applications* • Noncontact switching • Current measurement • Magnetic memory readout
Most common sensors • SQUID gradiometer • Optically pumped magnetometer	*Most common sensors* • Search-coil magnetometer • Flux-gate magnetometer • Magnetoresistive magnetometer	*Most common sensors* • Search-coil magnetometer • Hall-effect sensor

units (MIMUs), sun sensors, and MEMS magnetometers (MEMSMAGS). GSPS can also obtain the information about a satellite's attitude by means of difference, but the installation of two GSPS receivers for differential solution requires a long distance, which is not suitable for microsatellites. Infrared Earth profilers and star sensors use optical measurement, and the volume of the optical structure is too large to be used in the microsatellite field. The attitude determination accuracy can reach less than 0.5–2 degrees measuring by MEMSMAGS, and can reach less than 0.05 degree by the joint attitude determination technique of MIMU and MEMSMAGS [9–11].

Miniature magnetometers measure the instant geomagnetic field vector information under the satellite body coordinate system, and they need to establish the satellite attitude kinematics and dynamics model to be compared with the geomagnetic reference model (WMM: world magnetic model), and then through information processing the satellite attitude information can be obtained. The extended Kalman filter (EKF) or sampling Kalman filter (UKF) are generally used. The principles are shown in Fig. 9.2.

Here it is necessary to involve several aspects of content, which are mainly the status of the geomagnetic reference model and the research status of miniature magnetometers. Fig. 9.3 illustrates the current geomagnetic field model published by IAGSA (the International Airborne Geophysics Safety Association) (IGSRF2000: from 2000 to 2005; the model of 2005–2010 has also been published—IGSRF10.)

Table 9.2 Several Typical Magnetic Field Measurement Methods in Conventional Scales [1]

Measurement Methods	Performance Index			Advantages and Disadvantages	
	Resolution	Measurement Accuracy	Measuring Range	Advantages	Disadvantages
Magnetic resonance method	High resolution, up to pT stage	The accuracy of the simple device can reach 1×10^{-4}; the accuracy of sophisticated equipment can reach 1×10^{-5} after certain measures	The lowest magnetic field intensity that can be measured is 1×10^{-4} T, which can reach a limit of 1×10^{-7} T with the electron injection dynamic polarization method	Narrow resonance line; High stability; An order of magnitude can be improved when measuring weak magnetic field. The most accurate method in the absolute measurement methods of magnetic field at present	When measuring in the whole measurement range, it needs change to several kinds of different resonance frequency probes, which are not easy to carry out in continuous measurement. Complex structure; Large size
Hall effect method	Low, about 100 μT	1×10^{-4}	About 1×10^{-4} T to 10 T	Available to measure at extremely low temperatures (4.2K) and extremely high temperatures (573K). High sensitivity; Good linearity; Small probe size. Very small, IC process	Zero error, Hall voltage and internal resistance have temperature coefficient and can be affected by the input current and the uneven electrical potential and the self-excited zero potential when manufacturing
Electromagnetic induction method	High, about 10 nT	1×10^{-4} to 1×10^{-5}	From zero to the possible maximum	Available to measure the abnormal magnetic field; Wide measuring range; Good linearity; Available to carry out point measurement to a constant magnetic field	When measuring a constant magnetic field, the coil must move in the magnetic field. Large size

Continued

Table 9.2 Several Typical Magnetic Field Measurement Methods in Conventional Scales [1] *Continued*

Measurement Methods	Performance Index			Advantages and Disadvantages	
	Resolution	Measurement Accuracy	Measuring Range	Advantages	Disadvantages
Magnetic flux gate method	High, about 10 pT	1×10^{-3} (resolution of measuring the magnetic field below 10^{-3} can reach $10^{-8} - 10^{-9}$ T)	Measure the magnetic field under 0.01 T	High sensitivity; Simple structure; Reliable; Can be made very small	
Optical pumping method	Low, about μT	1×10^{-10}	The range of magnetic field is $5 \times 10^{-6} - 1 \times 10^{-3}$ T	High sensitivity; Small size; Continuous measurement and absolute measurement; Have direct reading and automatic recording features	Only able to measure the scalar intensity, but unable to measure the component value of the magnetic field vector
Magneto resistance method	Relatively high, 10 nT–1 nT	1×10^{-3}		High sensitivity; low power consumption; Simple structure; IC process (have potential to develop)	
Superconducting quantum interference device method	The resolution is 1×10^{-14} T hope to reach over 1×10^{-15} T			Magnetic field measurement method with the highest sensitivity up to date. It has a unique advantage in the weak magnetic field measurement	An of 10 Å is clamped between the two superconductors of the interferometer, which is difficult to achieve; Large size and structure; Need low temperature of 4K; High cost

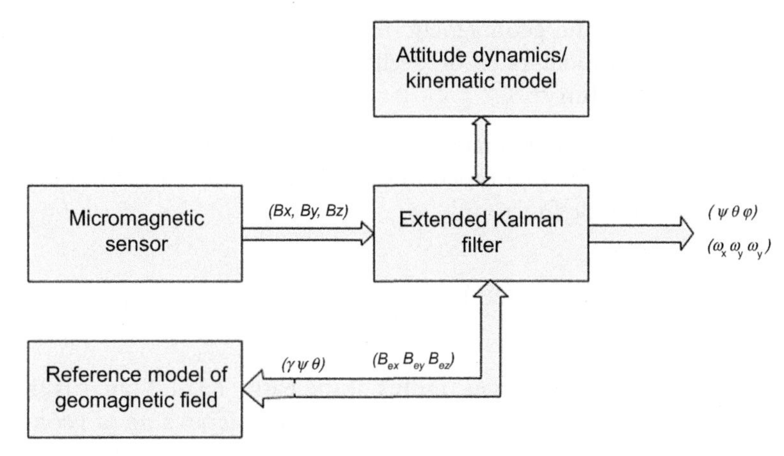

FIGURE 9.2
Principle diagram of miniature magnetometer attitude determination.

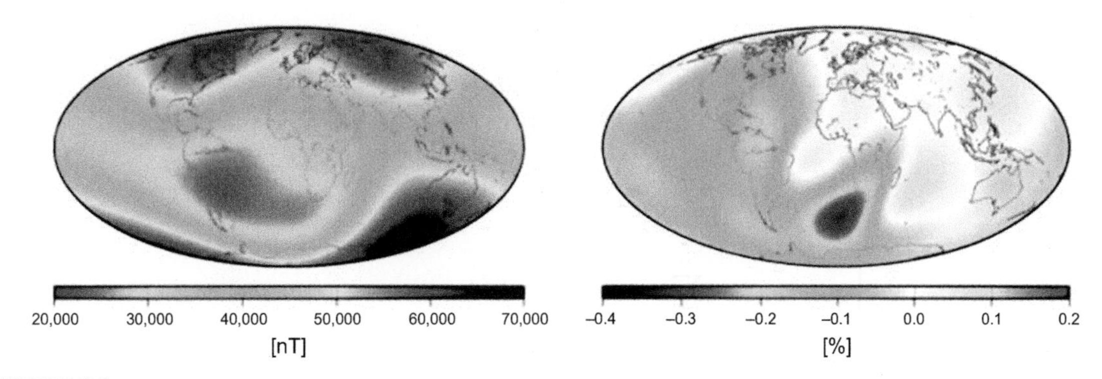

FIGURE 9.3
Current geomagnetic field model and geomagnetic field intensity.

At the same time, the IAGSA also published the calculation theory and method for the geomagnetic vector of an arbitrary point. The nominal magnetic field strength is 0.5 Gs (50,000 nT). While in the Earth's surface, the maximum magnetic field strength is about 0.7 Gs, the minimum is about 0.2 Gs. In the absolute position of 120 degrees east longitude, 34 degrees north latitude, the magnetic field strength for the year of 2000 was as follows: the total strength was 47,852 nT, the horizontal intensity was 30,713 nT, the declination was −5.16 degrees, and the magnetic inclination was 50.07 degrees. The predetermined height of the miniature space weapon platform is about 400–800 km, which requires the magnetometer measuring range to be roughly between −60 and +60 μT. The average annual rate of change of the geomagnetic field is approximately within ± 40 nT, and the average annual rates of declination and magnetic inclination are approximately within $\pm 3'$.

Supposing the geomagnetic field is caused by the internal source, the geomagnetic potential function can be expressed as the form of Gaussian spherical harmonic function:

$$V(r,\theta,\varphi,t) = R_E \sum_{n=1}^{N} \left(\frac{R_E}{r}\right)^{n+1} \sum_{m=0}^{n} \{g_{nm}(t)\cos(m\varphi) + h_{nm}(t)\sin(m\varphi)\} P_n^m(\theta)$$

where R_E is the average radius of the Earth, 6371.2 km; r is the geocentric distance, km; θ is the colatitude, complementary value of geographical latitude, 90 degrees latitude; φ is longitude; g_{nm}, h_{nm} is the Gaussian coefficient; $P_n^m(\theta)$ is the standardized Legendre polynomials of N order and M degree.

The magnetic flux density function $B(r,\theta,\varphi,t)$ of the geomagnetic field can be expressed as:

$$B(r,\theta,\varphi,t) = -\nabla V(r,\theta,\varphi,t)$$

In the spherical coordinate system (r,θ,φ), it can be expressed as:

$$\begin{cases} B_r(r,\theta,\varphi,t) = -\dfrac{\partial V(r,\theta,\varphi,t)}{\partial r} \\ \quad = \displaystyle\sum_{n=1}^{N}(n+1)\left(\frac{R_E}{r}\right)^{n+2} \sum_{m=0}^{n} \{g_{nm}(t)\cos(m\varphi) + h_{nm}(t)\sin(m\varphi)\} P_n^m(\theta) \\ B_\theta(r,\theta,\varphi,t) = -\dfrac{1}{r}\dfrac{\partial V(r,\theta,\varphi,t)}{\partial \theta} \\ \quad = \displaystyle\sum_{n=1}^{N}\left(\frac{R_E}{r}\right)^{n+2} \sum_{m=0}^{n} \{g_{nm}(t)\cos(m\varphi) + h_{nm}(t)\sin(m\varphi)\} \frac{dP_n^m(\theta)}{d\theta} \\ B_\varphi(r,\theta,\varphi,t) = -\dfrac{1}{r\sin\theta}\dfrac{\partial V(r,\theta,\varphi,t)}{\partial \varphi} \\ \quad = \displaystyle\frac{1}{\sin\theta}\sum_{n=1}^{N}\left(\frac{R_E}{r}\right)^{n+2} \sum_{m=0}^{n} m\{g_{nm}(t)\cos(m\varphi) + h_{nm}(t)\sin(m\varphi)\} P_n^m(\theta) \end{cases}$$

We obtain the B information based on the spherical coordinate system (r,θ,φ) from the above formula, but what we are interested in is the information value based on the geographic coordinate system (λ,φ,h) (geographic latitude, longitude, and altitude). Therefore, we need to perform coordinate conversion.

First, we can get the values of the point in (r, θ, φ) from those in (λ, φ, h):

$$\begin{cases} \cos\theta = \dfrac{\sin\lambda}{\sqrt{Q^2\cos^2\lambda + \sin^2\lambda}} \\ \sin\theta = \sqrt{1-\cos^2\theta} \\ Q = \dfrac{h\sqrt{a^2-(a^2-b^2)\sin^2\lambda}+a^2}{h\sqrt{a^2-(a^2-b^2)\sin^2\lambda}+b^2} \\ a\text{: semi-major axes, 6378.137 km} \\ b\text{: semi-minor axes, 6356.7523142 km} \end{cases}$$

$$r^2 = h^2 + 2h\sqrt{a^2-(a^2-b^2)\sin^2\lambda} + \frac{a^4-(a^4-b^4)\sin^2\lambda}{a^2-(a^2-b^2)\sin^2\lambda}$$

Get $(B_r, B_\theta, B_\varphi)$, then use:

$$\begin{cases} \cos\alpha = \dfrac{h+\sqrt{a^2\cos^2\lambda+b^2\sin^2\lambda}}{r} \\ \sin\alpha = \dfrac{(a^2-b^2)\cos\lambda\sin\lambda}{r\sqrt{a^2\cos^2\lambda+b^2\sin^2\lambda}} \\ \alpha = \lambda+\theta-\dfrac{\pi}{2} \end{cases}$$

The magnetic flux density component under the local coordinate system (geographical coordinates) is obtained:

$$\begin{cases} B_X(\lambda,\varphi,h,t) = -\cos\alpha B_\theta(r,\theta,\varphi,t) - \sin\alpha B_r(r,\theta,\varphi,t) \\ B_Y(\lambda,\varphi,h,t) = B_\varphi(r,\theta,\varphi,t) \\ B_Z(\lambda,\varphi,h,t) = \sin\alpha B_\theta(r,\theta,\varphi,t) - \cos\alpha B_r(r,\theta,\varphi,t) \end{cases}$$

We can also get:

1. Local horizontal magnetic flux density:

$$B_H(\lambda,\varphi,h,t) = \sqrt{B_X^2(\lambda,\varphi,h,t)+B_Y^2(\lambda,\varphi,h,t)}$$

2. Local total magnetic flux density of that point:

$$B_F(\lambda,\varphi,h,t) = \sqrt{B_H^2(\lambda,\varphi,h,t)+B_Z^2(\lambda,\varphi,h,t)}$$

3. Magnetic declination:

$$B_D(\lambda,\varphi,h,t) = \tan^{-1}\left(\frac{B_Y(\lambda,\varphi,h,t)}{B_X(\lambda,\varphi,h,t)}\right)$$

4. Magnetic inclination:

$$B_I(\lambda,\varphi,h,t) = \tan^{-1}\left(\frac{B_Z(\lambda,\varphi,h,t)}{B_H(\lambda,\varphi,h,t)}\right)$$

5. Raster change:

$$B_G(\lambda, \varphi, h, t) = \begin{cases} B_D(\lambda, \varphi, h, t) - \varphi, & \lambda \geq 0 \\ B_D(\lambda, \varphi, h, t) + \varphi, & \lambda < 0 \end{cases}$$

9.3 THE APPLICATION OF A MICROMAGNETOMETER IN NANOSAT

LAGSA published the theory and method of computation of the geomagnetic vector (including the three-axes components, magnetic declination, and magnetic inclination). The nominal geomagnetic intensity was 0.5 Gs (50,000 nT). The geomagnetic intensity ranged from about 0.2 Gs to about 0.7 Gs over the surface of the Earth. The total geomagnetic intensity, horizontal geomagnetic intensity, magnetic declination, and magnetic inclination were 47,852 nT, 30,713 nT, −5.16 degrees, and 50.07 degrees, respectively, in an absolute location on 30°N, 120°E, and 100 km altitude (Fig. 9.4).

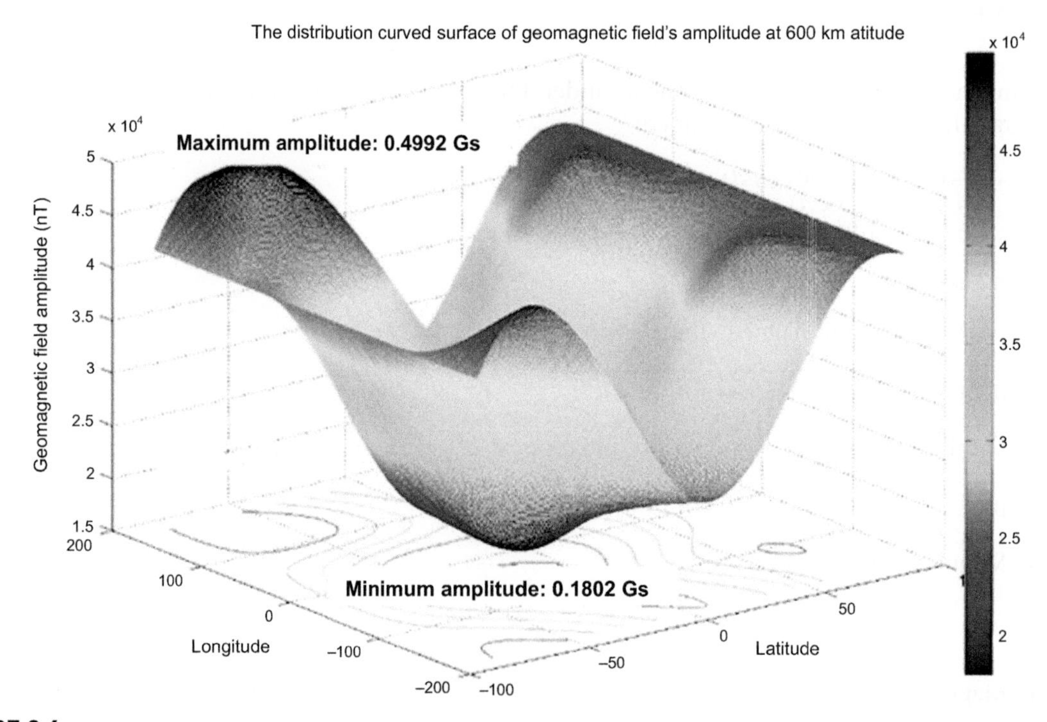

FIGURE 9.4
The magnetic field distribution at 600 km altitude.

The orbit height of a microsatellite is about 600 km. We can obtain the distribution of the geomagnetic field vector amplitude on the day of December 17, 2008, as shown in Fig. 9.1, using the WMM2005 model in MATLAB. The amplitude ranged from 0.1802 to 0.4992 Gs. The average annual rate of geomagnetic field was less than ± 40 nT, and the average annual rates of magnetic declination and magnetic inclination were less than $\pm 3'$.

The measurement of altitude of the magnetometer is designed between -60 and $+60$ μT based on the above model. According to the survey of existing commercial micromagnetometers, several types of micromagnetometers could barely reach the application conditions of the microsatellite.

Magnetometers are usually made up of a sensitive component, conversion component, and other components (including power supply part, signal conditioning part, etc.). The sensitive component is the kernel of the sensor. It perceives the magnetism of the environmental magnetic field directly, and outputs another physical quantity which has a relationship with the measured magnetism. The conversion component, also known as the converter, does not perceive the measured quantity generally, but converts the sensor's output quantities to electrical output. The signal conditioning part can convert the electrical parameter (voltage, current, etc.) into an electrical signal, which is useful, and can be easily displayed, recorded, processed, and controlled. Commonly used circuits include the weak signal amplifier, electric bridge, signal generator, modulation and demodulation circuit, filter and phase-shifter circuit, impedance matching, and conversion circuit, etc.

The APS Model 534 minifluxgate magnetometer from American Physical Society (APS) is shown in Fig. 9.5.

FIGURE 9.5
APS Model 534.

This minimagnetometer is machined by precision machining technology and has been used in aerospace application, such as in the NS-1 Nanosat. It can measure the three-axes geomagnetic field with an analogue signal output and has a temperature measurement model through reconstruction.

The technical index of this minimagnetometer is shown in the following table.

Measurement range	-1 Gs $\sim$ $+1$ Gs
Noise	$<1\times10^{-6}$ Gs RMS/√Hz
Bandwidth	DC to 400 Hz (-3 dB)
Linearity	$\pm0.1\%$ at full scale
Sensitivity	4.00 V/Gs
Dimensions	$20\times20\times70$ mm
Weight	30 gsm
Power	±5 to ±12 VDC at 20 mA

9.4 AMR MAGNETOMETER

9.4.1 The Principles and Realization of AMR Magnetometers

The system principle block diagram is shown in Fig. 9.6. The magnetometer is composed of single-axis and two-axes magnetometer chips based on the

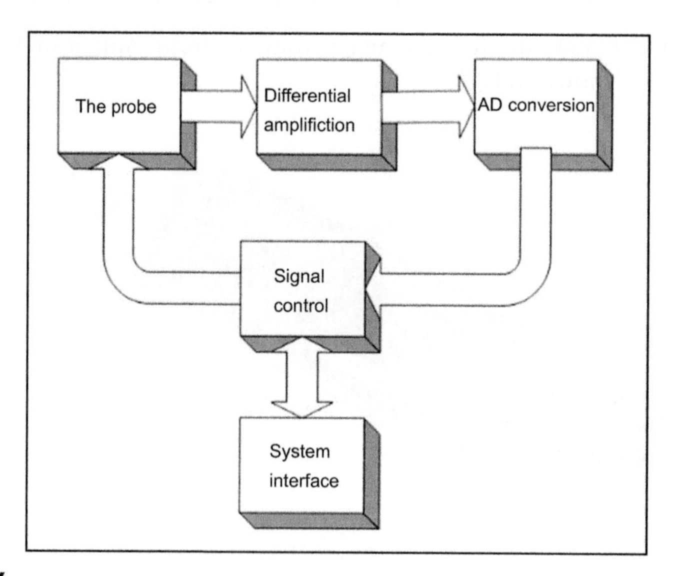

FIGURE 9.6
Principle diagram of the MAMRM three-axes magnetometer module.

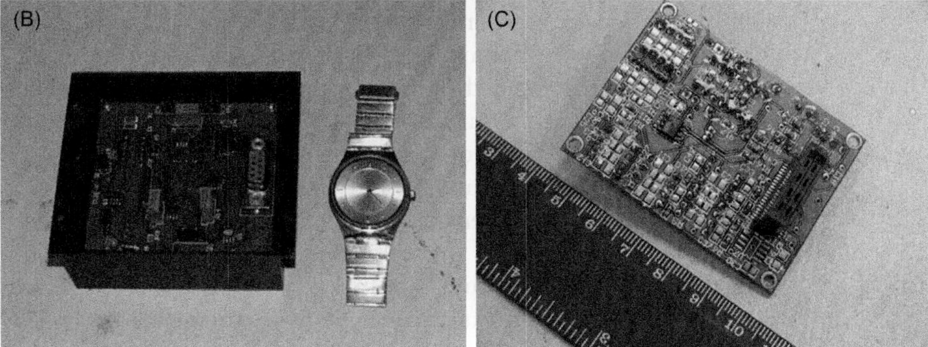

FIGURE 9.7
(A) The MAMRM three-axes magnetometer version 1. (B) The MAMRM three-axes magnetometer version 2. (C) The MAMRM three-axes magnetometer version 3.

AMR (anisotropic magnetoresistance) principle, which can be purchased, to measure the three-axes geomagnetic field intensity. It can amplify the analogue signal accurately through differential amplification and has a digital output through AD conversion.

Four round principle designs are performed according to the task demands of the microsatellite. Research on real objects is conducted in three rounds (Ver1, Ver2, Ver3), as shown in Fig. 9.7.

The magnetometer module is composed of a magnetic sensor component and other relevant electronics. Its circuit board is placed outside of the satellite through an installing box. Its function is for sensing the magnetic vector, generating an analogue voltage output corresponding to the magnetic vector. The circuit contains an internal resistance bridge using magnetic-resistance effect, converting the magnetic quantities to voltage output through a differential instrumentation amplifier, a low-pass filter, and an operational amplifier (op-amp) driver. In addition, a TTC (telecontrol telemetry subsystem) temperature measurement circuit is also placed in it. The power supply (+5 V) of the TTC temperature measurement circuit is provided by PDM, the analogue voltage output transfers to the TTC module directly through a wire.

The magnetometer circuit is composed of a TTC temperature measurement chip, a voltage regulator, an AMR chip, and an analogue signal conditioning device. The TTC temperature measurement chip used is an AD22100 chip. Because the AMR chip (HMC1001/1002 from Honeywell) has a resistance bridge structure inside which will affect the output if the power is not stabilized, it needs a voltage regulator to stabilize the voltage for the +5 V power supply. The signal conditioning device includes an instrumentation amplifier, low-pass filter, and op-amp. The differential analogue output of the AMR magnetometer single-ended outputs to the low-pass filter through preamplifying by an instrumentation amplifier to filter out noise, then outputs to the satellite's MUX port by an op-amp driven cable. Referencing to the magnetic field direction control circuit, the pulse signal (controlled by ADCS), which can make the magnetic sensor's magnetic field direction reverse, improves the measurement precision through twice measurement forward and inverse, respectively, to the same direction. In addition this can also prevent accidental magnetization by an external magnetic field reducing the magnetic measurement performance (Fig. 9.8).

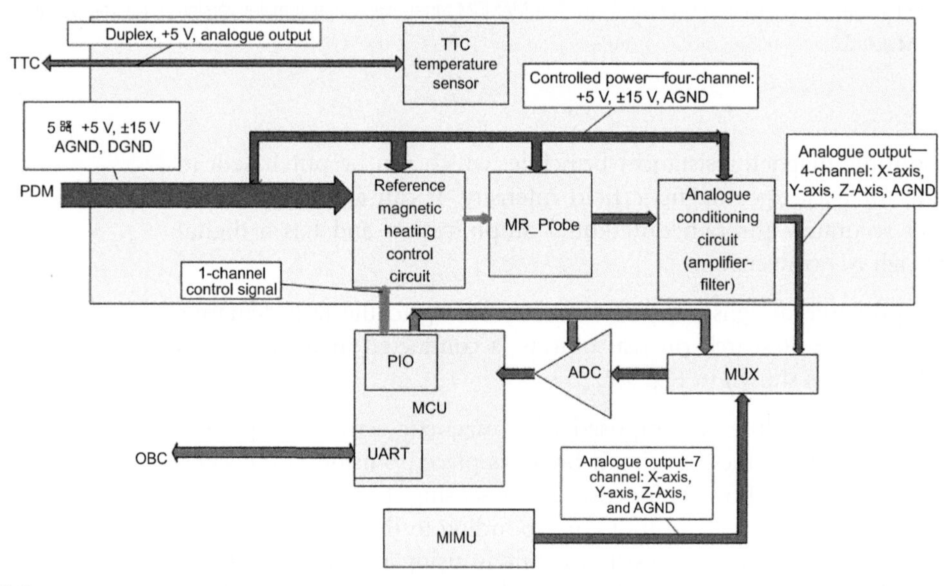

FIGURE 9.8
Schematic diagram of a magnetometer module.

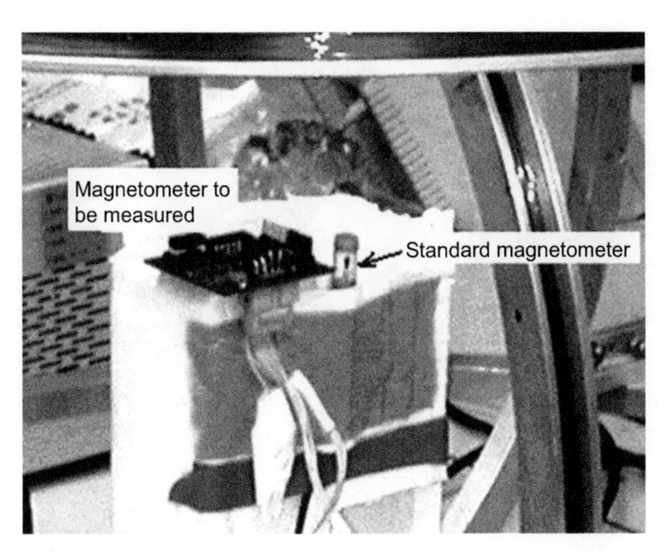

FIGURE 9.9
Parameter calibration test system.

9.4.2 Calibration of the System Main Parameters

The detection system can measure the main parameters of the magnetometer module, such as sensitivity and zero deviation. The detection system is mainly composed of a 3d Helmholtz coil and its drive and current monitoring circuit are shown in Fig. 9.9.

Using an existing fluxgate magnetometer whose precision is 1 nT as standard, the external magnetic field can be shielded by adjusting the value and direction of the electric current in three axes. And adjusting the current linear in one direction can change the value and direction of the drive magnetic field (this magnetic field can be recorded by a standard magnetometer). Then the output voltage of the magnetometer module is recorded and the three-axes calibration parameter of the magnetometer module can be obtained through linear fitting. The input and output curve of the X-axis are shown as Fig. 9.10:

X-axis: B(nT) = 1896*V(V) + 6475, linearity: 0.9997 (Fig. 9.11).

9.5 THE PRINCIPLES OF ORBIT AND ALTITUDE DETERMINATION USING A MAGNETOMETER

9.5.1 Orbit Determination Using a Magnetometer

The geomagnetism can be applied in navigation for near-Earth satellites, with improvement of the geomagnetic field model and the signal analysis

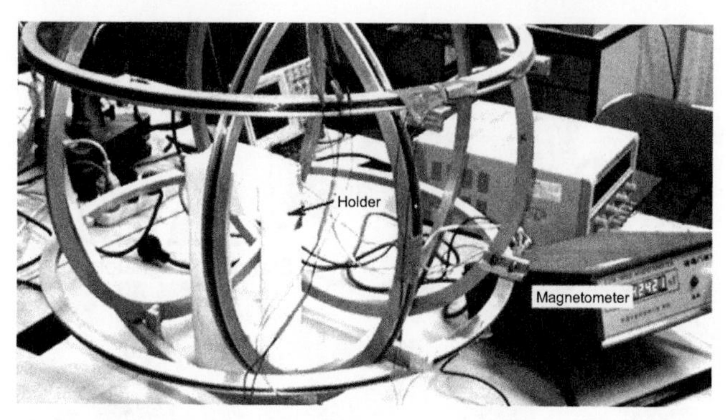

FIGURE 9.10
Calibration for magnetometer parameters.

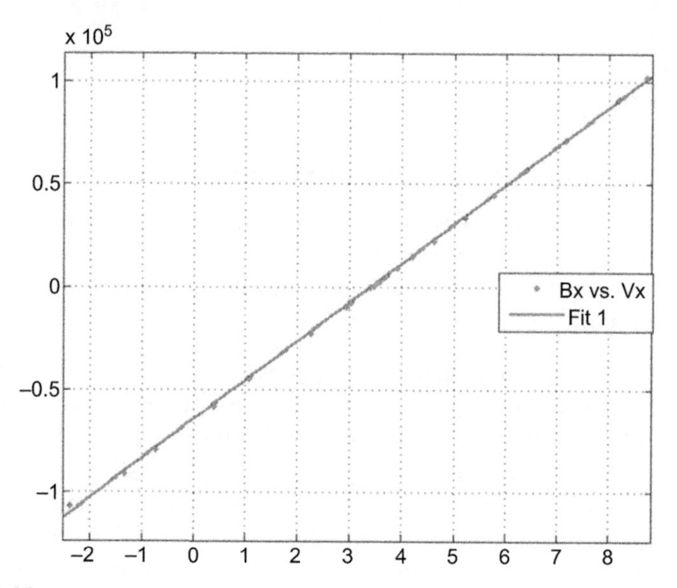

FIGURE 9.11
Linear fitting of measured data in the X-axis.

and processing technology [12]. Using a three-axes magnetometer to measure the three-dimensional component and the change rate of the geomagnetic field intensity vector, and using Kalman filtering technology to calculate the estimate of the position and speed of the satellite afterwards, the orbital elements can be acquired.

The system state equation is established by taking the position and velocity parameters $X = (x, y, z, \dot{x}, \dot{y}, \dot{z})^T$ of the satellite to the system state parameters:

$$\dot{X} = F(X,t)X + W$$

After linearization:

$$\Delta\dot{X} = \left.\frac{\partial F(X,t)}{\partial X}\right|_{x=\hat{X}(k/k-1)} \Delta X + W = A(X,t_k)\Delta X + W$$

The measurement model of the magnetometer:

$$y(k) = N[B(X(k),t_k)] + v(k)$$

After discretization:

$$Z(k) = \Delta y(k) = \left.\frac{\partial N(B)}{\partial X}\right|_{X=\hat{X}(k/k-1)} \Delta X(k) + V(k) = H(k)\Delta X(k) + V(k)$$

Therefore, the orbit determination module of a three-axes magnetometer is:

$$\begin{cases} \Delta X(k) = \Phi(k,k-1)\Delta X(k-1) + W(k) \\ Z(k) = H(k)\Delta X(k) + V(k) \end{cases}$$

where $\Phi(k,k-1)$ is the state-transition matrix, and $H(k)$ is the measurement matrix. $\Delta\hat{X}(k)$ can be acquired through a Kalman filter, therefore the $X(k)$ at time k is:

$$X(k) = \hat{X}(k/k-1) + \Delta\hat{X}(k)$$

The orbital elements can be ascertained when acquiring the position and velocity of the satellite in the geocentric inertial coordinate system at the moment t_k.

9.5.2 Attitude Determination Using a Magnetometer

Besides using a three-axes magnetometer to acquire position and velocity conditions with certain accuracy in order to determine satellite orbit elements, the magnetometer is mainly applied to determine satellite attitude. In order to acquire the satellite's attitude information, it is necessary to have information about the satellite orbit, namely the track number, to establish the current orbit coordinate system.

For MEMSSAT, there are several ways to get the above information:

1. Ground monitoring station and tracking system;
2. The three-axes magnetometer to estimate orbit elements;
3. The satellite-bone GSPS subsystem.

A rough method of acquiring satellite attitude is using the satellite's current location parameters to obtain the local geomagnetic field intensity vector, and comparing with the vector measured by the satellite-borne three-axes magnetometer, to acquire the altitude information at that moment.

A more accurate method is to use the orbit elements of the magnetometer and the vector measured by the satellite-borne three-axes magnetometer to obtain the best attitude estimation through a Kalman filter.

The coordinate systems used in determining altitude include:

1. The orbit coordinate system:
   $+X_0$ (the motion direction of the satellite), $+Y_0$ (the normal direction of the orbital plane), $+Z_0$ (the direction from the geo-center to the satellite).
2. The satellite coordinate system:
   $+X$ (the normal flight state) coinciding with the $+X_0$ direction, $+Y$ (the normal flight state) coinciding with the $+Y_0$ direction, $+Z$ (the stretch direction of the gravity gradient bar). The normal flight state coincides with the $+Z_0$ direction.
   The required direction the satellite points in when working normally is:
   The orbit coordinate $+X$, $+Y$, $+Z$ coinciding with the orbit coordinate $+X_0$, $+Y_0$, $+Z_0$.
   The required precision of the determining altitude is:
   Pitching: 1 degree (1σ), rolling: 0.5 degree (1σ), yaw: 1 degree (1σ)
   Because the MEMS satellite does not need to point to the Earth, and then considering the altitude precision (better than ± 1 degree), the poor quality and the low power consumption, the magnetometer can be used for determining altitude.
   Referring to the design of the NS $-$ 1 magnetometer, a three-axes flux-gate magnetometer can be chosen to measure the component of the Earth's magnetic field vector in the satellite coordinate system according to the required point precision (1 degree in three axes).
   In the period of low solar activity, using a magnetometer that has been calibrated, the measurement precision of satellite attitude angle can reach below 0.5 degree (1σ) for each axis. After the satellite launches into orbit and is in a state of rolling, we can still use the magnetometer and the fast Kalman filter to provide the satellite's three-axes attitude angular rate relative to the orbit coordinate system. The in-orbit calibration algorithm can correct the offset error and installation error of the magnetometer in real-time.
   The attitude dynamics equation of a satellite is:

$$I\dot{\vec{\omega}} = \vec{N}_{GG} + \vec{N}_M + N_D - \vec{\omega} \times (I\vec{\omega} + \vec{h}) - \dot{\vec{h}} \tag{9.1}$$

where $\boldsymbol{I}$ is the inertia tensor the satellite relative to the center of mass; $\vec{N}_{GG} = \frac{3\mu}{R^3}\left[I_{zz} - \frac{I_{xx} + I_{yy}}{2}\right](\vec{z}_0 \cdot \vec{z})(\vec{z}_0 \times \vec{z})$ is the gravity gradient torque vector; $\vec{N}_M = \vec{M} \times \vec{B}$ is the magnetic torque vector; $\vec{N}_D$ is attitude disturbance torque vector; and $\vec{h}$ is the angular momentum vector of the reaction wheel.

The attitude kinematics equation of a satellite is:

$$\begin{bmatrix} \dot{q}_1 \\ \dot{q}_2 \\ \dot{q}_3 \\ \dot{q}_4 \end{bmatrix} = \frac{1}{2} \begin{bmatrix} 0 & \omega_{oz} & -\omega_{oy} & \omega_{ox} \\ -\omega_{oz} & 0 & \omega_{ox} & \omega_{oy} \\ \omega_{oy} & -\omega_{ox} & 0 & \omega_{oz} \\ -\omega_{ox} & -\omega_{oy} & -\omega_{oz} & 0 \end{bmatrix} \begin{bmatrix} q_1 \\ q_2 \\ q_3 \\ q_4 \end{bmatrix} \tag{9.2}$$

where $\begin{bmatrix} \omega_{ox} & \omega_{oy} & \omega_{oz} \end{bmatrix}^T$ is the angular speed of the satellite relative to the orbit coordinate system; and $\begin{bmatrix} q_1 & q_2 & q_3 & q_4 \end{bmatrix}^T$ is the quaternion describing the altitude of the satellite.

The Kalman filter algorithm estimate for the altitude angular rate is:

Assuming the sampling period is T_s, making $\Gamma = T_s I^{-1}$, using $\vec{x}_k$ to represent the angular speed vector $\vec{\omega}(k)$ the satellite relative to the orbit coordinate system in the k-th sampling period, and defining the system state covariance matrix as $P_k = E\{\vec{x}_k \cdot \vec{x}_k^T\}$, the state vector spread as:

$$\hat{\vec{x}}_{k+1/k} = \hat{\vec{x}}_{k/k} + \Gamma u_k,$$

where u_k is the magnetometer measurement (the magnetic moment vector with the satellite loaded) in the k-th sampling period.

The disturbance covariance matrix spread is:

$$P_{k+1/k} = P_{k/k} + Q$$

where Q is the system noise covariance matrix.

The gain refresh is:

$$K_{k+1} = P_{k+1/k} H_{k+1}^T \left[H_{k+1} P_{k+1/k} H_{k+1}^T + R \right]^{-1},$$

where R is the measure noisy covariance matrix.

$$H_{k+1} = T_s \begin{bmatrix} 0 & -\omega_z(k) & \omega_y(k) \\ \omega_z(k) & 0 & -\omega_x(k) \\ -\omega_y(k) & \omega_x(k) & 0 \end{bmatrix}.$$

The system state refresh is:

$$\hat{\vec{x}}_{k+1/k+1} = \hat{\vec{x}}_{k+1/k} + K_{k+1}\left(\vec{y}_k - H_{k+1} \hat{\vec{x}}_{k+1/k} \right).$$

The covariance refresh is:

$$P_{k+1/k+1} = [1 - K_{k+1} H_{k+1}] P_{k+1/k}.$$

The real-time self-calibration algorithm of a magnetometer is shown as:

The vector error is defined as:

$$\vec{e}(k) = \vec{y}_{model}(k) - \vec{y}_{calib}(k) = A(k)\vec{B}_0(k) - \left[G(k)\vec{B}_m(k) + \vec{b}(k)\right]$$
$$= A(k)\vec{B}_0(k) - \vec{\phi}^T(k)\vec{\theta}(k)\vec{\phi}^T(t) = \begin{bmatrix} B_{mx} & B_{my} & B_{mz} & 1 \end{bmatrix},$$

where $A(k)$ is the altitude transform matrix from the orbit coordinate to the satellite coordinate in the k-th sampling period; $G(k)$ is the gain matrix; $\vec{B}_0(k)$ is the magnetic vector in the k-th sampling period; $\vec{B}_m(k)$ is the magnetometer measurement which has not been calibrated in the k-th sampling period; and $\vec{b}(k)$ is the partial adjustment to be ensured in the k-th sampling period.

The real-time self-calibration algorithm is:

Calculate $\vec{\phi}(k)$ and $\vec{e}(k)$

Update the gain vector: $\vec{K}(k) = P(k-1)\vec{\phi}(k)\left[\lambda + \vec{\phi}^T(k)P(k-1)\vec{\phi}(k)\right]^{-1}$

Update the parameter vector: $\vec{\theta}(k) = \vec{\theta}(k-1) + K(k)\vec{e}(k)$

Update the covariance matrix: $P(k) = \left[1 - K(k)\vec{\phi}^T\right]P(k-1)/\lambda$.

References

[1] Z.L. Mao, Magnetic Field Measurement, Atomic Energy Press, Beijing, 1985.

[2] Y.J. Zhao, K.C. Yang, Technology and Application of Amorphous Alloy Sensor, Huazhong University of Science and Technology Press, Wuhan, 1998.

[3] Z. You, Research on MEMS technology in Nanosat. The Final Report of 863 Project, 2001.

[4] Z. You, The Technology Report About MEMS Satellite., Tsinghua University, Beijing, 2001.

[5] R.S. Popovic, The future of magnetic sensors, Sens. Actuators A 56 (1996) 39–55.

[6] M.J. Caruso, T. Bratland, C.H. Smith, R. Schneider, A New Perspective on Magnetic Field Sensing, Honeywell, Inc., Morris Plains, NJ, 1998, p. 19.

[7] R. Rikpa, Review of fluxgate sensors, Sens. Actuators A 33 (1992) 129–141.

[8] J.E. Lenz, A review of magnetic sensors, in: Proceedings of IEEE, 1990, pp. 973–989.

[9] S. McLean, S. Macmillan, The US/UK World Magnetic Model for 2005–2010, British Geological Survey NOAA National Geophysical, Editor, 2005.

[10] S. Macmillan, J.M. Quinn, The derivation of World Magnetic Model 2000, United States GSeoloGsical Survey, Editor, 2000.

[11] J.M. Quinn, The Joint US/UK 2000 Epoch World Magnetic Model and the GEOMAGS/MAGSVAR Algorithm, U.S. Geological Survey MS 966 Geomagnetics Group, Federal Center, Denver CO 80225-0046, USA and British Geological Survey Geomagnetics Group, Editors, 1999.

[12] H.S. Cao, Designing of fluxgate magnetometer used by attitude test, J. Ballist. 14 (2002) 2.

CHAPTER 10

MEMS Microrelay

10.1 INTRODUCTION

10.1.1 Background and Significance of MEMS Relay Technology

Relay is a basic electronic control component, which plays an important role in aerospace, military electronic equipment, information technology, and other automation equipment. Relay has a control system (also known as the input circuit) and a controlled system (also known as the output circuit), and it is usually used in automatic control circuits to adjust automatically, protect security, convert the circuit, etc. According to the structure and working principle, relays are divided into EMR (electromechanical relay, mainly including electromagnetic relay and thermal dry reed relay) and solid-state relay (SSR), the essential difference between them being whether there is direct contact or not.

In recent years, with the development of semiconductors, the scale of IC chips has reached to well into the sub-micron level and the integration level is higher. In contrast, the progress of miniaturization technology in relays is far from adequate, its size and power have become the bottleneck of the system in attempts to further improve their working ability. It is predicted that the miniaturization of electromechanical relays will be widely demanded in the military and has a bright future.

10.1.2 Research Survey of MEMS Relay Technology at Home and Abroad

MEMS relay has gradually evolved from traditional relays. As early as 1978, Dr. K.E. Peterson developed the world's first MEMS relay [1]. To date, many domestic and foreign research institutes have conducted research on MEMS relays. Foreign institutes such as National Aeronautics and Space Administration (NASA), Defense Advanced Research Projects Agency (DAPAR), Germany Munich Polytechnic University, Microelectronics Center of North Carolina

Space Microsystems and Micro/Nano Satellites. DOI: http://dx.doi.org/10.1016/B978-0-12-812672-1.00010-2

(MCNC), United States Analog Device Corporation (United States AD), Northeast University, MIT Lincoln Laboratory, and others have different degrees of involvement in research into MEMS relays. There are also some domestic institutions that have researched MEMS relays, including Shanghai Jiao Tong University (SJTU), Beijing Jiaotong University (BJUT), and so on.

Compared with the traditional electromechanical relay, MEMS relay has small size, low power consumption, fast response, high isolation feature, and high load capacity. Also, MEMS relay can be integrated produced, similar to semiconductors. MEMS deposition and corrosion technology can be used to fabricate three-dimensional moving mechanical components, providing the basic process for MEMS relay fabrication.

There are two main classifications of MEMS relay. The first is divided according to the driving principle, mainly including electrostatic, electromagnetic, and thermomechanical types. In addition, there are also thermally controlled magnetization microrelay [2] and liquid-filled microrelay [3]. The other is divided according to the electrode switch pattern, including vertical and lateral switches. The MEMS relay using a vertical switch pattern is superior to the latter in switching frequency, contact resistance, insulation resistance. and other important parameters [4]. Parameter comparisons are shown in Table 10.1.

The most representative MEMS relay is the *ME-X* developed by Panasonic, which is a hybrid mechanical relay, adopting efficient magnetic circuits and MEMS technologies, integrating the minicontact and armature mechanical structures, as shown in Table 10.2. An ultra-small RF relay prototype product produced by OMRON can convert 20-GHz signal, the size of this micromachined relay (MMT) is 1.8 mm × 1.8 mm × 1 mm. TeraVicta Technologies have introduced a new SPDT MEMS relay, TT1244, whose working frequency is 0–26.5 GHz. The industrialization of MEMS relay is in the initial period, and the size of the MEMS relay is almost at the cm level—mm-level MEMS relays cannot yet be mass produced.

Table 10.1 The Main Performance Parameters of Vertical and Lateral Switches Relay

Index	Lateral Switches	Vertical Switches
Switching time (μs)	30	5
Contact resistance (kΩ)	1	3×10^{-3}
Insulation resistance (Ω)	–	10^{10}
Drive voltage (V)	50–260	20–100
Switching current (mA)	1	50
Contact electrode materials	Silicon	Metal
Moving electrode materials	Silicon	Multilayered materials

Table 10.2 The Parameters of MEMS Relay, *ME-X* Developed by Panasonic

Standard	Index	Details
Product number		AMEX1001
Coil specifications	Rated operational voltage	3 V DC (single coil self-lock)
	Rated power consumption	100 mW
Contact specifications	Contact form	1 alb
	Contact material	Au series
	Contact resistance (initial stage)	300 mΩ (typical)
	Contact allowed load	100 mA 1.5 V DC (resistance load)
High-frequency characteristics (initial stage, 50Ω series, ~6 GHz)	Insertion Loss	0.5 dB (typical)
	Isolation degree	28 dB (typical)
Usage times	Electrical usage times	More than 10 million times (10 mA, 1.5 V DC resistance load)
Voltage proof (10 mA test current)	Between contacts (initial stage)	100 V per minute

10.1.3 Overview of a Different Driving Method for MEMS Relay

According to the driving method, MEMS relay can be divided into electrostatic, electromagnetic, thermomechanical, and liquid-filled microrelays. The typical structures of various drive modes are introduced below.

10.1.3.1 Electrostatic MEMS Relay

The electrostatic drive method is commonly used, it has an easier fabrication process, low power consumption, and more rapid response time, also it has no special demands for driving materials. However, it demands a high driving voltage. There are two main structures of electrostatic MEMS relay, the cantilever type and the multifinger type. According to the electrode switch pattern, it can be divided into a vertical switch type or a lateral switch type.

A typical single-end fixed cantilever beam microrelay schematic is shown in Fig. 10.1. The upper plate is an active electrode, the bottom one is fixed. When switched on, it will build an electric field between the upper and bottom plates, the electric field force will move the upper plate downward contacting the bottom contact electrode, whereupon the relay is closed. When switched off, the electric field disappears and the upper plate disconnects from the contact electrode and recovers its original state because of the elasticity of the beam, and the relay is opened. The driving voltage of this kind of relay is generally between 20–90 V.

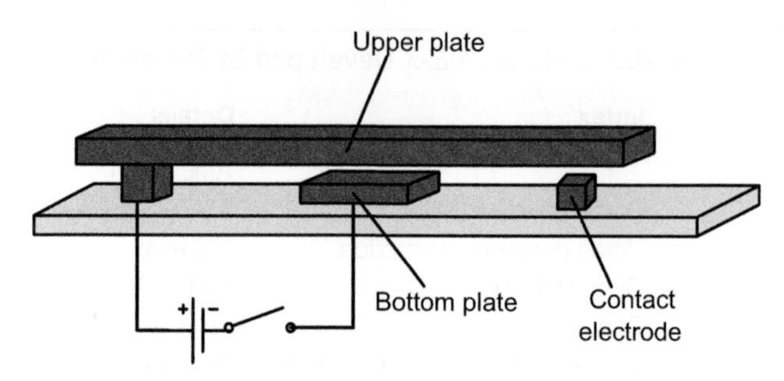

FIGURE 10.1
Single-ended fixed cantilever beam microrelay schematic.

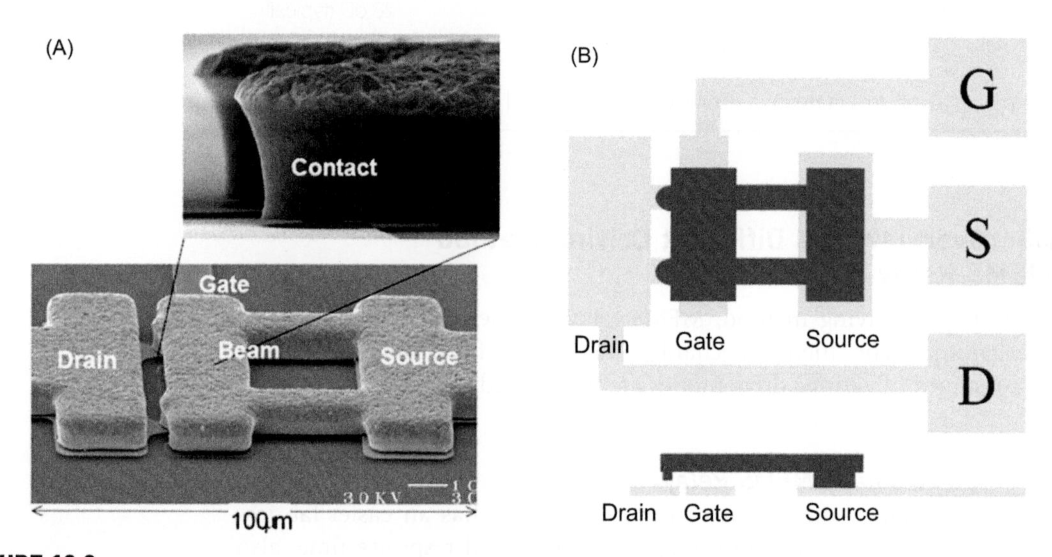

FIGURE 10.2
(A) Physical SEM photo; (B) structural schematic of a cantilever beam microrelay.

Fig. 10.2A is a physical SEM photograph of a single-end fixed cantilever beam microrelay fabricated by S. Majumder and others in America [5]. Fig. 10.2B shows its structural schematic.

The relay shown in Fig. 10.2 has two electrodes connecting to input and output signals, respectively, reducing the interference benefits of depart of switching part and driving part. When the upper plate moves downward, it will contact the two contact electrodes simultaneously, and the relay is closed. The relay's driving voltage is 80 V, contact resistance is less than 1 Ω, and the usage count can reach up to 10^{10} times.

Noryo Nishijima and Juo-Jung Hung proposed an improved single-ended fixed cantilever beam relay in 2004 [6]. As shown in Fig. 10.3, its plate is

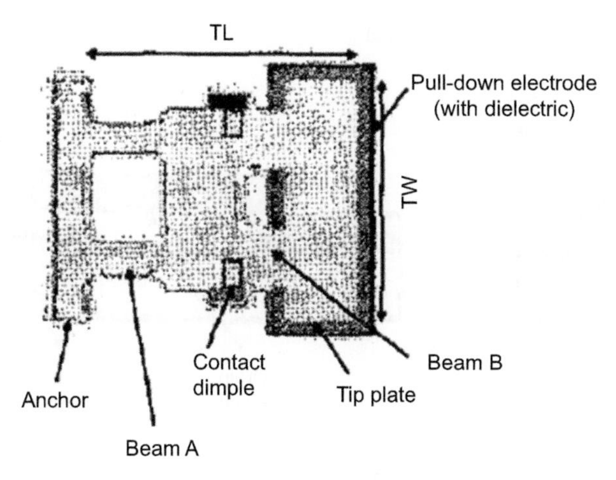

FIGURE 10.3
The top view of the relay (TL′ length is 190–220 μm; TW′ length is 160–220 μm).

divided into five parts, the fulcrum, beam A, contact part, beam B, and a driving electrode arranged from left to right, successively.

The pull-down electrode of the relay shown in Fig. 10.2 is located between the fulcrum and contact electrode; its schematic is shown in Fig. 10.4A. However, the pull-down electrode of the relay shown in Fig. 10.3 is located on the side of the contact electrode. When the former relay is closed, only its contact electrode switches on, as shown in Fig. 10.4C. However, when the latter relay is closed, its contact electrode switches on after its upper plate contacts the bottom plate, as shown in Fig. 10.4D.

The advantage of the latter design is reducing the pull-down voltage without reducing the releasing force. The releasing force is dependent on beam A, the movement of pulling-down is dependent on beams A and B, and the elastic coefficient of the beam is proportional to the cube of its length. Therefore, the force to bend two beams is less than that to bend one, and it will increase the contact force to make closure more stable.

Experiments indicate that the driving voltage of the driving force is 30 V, which is lower than that shown in Fig. 10.2, but also higher than the voltage a common circuit can provide, so it is not easy to integrate with the circuit.

As shown in Fig. 10.5, M.A. Gretillat and others proposed a double-end fixed MEMS relay in 1994, which can separate the driving part and working circuit [7]. Its driving voltage is 50–75 V, with a working frequency of 20–75 kHz, releasing time of 4 μs, and pull-in time of 1 μs. Comparing to the single-ended fixed relay, this kind of relay's releasing force is higher and the contact electrodes are not prone to stick to each other, however its drop voltage is higher.

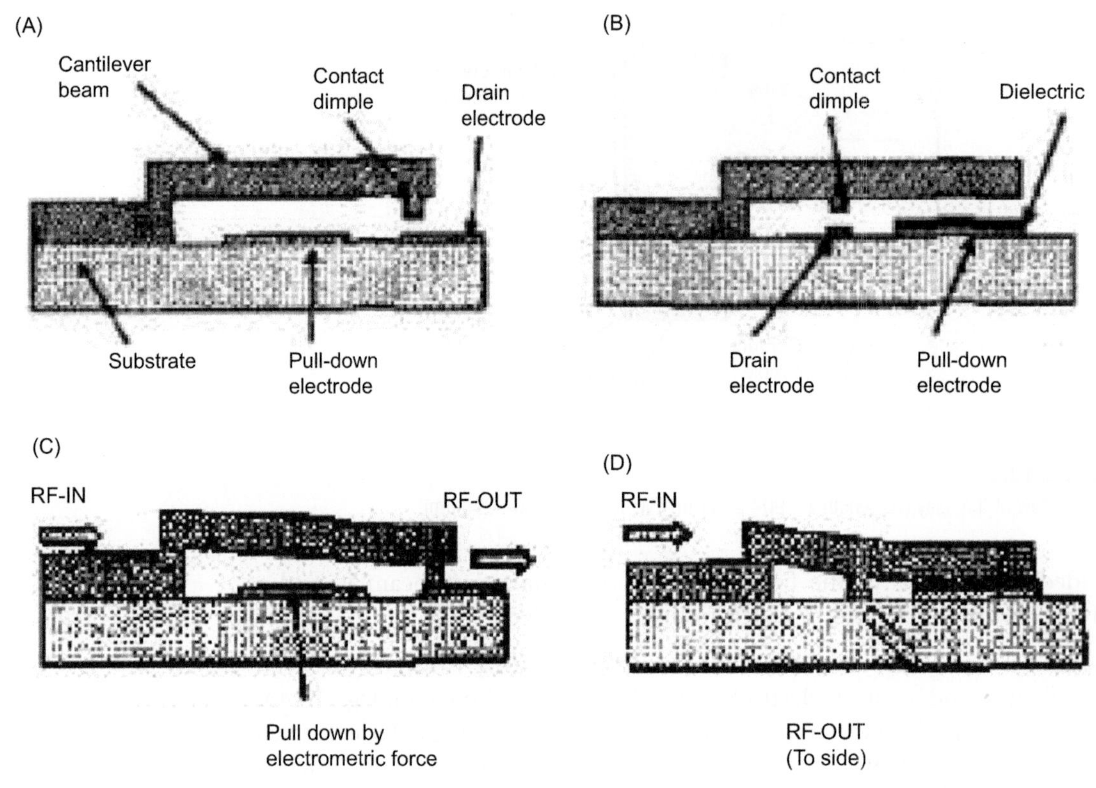

FIGURE 10.4
Schematics of two types of relays.

Torsion beam microrelay can bear less load, it is usually used in optical switching devices, such as optical modulators and three-dimensional scanning devices, etc. Torsion beam microrelay can realize a bistable state, but its fabricating process and service condition are too complex to be used.

A copper cantilever beam microrelay is shown in Fig. 10.6, its upper electrode is an entire cantilever, just as common is a single-fixed cantilever beam relay. However, this kind of microrelay has some drawbacks: (1) electric arc will make the contacts welded, affecting the usage time of the relay; (2) copper oxide will affect the switching period of the relay; and (3) high-contact resistance will reduce the performance of the relay. Low-contact resistance can ease the contact welding problem and reduce the consumption. Also, contact materials, contact force, and contact area will affect the contact resistance. In order to reduce the contact resistance, Han-Sheng Lee, Chi H. Leung, and others have designed a multibeam relay [8], which, as shown in Fig. 10.7, has four beams.

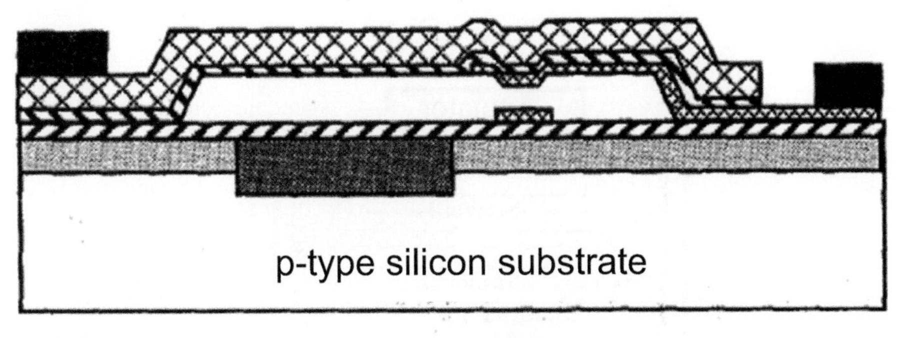

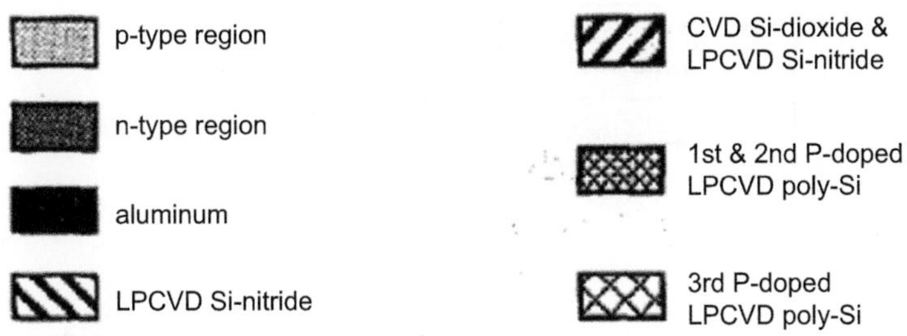

FIGURE 10.5

Double-ended fixed electrostatic MEMS relay with polycrystalline silicon structure.

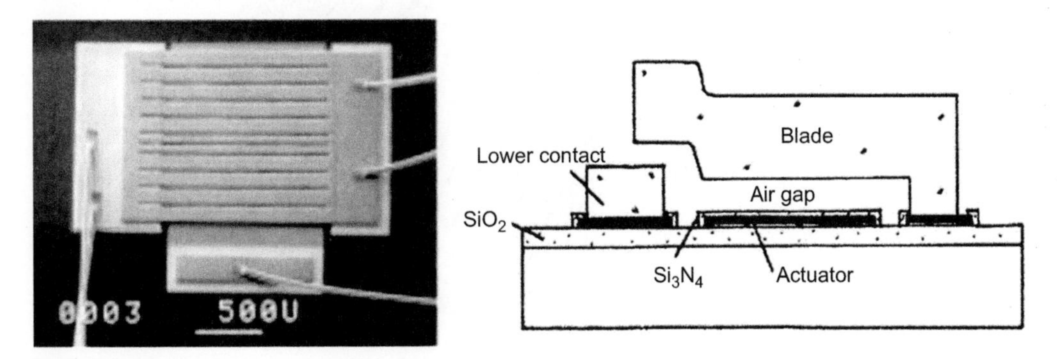

FIGURE 10.6

Copper cantilever beam microrelays: (A) physical photo, (B) relay sectional view.

The beam of the microrelay above is not connected rigidly, each of them has relative independence. Compared to the single cantilever beam, it has greater flexibility. The contact of this relay is 30 mΩ, which is 40 mΩ smaller than the single cantilever beam relay introduced above. It will appear contact welded phenomenon after 300 switching cycles under 1 A load.

Comb relay can increase the relative area of the parallel plate, and will create more electrostatic force under the same driving voltage. However, its

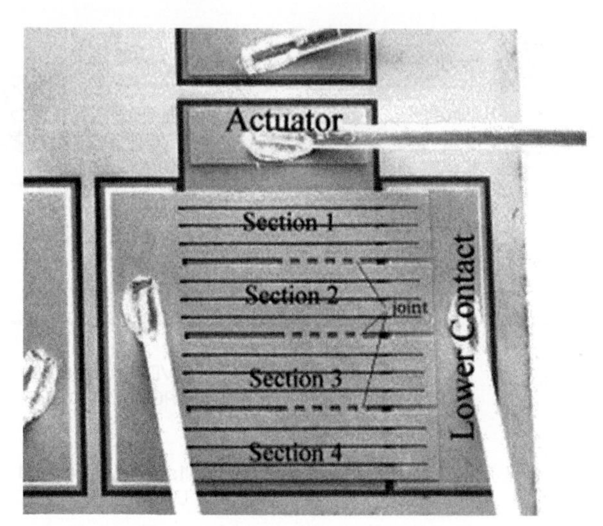

FIGURE 10.7
Microrelay of four beams.

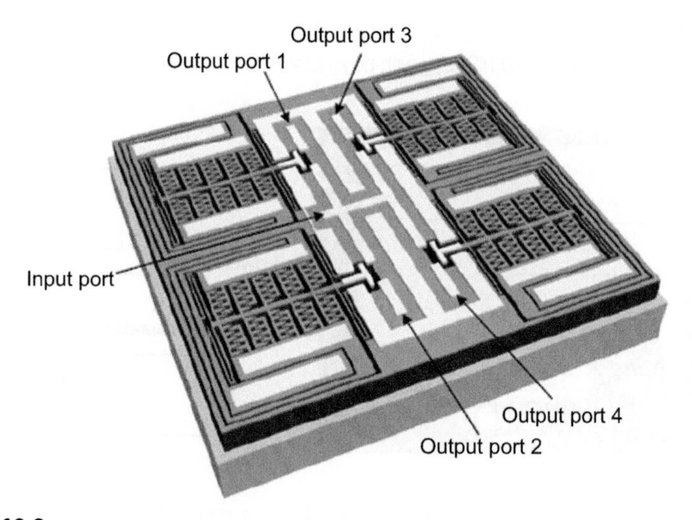

FIGURE 10.8
Single-pole, four-throw comb electrostatic relay.

structure and fabrication process are more complex than for the cantilever beam relay. The common drive failure of the comb relay is collisions between electrodes.

A single-pole, four-throw comb electrostatic relay is illustrated in Fig. 10.8 [9], its volume is $2.55\ \text{mm} \times 2.39\ \text{mm} \times 0.56\ \text{mm}$, and the driving voltage is

FIGURE 10.9
An electromagnetic microrelay with driving voltage and signal transmission depart.

only 15 V. It can be recycled as many as 1 million times under 100 mW load, and its insert loss is less than 0.31 dB at 60 GHz.

To summarize, electrostatic relays have two main types of structure: one has current flow through the beam, its control and load voltage are a unit, just like the microrelay proposed by Han-Sheng Lee (Fig. 10.6). The other cantilever beam is divided into a contact area and an upper plate, the contact area conducts the input and output parts to the relay; the driving voltage is only put on the plate area to control the relay switch, as shown in Fig. 10.9.

10.1.3.2 Electromagnetic MEMS Relay

The driving force of the electrostatic actuator is small, it is necessary to provide higher driving voltage to get higher driving force. The electromagnetic actuator has a high driving force under a small driving voltage (5 V) and is easy to integrate with the IC circuit. However, its power consumption is high and the fabrication process is complex, furthermore it cannot be used in magnetic-sensitive circuits.

In order to reduce power consumption, M. Ruan and J. Shed designed a bistable electromagnetic microrelay in 2001, as shown in Fig. 10.10 [10]. Power is only needed at the state switching stage, the stable state will not consume any power. This relay has been used in handheld devices.

In 2007, Shi Fu, Guifu Ding, and others proposed a new bistable electromagnetic microrelay in SJTU [11], as shown in Fig. 10.11. Compared to a traditional cantilever beam relay, this kind of relay does not need to consider the damping generated by air when the beam is close to or leaves from the basement, thus its response time is low. However, the fabrication process is complicated and the rotation will cause friction and wear to the beam. Experiments reveal that this relay has a good current-carrying capacity up to 2 A.

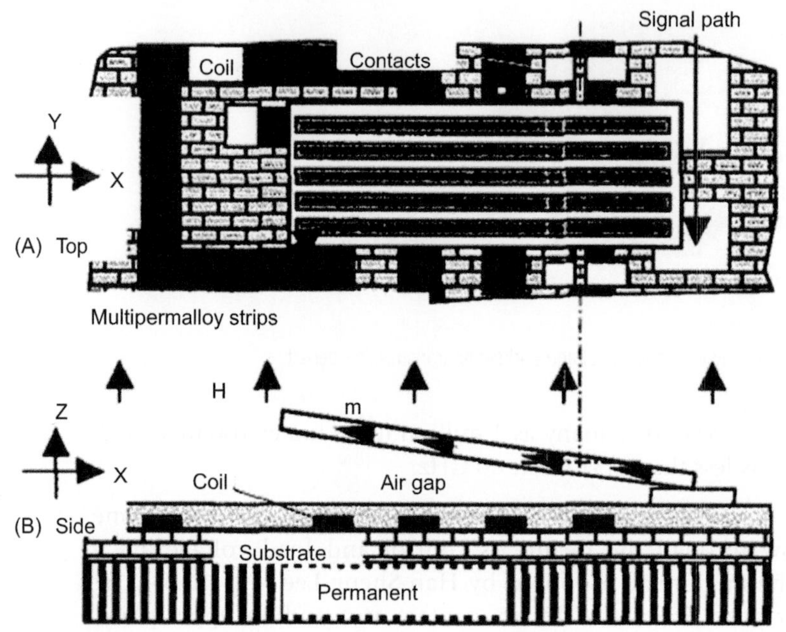

FIGURE 10.10
Bistable electromagnetic microrelay.

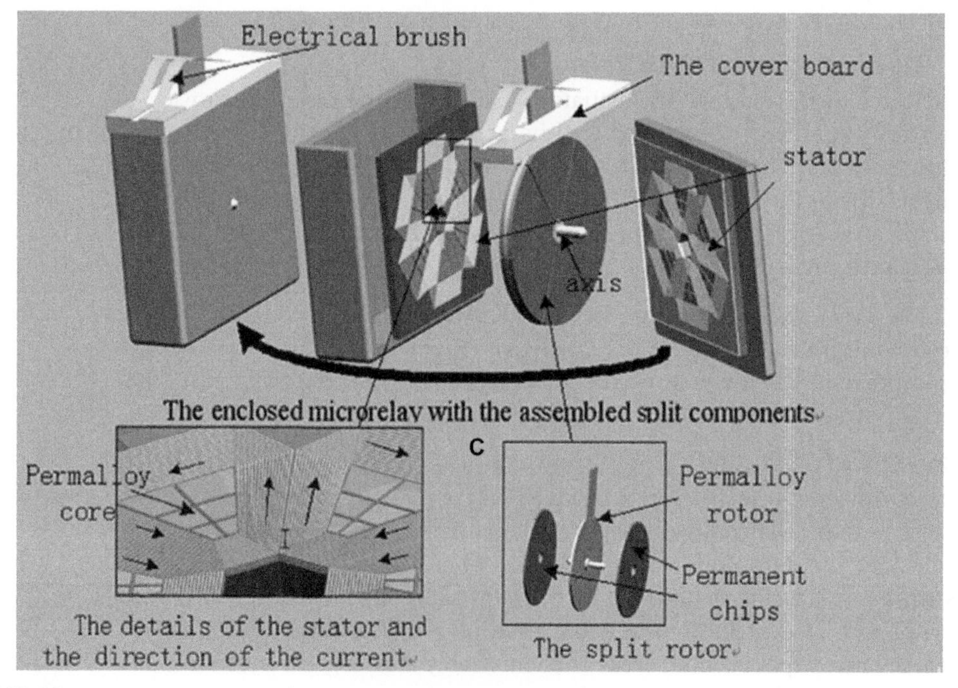

FIGURE 10.11
Structure diagram.

10.1.3.2.1 Thermodynamic Driving MEMS Relay

When the driving material is electrified, the expansion after heating will cause a displacement. The thermodynamic driving mode needs less voltage, but it costs high power consumption and its driving frequency is low.

Jin Qiu and Jeffrey H. Lang designed a bulk-micromechanical bistable thermodynamic driving relay in 2005 [12], as shown in Fig. 10.12. This relay only needs power in the state-switching stage, and its carrying current can be up to 3 A, but its switching frequency is low, with a maximum of 5 Hz.

10.1.3.2.2 Liquid MEMS Relay

The operating principle of liquid MEMS relay is to heat the gas, pushing the conductive liquid droplet to move a displacement to switch the relay.

A typical liquid MEMS relay is shown in Fig. 10.13 [13]. The moving of mercury droplets controls the circuit switching, the friction force between two signal electrodes and contact loss are low because of the use of this liquid metal. The critical problems of this kind of relay are bubbles, mercury droplet control, and sealing.

10.1.3.2.3 Hybrid Driving Mode MEMS Relay

Each single driving mode relay has its advantages and disadvantages, and hybrid driving mode relays can adopt the advantages of multiple relays and overcome each other's shortcomings.

A gold electroplated electromagnetic and electrostatic hybrid MEMS relay is shown in Fig. 10.14, composited by a movable beam, contact, and a

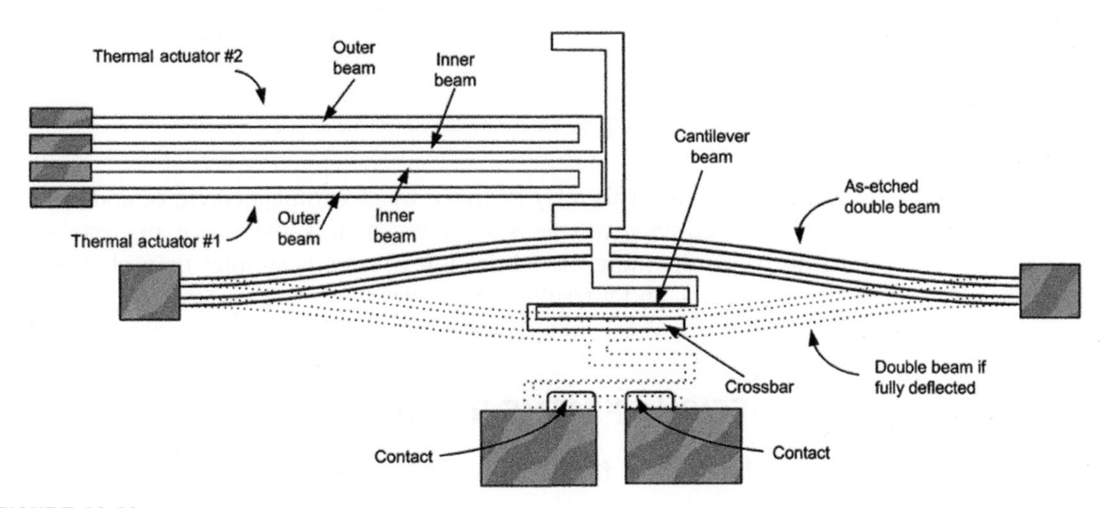

FIGURE 10.12
Bulk-micromechanical bistable thermodynamic driving relay.

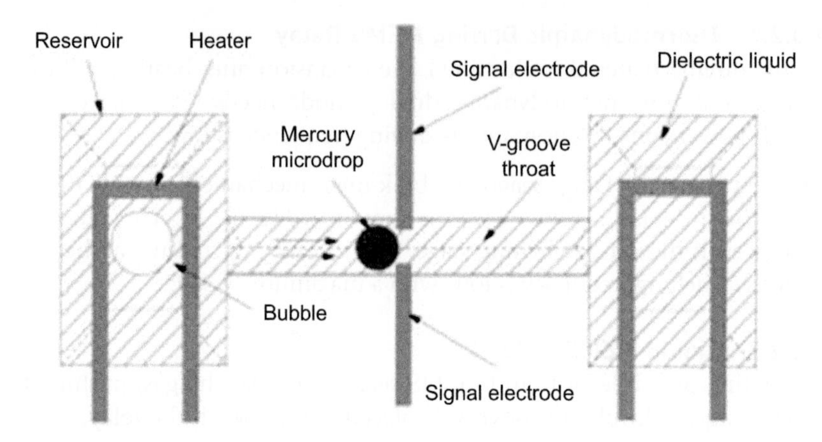

FIGURE 10.13
Liquid MEMS relay.

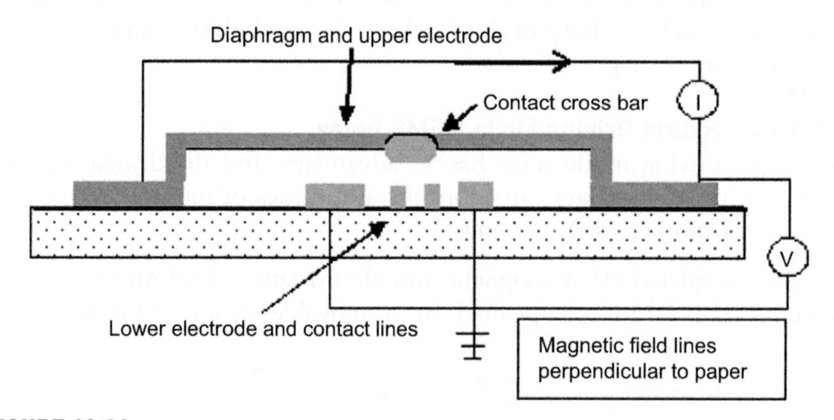

FIGURE 10.14
A gold-electroplated electromagnetic and electrostatic hybrid MEMS relay.

permanent magnet. This relay is driven by electromagnetic force and electrostatic force, its driving voltage is 6–12 V, which is lower than an electrostatic microrelay. Furthermore, its fabrication process is simpler than electromagnetic microrelays, and it results in no coils.

10.2 DESIGN OF MEMS RELAY

10.2.1 Materials for MEMS Electromagnetic Relays

The main materials of bistable MEMS relays are coils, soft magnets, and permanent magnet thin-film materials. To improve the relay's switch response speed and stability, soft and permanent magnets should have small coercivity, small remanence, and a linear demagnetization curve. A soft magnet with high magnetic permeability and a permanent magnet with high magnetic energy product should be chosen in order to miniaturize the device. The complex structure and difficulty in fabricating are the main problems of electromagnetic MEMS relays, because of the incompatibility of the fabrication process for soft and permanent magnets. Hence, the main content of this section is the fabrication process of soft and permanent magnets.

10.2.1.1 Soft Magnetic Thin-Film Materials

The fabrication process of soft magnetic thin-film materials includes the selection of materials and process, manufacturing of thin film, and testing.

Typical magnetic hysteresis and normal magnetization curves are shown in Fig. 10.15, H_c, M_r, H_s, and M_s represent the coercivity, remanence, saturation magnetization, and saturation magnetization, respectively.

The magnetic hysteresis curve of a soft magnet is narrow, and its coercivity, remanence, and hysteresis loss are low. The curve connected by different magnetic hysteresis vertices, whose starting value is less than H_s, is called a normal magnetization curve. The slope of the curve at any point is the permeability of materials at the corresponding magnetic density. This curve will represent the material's magnetic properties, and the common magnetic parameters, such as μ,B_s, are obtained from it [14].

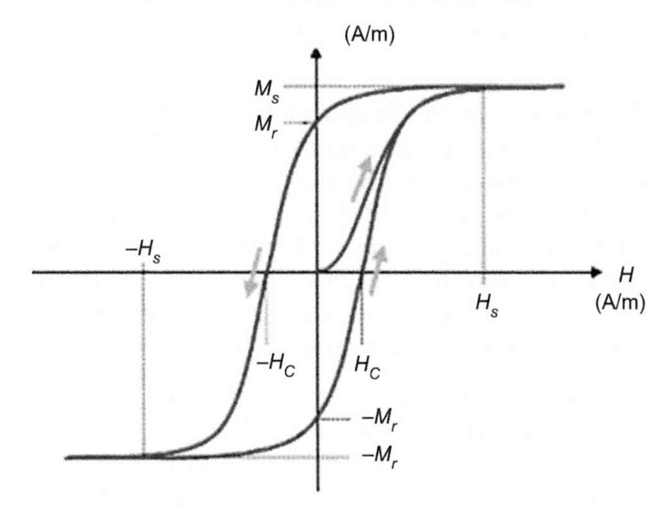

FIGURE 10.15
Magnetic hysteresis and normal magnetization curve.

Soft magnetic materials can be divided into three categories: Fe alloy, Ni alloy, and Co alloy. The development of soft magnetic materials has gone through three main stages: traditional soft magnetic materials made by melting technology (such as permalloy) and ferrite (such as MnZn, NiZn series, etc.), amorphous, and nanocrystalline soft magnetic alloy (such as Finement). The fabrication process of amorphous and nanocrystalline soft magnetic alloy film materials adopts liquid quenching technology, dozens of micron-thick films built by molten alloy rotating on a roll surface, whose thickness is hard to control and the surface of which is rough [15]. Furthermore, the processes of amorphous and nanocrystalline soft magnetic alloy and MEMS production are incompatible, traditional materials still need to be selected.

It is very common to obtain the soft magnetic functional structure by microelectroforming in the MEMS process. Foreign researcher William PT and others have studied the manufacture and testing of Ni–Fe alloy, Ni–Co alloy, and Ni–Co–Mo alloy through the microelectroforming process [16]. Through the comparison of magnetic parameter measurement, the Ni–Fe alloy film has the highest relative magnetic permeability (500–1000) and smallest coercivity (1–5 Oe, about 80–400 A/m). Although the relative magnetic permeability of Ni–Fe alloy film is much less than that of its bulk materials, the Ni–Fe alloy is also the most suitable material applied to MEMS electromagnetic switching devices, and its consumption is too low to significantly affect the cost.

The Ni–Fe alloy film adopted microelectroforming technology has uniform thickness, less membrane stress, and better magnetic properties by repeatedly adjusting the parameter settings.

10.2.1.1.1 Manufacture of Ni–Fe Alloy Film Sample

The principle of alloy microelectroforming is shown in Fig. 10.16A. Alloy microelectroforming is using electrochemical methods to deposit two or more metals on the cathode (plating), which is an oxidation–reduction reaction. The atomic structures of Ni and Fe are similar and they are iron subgroup elements. Their standard potentials are 0.25 V and −0.44 V, respectively, judging from the electrochemical potential, they are able to coprecipitate in single-salt solutions. Fig. 10.16B illustrates a microelectroforming device.

The solution of Ni–Fe alloy microelectroforming consists of Ni^{2+}, Fe^{2+}, SO_4^{2-}, and H^+, and the recipe is shown in Table 10.3. The basic operating conditions for getting a smooth and bright Ni–Fe alloy repeatedly are as follows: solution pH value between 3.4–3.6, temperature 63°C, direct current, density of cathode current between 3–4 A/dm^2, and medium magnetic stirring speed (30–40 r/min).

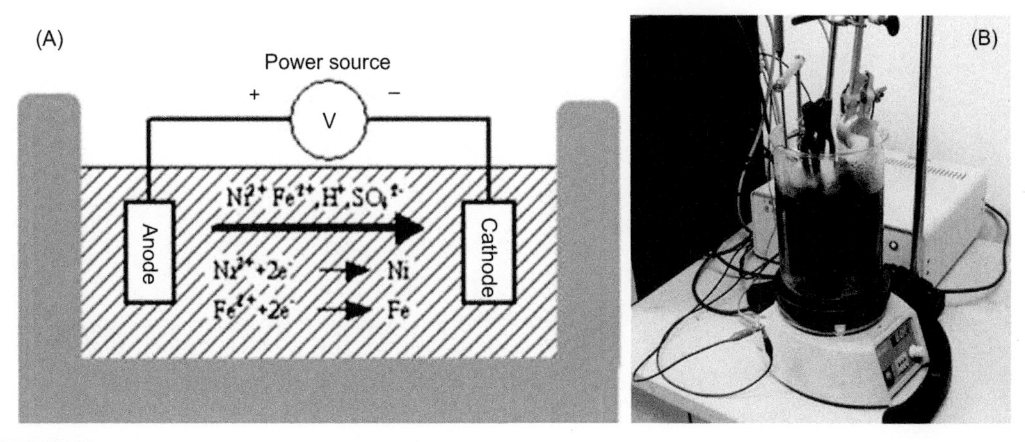

FIGURE 10.16
(A) Ni–Fe alloy microelectroforming principle and (B) device.

Table 10.3 Recipe for the Ni–Fe Alloy Solution

Composition	Content	Composition	Content
$NiSO_4 \cdot 6H_2O$	200 g/L	$NaC_{12}H_2SO_4$	0.1 g/L
$FeSO_4 \cdot 7H_2O$	15 g/L	$H_3C_6H_5O_7$	10 g/L
NaCl	25 g/L	H_3BO_3	40 g/L
$C_4H_4O_3NSNa \cdot 2H_2O$	4 g/L	$NaC_6H_5O_7 \cdot 2H_2O$	20 g/L

10.2.1.1.2 Condition Parameter Control and Operation Essentials of Microelectroforming

The pH value of the electrolytic solution has a great influence on the quality of the deposition layer. A high pH value will accelerate the formation of Fe^{3+} and precipitate the $Ni(OH)_2$, however, a low-pH solution will accelerate the chemical dissolution of the iron anode and produce a lot of bubbles on the cathode, which would cause the hydrogen evolution reaction to reduce the current efficiency. The solution pH value should be set at about 3.5 to obtain a bright Ni–Fe alloy. During the process of microelectroforming, the pH value will increase continuously. The pH value should be tested by pH test paper or a pH meter at intervals and dilute sulfuric acid or dilute hydrochloric acid added according to the obtained value.

The low temperature of the solution will reduce the brightness, smoothness, and deposition rate of the deposition layer. High temperatures will decompose the stabilizer solution and oxidate the Fe^{2+} to Fe^{3+}, resulting in an increased brittleness of deposition layer and reduced mechanical performance. For a bright Ni–Fe alloy the microelectroforming temperature should be set at 63°C.

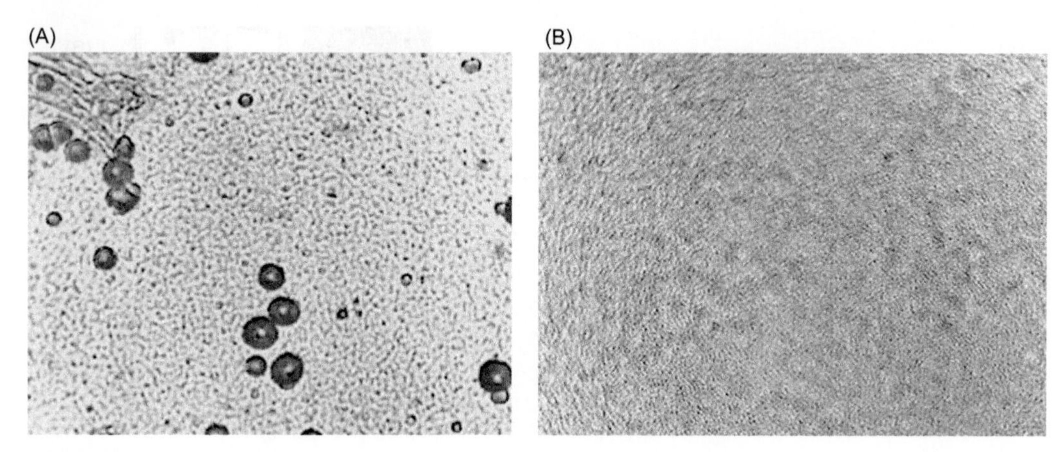

FIGURE 10.17
The effect of current density on the Ni–Fe alloy surface quality: (A) current density of 20 A/dm^2; (B) current density of 1 A/dm^2.

High current density will increase pinholes and pits, and enlarge particles in the deposition layer, reducing the current density could solve these problems. As shown in Fig. 10.17A and B, compared to a current density of 20 A/dm^2, which causes a lot of pinholes and pits on the surface, the deposition layer with 1 A/dm^2 current density has virtually no pinholes or pits, and its particles are smaller. In order to take the quality of the deposition layer and efficiency into consideration, the current density should be set at about 5 A/dm^2.

Reasonable stirring can accelerate the speed of microelectroforming solution getting into tiny structures and distribute the pH value uniformly to decrease the possibility of a hydrogen evolution reaction. When there are some impurities, such as iron and copper, in the solution, accelerating the speed of stirring will oxidate the iron and copper ions, causing the impurities to precipitate. Therefore, filtering the solution after electroforming each time is as important as controlling the stirring speed.

10.2.1.1.3 Stress of the Microelectroforming Film and Hydrogen Evolution Reaction

In the process of Ni–Fe alloy microelectroforming, two key points should be focused: stress of the microelectroforming film and hydrogen evolution reaction.

1. Film stress
   During the microelectroforming process, stress will be introduced in the film; the thicker the film is, the greater the stress is, and when it reaches a certain point, the film may drop from the substrate. Stress

will limit the application of soft magnetic film in MEMS. After microelectroforming, the level of stress can be determined by observing the surface of the photoresist around the metal. If there are obvious cracks on the surface, the stress of the deposition layer is high, and some brightener needs to be added to regulate this stress.

During the electrodeposition process, adding brightener can adjust the deposition layer stress and improve the surface smoothness, using saccharin or crystallox, 1,4-butynediol or its derivatives such as "791" brightener, etc. For MEMS processing, adding brightener should be guided by the following principles: (1) more feeding times and a small feeding amount; (2) minimize the brightener types. If two or more brighteners are needed, the consumption rate of each brightener should be examined, otherwise, the imbalanced composition of the brightener will increase the internal stress. Generally, the collocation of primary and secondary brighteners is appropriate, and they should be added on time. The film stress cannot be regulated by adding brightener, otherwise the effect is reversible.

For electrodeposition, hydrogen theory is widely recognized, that is, during deposition, hydrogen ions get into the crystal lattice of the deposition layer and expand the crystal lattice; after electroforming, the hydrogen ions leave the lattice and shrink the crystal lattice, resulting in stress. In addition, the pH value, current density, temperature, and impurities also have an effect on the internal stress.

2. Hydrogen evolution reaction
The standard electrode potential of nickel is lower than hydrogen, H^+ will get the electron firstly to generate the gas. Therefore, the hydrogen evolution reaction is inevitable in the process of electrodeposition of Ni or Ni–Fe alloy. There are two consequences of the hydrogen evolution reaction, one is increasing number of pinholes and pits on the surface, and the other is hydrogen embrittlement, which reduces toughness.

For the MEMS switch structure, the impact caused by the hydrogen evolution reaction is not limited to these, one is to reduce the yield, and the second is to reduce the reliability of the device. MEMS structures are generally in the micrometer size, a lot of bubbles will attach to the cathode surface during a serious hydrogen evolution reaction, the diameter of the bubbles and the microelectroforming pattern size are generally comparable. Therefore, a lot of units is invalided and by attached bubbles, and the actual electroforming area is reduced, so the unit thickness without bubbles attached is not controllable, which decrease the actual yield greatly. For an active structure, cast layer toughness decreases directly, leading to reduced reliability and service life.

10.2.1.2 Permanent Magnetic Film Material

A hysteresis curve of a permanent magnetic material is wide, and its coercive force and remanence induction are high. Ferrite permanent magnetic materials, rare earth permanent magnet materials, transition metal-based permanent magnet materials, and other types of permanent magnetic materials are widely applied in MEMS electromagnetic devices [17].

The curve segment located in the second quadrant of the hysteresis curve is called the demagnetization curve; mostly permanent magnetic materials work on this curve. Demagnetization curves and recoil lines are the basic characteristics of permanent magnetic materials curves; as shown in Fig. 10.18A, the permanent magnetic material is not working along the demagnetization curve $B_r mm' H_c$ generally, but is working along the recoil line mr, the operating point in the recoil line. The slope of the recoil line is called the recoil permeability, and is generally expressed by μ_{rec}; apparently μ_{rec} characterizes magnetic stability under dynamic operating conditions. It is generally desirable to coincide the recoil line with the demagnetization curve of permanent magnets in a linear mode, and stable device performance parameters can be measured. As shown in Fig. 10.18B, the operating point D (B_d, H_d) also found periodic changes back and forth, definite the trajectory of point D dynamic recoil line of magnet, the smaller μ_{rec} is, the more stable the magnet performs under dynamic operating conditions.

Millimeter-scale permanent magnetic film may be obtained by processing the permanent magnetic block appropriately, however, film with thickness under

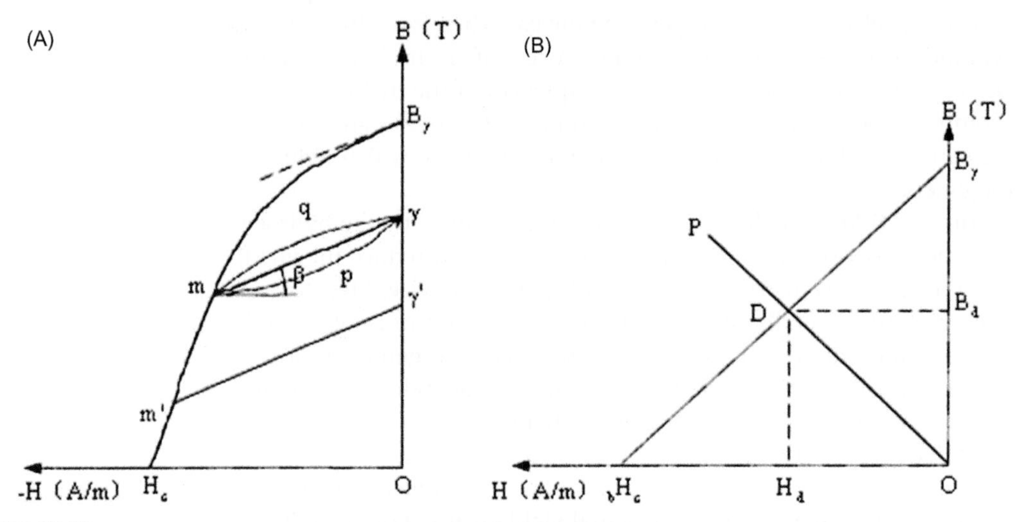

FIGURE 10.18

The basic characteristic curve of permanent magnetic material. (A) Demagnetization curve of permanent magnetic materials and recoil line and (B) their coincidence.

100 microns cannot be processed by a permanent magnetic block. Because conventional high-performance permanent magnetic block materials are relatively brittle, the split size limit is generally about 100 μm. Permanent magnetic film of micrometer thickness must be directly deposited to form a thin film on the components requiring a magnetic field, in order to adapt to a variety of microelectronic and micromechanical systems with high performance and high reliability requirements. Currently permanent magnetic materials deposited mainly in the form of a thin film are transition metal-based permanent magnetic materials such as AlNiCo, FeCrCo, MnAl, PtCo, and PtFe, which are mainly produced by sputtering or microelectroforming. An advantage is oxidation of corrosion inhibitors is compatible with MEMS process, but disadvantages are the low coercivity and remanence flux density, the energy product is small, and demagnetization curves generally do not coincide with the recoil line. For example, domestic scholars Hongchuan Jiang and others have carried out research on CoNiMnP microelectroforming conducted at room temperature, but the coercivity of the vertical direction on the film is only 59.7 kA/m, remanence flux density 0.57 T, and energy product 11.3 kJ/m^3.

10.2.2 Structural Design

It is required to consider a variety of factors on structural design. Two factors should be emphasized especially: one is the design itself is reasonable, another is the process is able to meet the design requirements.

10.2.2.1 Electrostatic Drive Cantilever

A cantilever contact-type switch is shown in Fig. 10.19, one end of the switch is fixed and the other is free-floating. A capacitor is formed between the cantilever and the lower electrode, and a drop-down electrostatic force will be generated on the cantilever when the bias voltage is applied between the cantilever and the lower electrode. Setting the width of the cantilever (also the

FIGURE 10.19

Driving voltage and signal transmission separate microelectromagnetic relay schematic.

width of upper plate) as w, and the lower electrode width W, the value of the parallel plate capacitor can be expressed as

$$C = \frac{\varepsilon_0 wW}{g + \frac{t_d}{\varepsilon_r}} \tag{10.1}$$

where t_d and ε_r are the insulating layer thickness and relative dielectric constant, respectively, and g is the spacing between the cantilever and the electrode.

Therefore the electrostatic force between the parallel capacitor is

$$F = \frac{1}{2}V^2\frac{dC(g)}{dg} = -\frac{\varepsilon_0 \varepsilon wWV^2}{2\left(g + \frac{t_d}{\varepsilon_r}\right)^2} \tag{10.2}$$

where ε is a scale coefficient expressing the capacitance reduction caused by the unclosed attachment between the lower electrode and the underlying silicon dioxide insulation layer

$$\varepsilon = \begin{cases} 1 \ (g = 0) \\ 0.4 \sim 0.8 \ (g \neq 0) \end{cases} \tag{10.3}$$

where g represents the spacing between the lower electrode and the underlying silicon dioxide insulation layer after the bias voltage is pulled down the cantilever.

When the electrostatic force is greater than or equal to the restoring force of the cantilever, $F = -kx$,

$$F = -\frac{\varepsilon_0 \varepsilon wWV^2}{2\left(g + \frac{t_d}{\varepsilon_r}\right)^2} = -k(g_0 - g) \tag{10.4}$$

The driving voltage is

$$V = \sqrt{\frac{2k}{\varepsilon_0 wW}(g_0 - g)\left(g + \frac{t_d}{\varepsilon_r}\right)^2} \tag{10.5}$$

where g_0 is the spacing between the cantilever and the lower electrode with no bias voltage, for the cantilever, its elastic coefficient k can be expressed as

$$k = \frac{2}{3}Ew\left(\frac{t}{l}\right)^2 \tag{10.6}$$

where E is the elastic modulus of the cantilever material, t is the thickness of the cantilever, and l is the length of the cantilever.

When the cantilever is pulled to

$$g = \frac{2}{3}g_0 \tag{10.7}$$

At this point, the closing voltage is

$$V_{th} = \sqrt{\frac{2k}{3\varepsilon_0 wW} g_0 \left(\frac{2}{3} g_0 + \frac{t_d}{\varepsilon_r}\right)^2} \tag{10.8}$$

Thereafter the cantilever rapidly falls, until contact is made with the lower electrode, and the signal between the contact metal and the signal line is conducted.

10.2.2.2 Electromagnetic Drive Structure

For reasonable electromagnetic drive structure design, the following aspects should be followed:

1. An Ni–Fe alloy film should work in an external magnetic field with high permeability. As shown in Fig. 10.20, Ni–Fe alloy film magnetic characteristic parameters of the test results show that the relative permeability of the nickel–iron alloy thin film changes when the

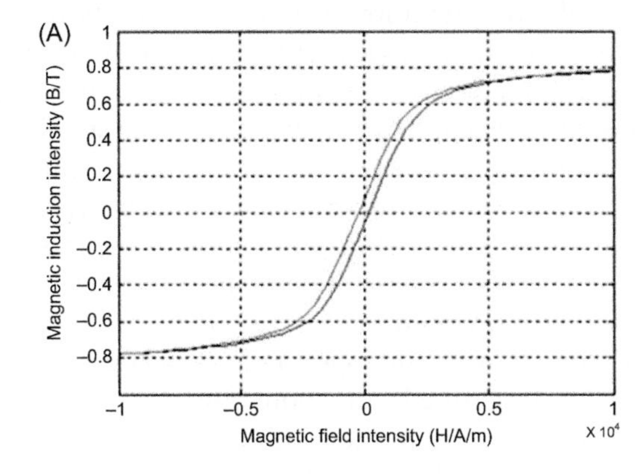

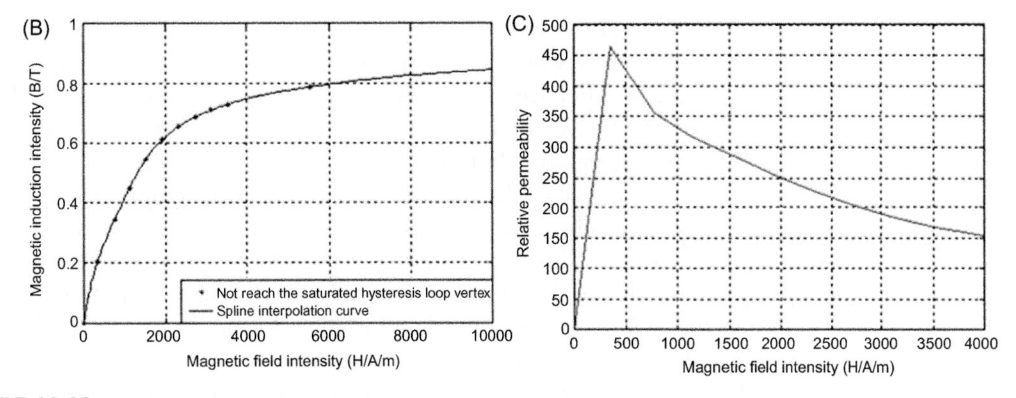

FIGURE 10.20
Characteristic curve of an Ni–Fe alloy thin-film sample. (A) Hysteresis curve; (B) basic magnetization curve; (C) changes in the relative permeability with the external magnetic induction.

external magnetic field changes, after increasing rapidly, it will decrease until near vacuum permeability. In order to make full use of Ni–Fe alloy film and reduce the size of the movable electrode, the film should work in the range of high permeability. According to the law, shown in Fig.10.20, about relative permeability changes with the variation of the external magnetic field, selected 500 ~ 1500 A/m as the working electrode of the reference magnetic field, in this magnetic field range, the relative permeability of nickel–iron film changes between 420 and 280, nickel–iron alloy film has a high utilization rate.

2. Appropriate coil center line spacing d_{cc}. On the limited area extent, the greater the d_{cc} is, the fewer the number of turns the coil should have, and the greater the fluctuation value of the magnetic field strength the coil generates; the smaller the d_{cc} is, the greater the number of turns the coil should have to cover the electrode, which will increase the total length of wire and power consumption. Therefore, to insure the process runs smoothly, the distribution of coils should be limited to a rational range, with lower power consumption.
3. Appropriate twist angle ∂ of the twist beam. Generally speaking, when the angle between two planes is small, the two planes can be considered as approximately parallel. The smaller the ∂ is, the lower the utilization efficiency is, increasing the volume of the permanent magnet; however, a much greater ∂ will increasing the component along the direction of magnetization of the movable electrode, which can reduce the volume of permanent magnet, but the power consumption would increase.
4. Appropriate movable aspect ratio. The larger the aspect ratio of the movable electrode is, the more obvious the demagnetization of soft magnetic material in the width direction, and the more insignificant it is in the longitudinal direction. When the movable electrode contacts the fixed electrode, it will make contact with two independent fixed electrodes, the greater the contact area, the smaller the contact resistance is. Therefore, the width of the movable electrode should be designed to be as small as possible, but the contact area and magnetic moment it generates should also be considered.
5. Minimize the size of the permanent magnet. The permanent magnet size directly affects the degree of miniaturization of the relay, therefore, to maximize the utilization of permanent magnets, the most direct way is to reduce size of the movable electrode and the permanent magnet pole distance, which are mainly related to the package.
6. The planar coil design should not affect the size of the torsion structure. In the MEMS process, it is necessary to consider all the structures on the same substrate with the mutual influences between the front and rear steps, as there is a corresponding relationship between their position, which is embodied well in mask design.

Table 10.4 Characteristic Parameters of the Relay

Structure	Parameter	Structure	Parameter
Moving spacing g (μm)	4–16	Coil width w_c (μm)	15
Active electrode length l_m (μm)	700–800	Coil spacing d_{cc} (μm)	20
Active electrode width w_m (μm)	170–200	Coil thickness t_c (μm)	1–1.5
Torsion beam length l (μm)	180–200	Fixed contact electrode area (μm^2)	<100 × 100
Torsion beam width w (μm)	100–150	Magnet diameter Φ (μm)	<10,000
Torsion beam thickness t (μm)	2–8	Magnet thickness (μm)	<1500
Soft magnet thickness t_m (μm)	4–25	Spacing between soft and permanent magnet h_O (μm)	500–700

Based on the above six design principles and taking realizable microprocessing technology into consideration, a typical set of design parameters for the relay is shown in Table 10.4 [18], wherein the plane coil uses a regular octagonal structure.

10.2.3 Contact Design

10.2.3.1 Contact Material Selection

Select the electrical contact material is a key to relaying electrical contact properties, selecting contact materials with excellent properties is the most obvious way to improve the electrical performance.

Electrical contact materials have nearly 100 years of development history; the initial materials used were gold, silver, and platinum as pure material nodes; from the 1940s, Ag–Cu, Au–Ag, Pt–Ir, and Pt–Ag alloys were used; a diverse variety of precious metals and precious metal composites have been developed since the 1960s. The best contact material processing technology is not compatible with MEMS processing technology, e.g., silver and silver alloy contact electrodes commonly used in the production of sintered solution infiltration, extrusion, and high-temperature annealing processes (up to 1000°C or more), have a negative effect on MEMS devices, therefore, the range of MEMS relay materials is narrow.

For MEMS relay, as the voltage of the contact between the electrodes and the current through the contact electrode increases, the welding and material transfer phenomenon is more obvious, leading to a rapid decline in the electrical life of the relay. When selecting contact materials, the following basic requirements should be followed: good electrical and thermal conductivity and resistance to arc burning, antiwelding, wear resistance, low and stable contact resistance, no chemical reaction with media, a certain strength continuity, and ease of machining.

Throughout the development history of MEMS relay, precious metals and their alloys have been the preferred MEMS relay contact materials; precious metals are easily oxidized, with high electrical and thermal conductivity, and the quantities are small so that the cost is negligible. Gold is an ideal MEMS

Table 10.5 Contact Characteristics of Au, $AuNi_5$, and Rh

Characteristics	Au	$AuNi_5$	Rh	Unit
F_{min}	<0.1	0.3	0.6	mN
R (F_{min})	<30	<100	<1000	mΩ
F_{ahdh}	2.7	0.3	<0.1	mN

relay contact material as it has a high melting point, soft material quality, and does not easily generate an oxide film.

The German scholar Schimkat has conducted comparison research on the contact performance of three metals (Au, AuNi5, and Rh), which have increasing hardness sequentially [19]. The results showed that the minimum force to maintain stable contact and the contact resistance gradually increases, however the suction force decreases gradually, as shown in Table 10.5. Obviously, AuNi5 and Rh are more suitable contact materials than Au. Due to limited technology and other factors, the study process for MEMS relay contact materials has been relatively slow, so that the main relay material is still Au.

As contact materials used in conventional electromechanical relays, adding small amounts of other materials to generate alloy materials can significantly improve the contact characteristics of pure precious metals, such that their mechanical properties, arc-quenching ability, and hardness are improved. In the future, precious metal-based alloys will be an important development direction for MEMS relay contact materials, and the application of these materials will depend on the compatibility of MEMS technology.

10.2.3.2 Contact Electrode Design

The working conditions of the contact electrode directly affect the value of the contact resistance and effect the relay electrical life, therefore, the contact electrode design is particularly important. For ordinary electromechanical relays, the Holm model is commonly used to describe the relationship between the contact resistance and the contact force.

According to the Holm model, the value of the contact resistance is significantly influenced by contact area. Contact resistance is a function of the contact force, the contact area, and the contact surfaces of cleanness. In a stable region, the contact resistance increases when the contact force decreases in inverse proportion to the contact area, and the cleaner the contact surface, the smaller the contact resistance is. Therefore, as far as possible under the premise of increasing the contact area of the electrode, distribution should be rational.

Compared with ordinary electromechanical relays and diverse forms of contact, in the MEMS relay, contact structure, planar structures between the active contact electrode and the fixed contact electrode are used mostly; this process is also a planar process. As shown in Figs. 10.2 and 10.5, the process is not very complicated and is widely applied.

10.3 DYNAMICS MODELING AND SIMULATION ANALYSIS FOR MEMS RELAY

10.3.1 Electrostatically Actuated MEMS Relay

10.3.1.1 Driving Voltage of Electrostatic Relay

For the structure shown in Fig. 10.21, the beam width is w, which is also the width of electrostatic relay, and the lower electrode width is W, when the spacing between the upper and lower electrodes is h and the driving voltage is U, the electrostatic force is:

$$F_e = \frac{1}{2}U^2 \cdot \frac{dC(h)}{dh} = -\frac{1}{2} \cdot \frac{\varepsilon_0 A U^2}{h^2} \tag{10.9}$$

($A = w \times W$, is the relative area between the upper and lower plates.)

With the increase of bias voltage U, electrostatic force F_e between the upper electrode and the lower electrode also increases, meanwhile the spacing h decreases,

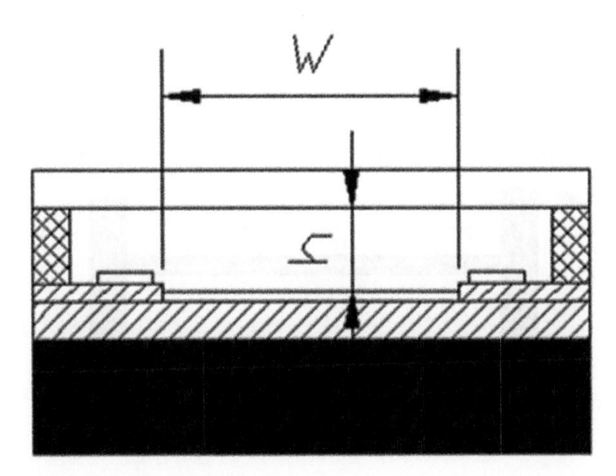

FIGURE 10.21
Diagram of double-ended fixed microrelay structure.

which makes F_e increase also. When the upper electrode is pulled down to $2/3h$, the increasing electrostatic force is much greater than the increasing elastic restoring force, the balance is broken and the upper electrode is pulled down.

The bias voltage that make the upper electrode be pulled down to $2/3h$ is called driving voltage and the expression is

$$V_{pull-down} = \sqrt{\frac{8k}{27\varepsilon_0 A} h_0^3} \tag{10.10}$$

where k is equivalent to the elastic coefficient of the beam, ε_0 is vacuum permittivity with a value of 8.854×10^{-12} F/m, and h_o is the initial value of h.

10.3.1.2 Elasticity of the Double-Ended Fixed Beam

The elasticity of the double-ended fixed beam consists of two elements, one is caused by the stiffness of the beam, and its material properties such as Young's modulus and moment of inertia about; the other is caused by the average residual stress of the beam, the expression of which can be expressed as

$$k = k_c + k_\sigma \tag{10.11}$$

where k_c is the coefficient of elasticity decided by the material and structure, and k_σ is the coefficient of elasticity caused by the residual stress.

10.3.1.2.1 Coefficient of Elasticity k_c

For vertical concentrated load double-ended fixed beams under the load of F shown in Fig. 10.22, the deflection of the beam can be expressed as

$$y = \frac{M_A x^2}{2EI} + \frac{R_A x^3}{6EI} (x \leq a) \tag{10.12}$$

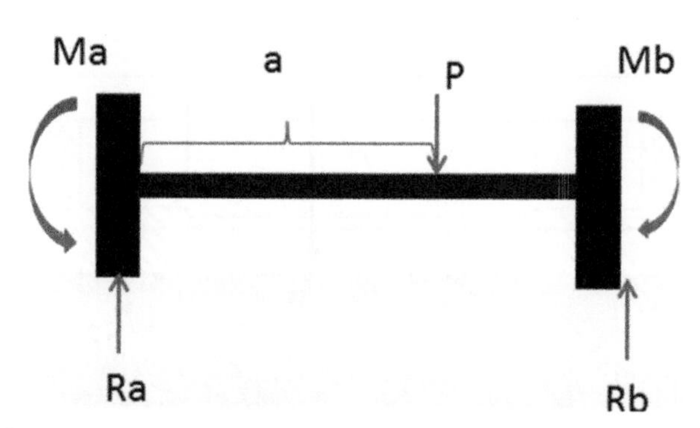

FIGURE 10.22
Concentrated-load double-ended fixed beams.

where

$$\begin{cases} R_A = \frac{F}{l^3}(l-a)^2(l+2a) \\ M_A = -\frac{Fa}{l^2}(l-a)^2 \\ I = \frac{wt^3}{12} \end{cases} \tag{10.13}$$

y is the deflection at x ($x \le a$), l is length of the beam, M_A and R_A are the torque and vertical force of the A end of beam, and the units are $N \cdot m$ and N, respectively. I is the moment of inertia, w and t are the width and height of the rectangle beam section, respectively, and E is the Young's modulus of the beam material.

Since the load of the beam is distributed, take $x = l/2$ into Eq. (10.12) to obtain the deflection of the center of the beam when the load is pulled at a point.

For the design of the relay its drop-down electrode is located directly below the beam, the deflection of the center can be calculated by integration through the principle of superposition, which can be expressed as

$$y = \frac{2}{EI}\int_{l/2}^{x} \frac{\zeta}{48}(l^3 - 6l^2a + 9la^2 - 4a^3)da \tag{10.14}$$

where ζ is the load per unit length, the total load is $F = \zeta W$; $x = (W + l)/2$, the equivalent modulus of elastic coefficient can be expressed (Fig. 10.23).

$$k_c = -\frac{F}{y} = -\frac{\zeta W}{y} = 32Ew\left(\frac{t}{l}\right)^3 \frac{1}{8(x/l)^3 - 20(x/l)^2 + 14(x/l) - 1} \tag{10.15}$$

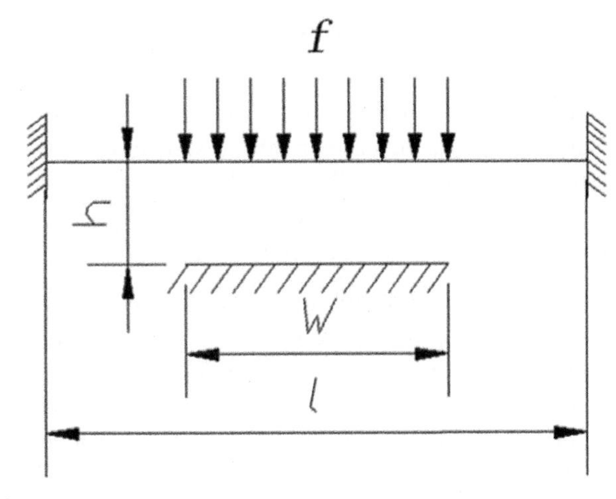

FIGURE 10.23
Diagram of double-ended fixed-beam load distribution.

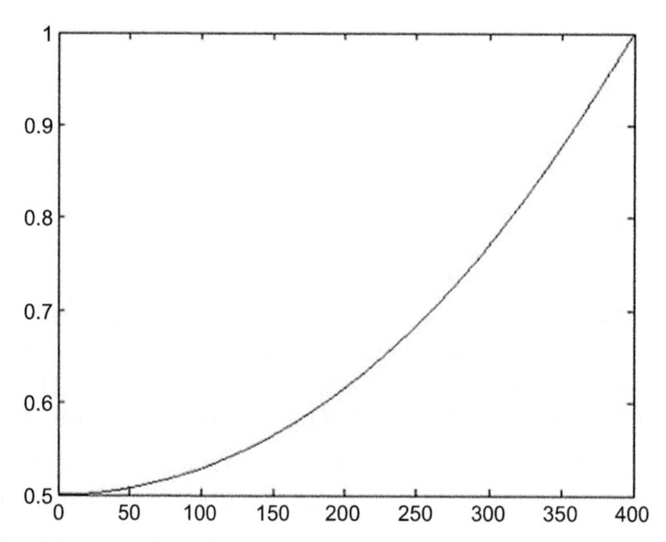

FIGURE 10.24
Relationship of k_c–W.

Take the $x=(W+l)/2$ into Eqs. (10.3)–(10.7) can obtain the equivalent elastic coefficient.

$$k_c = 32Ew\left(\frac{t}{l}\right)^3 \frac{1}{(W/l)^3 - 2(W/l)^2 + 2} \tag{10.16}$$

Using Eq. (10.16), the relationship between k_c and W is shown in Fig. 10.24.

As can be seen from Fig. 10.24, the increase in the concentration of the force, that is reducing W, will reduce the elastic coefficient of the beam, but for electrostatic relays, reducing W will reduce the electrostatic force.

When $l=400\ \mu m$, $w=50\ \mu m$, and $t=1\ \mu m$, the influence of driving voltage caused by h_0 and W is shown in Fig. 10.25. As can be seen from this figure, the driving voltage $V_{\text{pull-down}}$ increases as the distance h_0 between the upper and lower electrodes increases rapidly, and decreases with increasing width W. When $W>250\ \mu m$, the impact of increasing W to $V_{\text{pull-down}}$ is very small, therefore, the value of W is 250 μm and h_0 is 4 μm.

When $W=250\ \mu m$ and $h_0=4\ \mu m$, the influence of driving voltage caused by l and t is shown in Fig. 10.26. When the upper electrode length l increases and the upper electrode thickness t is reduced, the driving voltage $V_{\text{pull-down}}$ declines. However, the decreasing rate of driving voltage becomes slow as l increases gradually. When $l>500\ \mu m$, the impact on driving voltage is relatively small. When l is too large, it will affect the size limit of the relay, too small t will greatly improve the sensitivity of the upper electrode to stress gradients, and thereby a warping problem may happen. Taking the possibilities of the process and the settings of W and h_0 into consideration, $l=400\ \mu m$ and $t=1\ \mu m$.

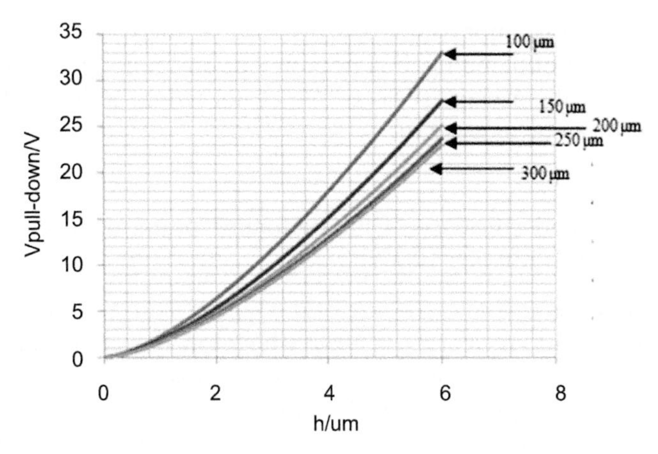

FIGURE 10.25
Relationship of $h/W - V_{pull\text{-}down}$.

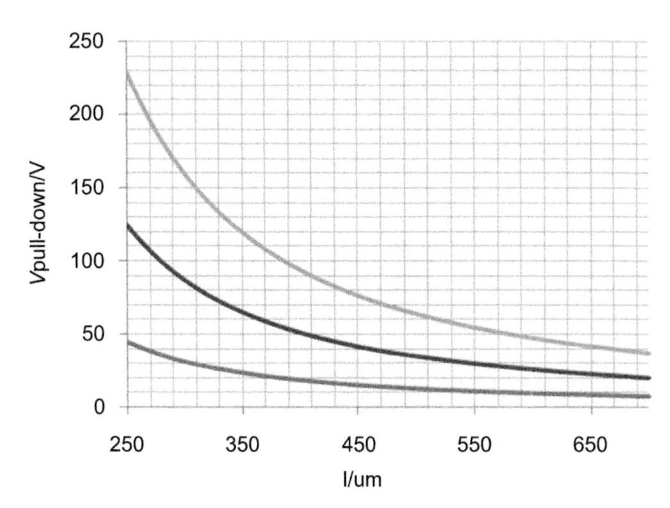

FIGURE 10.26
Relationship of $l/t - V_{pull\text{-}down}$.

10.3.1.2.2 Coefficient of Elasticity k_σ

The elastic coefficient k_σ of a double-ended fixed beam is mainly influenced by the average stress σ_0 within the beam, the relationship is shown in following expression:

$$k_\sigma = 8\sigma_0(1-v)w\left(\frac{t}{l}\right)\frac{1}{3-2(x/l)} \tag{10.17}$$

where v is Poisson's ratio for the material.

Generally, in order to improve the speed of the sacrificial layer releasing and reducing the effect of damping, it is necessary to drill holes in the beam. The holes can reduce the average stress of the beam, reducing the elastic coefficient k_σ, which can be reduced to

$$\sigma' = (1-\mu)\sigma_0 \tag{10.18}$$

where μ is the band efficiency of the perforated beam, defined as: $\mu = l/p$, with l and p as shown in Fig. 10.27.

In the design of this example, the size of the hole is 5 μm × 5 μm, $p = 20$ μm, $l = 10$ μm, and band efficiency $\mu = 50\%$, which can reduce the stress of the beam to half of the original stress. In addition, stress is also greatly different between the different processes; in the three processes usually used to produce metal thin films, the stress of evaporation coating is highest, followed by sputtering, and finally electroplating. The inertial stress within the metal in this design is about 50–60 Mpa.

10.3.1.3 Design Example

10.3.1.3.1 Parameter Selection

Parameters are determined using the analysis in Section 10.3.1.2, they are also shown in Table 10.6.

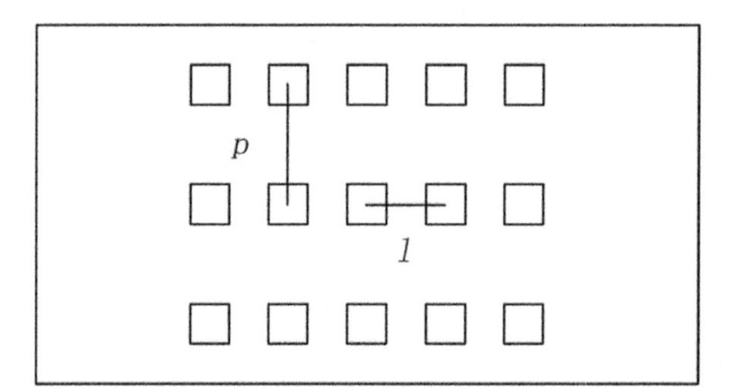

FIGURE 10.27
Schematic of *p/l*.

Table 10.6 Parameters of Double-Ended Fixed Electrostatic Relay

Parameter	Value
Length of upper electrode *l*/μm	400
Thickness of upper electrode *t*/μm	1
Length of lower electrode *W*/μm	250
Width of electrode *l*/μm	50
Spacing between upper and lower electrode h_0/μm	4

10.3.1.3.2 Parameter Testing

Taking the parameters in Eq. (10.16) to obtain

$$k_c = 32Ew\left(\frac{t}{l}\right)^3 \frac{1}{(W/l)^3 - 2(W/l)^2 + 2} = 0.9741\text{N/m} \tag{10.19}$$

Taking the parameters in Eq. (10.17), $\sigma_0 = \sigma_{0\max} = 2.2$ Gpa,

$$k_{\sigma\max} = 8\sigma_{0\ \max}(1-\nu)w\left(\frac{t}{l}\right)\frac{1}{3-2(x/l)} = 14.18\text{N/m} \tag{10.20}$$

Therefore,

$$k = k_{\sigma\max} + k_c = 15.16\text{N/m} \tag{10.21}$$

Taking k in Eq. (10.10)

$$V_{\text{pull-down}} = \sqrt{\frac{8k}{27\varepsilon_0 A} h_0^3} = 50.97\ \text{V} \tag{10.22}$$

10.3.1.3.3 Structural Modeling

Using Ansys to simulate the design structure, since it is complicated to simulate the upper plate because of holes, we can simulate the situation of a plate without holes. If the upper plate without holes can meet the drop-down conditions, so does the plate with holes. We simulate and analyze the displacement and models after modeling.

1. Displacement analysis

   The first coupling field model is shown in Fig. 10.28.

   The obtained displacement is shown in Fig. 10.29.

   As can be seen from Fig. 10.29, the center displacement of the plate is 2.88 μm, which is more than two-thirds of 4 μm, therefore it can be pulled down.

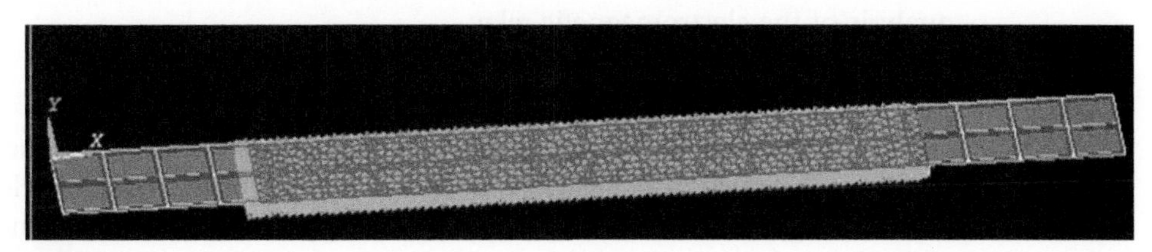

FIGURE 10.28
Coupling field.

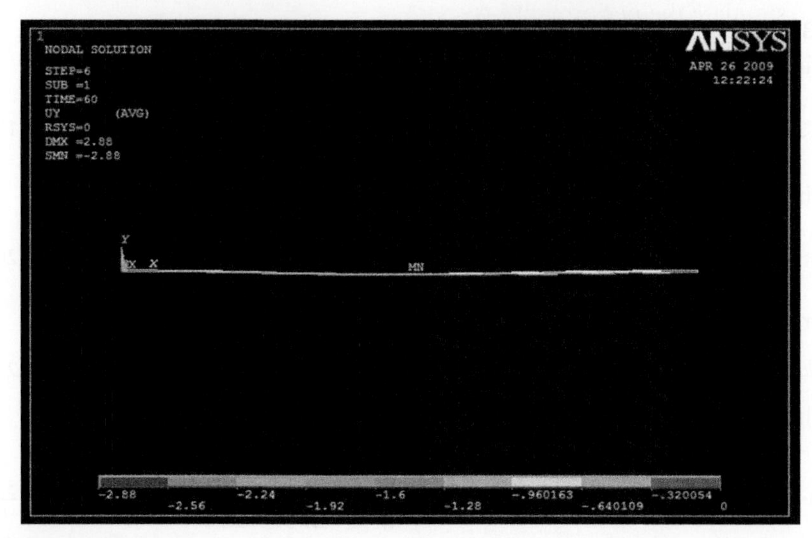

FIGURE 10.29
Displacement of upper plate.

2. Models analysis

   As expected, the relay can also be used in the case of high frequency, therefore study of the model analysis was carried out to observe the vibration mode of relay in high frequency. The results are shown in Fig. 10.30.

   As can be seen from this figure, the relay is unable to work on the second-order model or more, therefore the working frequency of the relay should be controlled at under 50 kHz to maintain normal use.

10.3.2 Electromagnetic Actuated MEMS Relay

In this section, we use the parameters of electromagnetic relay in Table 10.4 as the basis for modeling of dynamics analysis, air damping, magnetic field strength, and modality, which reflects the general method for the design and analysis of the electromagnetic relay.

10.3.2.1 Dynamic Modeling and Analysis

Fig. 10.31 is a geometric model of the movable electrode kinetics. The origin of the Cartesian coordinate system is at the centroid of the movable electrode and the angle between the x-axis and the longitudinal direction of the movable electrode is θ, the y-axis coincides with the axis of the twist beam. A movable electrode rotates clockwise or counterclockwise with angular velocity ω between the two stable states and the angle of rotation is α. The direction of the permanent magnetic field is parallel to the axis.

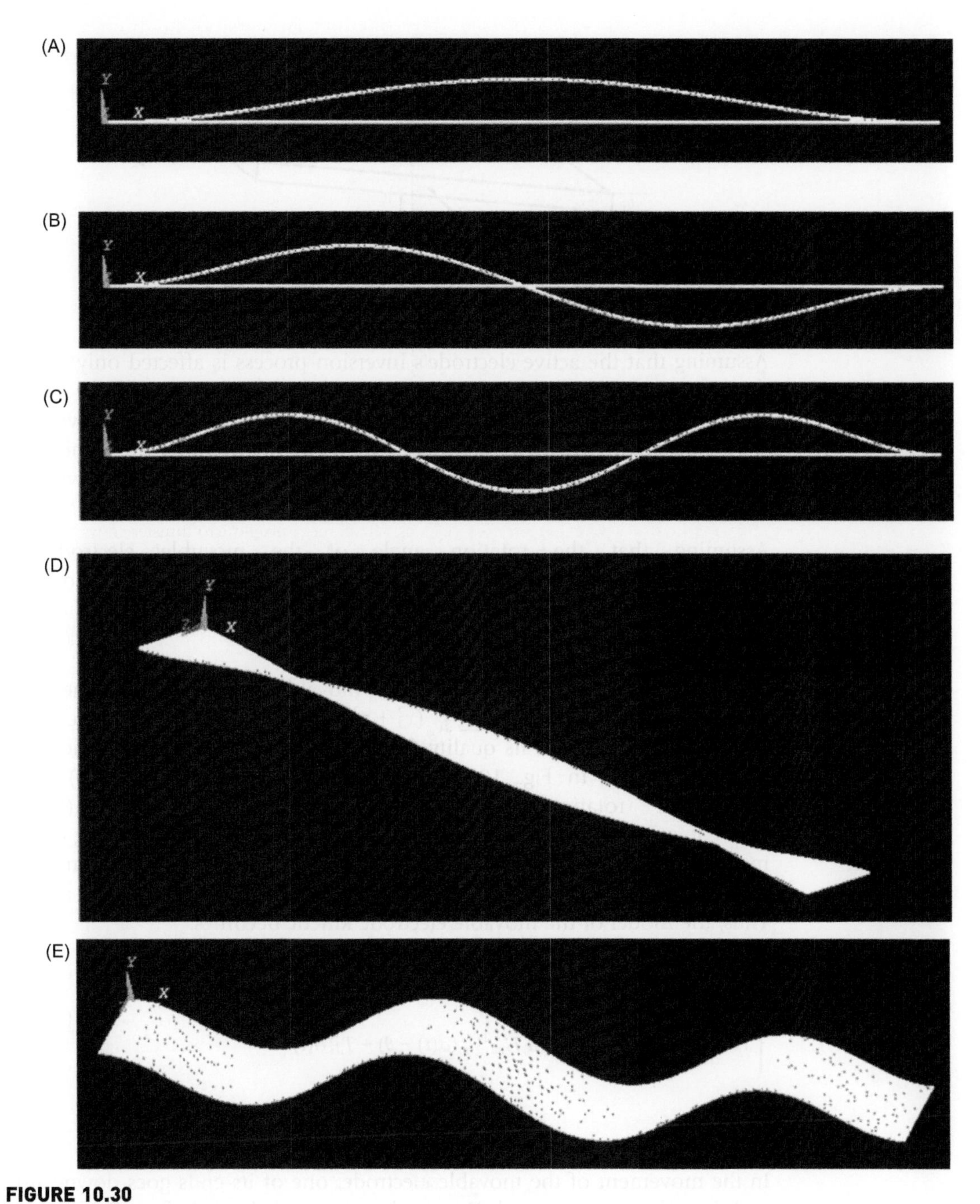

FIGURE 10.30

Schematic of different models. (A) First-order model, $f = 19,312$ Hz; (B) second-order model, $f = 52,853$ Hz; (C) third-order model, $f = 102,980$ Hz; (D) fourth-order model, $f = 129,540$ Hz; (E) fifth-order model, $f = 169,300$ Hz.

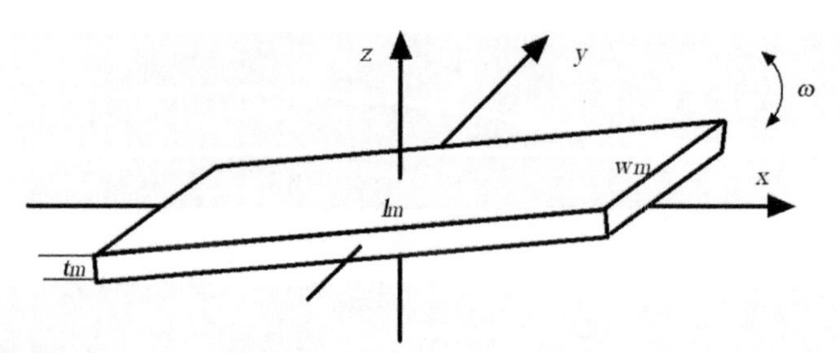

FIGURE 10.31
The positional relationship between the movable electrode and the axis.

Assuming that the active electrode's inversion process is affected only by the permanent magnetic field, we can use a typical second-order dynamic model to describe the movement of the movable electrode, excluding the movement that is caused by being picked up to the contact electrode. And when the contact is closed, the kinetic energy of the movable electrode is zero with no rebound.

Assuming that the rotation angle of the movable electrode is $\alpha(t), 0 \le \alpha(t) \le 2\theta$, the kinetic model of the movable electrode is as follows:

$$I_y \frac{d^2\alpha(t)}{dt^2} + 2K_\alpha \alpha(t) = T_{mag} + T_d \tag{10.23}$$

where I_y is the moment of inertia of the movable electrode around the twist beam's axis of rotation, $I_y = \sum \int_{V_m} (x^2 + z^2) dm$. $V_m$is the volume of the movable electrode and m is its quality. When the rotation angle of the movable electrode shown in Fig. 10.31 $\alpha(t) \le \theta$, $T_{mag} = \mu_0 \mu_r H_{pz}{}^2 V_m(\theta - \alpha(t))$. And when the rotation angle of the movable electrode $\alpha(t) > \theta$, $T_{mag} = \mu_0 \mu_r H_{pz}{}^2 V_m(\alpha(t) - \theta)$. T_d is the damping torque by air damping of the movable electrode during the descent process, $T_d = C\frac{d\alpha}{dt}$, C is the damping coefficient, the other parameters have the same meanings as defined above. Thus, the model of the movable electrode kinetic becomes:

$$\begin{cases} I_y \dfrac{d^2\alpha(t)}{dt^2} + 2K_\alpha \alpha(t) = \mu_0 \mu_r H_{pz}{}^2 V_m(\theta - \alpha(t)) + T_d, \alpha(t) \le \theta \\ I_y \dfrac{d^2\alpha(t)}{dt^2} + 2K_\alpha \alpha(t) = \mu_0 \mu_r H_{pz}{}^2 V_m(\alpha(t) - \theta) + T_d, \alpha(t) > \theta \end{cases} \tag{10.24}$$

10.3.2.2 Analysis of Air Damping

In the movement of the movable electrode, one of its ends goes downward, and the other goes upward. Due to the presence of an air flow field, when the movable electrode rotates around the axis of the twist beam, the air pressure between the end that goes downward and the plane of the fixed electrode changes in distribution, resulting in a squeeze-film air damping effect from the presence of the air damping torque generating the pressure distribution on the surface of the movable electrodes [20]. Assuming that during the twisting of the movable electrodes, the air near the end that goes downward is incompressible and has constant viscosity. The air damp in MEMS devices could usually linearize nonlinear squeeze-film Reynolds equation [21], and get the squeeze-film Reynolds equation:

$$\frac{\partial}{\partial x}\left(h^3\frac{\partial P}{\partial x}\right)+\frac{\partial}{\partial y}\left(h^3\frac{\partial P}{\partial y}\right)=12\eta_0\frac{\partial h}{\partial t} \tag{10.25}$$

where $P$is the air damping pressure on the movable electrode. $h$is the average thickness of the air film, referring to the film thickness of the air between the plane of the lower surface of the movable electrode and the contact electrode and is influenced by the rotation angle of the movable electrode. η_0 is the dynamic viscosity of air, which is usually 1.79×10^{-5} Pa.s at room temperature. As shown in Fig. 10.32, suppose that the initial state of the movable electrode is stable state 1, for example, when the movable electrode twist from stable state 1 to stable state 2, the end that goes downward compresses the air below. In the figure, h_0 is the height of the axis of the twist beam from the fixed electrode, and ω is the rotation angular velocity of the movable electrode.

Since the movable electrode is rotating around the y-axis, the air damping pressure P is distributed consistently in the y-axis direction. Thus, $\partial P/\partial y=0$. The initial angle between the movable electrode and the y-axis is θ, therefore $h=h_0+(\theta-\alpha(t))x$, and the squeeze-film Reynolds equation is:

$$\frac{\partial^2 P}{\partial x^2}+\frac{3}{h}(\theta-\alpha(t))\frac{\partial P}{\partial x}=-\frac{12\eta_0}{h^3}\omega x \tag{10.26}$$

The angular velocity $\omega=\alpha'(t)$ and $P(0)=P(0.5l_m)=0$. Therefore, we get the air-damping pressure $P(x)$. Due to the rotation around the y-axis of the twist beam, the air-damping torque on the end that goes downward of the movable electrode is:

$$T_d=\iint_A P(x)x\,dA=\iint_A P(x)x\,dx\,dy \tag{10.27}$$

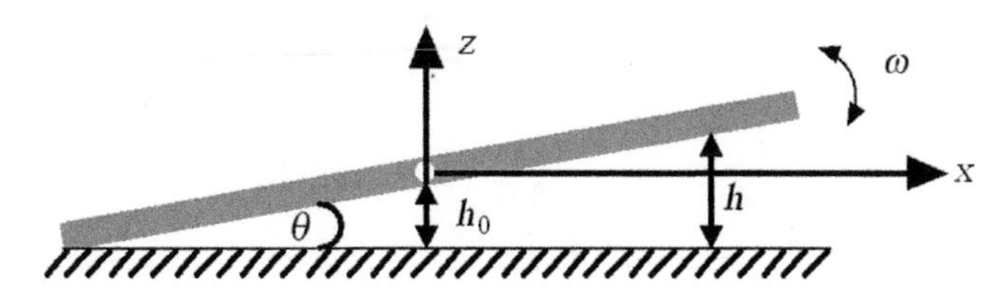

FIGURE 10.32
Model of air compression.

where A is the surface of the movable electrode dropping side, and we obtain the damping coefficient $C = \frac{T_d}{\omega}$.

The air-damping influence on the dynamic response of the movable electrode is mainly reflected in the impact of air-damping torque on the switching time and moving speed generated by the movable electrode's downward movement, which depends mainly on the ratio of the damping torque of the air and the driving torque. When the driving torque and air pressure damping torque are almost the same magnitude, the influence of the squeeze-film air damping effect on the movement of the movable electrode can not be ignored. However, when the driving torque is much larger than the air-damping torque, the air-damping torque can be ignored.

10.3.2.3 The Effect of a Permanent Magnetic Field and Torsional Stiffness of the Movable Electrode Beam on Switching Time

In the following section, the impact of magnetic moment on the movable electrode switching time by increase the permanent magnetic field, as shown in Fig. 10.33, and the relationship between the switching times of movable electrode and the strength of permanent magnetic field are studied. When

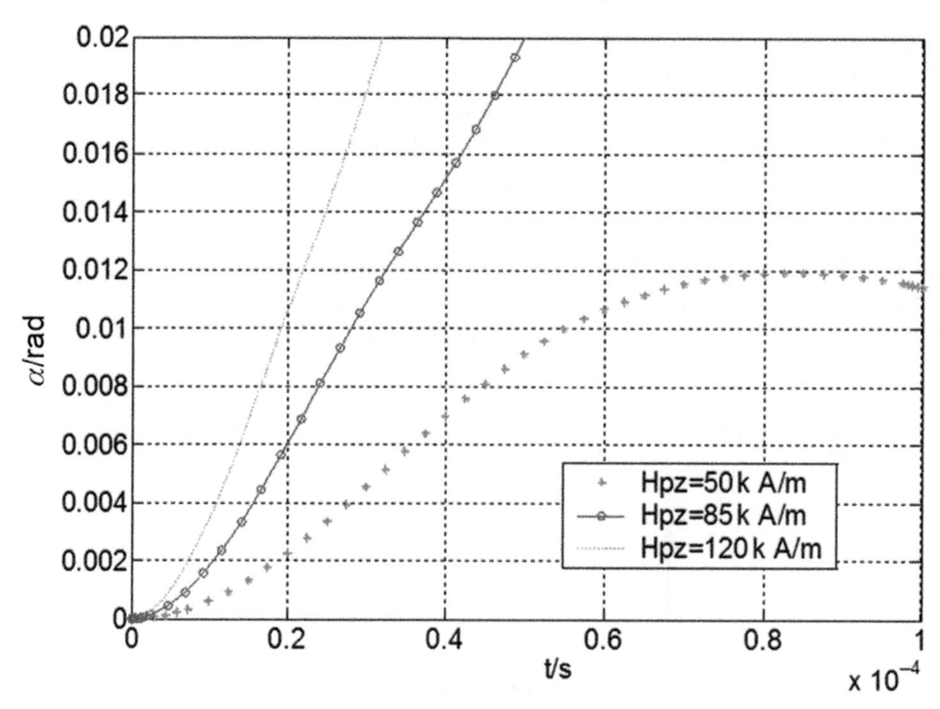

FIGURE 10.33
The impact of magnetic moment on the switching time of the movable electrode.

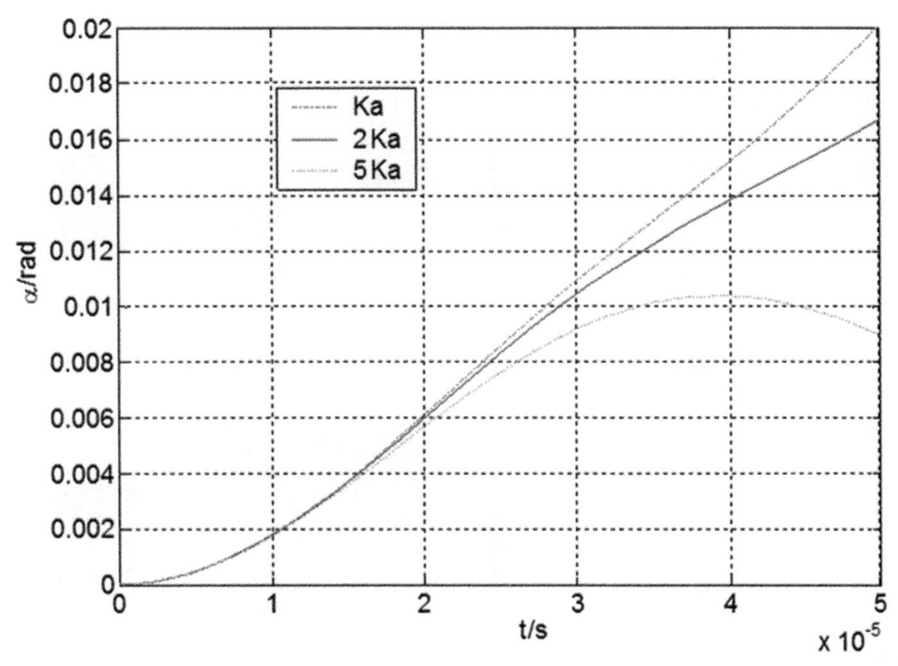

FIGURE 10.34
Effect of the stiffness of the twist beam torsional on the angular velocity.

the magnetic moment increases, the switching speed of the movable electrode will also increase, and the speed increase ratio is almost equal to that of the magnetic field. If the magnetic moment is too small, the movable electrode will not be able to reach the fixed electrode. Thus, an appropriate increase in the permanent magnetic field can effectively improve the switching speed of the movable electrode. However, the faster the movable electrode is, the shorter the life of the electrode will be, and it will also be easier to rebound, which is contradictory to the pursuit of a fast response.

As shown in Fig. 10.34, the stiffness of the torsion beam,K_α, has a direct impact on the switching time of the movable electrode. We can see from the figure that the greater the stiffness is, the longer the response time that the movable electrode will have. Thus, under the premise of ensuring the torque of the torsion beam is less than the maximum allowable torque, we should reduce the stiffness to improve the response speed of the movable electrode.

10.3.2.4 Modal Analysis

Mode shape is a characteristic of the mechanical structure's natural vibration. Each modal has its own specific natural frequency, damping ratio, and mode shape. Modal analysis is an important method of dynamic design and fault

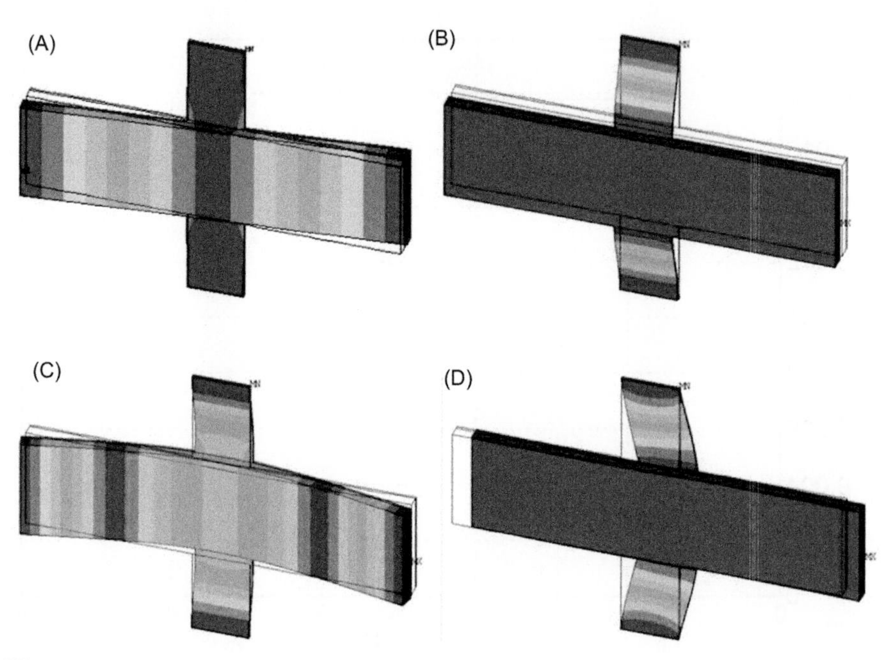

FIGURE 10.35
Images of the first- to fourth-order mode shape of the movable structure. (A) The first-order mode shape, (B) the second-order mode shape; (C) the third-order mode shape; (D) the fourth-order mode shape.

Table 10.7 The Frequencies of the First- to Fourth-Order Mode Shapes (Hz)

First-Order Mode Shape	Second-Order Mode Shape	Third-Order Mode Shape	Fourth-Order Mode Shape
4525	14088	0.16944×10^6	0.19245×10^6

diagnosis of the structure to be analyzed to determine order resonance frequencies and mode shapes. We used finite element software ANSYS to analyze and used unit Solid98 to build an analysis model. We took the structural parameters in Table 10.4 as an example for finite element analysis without damping, as shown in Fig. 10.35, and the frequencies of each mode are shown in Table 10.7.

We can see from the image of the modal shape, the first-order mode shape is torsion of the twist beam along the longitudinal direction (y-axis) without movement in another direction, which also verified that the bending deflections caused by the gravity and electromagnetic force are negligible in the analysis.

The frequency of the second-order mode is 14,088 Hz, which is almost four times the frequency of the first order. A translational motion of movable electrode parallel to the direction along its thickness bends the twist beam. The resonant frequencies of second- and higher-order modes are much larger than that of the first order, so the target mode and nontarget modes are well-isolated, which is to say that the first-order mode is not affected by other higher-order modes.

10.3.2.5 Transient Analysis

Let's take the structural parameters in Table 10.4 as an example. We used the MEMS simulation software CoventorWare for transient simulation of the movable electrodes, and the analysis results are shown in Fig. 10.36. The strength of the permanent magnetic field $H_{pz} = 85$ kA/m, the pick-up current is about 100 mA, and the switching time is approximately 100 μs.

Input pulse voltage signal into the input terminal of the coil to get the magnetization in three directions near the tip of the movable electrode and the comparison of the input and the output signal. Here, *i* is the current signal input in amperes (A), and *mx*, *my*, and *mz* represent magnetization along the direction of the length (*x* axis), width (*y* axis), and thickness (*z* axis) of the movable electrode in Tesla (T), respectively. *hx* represents the induction density generated by the coil along the direction of *x*, in amperes per meter (A/m). *ry* is the output rotation angle of the movable electrode around the *y*-axis in radians (rad).

As can be seen from Fig. 10.36, the coil generates a pulse magnetic field along the *x*-direction when applied on an input pulse signal, and the state of the movable electrode is reversed, and continues to maintain a constant twist angle after the end of the pulse signal until the arrival of the next switching pulse.

During the pulses, the maximum amplitude of *mx* is about 0.8 T, the order of magnitude of the amplitude of *my* is 10^{-6}, and the amplitude of *mz* is about 0.12 T. As can be seen from Fig. 10.36, during the pulses, H_{pz} and $H_{coil\text{-}x}$ have a superimposed effect on *mx*. When the pulse ends, there is a weak decrease in *mx*, as a result of $H_{coil\text{-}x} = 0$. The magnetization in the direction of the *y*-axis is only related to $H_{coil\text{-}y}$. This is zero when the pulse ends. Although the thickness of the movable electrode is much smaller than its length and width, *mz* can not be ignored compared to *mx*. As the length is much larger than thickness, the easy magnetization direction of the movable electrode is still in the longitudinal direction, which is to say in the direction of the *x*-axis; *mx* is much larger than *mz*.

Simulation results show that the magnetization direction of soft magnetic material is affected by many factors. The size and amplitude of the magnetic field in the direction of the material are both important determinants.

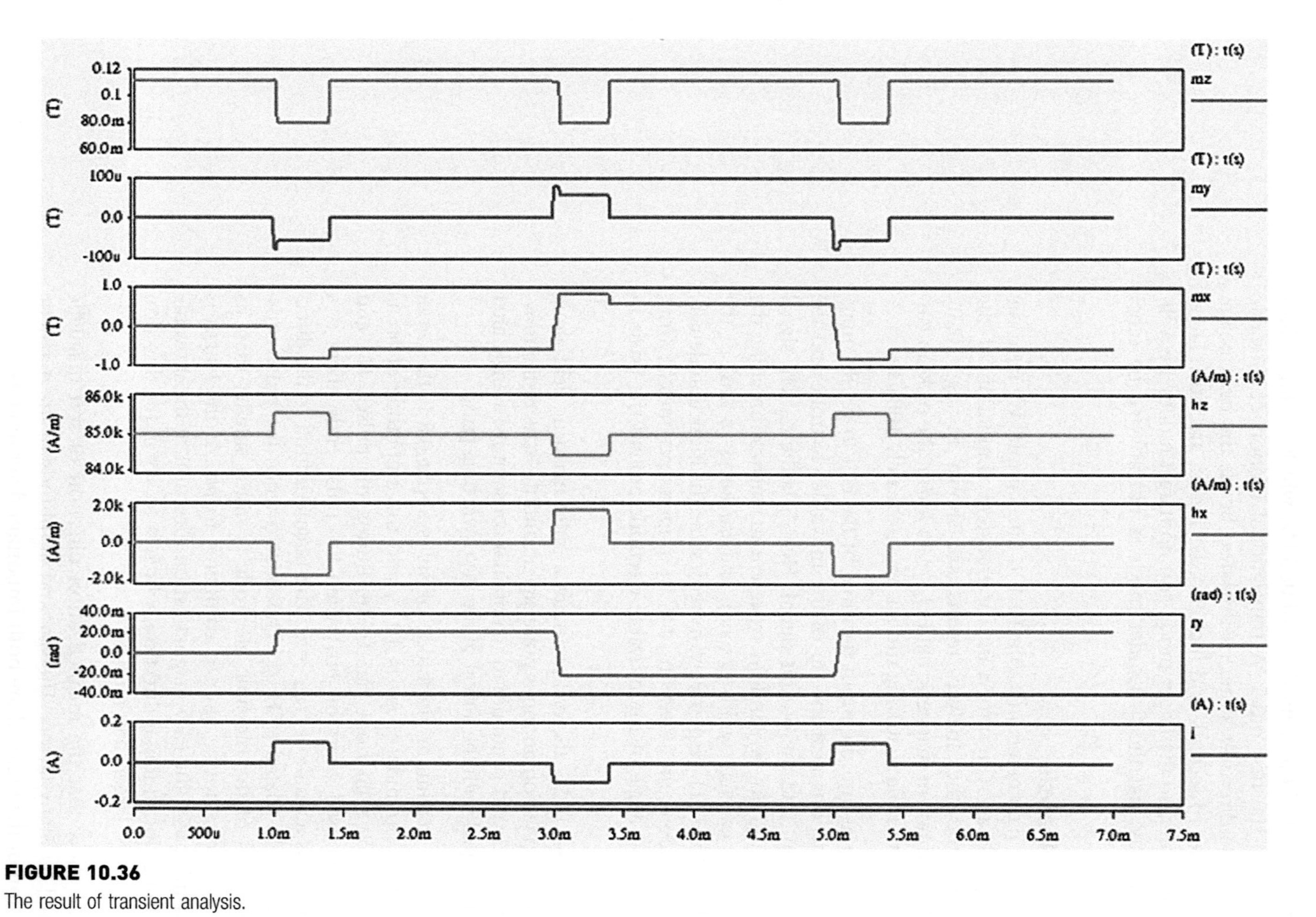

FIGURE 10.36

The result of transient analysis.

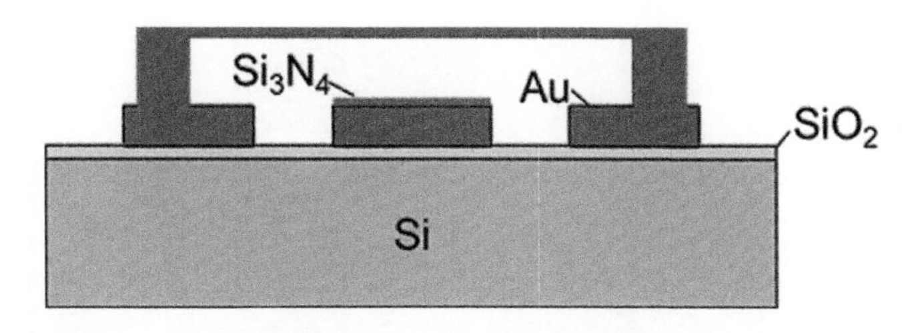

FIGURE 10.37
The process design of the high-resistivity Si substrate.

FIGURE 10.38
Layout design.

10.4 PROCESSING TECHNOLOGY FOR MEMS RELAY

10.4.1 Process of Electrostatic-Driven Relay

Use the release process of the sacrificial layer for the metal membrane structure [22] (Figs. 10.37–10.39).

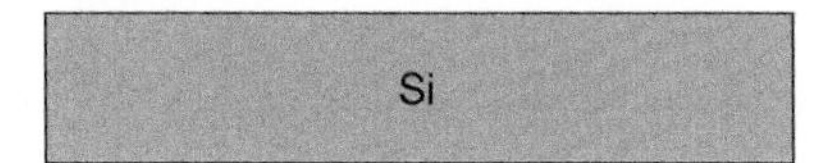

Use high-resistivity silicon (Si) as the substrate (layer 1).

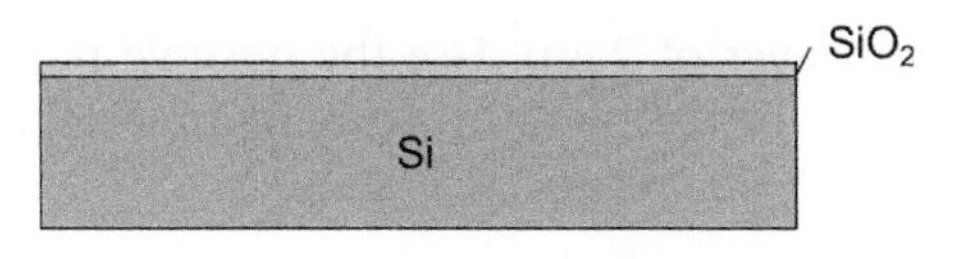

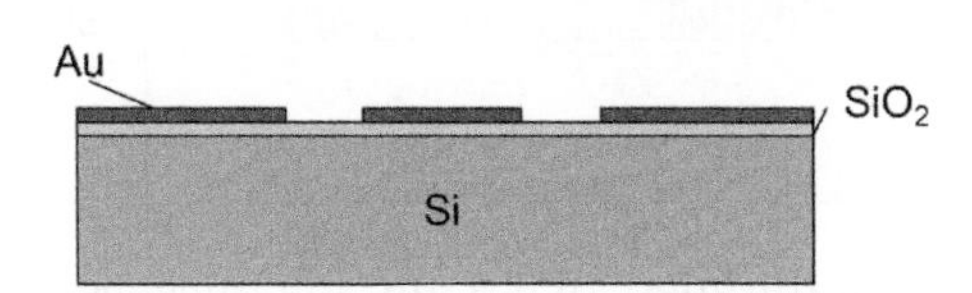

Produce an SiO_2 layer of 5000 angstroms by thermal oxidation as a passivation layer. Sputter a Ti layer of 1000 Å as an adhesion layer, then sputter Au

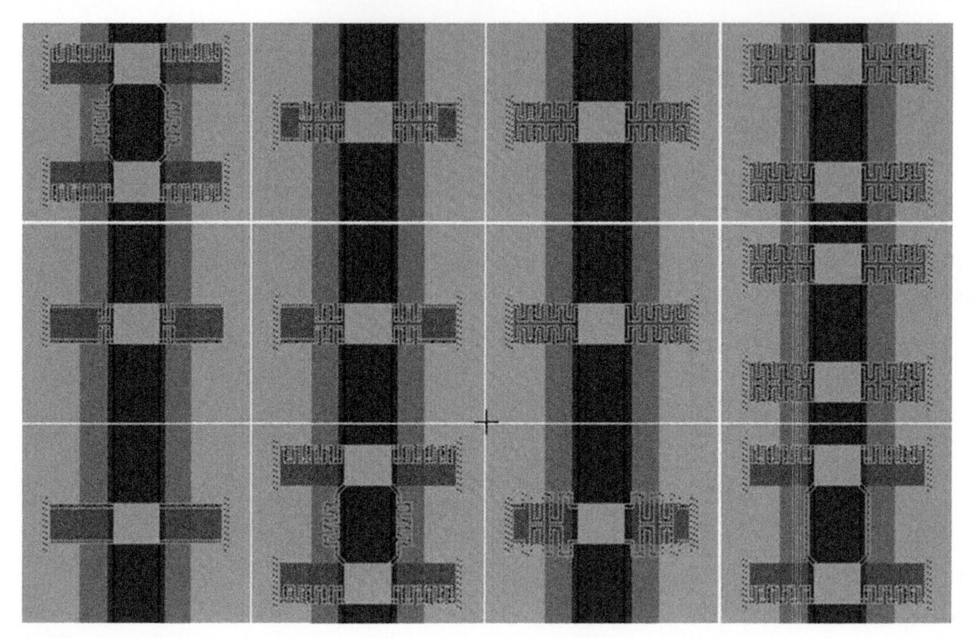

FIGURE 10.39
Layout that contains multiple structural beams.

with a thickness of 1000 Å, and get a seed layer for electroplating by a positive photoresist stripping process.

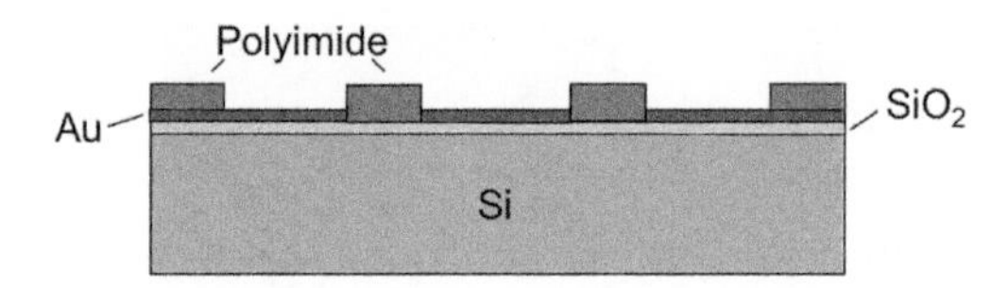

Spin-coat a polyimide layer of 2 μm. Use the positive resistant developer to etch the polyimide layer in the photolithography process, to expose the bottom layer of Au as the plating mold.

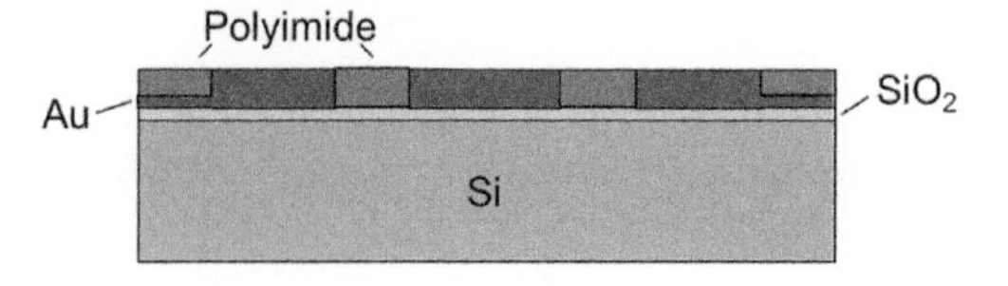

Electroplate the Au to a thickness of 1.8 μm, the thickness of the plated layer is close to that of the polyimide layer.

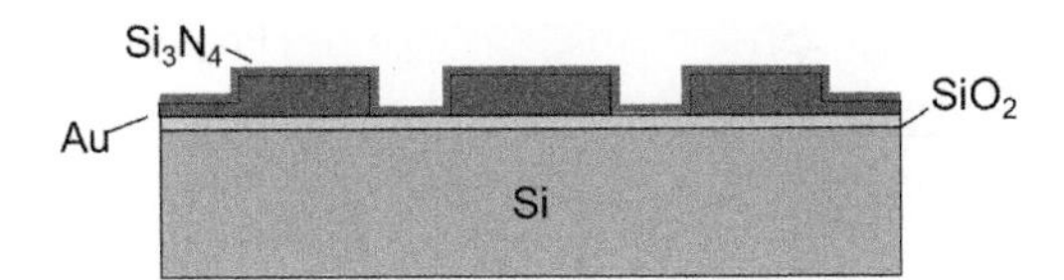

Remove the polyimide by etching, use solution ($KI + I_2$) to etch to Au seed layer on the edge and remove the TiW adhesion layer by hydrogen peroxide immersion. Deposit Si_3N_4 by PECVD (with a reference temperature of 300°C), and make sure the thickness is 3000 Å with a flat upper surface whose surface roughness is lower than 10 Å, to get a dielectric layer.

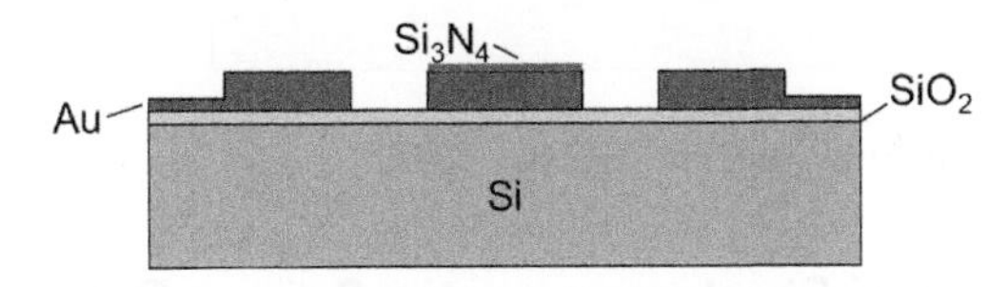

Remove the peripheral portion of Si_3N_4 by etching to expose the Au layer.

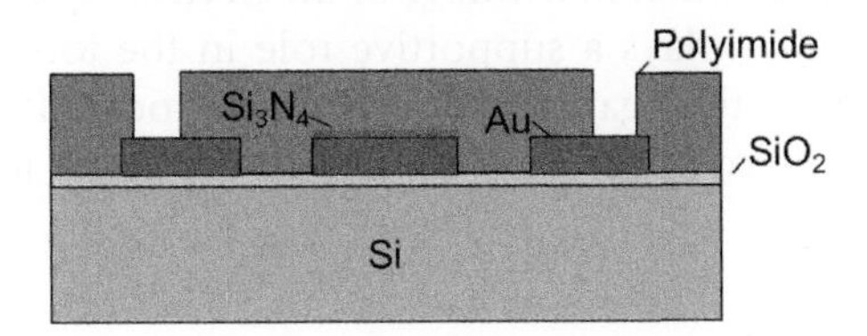

Spin coat polyimide (6.3 μm), the upper surface of which needs to be flat, and patternize the polyimide to prepare for the sputtering of the electrode.

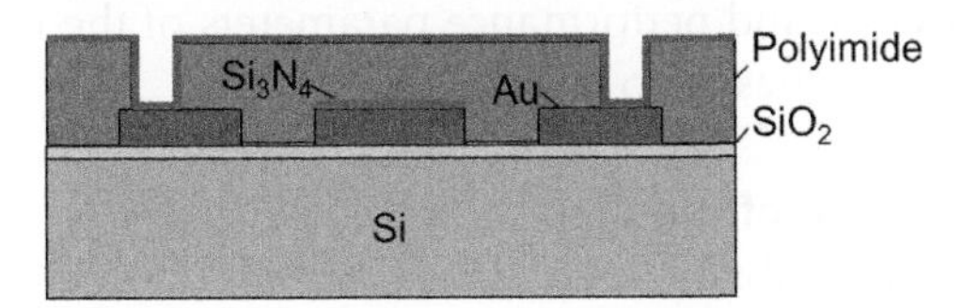

Step sputter Au, 0.25 μm per step with a total thickness of 1 μm. The stress of the sputtered metal layer should be as low as possible to avoid buckling after the structure releases.

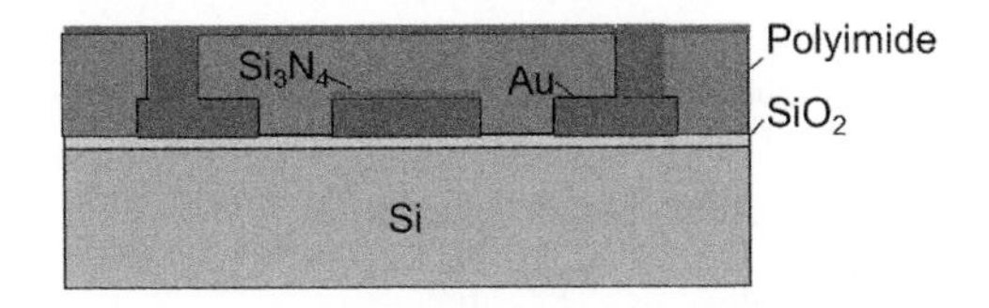

Electroplate the anchor area (the plating area expands by 2 μm), and bond with the sputtered layer. The anchors need to be firmly connected with the structure layer.

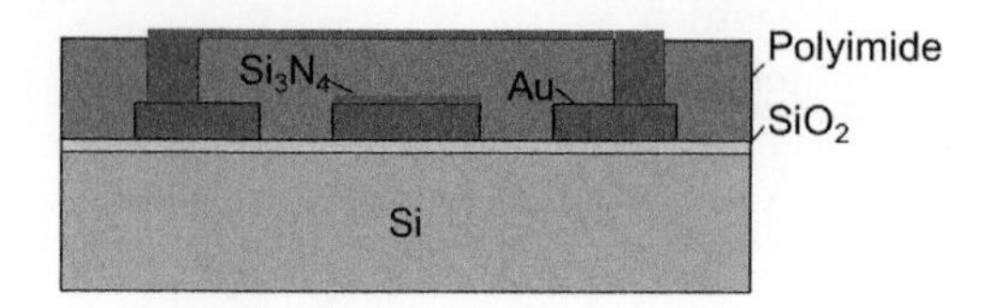

Pattern the Au layer by lithography.

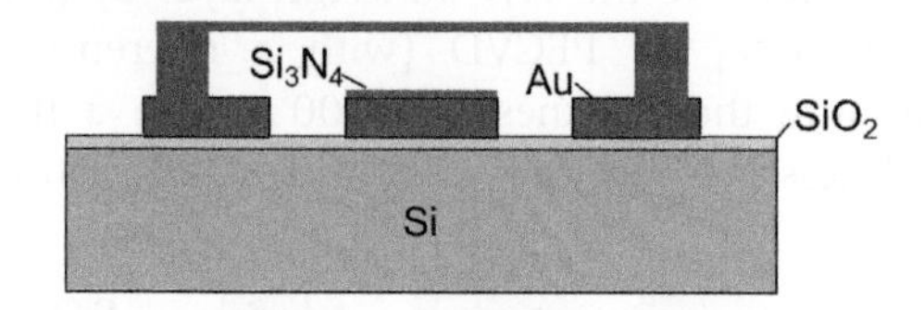

Remove the polyimide to release the structure.

10.4.2 Process of Electromagnetic-Driven Relay

Fig. 10.40 illustrates the micromachining of an electromagnetic-driven MEMS relay, wherein the pillar plays a supportive role in the torsion structure and thereby generates a motion gap for movable electrodes. It is an important auxiliary structure. The coil and fixed contact electrode are located below the torsion structure.

10.4.2.1 Key Process

Throughout the process, the concrete realization of the coil and soft magnetic material processing and other critical processing steps determine the overall processing performance and performance parameters of the relay. The following subsections discuss these two key processes.

10.4.2.1.1 Processing of the Coil

To pattern the metal membrane, a common method is wet etching, resist stripping, and electroplating. To increase the current capacity of the coil's wire, it is necessary to improve the cross-sectional area of the wire. The increase in the wire width will increase the coil size, resulting in an increase in the relay structure and the inhomogeneity of the magnetic field, which is not conductive to

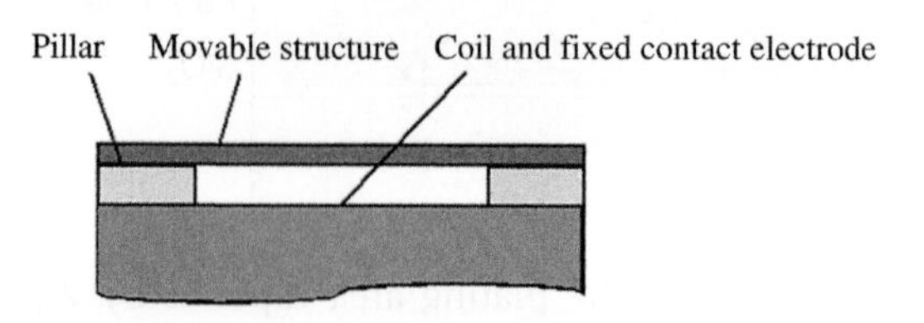

FIGURE 10.40
Relay structure micromachining diagram.

the miniaturization of the structure and the polarization of movable electrodes. Therefore, a suitable method is to increase the wire thickness.

Due to the large thickness of the coil, it is easy to take the metal coil off the substrate totally when stripping the photoresist. Therefore, methods like wet etching and electroplating are used to process the coil, and then assess and compare the processing quality. On this basis, we chose electroplating as the method for processing a planar coil and the fixed contact electrodes.

1. Wet Etching
   In the MEMS process, wet etching is widely used in a variety of structures for its simplicity, fast etching rate, and other features. Thus, wet etching is first used to processing the coil. The process is as follows: first, sputter Cr/Au sequentially by magnetron sputtering on the surface of the glass as the seed layer for plating, whose thicknesses are 30/150 nm each. Second, prepare the Au layer by electroplating, with a thickness t_c of ~ 1 μm and a line width w of 12 μm, while the centerline spacing d_{cc} is 26 μm, and wire segment pitch is 14 μm. Third, coat photoresist, pattern the coil and pads, and use the wet etching method to etch the coil.
   As shown in Fig. 10.41A, after the end of the corrosion, the line width w is less than 10 μm, while the spacing for wires s is close to 16 μm. According to the characteristics of the coil when generating a magnetic field, the increase in the centerline spacing under the same current leads to a sparse magnetic field distribution. Thus, such a process affects the load capacity of the relay.

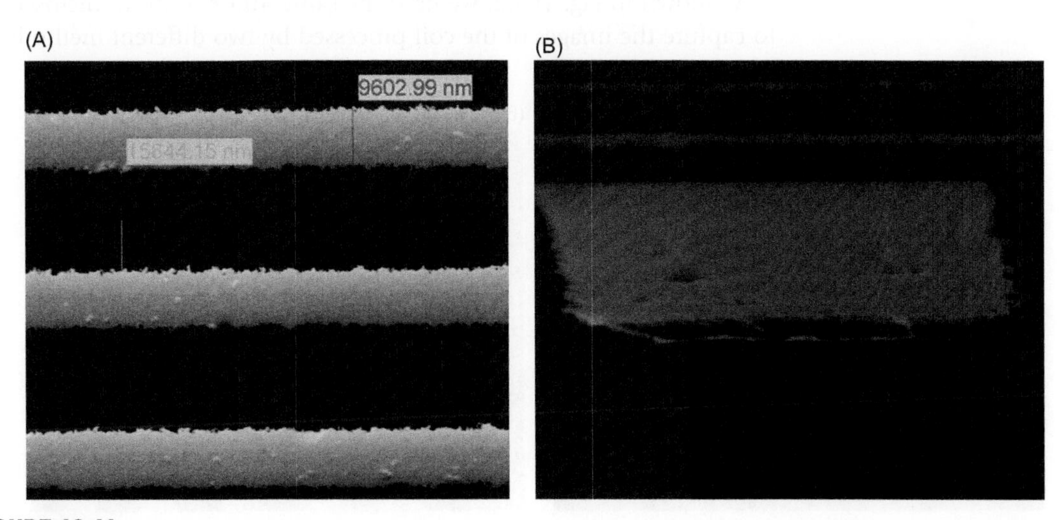

FIGURE 10.41
SEM image of the planar coil after wet etching process. (A) The actual size of the coil; (B) etched coil sections.

Fig. 10.41B shows a cross-sectional side cut of the coil. We analyze and discuss such process effectiveness from the aspect of the mechanics of the materials. In this process, we sputter the seed layer first, then electroplate, and finally wet etch, in order to obtain the coils. As the bottom Au layer is made by sputtering, its grain size is small and the density of the grain boundary is high, resulting in many intergranular boundary defects. As the energy needed when etching is low, the etching speed is high. When the top Au layer is made by plating, its grain size is large with less grain boundaries, which means less defects. Thus, the etching rate is low.

2. Electroplating

   Electroplating is commonly used in metal patterning in MEMS, the biggest advantage is obtaining a desired pattern fast and accurately. The processing order when processing coil by electroplating is slightly different from that by wet etching. The main steps are: (1) sputter the seed layer of Ti/Au, to a thickness of 50/100 nm; (2) coat photoresist, expose, and develop; (3) electroplate and remove the resist; and (4) physical sputter argon ions to remove the seed layer, which is usually referred to as the antisputtering process.

   Electroplating can control the size and precision of the coil accurately, and is a metallization forming method to obtain high dimensional accuracy. However, such a method has its disadvantages, i.e., when physical sputtering the argon ion, the surface of the coil and the fixed contact electrode also suffer from high-speed ion bombardment, and surface materials are consequently lost. Thus, the timing of the antisputtering process requires strict control.

   As shown in Fig. 10.42, we used the same 400 × optical microscope to capture the images of the coil processed by two different methods;

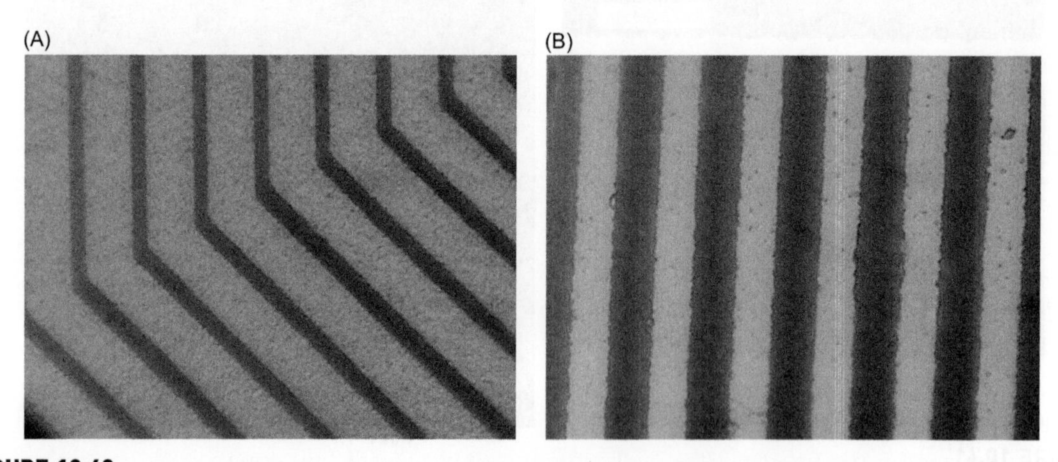

FIGURE 10.42
Comparison of the planar coil processed by electroplating and wet etching: (A) electroplating; (B) wet etching.

Fig. 10.42A shows the coil wire layout width of 15 μm, and Fig. 10.42B coil wire layout width of 12 μm. Fig. 10.42A is the image of coil processed by plating, and Fig. 10.42B by a wet etching method. Obviously, the coil in Fig. 10.42A has neater edges and a better-quality surface than that in Fig. 10.42A. Although improved electroplating solution formulation could help improve the coil's surface quality, it is relatively difficult to avoid the side cut phenomenon.

Based on the above analysis, the electroplating method is more suitable for the structure of the relay in this embodiment, and thus the ultimate method for making a planar coil is electroplating.

10.4.2.1.2 Soft Magnetic Materials Processing

MEMS magnetic sensors or actuators usually use the microelectroforming method to deposit soft magnetic material, which is several microns to tens of microns in thickness. In the microelectroforming process, photoresist such as positive photoresist AZ1500, AZ4620, negative photoresist SU-8, etc., are commonly used, the latter of which may be up to a thickness of about 500 μm. AZ4620, generally used for construction of about 10 μm, whose prominent feature is being able to obtain a uniform thick film structure, has consequently been widely used by researchers.

Usually in the microelectroforming process, it is required that the thickness of the photoresist is slightly greater than that of the microelectroforming membrane, in order to achieve better electroplating effects. Without considering some weak deformation of the edge of graphics of the photoresist caused by various factors in the lithography and microelectroforming process, the pattern of metal membrane after plating and photoresist is consistent in size. In this embodiment, the movable electrode pattern is relatively simple, the requirement is uniform deposition layer and thickness to meet the conditions. The following focuses on the process experiments using photoresist of the AZ1500 series.

In the AZ series photoresist, the 1500 series is most widely used in the thin-membrane lithography process, with common thicknesses of about 2 μm, but the thickness of the nickel–iron film herein is larger than 2 μm. Thus, in order to verify the feasibility of the process design and to measure and compare the growth rate of the microelectroforming membrane on both sides and upward, we designed a test process in which the resist pattern's thickness is less than that of the target film.

When microelectroforming, as the thickness of the membrane exceeds the thickness of the photoresist material, the membrane continues to grow and is not restricted by the photoresist; it will grow not only upward but along the sides as well, which means the final membrane's pattern sectional shape

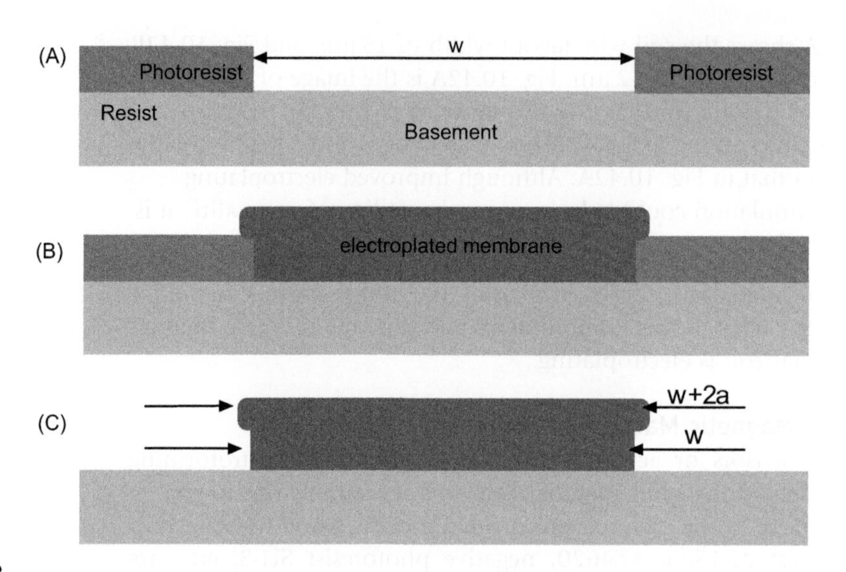

FIGURE 10.43
Microelectroforming process using AZ1500 photoresist. (A) Width section of lithographic pattern before microelectroforming; (B) width section of membrane with photoresist after microelectroforming; (C) width section of membrane without photoresist after microelectroforming.

will be mushroom-like, as shown in Fig. 10.43. w is the cross-sectional width dimension of the resist pattern, a is the width dimension of the membrane's lateral growth. In addition, in order to facilitate the description, a' is the lateral width of the longitudinal direction of the membrane's growth.

The process is as follows: (1) the thermally oxidized silicon has an oxide thickness of about 4300 Å. (2) Sputter Cr/Au on the wafer surface after oxidation. The sputtering thickness is 200/900 Å, the sputter time is 3/8 min, respectively. (3) Coat photoresist and then lithograph. If the coating process needs to be done more than once, in order to improve the affinity of the substrate, it should be dried for 2 min at 100°C each time after lithography, and RF plasma-based dry cleaning should be used to strip the resist for 2 min for descumming. (4) Use the stylus profiler to measure the thickness of the photoresist, and backup.

In order to investigate the microelectroforming effect when the thickness of the microelectroforming membrane is larger than or equal to that of the photoresist, we carried out four sets of experiments in total, using four groups of silicon wafers (No. 1–4). For each group, the photoresist coating thickness of wafers is different from the other group. The target thicknesses for No. 1–4 were 2, 3, 5, and 7 μm, respectively. To verify the actual design size as accurately as possible, considering that the

Table 10.8 Some Photoresist Coating Parameters of the Soft Magnetic Material Microelectroforming Experiment Using AZ1500

No.	Resist Thickness (μm)		Method of Coating	Rotation Speed (Rad/min)	Time (min)	Number of Coatings
1	Edge 2.28	Center 2.30	Spin coating	3000	5	1
2	Edge 3.02	Center 2.85	Spin coating	3000	5	1
3	Edge 5.40	Center 5.46	Spin coating	3000	8	2
4	Edge 8.83	Center 7.54	Spin coating	3000	10	3

membrane has lateral growth when exceeding the lithograph thickness, the pattern size used by the experiment was 780 μm × 180 μm, 840 μm × 180 μm, and 850 μm × 190 μm, respectively. The above steps and the specific parameters after lithography are shown in Table 10.8.

As can be seen from Table 10.8, for the thickness of photoresist, as the number of coating times increases, the difference between the center and edge of the wafer grows. There are many reasons for this thickness difference, including human factors in process operation, test point selection when using the stylus profiler, the balance of the rotation platform, and so on. However, the main cause was the accumulation of surface undulation caused by repeated photoresist coating. Thus, as the thickness of the photoresist and the number of coating times increases, the thickness uniformity of the photoresist becomes increasingly poor.

Before the microelectroforming process, the wafer was dried for 20 minutes at 100°C, and then descummed. Use the microelectroforming formulations and techniques mentioned in section 10.2 on microelectroforming, wherein the temperature is 60°C and the current density is 10 mA/cm^2. The time of microelectroforming is about 66 minutes and the pH is 3.25.

As shown in Fig. 10.44, we use sample No. 3 with the best photoresist consistency to illustrate the sectional shape and size of the microelectroforming sample. The pattern size of the photoresist is 780 μm × 180 μm and 840 μm × 180 μm.

Since the pattern size of the microelectroforming membrane is tiny, it is difficult to get the cross-section of the sample membrane vertical to the SEM imaging direction, so the measurements will be slightly different from reality, however the data are basically credible. As shown in Fig. 10.44(A) and (C), the cross-sectional shape of the microelectroforming membrane is mushroom-like, the cross-sectional width w of the resist pattern on the bottom part of the membrane is approximately 180 and 179 μm, while the cross-sectional width of the resist pattern is 180 μm. Considering human measuring errors, the thickness of the microelectroforming membrane can be

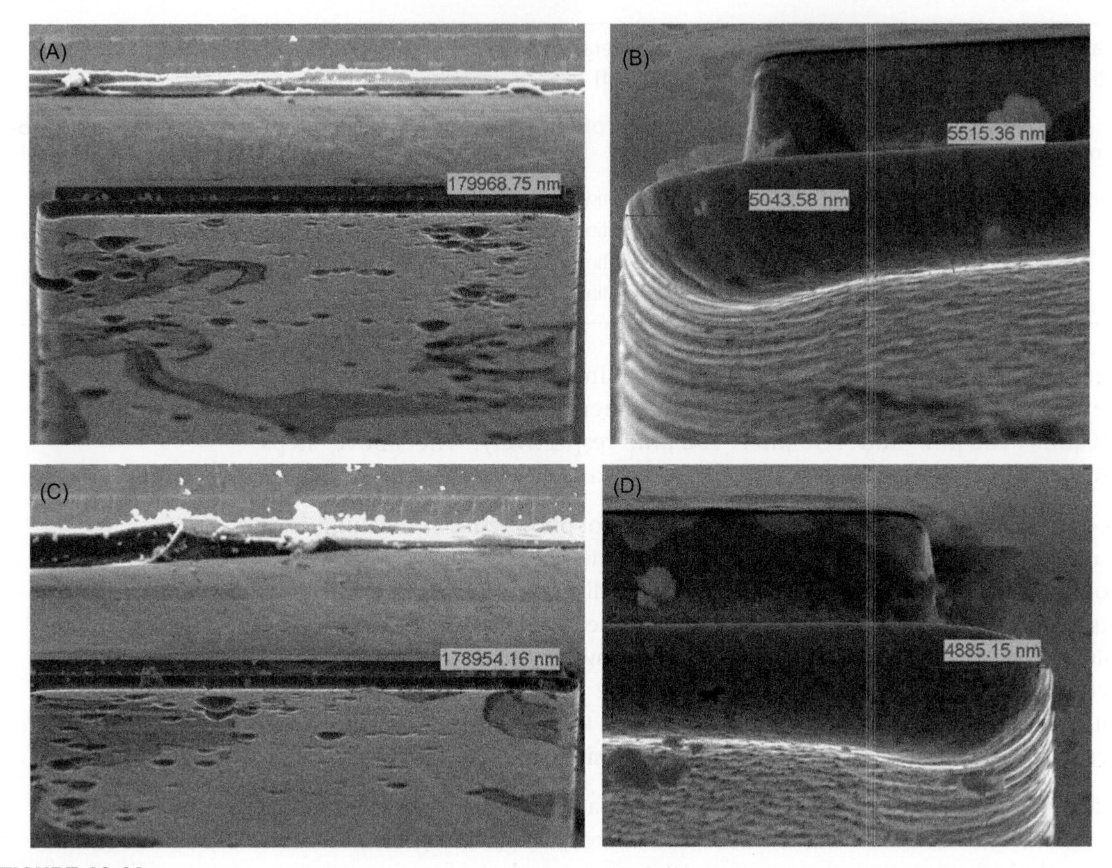

FIGURE 10.44
Cross-sectional SEM images of microelectroforming soft magnetic material sample. (A) General cross-sectional SEM image (780 μm × 180 μm); (B) cross-sectional SEM image of membrane (780 μm × 180 μm); (C) general cross-sectional SEM image (840 μm × 180 μm); (D) cross-sectional SEM image of membrane (840 μm × 180 μm).

assumed to be within the resist pattern coinciding with the photoresist thickness. As shown in Fig. 10.44(B) and (D), the thickness of the microelectroforming membrane above the resist pattern is about 5.5 μm, and (A) is about 5.04 and 4.9 μm in each figure. Their sizes of are almost the same.

Fig. 10.44 shows only relevant dimensions of cross-sectional width with the width direction for comparison. For a more comprehensive study to compare the growth rate of microelectroforming membrane over the photoresist, the sizes of a' are needed. As the membrane is tiny, it is difficult to get the cross-section along the length direction, so we need to obtain a' in an indirect way by measuring the total size of the membrane to estimate the approximate

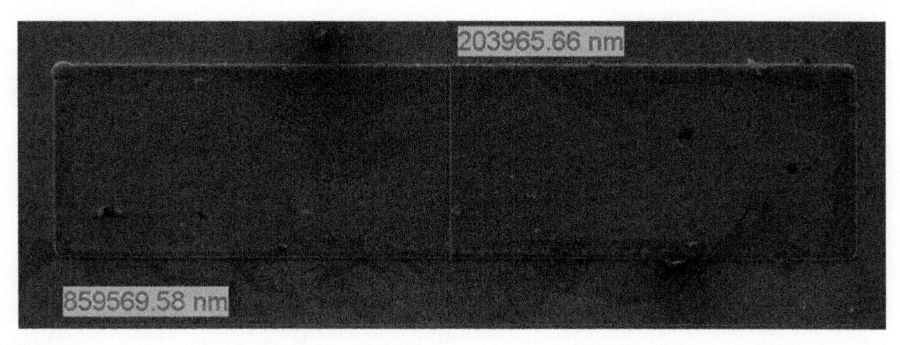

FIGURE 10.45
The SEM image of the total membrane size (850 μm × 190 μm).

range. As shown in Fig. 10.45, a complete vertical view of the nickel–iron membrane, the pattern size of photoresist is about 850 μm × 190 μm. And the membrane is about 860 μm × 204 μm, which is 14 μm larger along the width direction and 10 μm larger along the longitudinal direction. Thus, a is 7 μm and a' is 5 μm. Considering the error caused by the imaging angle and human error, we can conclude that the lateral growth rate of the membrane along the width direction and the longitudinal direction should be very close when it exceeds the photoresist in thickness.

10.4.2.2 Micromachining Process

Under the premise of the critical process being implemented, we conducted a feasibility analysis of the process, to clarify the tasks of various stages of the whole process. We took the use of mature technology as a criterion, combined with the current status of the domestic MEMS technology, and designed the detailed process. Each major step process is as follows:

1. Preparation of the substrate
   Use 7740 4-inch glass as the substrate, with a thickness of 500 μm, wash and dry the glass, as shown in Fig. 10.46A.
2. Sputtering seed layer # 1
   Sputter a Ti/Au composite metal layer on glass with a substrate thickness of 50/100 nm as the pad plating fixed-contact electrode and a seed layer for plane coil, as shown in Fig. 10.46B:
3. Lithography No. 1
   Coat positive photoresist AZ4330, expose, and develop, to form a fixed contact electrode pad pattern.
4. Plating fixed-contact electrode pads
   Plate Au fixed-contact electrode pad. The thickness is 0.3 μm. Remove the resist, as shown in Fig. 10.46C.

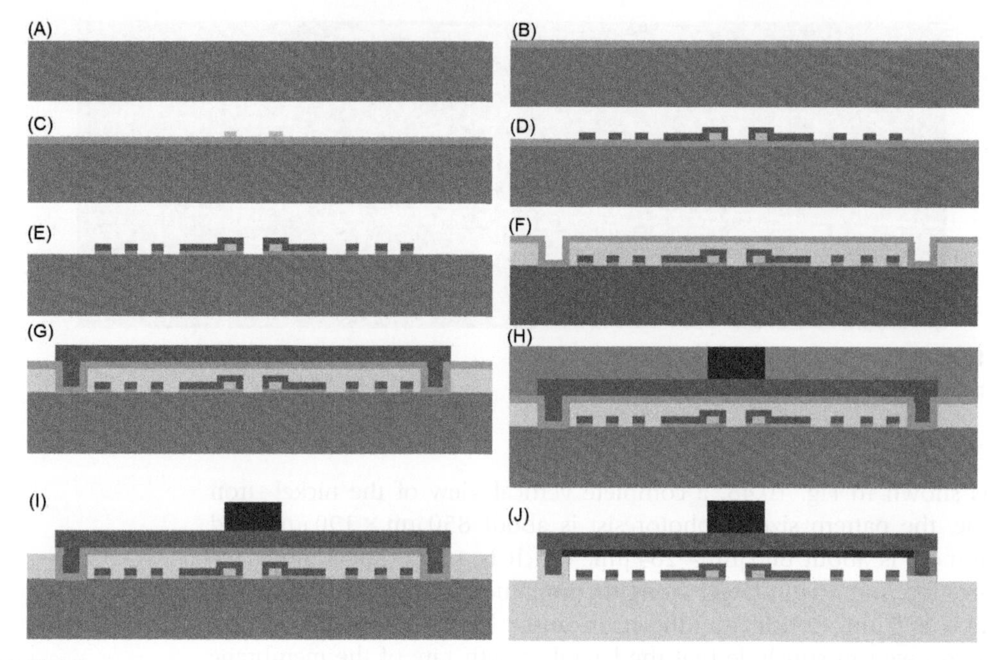

FIGURE 10.46

Micromachining process of relay. (A) Preparation of the substrate; (B) sputtering seed layer 1; (C) plating fixed contact electrode pads; (D) plating planar coil and fixed contact electrodes; (E) remove the seed layer 1; (F) sputtering seed layer 2; (G) electroplating Au; (H) microelectroforming nickel–iron alloy; (I) removing the seed layer 2; (J) dicing and releasing the sacrifice layer.

5. Lithography No. 2
   Coat positive photoresist AZ4330, expose, and develop, to get the fixed-contact electrode pattern and the planar coil.
6. Plating planar coil and fixed-contact electrodes
   Plate Au planar coil and fixed-contact electrodes. The thickness is 1.2 μm. Remove the resist, as shown in Fig. 10.46D.
7. Remove seed layer No. 1
   Antisputter to etch the seed layer of No. 1, and form the planar coil and fixed-contact electrodes, as shown in Fig. 10.46E.
8. Lithography No. 3
   Coat photosensitive polyimide with a thickness of 5.5 μm, expose, and develop to get the pattern of the twist beam pillar.
9. Sputtering seed layer No. 2
   Sputtering a Ti/Au composite metal layer and the layer for the contact electrodes with a thickness of 50/100 nm, as shown in Fig. 10.46F.
10. Lithography No. 4
    Coat positive photoresist AZ4330 with a thickness of 30 μm, expose, and develop, to form the pattern of the movable structure.

11. Electroplating Au
    Plate Au to form the movable structure with a thickness of 5.5 μm, as shown in Fig. 10.46G.
12. Lithography No. 5
    Coat positive photoresist AZ4620, expose, and develop, to get the fixed movable electrodes pattern.
13. Microelectroforming nickel–iron alloy
    Microelectroform nickel–iron alloy with a thickness of 24 μm, using acetone to remove the resist, as shown in Fig. 10.46H.
14. Remove seed layer No. 2
    Antisputter to etch seed layer No. 2, to get the polyimide exposed, as shown in Fig. 10.46I.
15. Dicing and releasing the sacrifice layer
    Dice and remove the photosensitive polyimide to release the movable structure, as shown in Fig. 10.46J.
16. Packaging
    Bond the wires and package.
17. Adhesion of the permanent magnets
    Adhere the precision machined permanent magnetic membrane directly above the envelope.

10.4.2.3 The Layout Design

After designing the process, we need to design and draw the mask.

The mask design must minimize the adverse impact on the parameter of image size due to the characteristics of the process, which is the basic criterion of mask design. Meanwhile, the mask design needs to take full account of the different requirements and characteristics of every step in the process, such as the type and characteristics of the resist, process type, pattern transfer, overlay operation, dicing, wire bonding, and so on, as well as positive and negative mask, and the corresponding relationship between the different layers.

To make it easy for operators to quickly align the adjacent overlay relationship between the two layers, the mask uses a cross symbol as an alignment symbol, as shown in Fig. 10.47, with the pattern of the alignment symbol overlapping. Considering a variety of factors in the process and according to the photolithography sequence, beginning from the second layout, the edge of the alignment mark on each layout is widened for 10 μm compared with the previous layout. As shown in Fig. 10.47, the color of the cross edge is faded from the inside to the outside, and the graphic with the deepest color is the alignment symbol on the first mask.

On the basis of the above factors, we used the Tanner EDA L-Edit software as the tool for mask drawing. According to this process, to complete the

FIGURE 10.47
The alignment symbol.

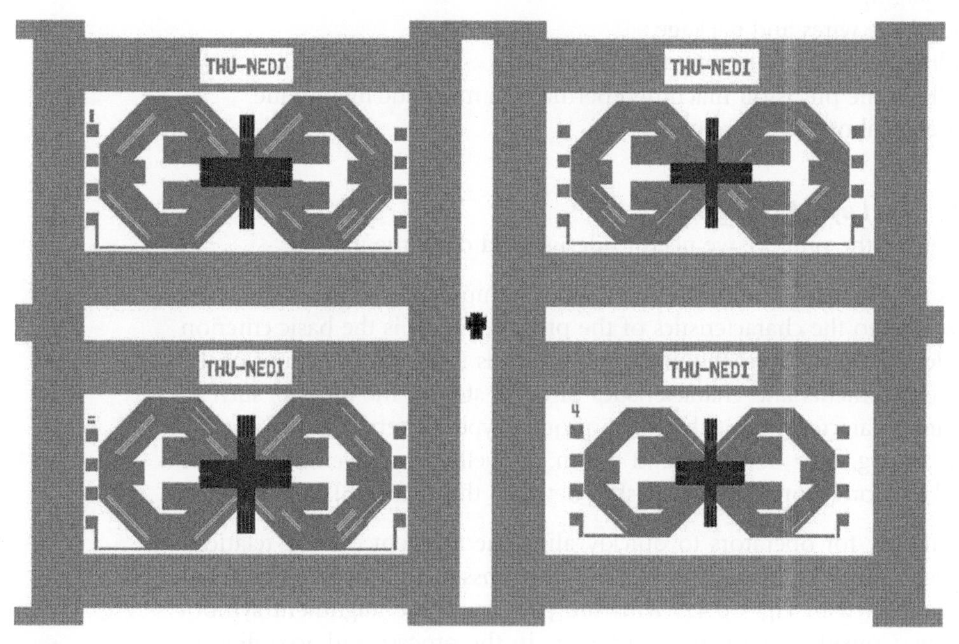

FIGURE 10.48
Overall layout graphics.

micromachining of relay, a total of five times lithography and a total of five different designs of the masks are needed. We also designed four groups of graphics with different structure sizes in each, as shown in Fig. 10.48. Each group's structures are distinguished from the others by the scribe line. After the completion of processing, each group is packaged individually.

10.5 TEST TECHNIQUES FOR MEMS RELAY

10.5.1 Test Target and Equipment

1. The test target
   (1) Pick-up current; (2) release current; (3) switching time; (4) contact resistance; (5) switching current.
2. Test equipment
   (1) Welding the test circuit board with related components; (2) signal generator, Agilent 33120A; (3) oscilloscope, TDS2024B; (4) computer power, WD990A; (5) multimeter.

10.5.2 Test Circuit and Parameters

The signal for the relay test should have the following characteristics: (1) containing positive and negative pulses; and (2) having zero signal between positive and negative pulses. Input the square wave into the RC differentiating circuit. When the time constant $\tau = RC$ is much smaller than the pulse width of the input square wave signal, we can get the test signal with the two output characteristics above. As shown in Fig. 10.49, we use the square wave signal U_s as the input U_i of the RC differentiating circuit, and the pulse signal U_R is the output. Generally, when τ is less than one-fifth to one-tenth the pulse width t_w of the square wave signal, the differential circuit will be set up, as shown in Eq. (10.28).

$$U_R = RC\frac{dU_i}{dt} \tag{10.28}$$

For example, connect the electromagnetic relay to the test circuit shown in Fig. 10.50. The irregular curve represents the golden wire connecting the pads.

The resistance of a single coil is about 34–60 Ω, and the parallel resistance for two coils is generally between 17–30 Ω. As the resistance of the signal generator is about 50 Ω, if R is less than the parallel resistance of the coils, it will reduce the utilization of the generator. So we ignored R, as shown in Fig. 10.50. In the

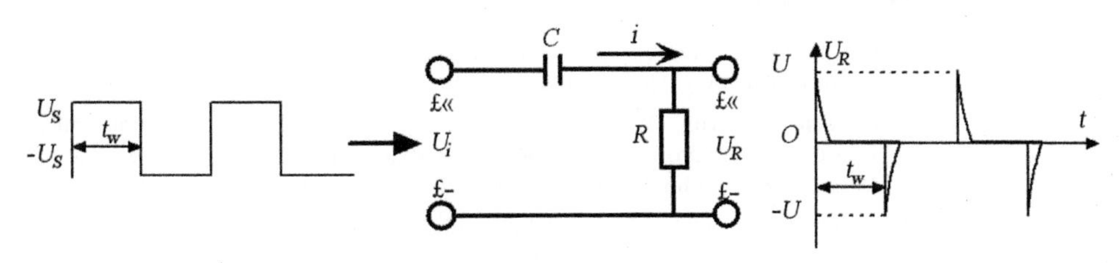

FIGURE 10.49
Using the RC differentiating circuit to generate the pulse signal needed.

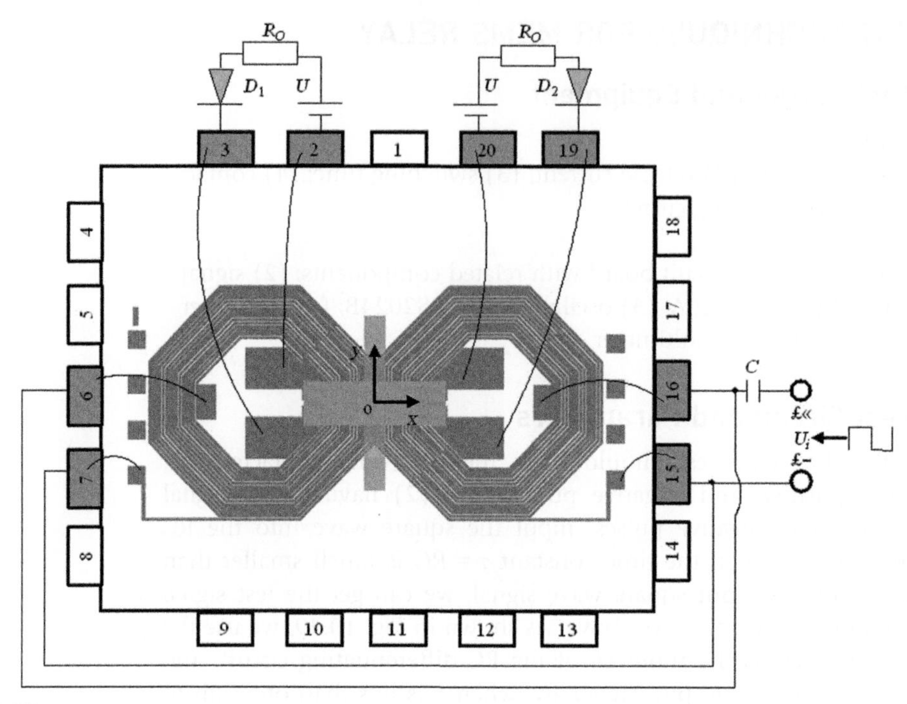

FIGURE 10.50
Package pins and the test circuit connection diagram of LCC20.

ideal case, the simulation results of the relay switching time are several tens of microseconds to one hundred milliseconds. In reality, many factors have led to the extension of the switching time. For Capacitor C, we use an electrolytic capacitor of 100 and 220 μF whose value of permissible voltage is 100 V. And the time constant for the RC differentiating circuit is about 3.7–6.6 ms. The multimeter is used to measure the voltage drop of the light-emitting diode, which is about 1.68–1.85 V. The resistance values of the dividing resistors *Ro* in the contacting loop were 100, 3.3, and 5.6 kΩ.

References

[1] H. Junjun, L. Desheng, Z. Yufeng, Research status and development of micro relays, J. Heilongjiang Aug. First Land Reclam. Univ. 12 (1) (2000) 59–64.

[2] H. Etsu, U. Yuji, W. Akinori, Thermally controlled magnetization microrelay, in: Transducers'95, Eurosensors IX, the 8th International Conference on Solid-State Sensors and Actuators, Stockholm, Sweden, June 25–29, 1995, pp. 361–364.

[3] S. Jonathan, S. Scott, C.J. Kim, A liquid-filled microrelay with a moving mercury microdrop, J. Microelectromech. Syst. 6 (3) (1997) 208–216.

[4] S. Ignaz, H. Bernd, Comparison of lateral and vertical switches for application as microrelays, J. Micromachin. Microeng. 9 (2) (1999) 146–150.

[5] S. Majumder, J. Lampen, R. Morrison and J. Maciel, A packaged, high-lifetime ohmic MEMS RF switch, in: Microwave Symposium Digest, 2003.

[6] N. Nishijima, J.-J. Hung, G.M. Rebeiz, A low-voltage high contact force RF-MEMS switch, 2004.

[7] M.-A. Gretillat, P. Thiebaud, N.F. de Rooij, C. Linder, Electrostatic polysilicon microrelays integrated with MOSFETs, in: Proc. IEEE MEMS Workshop 94, 1994.

[8] H.-S. Lee, C.H. Leung, J. Shi, S.-C. Chang, Electrostatically actuated copper-blade microrelays., Sens. Actuat. 100 (2002) 105–113.

[9] S. Kang, H.C. Kim, K. Chun, A low-loss, single-pole, four-throw RF MEMS switch driven by a double stop comb drive, J. Micromech. Microeng. 19 (2009) 035011.

[10] M. Ruan, J. Shed, C.B. Wheeler, Latching microelectromagnetic relays, Sens. Actuat. 10 (2001) 511–517.

[11] F. Shi, D. Guifu, W. Yan, Development of electromagnetic micro relay for a novel bistable microelectromechanical system, J. Shanghai Jiaotong Univ. (2006).

[12] J. Qiu, J.H. Lang, A.H. Slocum, A bulk-micromachined bistable relay with U-shaped thermal actuators, J. Microelectromech. Syst. 14 (2005) 1099–1109.

[13] J. Simon, S. Saffer, C.-J. Kim, A liquid-filled microrelay with a moving mercury microdrop, J. Microelectromech. Syst. 6 (1997) 208–216.

[14] W. Baoling, Design Basis of Electromagnetic Apparatus, National Defense Industry Press, Beijing, 1989 32, 34, 46

[15] G. Zhancheng, L. Yuxing, L. Meifeng, Microstructure and magnetic properties of electrodeposited Fe and Ni-based alloy foils, Chin. J. Nonferr. Metals 14 (2) (2004) 275–279.

[16] P.T. William, S. Michael, B. Henry, et al., Electroplated soft magnetic materials for microsensors and microactuators, in: Transducer'97, Chicago, 1997, pp. 1445–1449.

[17] T.S. Chin, Permanent magnet films for applications in microelectromechanical systems, J. Magnet. Magnet. Mater. 209 (2000) 77.

[18] L. Huijuan, Design and Technology of MEMS Bistable Electromagnetic Relay [D], Tsinghua University, Beijing, 2008.

[19] J. Schimkat, Contact materials for microrelays, in: IEEE, 1998, pp. 190–194.

[20] M. Andrews, G. Turner, G. Turner, A comparison of squeeze-film theory with measurements on a microstructure, Sens. Actuat. A 36 (1993) 79–87.

[21] F. Pan, J. Kubby, E. Peeterst, et al., Squeeze film damping effect on the dynamic response of a MEMS torsion mirror, J. Micromech. Microeng. 8 (1998) 200–208.

[22] C. Junshou, Study on Ka-band MEMS Capacitive Switch of MOS Coplanar Waveguide, Tsinghua University, Beijing, 2012.

Further Reading

S. Guan, K. Vollmers, A. Subramanian, B.J. Nelson, Design and fabrication of a gold electroplated electromagnetic and electrostatic hybrid MEMS relay, J. Appl. Phys. 97 (2005) 10R506.

Index

Note: Page numbers followed by "*f*" and "*t*" refer to figures and tables, respectively.

A

E

F

G

H

I

K

L

M

N

T

U

W

Z

Printed and bound by CPI Group (UK) Ltd, Croydon, CR0 4YY
08/05/2025
01864801-0001